Thioredoxin and Glutaredoxin Systems

Thioredoxin and Glutaredoxin Systems

Special Issue Editors

Jean-Pierre Jacquot
Mirko Zaffagnini

MDPI • Basel • Beijing • Wuhan • Barcelona • Belgrade

MDPI

Special Issue Editors

Jean-Pierre Jacquot
Université de Lorraine
France

Mirko Zaffagnini
University of Bologna
Italy

Editorial Office
MDPI
St. Alban-Anlage 66
4052 Basel, Switzerland

This is a reprint of articles from the Special Issue published online in the open access journal *Antioxidants* (ISSN 2076-3921) from 2018 to 2019 (available at: https://www.mdpi.com/journal/antioxidants/special_issues/Thioredoxin_and_Glutaredoxin_Systems)

For citation purposes, cite each article independently as indicated on the article page online and as indicated below:

LastName, A.A.; LastName, B.B.; LastName, C.C. Article Title. *Journal Name* **Year**, *Article Number, Page Range.*

ISBN 978-3-03897-836-7 (Pbk)
ISBN 978-3-03897-837-4 (PDF)

Cover image courtesy of Jean-Pierre Jacquot.

Fog rolling over the french Ormont mountain in the Vosges near Nayemont les Fosses in the fall of 2018. This was taken on a clear and very cold and sunny day, conditions favorable for generating reactive oxygen species in plants. The article of Dreyer and Dietz in this issue deals with cold and light stress in plants and the research done in the Rouhier group in Nancy addresses the physiology of trees in relation with their interacting with microorganisms and also in response to stress.

Contents

About the Special Issue Editors

Jean-Pierre Jacquot, Prof., works on redox regulation in plants via the thioredoxin and glutaredoxin systems. His early work elucidated the regulatory cascade of the ferredoxin–thioredoxin system and the physicochemical properties of its components (ferredoxin, thioredoxin reductase, and thioredoxins). He has also studied the molecular properties of its target enzymes (NADP-MDH and FBPase). More recent work of his concerns the study of plant glutaredoxins and their involvement in stress response and iron–sulfur assembly processes.

Mirko Zaffagnini, Dr., currently works at the Department of Pharmacy and Biotechnology (FaBiT), University of Bologna. Dr. Zaffagnnini conducts research in plant physiology and biochemistry. His current projects are mainly focused on the role of thiol-based redox modifications in photosynthetic organisms.

Fog rolling over the french Ormont mountain in the Vosges near Nayemont les Fosses in the fall of 2018. This was taken on a clear and very cold and sunny day, conditions favorable for generating reactive oxygen species in plants. The article of Dreyer and Dietz in this issue deals with cold and light stress in plants and the research done in the Rouhier group in Nancy addresses the physiology of trees in relation with their interacting with microorganisms and also in response to stress.

Jean-Pierre Jacquot
Guest Editor

antioxidants

MDPI

Editorial

Thioredoxin and Glutaredoxin Systems Antioxidants Special Issue

Jean-Pierre Jacquot [1,*] and Mirko Zaffagnini [2]

[1]   Université de Lorraine, Inra, IAM, F-54000 Nancy, France
[2]   Laboratory of Molecular Plant Physiology, Department of Pharmacy and Biotechnology,
      University of Bologna, via Irnerio 42, 40126 Bologna, Italy; mirko.zaffagnini3@unibo.it
*   Correspondence: j2p@univ-lorraine.fr

Received: 14 March 2019; Accepted: 16 March 2019; Published: 18 March 2019

The special issue on Thioredoxin and Glutaredoxin systems (http://www.mdpi.com/journal/antioxidants/special_issues/Thioredoxin_and_Glutaredoxin_Systems) was initiated in response to solicitations from Antioxidants after discussing with colleagues at two successive redox meetings sponsored by European Molecular Biology Organization (EMBO) and held in July 2017 in Moscow/St. Petersburg (http://redox.vub.ac.be/events/embo-redox-biology-conference.html) and in September of the same year in San Feliu de Guixols (Spain) (http://meetings.embo.org/event/17-thiol-ox). We could then submit the idea to long time collaborators and redox friends but also to other colleagues with whom we had the chance to get in touch with at these meetings. In general, although Antioxidants is a rather recent creation and its credentials were at the time not so well known, the idea of participating in a special issue was very well received and many of the contacted colleagues have answered positively. Of course, as our background is in plant sciences, this special issue mostly contains papers dealing with oxygenic phototrophs but other experimental model organisms are also addressed (bacteria, mammals, zebrafish, etc.). Overall the special issue contains 16 papers, 12 of those reporting experimental research data, and 4 others being more review-like although some of them also contain original bioinformatics data.

The two volume editors (J.P.J. and M.Z.) wish to testify that the reviewing process has been done in a very professional way by Antioxidants. We have been phased out of the few papers that presented a conflict of interest, but asked to give a final approval for those. All papers were reviewed by at least two international experts, very often three and more rarely four. Occasionally the participants were asked to cross review papers of the special issue but on average this happened quite rarely. This very thorough evaluation system has helped improve the quality of several of the papers by pointing out some weaknesses that were fixed in a second or third round of evaluation.

Thus overall, we are very pleased with the outcome of this editorial effort and we wish here to give a brief summary of its content and most exciting features.

The first article is a review by Sophie Vriz and colleagues [1]. It deals with Reactive Oxygen Species (ROS) signaling, the very dynamic variation of those species and the morphogenetic, embryogenesis, regeneration, and stem cell differentiation consequences of these molecules. Of course the interaction with reducing molecules as those of the thioredoxin (TRX) and glutaredoxin (GRX) systems can modulate the oxidant signaling. As Vriz and colleagues have extensive experience with zebrafish this experimental model is especially well treated (most other papers deal essentially with bacteria and plants). The two next papers by the laboratories of Pascal Rey and Bertrand Friguet describe properties of methionine sulfoxide reductases (MSRs, enzymes that are able to reduce methionine sulfoxide back to methionine. In the context of increased generation of oxidant molecules, damages can occur on macromolecules including lipids and proteins and thus the function of MSRs is very important in the cell. More precisely Rey and Tarrago [2] describe the relative properties of MSRA and MSRB in plants and the interplay of MSRs with Ca^{2+}- and phosphorylation-dependent cascades, thus transmitting

ROS-related information in transduction pathways. Lourenço dos Santos, Petropoulos and Friguet [3] discuss the properties of MSRs essentially in bacteria but also detail the generation of cysteine sulfenate (-SOH) leading to the buildup of disulfide bonds that can be reduced back to dithiols via the TRX and GRX systems.

The next paper by Dreyer and Dietz [4] discusses the regulatory network of the cell leading to cold stress adaptation and provides short-term transcriptome and metabolome analyses that help understand the physiological responses of plants to cold adaptation. Undeniably, in agronomy the resistance of plants to cold is essential for maintaining a high yield required for animal and human nutrition.

The next paper in the series is a contribution by Monica Balsera and colleagues [5]. Buey, Schmitz, Buchanan and Balsera report the structure of an nicotinamide adenine dinucleotide phosphate reduced (NADPH) TRX reductase (NTR) from an archaeon producing methane, *Methanosarcina mazei*. Interestingly, the protein crystallizes in the apo form lacking flavin adenine dinucleotide (FAD). The apo NTR displays the same dimeric head to tail organization than previously characterized NTRs despite lacking the flavin coenzyme. They discuss the significance of this weaker FAD binding compared to NTR from other organisms including bacteria, animals and plants.

The next six papers discuss structural and physiological properties of TRXs and GRXs in plants and bacteria. The first of these by the Frendo group [6] discusses the interactions between *Medicago truncatula–Sinorhizobium meliloti* which are extremely important in nitrogen fixation for leguminous plants. The symbiotic interaction leads to the formation of a new organ, the root nodule, where a coordinated differentiation of plant cells and bacteria occurs. The crucial role of glutathione in redox balance and sulfur metabolism is presented. They also highlight the specific role of some TRX and GRX systems in bacterial differentiation. Transcriptomics data concerning gene encoding components and targets of TRX and GRX systems in connection with the developmental step of the nodule are considered in this contribution. The paper by Mariam Sahrawy and colleagues [7] follows an interesting approach that has not been used thus far: determining by proteomics the relative abundance of a large panel of proteins in plants lacking either TRX *f* or TRX *m*. Their results revealed a quantitative alteration of 86 proteins and demonstrate that the lack of both the *f*- and *m*-type TRXs have diverse effects on the proteome. Most of the differentially expressed proteins fell into the categories of metabolic processes, the Calvin–Benson cycle, photosynthesis, response to stress, hormone signaling, and protein turnover. Photosynthesis, the Calvin–Benson cycle and carbon metabolism are the most affected processes. Notably, a significant set of proteins related to the answer to stress situations and hormone signaling were affected. Overall, this suggests that the TRX systems may regulate transcription and translation in plants.

The paper by Stéphane Lemaire et al. [8] reports the crystal structure of TRX *f*2 from *Chlamydomonas reinhardtii*. The systematic comparison of its atomic features to other *f*-type TRXs reveals a specific conserved electropositive crown around the active site, complementary to the electronegative surface of their targets. They postulate that this surface provides specificity to the various type of TRX. The following article of the Javier Florencio group [9] provides information about TRX C, an atypical TRX present exclusively in cyanobacteria. TRX C has a modified active site (WCGLC) instead of the canonical (WCGPC) present in most TRXs and is not active in the classical TRX in vitro tests. Nevertheless, the Δ*trxC* mutant, although growing at similar rates to WT in all conditions tested, showed an increased carotenoid content especially under low carbon conditions. Their data suggest that TRX C might have a role in regulating photosynthetic adaptation to low carbon and/or high light conditions. Marchand et al. [10] provide a refined 3D structure of *C. reinhardtii* TRX *h*1 in the reduced and oxidized form as well as the one of cysteine mutants. This paper also features data concerning the pK_a values of both catalytic cysteines by means of iodoacetamide-based mass spectrometry analysis. The next contribution by the Nicolas Rouhier group [11] describes physico-chemical and catalytic properties of the poorly characterized mitochondrial TRX *o*. Very interestingly, they show for the first time that this isoform can in vitro bind iron sulfur centers (ISCs) of the [4Fe-4S] type likely ligated by the classical catalytic cysteines present in the conserved WCGPC signature. This situation is somewhat similar to those of various GRXs which also bind ISCs in a homodimer although the nature of the center is different. Remarkably, their results suggest that a novel regulation mechanism may prevail for mitochondrial *o*-type TRXs, possibly existing as a redox-inactive Fe–S cluster-bound form that could be rapidly converted in a redox-active form upon cluster degradation under specific physiological conditions. NFU proteins could be target of the catalytically active form. It remains to be seen whether the ISC-replete form could be involved in ISC transfer/assembly as suggested for GRXs in many species and sub-compartments.

In the next four papers, we now turn our attention to target enzymes of the TRX systems in plants. Yoshida and Hisabori [12] investigate the rate limiting step of enzyme light activation in chloroplasts. They found that the catalytic subunit of ferredoxin:TRX reductase (FTR) and *f*-type TRX are rapidly reduced after the drive of reducing power transfer, irrespective of the presence or absence of their downstream target proteins. By contrast, three target proteins, fructose 1,6-bisphosphatase (FBPase), sedoheptulose 1,7-bisphosphatase (SBPase), and RuBisCO activase (RCA) showed different reduction patterns; in particular, SBPase was reduced at a low rate. The in vivo study using *Arabidopsis* plants showed that the TRX family is commonly and rapidly reduced upon high light irradiation, whereas FBPase, SBPase, and RCA are differentially and slowly reduced. Among the GRX targets is cytosolic isocitrate dehydrogenase (cICDH), the activity of which is controlled by S-nitrosylation upon nitrosoglutathione (GSNO) treatment as shown by Reichheld and colleagues [13]. In particular, they have shown that GRXs are able to rescue the GSNO-dependent inhibition of cICDH activity, suggesting that they can act as a denitrosylation system in vitro. They observe that the GRX system, contrary to the TRX system, is able to remove S-nitrosothiol adducts from cICDH and they have specified on which specific cysteine this is occurring. The report by Vanacker et al. (Emmanuelle Issakidis lead author) investigates the redox regulation of monodehydroascorbate reductase (MDHAR) [14]. They found that the activity of leaf extracted or the recombinant plastidial *Arabidopsis thaliana* MDHAR isoform 6 was specifically and strongly increased by reduced TRX *y*, and not by other plastidial TRXs. In addition, TRX *y* mutant plants showed reduced stress tolerance in comparison with wild-type (WT) plants that correlated with an increase in their global protein oxidation levels. The last of the papers dealing with redox regulated enzymes provides results concerning chlorophyll biosynthesis. It is well known that chlorophyll synthesis requires light and one key regulatory enzyme is aminolevulinate dehydratase (ALAD). Berhanrd Grimm and colleagues [15] show that this enzyme interacts with TRX *f*, TRX *m* and NTRC in chloroplasts. The reduced and oxidized forms of ALAD differed in their catalytic activity and they conclude that (i) deficiency of the reducing power mainly affected the in planta stability of ALAD; and (ii) the reduced form of ALAD displayed increased enzymatic activity.

The last paper of this special issue is by Nicolas Navrot et al. [16]. This contribution of the Svensson lab has investigated the biotechnological potential of the NTR/TRX system (NTS), and especially of NTR, as a useful tool in baking for weakening strong doughs, or in flat product baking. They have shown that the barley NTS is capable of remodeling the gluten network and weakening bread dough.

In conclusion, we believe that this special issue had indeed provided new information about a number of proteins involved in the redox regulatory networks, either oxidants, reductants, TRXs, or targets. Of course, we would have liked to cover additional topics, in particular the field of peroxiredoxins and glutathione peroxidases was not addressed at all. Likewise, two recent papers reported on the interesting diversity of protein disulfide isomerases in plants [17] and this will certainly require a more thorough investigation in the future. Many other examples could be provided for sure, so there are many opportunities for producing future special issues as highlighted by the recent paper uncovering the importance of Cys-based redox mechanisms in plant development, growth and adaptation to stress [18].

Funding: This research received no external funding.

Conflicts of Interest: The authors declare no conflict of interest.

References

1. Rampon, C.; Volovitch, M.; Joliot, A.; Vriz, S. Hydrogen Peroxide and Redox Regulation of Developments. *Antioxidants* **2018**, *7*, 159. [CrossRef] [PubMed]
2. Rey, P.; Tarrago, L. Physiological Roles of Plant Methionine Sulfoxide Reductases in Redox Homeostasis and Signaling. *Antioxidants* **2018**, *7*, 114. [CrossRef] [PubMed]
3. Lourenço dos Santos, S.; Petropoulos, I.; Friguet, B. The Oxidized Protein Repair Enzymes Methionine Sulfoxide Reductases and Their Roles in Protecting against Oxidative Stress, in Ageing and in Regulating Protein Function. *Antioxidants* **2018**, *7*, 191. [CrossRef] [PubMed]
4. Dreyer, A.; Dietz, K. Reactive Oxygen Species and the Redox-Regulatory Network in Cold Stress Acclimation. *Antioxidants* **2018**, *7*, 169. [CrossRef] [PubMed]
5. Buey, R.; Schmitz, R.; Buchanan, B.; Balsera, M. Crystal Structure of the Apo-Form of NADPH-Dependent Thioredoxin Reductase from a Methane-Producing Archaeon. *Antioxidants* **2018**, *7*, 166. [CrossRef] [PubMed]
6. Alloing, G.; Mandon, K.; Boncompagni, E.; Montrichard, F.; Frendo, P. Involvement of Glutaredoxin and Thioredoxin Systems in the Nitrogen-Fixing Symbiosis between Legumes and Rhizobia. *Antioxidants* **2018**, *7*, 182. [CrossRef] [PubMed]
7. Fernández-Trijueque, J.; Serrato, A.; Sahrawy, M. Proteomic Analyses of Thioredoxins f and m Arabidopsis thaliana Mutants Indicate Specific Functions for These Proteins in Plants. *Antioxidants* **2019**, *8*, 54. [CrossRef] [PubMed]
8. Lemaire, S.; Tedesco, D.; Crozet, P.; Michelet, L.; Fermani, S.; Zaffagnini, M.; Henri, J. Crystal Structure of Chloroplastic Thioredoxin f2 from Chlamydomonas reinhardtii Reveals Distinct Surface Properties. *Antioxidants* **2018**, *7*, 171. [CrossRef] [PubMed]
9. López-Maury, L.; Heredia-Martínez, L.; Florencio, F. Characterization of TrxC, an Atypical Thioredoxin Exclusively Present in Cyanobacteria. *Antioxidants* **2018**, *7*, 164. [CrossRef] [PubMed]
10. Marchand, C.; Fermani, S.; Rossi, J.; Gurrieri, L.; Tedesco, D.; Henri, J.; Sparla, F.; Trost, P.; Lemaire, S.; Zaffagnini, M. Structural and Biochemical Insights into the Reactivity of Thioredoxin h1 from Chlamydomonas reinhardtii. *Antioxidants* **2019**, *8*, 10. [CrossRef] [PubMed]
11. Zannini, F.; Roret, T.; Przybyla-Toscano, J.; Dhalleine, T.; Rouhier, N.; Couturier, J. Mitochondrial Arabidopsis thaliana TRXo Isoforms Bind an Iron–Sulfur Cluster and Reduce NFU Proteins In Vitro. *Antioxidants* **2018**, *7*, 142. [CrossRef] [PubMed]
12. Yoshida, K.; Hisabori, T. Determining the Rate-Limiting Step for Light-Responsive Redox Regulation in Chloroplasts. *Antioxidants* **2018**, *7*, 153. [CrossRef] [PubMed]
13. Niazi, A.; Bariat, L.; Riondet, C.; Carapito, C.; Mhamdi, A.; Noctor, G.; Reichheld, J. Cytosolic Isocitrate Dehydrogenase from Arabidopsis thaliana Is Regulated by Glutathionylation. *Antioxidants* **2019**, *8*, 16. [CrossRef] [PubMed]
14. Vanacker, H.; Guichard, M.; Bohrer, A.; Issakidis-Bourguet, E. Redox Regulation of Monodehydroascorbate Reductase by Thioredoxin y in Plastids Revealed in the Context of Water Stress. *Antioxidants* **2018**, *7*, 183. [CrossRef] [PubMed]
15. Wittmann, D.; Kløve, S.; Wang, P.; Grimm, B. Towards Initial Indications for a Thiol-Based Redox Control of Arabidopsis 5-Aminolevulinic Acid Dehydratase. *Antioxidants* **2018**, *7*, 152. [CrossRef] [PubMed]
16. Navrot, N.; Buhl Holstborg, R.; Hägglund, P.; Povlsen, I.; Svensson, B. New Insights into the Potential of Endogenous Redox Systems in Wheat Bread Dough. *Antioxidants* **2018**, *7*, 190. [CrossRef] [PubMed]
17. Selles, B.; Zannini, F.; Couturier, J.; Jacquot, J.P.; Rouhier, N. Atypical protein disulfide isomerases (PDI): Comparison of the molecular and catalytic properties of poplar PDI-A and PDI-M with PDI-L1A. *PLoS ONE* **2017**, *12*, e0174753. [CrossRef] [PubMed]
18. Zaffagnini, M.; Fermani, S.; Marchand, C.H.; Costa, A.; Sparla, F.; Rouhier, N.; Geigenberger, P.; Lemaire, S.D.; Trost, P. Redox Homeostasis in Photosynthetic Organisms: Novel and Established Thiol-Based Molecular Mechanisms. *Antioxid Redox Signal.* **2019**. [CrossRef] [PubMed]

antioxidants

MDPI

Review

Hydrogen Peroxide and Redox Regulation of Developments

Christine Rampon [1,2], **Michel Volovitch** [1,3], **Alain Joliot** [1] **and Sophie Vriz** [1,2,*]

1   Center for Interdisciplinary Research in Biology (CIRB), College de France, CNRS, INSERM, PSL Research
    University, 75231 Paris, France; Christine.rampon@college-de-france.fr (C.R.);
    Michel.volovitch@ens.fr (M.V.); alain.joliot@college-de-france.fr (A.J.)
2   Sorbonne Paris Cité, Univ Paris Diderot, Biology Department, 75205 Paris CEDEX 13, France
3   École Normale Supérieure, Department of Biology, PSL Research University, 75005 Paris, France
*   Correspondence: vriz@univ-paris-diderot.fr

Received: 19 September 2018; Accepted: 10 October 2018; Published: 6 November 2018

Abstract: Reactive oxygen species (ROS), which were originally classified as exclusively deleterious compounds, have gained increasing interest in the recent years given their action as *bona fide* signalling molecules. The main target of ROS action is the reversible oxidation of cysteines, leading to the formation of disulfide bonds, which modulate protein conformation and activity. ROS, endowed with signalling properties, are mainly produced by NADPH oxidases (NOXs) at the plasma membrane, but their action also involves a complex machinery of multiple redox-sensitive protein families that differ in their subcellular localization and their activity. Given that the levels and distribution of ROS are highly dynamic, in part due to their limited stability, the development of various fluorescent ROS sensors, some of which are quantitative (ratiometric), represents a clear breakthrough in the field and have been adapted to both ex vivo and in vivo applications. The physiological implication of ROS signalling will be presented mainly in the frame of morphogenetic processes, embryogenesis, regeneration, and stem cell differentiation. Gain and loss of function, as well as pharmacological strategies, have demonstrated the wide but specific requirement of ROS signalling at multiple stages of these processes and its intricate relationship with other well-known signalling pathways.

Keywords: H_2O_2; redox signalling; development; regeneration; adult stem cells; metazoan

1. Introduction

For a long time, reactive oxygen species (ROS), including hydrogen peroxide (H_2O_2), were considered deleterious molecules. Emphasis was given to their role in neutrophils where they are produced to contribute to anti-microbial defence [1], and extensive studies have been performed on ROS over-production due to mitochondrial dysfunction in neurological disorders or cancer progression [2–4]. Consistent with these detrimental functions, attention has been almost exclusively focused on their toxicity, and many studies strengthened this aspect of redox biology. However, pioneer works highlighted a new role of ROS in signalling, which led to the emergence of the redox signalling field [5,6]; recent reviews in [7,8]. Redox signalling soon also proved to be important during animal development for review [9,10]. In 2017, Helmut Sies, a pioneer in redox biology, reviewed the topic and developed the concept of oxidative eustress (physiological redox signalling) and oxidative distress (pathophysiological disrupted redox signalling), bringing the two faces of ROS back together [11]. As recently noted [12], a new reading of the past literature might shed a new light on the tenets of redox signalling. Relevant issues are the nature of the ROS invoked, the accurate localization of its site of production, and its concentration, spreading and dynamics in the context of a defined physiological process. The present review focuses on H_2O_2, a central ROS in redox signalling during development and regeneration in metazoans, and its interplay with the redox machinery. We will not address the role of other reactive species, and readers are referred to excellent reviews on Reactive Nitrogen Species (RNS) or oxidized lipids recently published [13,14].

H_2O_2 is the major ROS produced by cells that acts in signalling pathways as a second messenger [11,15–17]. H_2O_2 is a by-product of many oxidative reactions, such as oxidative protein folding in the endoplasmic reticulum (ER) and peroxisomal enzyme activities. For signalling purposes, the main sources of H_2O_2 are the mitochondrial respiratory chain and NADPH oxidases (NOXs) [18]. NOXs are trans-membrane proteins that use cytosolic NADPH as an electron donor. NOXs belong to multi-component complexes that generate either O_2^- (NOX 1, 2, 3 and 5) or H_2O_2 (NOX 4, DUOX 1 and 2) upon appropriate stimulation (by growth factors, cytokines ...) [19,20]. Even when the primary product of NOX activity is O_2^-, it is largely and immediately transformed into H_2O_2 by a superoxide dismutase (SOD) enzyme physically associated with NOX, or it dismutates spontaneously at low pH levels. Several NOXs are located at the plasma membrane, which is a hub for cell signalling. In this case, H_2O_2 is delivered in the extracellular space, a somehow puzzling situation considering that most known H_2O_2 targets localize in the cell interior. It was first thought that H_2O_2 could pass from the extracellular to the intracellular milieu by passive diffusion through the plasma membrane, but it was later shown that H_2O_2 has poor lipid membrane diffusion capacities and crosses into cells via aquaporin channels [21–23]. This facilitated transport of H_2O_2 across the plasma membrane is itself subject to redox regulation [24], and further investigations are needed to better understand the role of aquaporins in redox signalling. The unique and specific enzyme for H_2O_2 degradation into H_2O is the ubiquitously expressed protein catalase. It mainly localizes in the peroxisome where it is devoted to the reduction of excess H_2O_2 produced there. However, it can also be secreted by an unknown mechanism and associate with the plasma membrane [25–27] or spread in the extracellular milieu [28].

The main physiological target of H_2O_2 action is the reversible oxidation of cysteine residues in proteins. Modification only occurs on the thiolate anion form (S^-). However, at physiological pH, most cysteines are protonated and thus react weakly with H_2O_2. However, the pK_a of cysteine greatly depends on its protein environment and can reach several units below ~8.5, the approximate value of cysteine alone [29], making these residues ionized and reactive. H_2O_2 oxidizes the thiolate anion to produce sulfenic acid, which is highly reactive and readily forms a disulfide bond in contact with accessible –SH group. Reciprocally, in reducing conditions, disulfide bonds can be easily cleaved to restore the thiol functions. As oxidative condition increases, sulfenic acid will further oxidize to sulfinic and ultimately sulfonic derivatives. These two reactions are generally irreversible deleterious modifications; however, exceptions were reported for sulfinic derivatives (see below). Redox signalling depends both on the local concentration of H_2O_2 and the state (protonated or deprotonated) of the cysteine. Although some cysteines can be directly oxidized by H_2O_2, most of them require prior activation to be deprotonated, involving additional redox-sensitive relays. The best candidates for this relay function appear to be proteins first identified as antioxidant safe-guarders [30–36] reviews in [37–41], and they will be discussed below. It is now clear that the role of H_2O_2 signalling in oxidative eustress has to integrate the entire redox machine.

2. The Redox Machine

The central redox machine contains at least six main protein families: thioredoxin reductases (TrxRs), thioredoxins (Trxs), peroxiredoxins (Prxs), glutathione reductases (GRs), glutaredoxins (Grxs) and glutathione peroxidases (Gpxs) (Figure 1) [for general reviews, see [42–45]. Moreover, as schematized in Figure 1, the activities of all enzymes in the redox machine are interconnected (some additional branches between cycles have been omitted), and the final outcome of thiol-oxidation reactions depends on many parameters, making computational modelling useful but hampering genetic approaches.

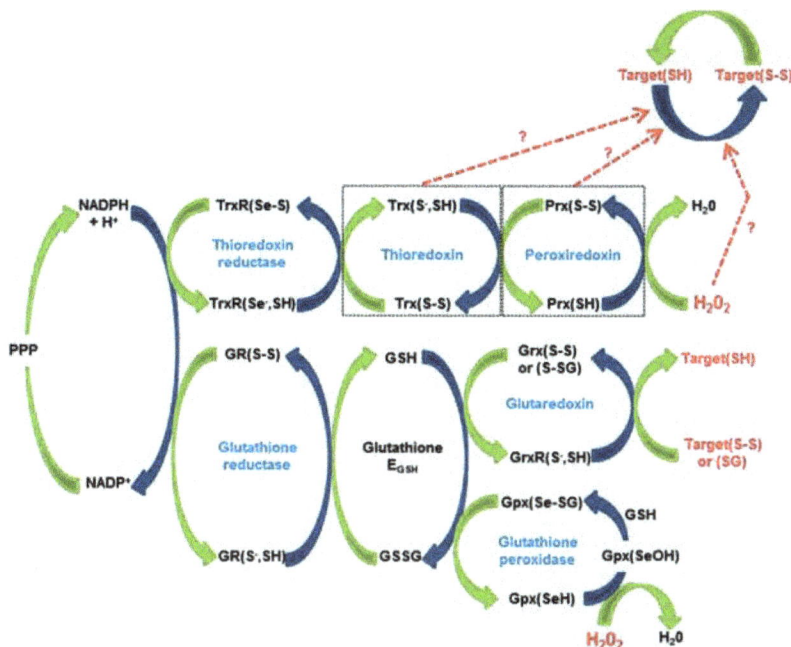

Figure 1. The redox machinery. Interconnection of redox couples from H_2O_2 to thiol targets are represented. H_2O_2 is a by-product of oxidative reactions. Major sources include mitochondrial respiratory chain and NOXs for review [18]. PPP: Pentose Phosphate Pathway.

As mentioned in the introduction, the central redox machine has pleiotropic functions. In addition to detoxification of harmful amounts of ROS, they also act as sensors of oxidant concentrations and can even acquire new functions, such as chaperones activity for some Prxs [46] for a review. This sensor and/or transducer functions are very important given that the vast majority of redox-sensitive proteins are poorly sensitive to direct oxidation by H_2O_2 (a possible exception, PTP1B, is discussed in [39]). Prxs have attracted considerable attention as potent mediators redox signals, as first established in yeast [31,47–50], and some years after in mammals. Ledgerwood and colleagues demonstrated that Prx1 participates in the propagation of peroxide signals via disulfide exchange with the target kinase ASK1 [51], and the group of Tobias Dick showed that Prx2 forms a redox relay for H_2O_2 signalling together with the transcription factor STAT3 [36]. Very recently, the same group demonstrated that the relay activity of cytosolic Prxs (1 and 2) is not dependent on Trx1 or TrxR1 but is based on transient disulfide conjugates with protein targets and occurs mainly in conditions of fast response to small variations in H_2O_2 [52].

3. Seeing Is Believing

A critical step to model redox signalling is to determine the spatiotemporal localization and amount of the different protagonists. Several synthetic dyes were actively used to measure ROS and RNS [53–55]. However, these dyes are often poorly specific, do not penetrate in tissue, or are unstable. Moreover, their reaction with ROS/RNS is irreversible. In the last decade, a major effort was devoted to develop genetically encoded fluorescent biosensors for the redox machine elements.

3.1. H_2O_2 Sensors

For all ROS, ex vivo and in vivo measurements of H_2O_2 concentration are challenging due to its short half-life, fast-spreading and high reactivity. The development of a genetically encoded fluorescent biosensor specific for H_2O_2 revolutionized the field. It provides access to the dynamics of H_2O_2 concentration in living systems and its modulation by genetic or chemical approaches. This goal was first achieved by Vsevolod Belousov who designed the HyPer probe [56]. The HyPer biosensor is based on the fusion of a circularly permutated fluorescent protein (cpYFP) with the H_2O_2-sensing domain of *E. coli* OxyR. Two cysteines of OxyR moiety form a disulfide bond in the presence of H_2O_2 and the resulting conformational change induces a modification of cpYFP spectra, which allows a ratiometric measurement of H_2O_2 levels. Advantages of this probe are its high sensitivity (nanomolar), its reversibility and its fast reaction rate constant. Moreover, ratiometric measurement is independent of the expression level. The main drawback of this sensor is its sensitivity to pH. To circumvent this problem, a cysteine-mutated form of HyPer (SypHer), which is still sensitive to pH but no longer to H_2O_2, can be used as a control or to measure pH in vivo [57]. Since the initial version, HyPer probe has evolved, and the HyPer family currently includes members with different spectral and redox properties [58].

When expressed in *Xenopus laevis* oocytes, HyPer revealed an oscillating production of H_2O_2 induced by fertilization. This production of H_2O_2 is of mitochondrial origin, dependent on calcium waves initiated by fertilization and involved in cell cycle progression at the beginning of development [59]. HyPer was also expressed by transgenesis in two animal models (nematode and fish), where it revealed a highly dynamic fluctuation in H_2O_2 levels during embryonic and post-embryonic development. In *Caenorhabditis elegans* (where HyPer expression was under the control of the ubiquitous RPL-21 promoter), H_2O_2 levels were high during larval development (in the head, notably in the pharynx and neurons), strongly decreased at the transition to the adult stage, and remained low during most of the reproductive period [60]. A similar pattern was observed in *Danio rerio* transgenic animals with high levels of H_2O_2 during development and a massive reduction at 3 days post fertilization (dpf) when most of the developmental programmes have ended. Notably, in fish and nematode, HyPer revealed a highly dynamic pattern of H_2O_2 levels in the developing nervous system [61] (Figure 2).

Another type of H_2O_2 sensor was developed from a fusion between roGFP2 (a redox-sensitive GFP) and Orp1, the yeast H_2O_2 sensor and modulator of redox-sensitive transcription factor Yap1 [34]. Orp1 is sensitive to H_2O_2; once oxidized, Orp1 promotes the nearby oxidation of roGFP2 (as it does for Yap1), resulting in a shift of roGFP2 spectral properties. Compared with HyPer, this biosensor is insensitive to pH but less sensitive to H_2O_2. This lower affinity for H_2O_2 was overcome by fusion of roGFP2 to the yeast Prx Tsa2 (Tsa2ΔC_R) [63]. roGFP2-Orp1 has been successfully used to measure H_2O_2 in developing and adult *Drosophila* [64]. One of the advantages of genetically encoded sensors is their ability to be addressed to a cell-specific compartment upon fusion with appropriate targeting sequences. Differential targeting into either the cytosol or the mitochondria allowed Albrecht et al. to demonstrate the heterogeneity of H_2O_2 levels depending on the tissue and that H_2O_2 level is not coupled with the redox state of glutathione during development [64]. Cytosol/mitochondria expression of roGFP2-Orp1 in the germline of *Caenorhabditis elegans* revealed an increase in H_2O_2 levels in the proximal side of the germline and a peak within the oocytes and in the zygote [65]. An elegant approach preserving the redox status of roGFP2-Orp1 during tissue cryo-section allowed H_2O_2 measurements in mammalian development and adult tissues [66]. This strategy is very promising to acquire redox maps of non-optically accessible tissue. roGFP2-Orp1 was also targeted to zebrafish cardiomyocytes in different compartments (nucleus, mitochondria and cytosol) to follow H_2O_2 level variations during cardiac function and upon pharmaceutical treatments, demonstrating the interest of this H_2O_2 probe to score oxidant or antioxidant molecules [67].

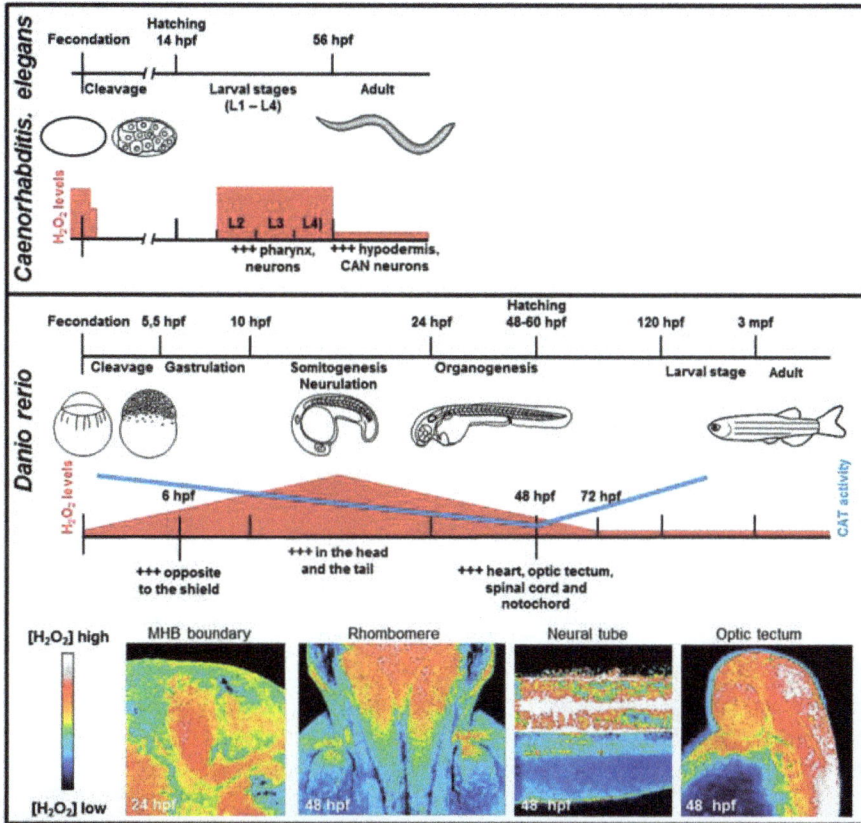

Figure 2. H_2O_2 detection during development. Upper panel: H_2O_2 detection during *C. elegans* development. Adapted from [62] and [60]. Middle panel: H_2O_2 levels and catalase activity during *Danio rerio* development. Adapted from [61]. Lower panel: HyPer fish reveal spatio-temporal dynamic and gradients of H_2O_2 during neural development. hpf: hours post fertilization, mpf: month post fertilization.

3.2. Glutathione Redox Potential Sensors

Glutathione plays a key role in cellular thiol-disulfide exchange reactions, and the GSH/GSSG ratio is considered a good indicator of redox balance (E_{GSH}) (Figure 1). A fluorescent sensor for E_{GSH} was generated by fusion of the roGFP2 with the human glutaredoxin-1 (Grx1). roGFP2 alone exhibits a slow response to redox changes. Grx1 fusion to roGFP2 resulted in a rapid equilibrium between the GSH/GSSG couple and the reporting redox couple (roGFP2red and roGFP2ox), thus reflecting the level of E_{GSH} [68]. This sensor was introduced in several species. In *Drosophila*, it was addressed to mitochondria and cytosol to compare E_{GSH} with H_2O_2 levels in developing structures and adults [64]. Live imaging of the third-instar larvae revealed high variations in mitochondrial E_{GSH} amongst different tissues, whereas the cytosolic E_{GSH} was almost constant [64]. Transgenic *C. elegans* expressing a cytosolic form of Grx1-roGFP2 under a ubiquitous promoter was used to analyse E_{GSH} during development [62]. E_{GSH} decreases globally during development and then remains constant in adult except in the spermathecae where fertilization occurs [62].

3.3. NADPH Sensor: iNap

NADPH is a key element in the redox machine as a final electron donor for thiol oxidation by H_2O_2. A genetically encoded fluorescent indicator for NADPH (iNap sensor) was developed by mutagenesis of the ligand binding site of the NADH/NAD+ sensor SoNar to switch the selectivity from NADH to NADPH [69]. iNap can be used in vivo, and a proof of concept experiment was performed in a wound healing assay in zebrafish larvae. In combination with a red version of HyPer, the iNap biosensor revealed the concomitant decrease of NADPH levels after tissue wounding with an increase in H_2O_2 levels, which is consistent with NADPH consumption by NOXs during wound healing [69].

All these sensors provide invaluable information on cellular redox status in vivo. However, it is worth reminding that all H_2O_2 sensors consume H_2O_2 to measure its concentration. Thus, there is room to improve these systems in particular through increased sensitivity. Only one molecule of H_2O_2 is consumed to modify each molecule of sensor for both HyPer and roGFP2-Orp1. However, for the later, it also depends on the local redox potential. This is a clear illustration of a common difficulty: seeing is modifying.

4. Redox Signalling in Animal Development and Regeneration

H_2O_2 is generated in response to many stimuli, including cytokine or growth factors, which are involved in embryonic development and adult homeostasis. Given redundancy in the redox machinery components, most single Knock out (KOs) in mice are viable, and embryonic development generally occurs normally [70]. Metazoan development can be divided into 3 phases: (1) fertilization and cleavage period, (2) gastrulation, and (3) morphogenesis. At the adult stage, most tissues are continuously renewed and, in some species, rebuilt after amputation. This section will describe the state of the art for the role of redox signalling during these processes. Many molecular targets of redox signalling are now known, as well as the mechanisms by which their redox balance influences the pathways they belong to, for which excellent recent reviews exist [7–10,71]. The role of RNS and oxidative stress in pathologies of the nervous system, which are not discussed here, are extensively reviewed in [72–74]. Finally, it is worth mentioning that some developmental effects of redox signalling in the brain only become apparent much later (as for instance in the case of critical periods [75], mental illnesses, such as schizophrenia [76], or autism-like behaviours in mouse [77]). These effects have only been analysed in the context of their dysregulation, but their normal progress certainly warrants better examination.

4.1. Embryonic Development

4.1.1. NADPH Oxidase Complexes in Embryonic Development

Given the importance of NOXs in anti-microbial defence, inflammation, disorders including cancers, and more generally in the maintenance of redox balance, most studies on these enzymes in metazoans focus on adult expression. Some NOXs exhibit broad distribution among tissues (Nox2 was first believed to be exclusively expressed in neutrophils and macrophages and is currently known to present the largest distribution). Others are more restricted (Nox3 is predominantly expressed in the inner ear), but none of them is ubiquitously expressed. Constitutive expression coexists with induction phenomena (for review see [19,78]. NOX expression also greatly varies during development (Table 1). The most detailed study was performed by qPCR and in situ hybridization in developing zebrafish [79]. Unlike the uniform and homogeneous expression of Nox2, Nox1 and Nox5 expression is high during gastrulation and then decreases to a basal level upon morphogenesis. During this period of development, Nox1 expression is increased in the brain. The expression of dual oxidase (Duox), which is a member of the NOX family, is increased during late morphogenesis [79]. In rodents, Duox expression patterns were described in embryonic thyroid [80] and Nox2 and 4 were looked at during limb formation [81]. Though useful, these expression data did not provide a clear picture of the physiological function of these enzymes. Till the advent of tools enabling to study redox biology in live organisms (see below), the contribution of H_2O_2 signalling to developmental processes was essentially analysed in embryonic stem (ES) cells. More than 10 years ago, it was proven that ES cell differentiation into cardiac lineage was dependent on NOX enzymes [82,83]. More examples of ES cell sensitivity to redox potential are now known, which can be found in recent reviews on the subject [84,85].

Table 1. Expression of the redox machinery genes during *Danio rerio* development.

Enzyme	Gene	Cellular Localization	Gene Expression	Reference
Catalase	cat	mitochondria, peroxisome	brain, digestive system, gill, muscle, sensory system	[86]
Glutaredoxin	glrx2	cytoplasm	whole organism	[87]
	glrx3	cytoplasm	brain, heart, sensory system	[86,88]
	glrx5	cytoplasm	blood island, digestive system, heart, sensory system	[89]
Glutathione Peroxidase	gpx1a	nd	digestive system, muscle, sensory system	[90,91]
	gpx1b	cytoplasm	digestive system, sensory system	[86,92]
	gpx4a	nd	digestive system, peridermis	[90,93]
	gpx4b	nd	blastoderm, digestive system, epidermis, epiphysis, muscle, pharyngeal arch, pronephric duct, sensory system	[90,93,94]
	gpx7	nd	notochord, splanchnocranium	[86]
	gpx8	membrane	notochord, pharyngeal arch, sensory system	[86]
Glutathione Reductase	gsr	cytoplasm	digestive system, macrophage	[86]
NADPH Oxidase	nox1	membrane	brain, spinal cord, sensory system	[79]
	nox2/Cybb	membrane	blood, brain, spinal cord, sensory system	[79,86]
	nox5	membrane	brain, spinal cord, sensory system	[79]
	duox	membrane	brain, digestive system, epidermis, spinal cord, sensory system, swim bladder, thyroid,	[79,95,96]
Peroxiredoxin	prdx1	cytoplasm	brain, neural crest derivatives, vessels	[97,98]
	prdx2	nd	blood, CNS, digestive system, pharyngeal arch, sensory system	[90]
	prdx3	nd	blood, digestive system, myotome, pharyngeal arch, sensory system	[90]
	prdx4	nd	digestive system, hatching gland, pharyngeal arch, sensory system	[86,99]
	prdx5	nd	macrophage, pronephric duct, sensory system	[86]
	prdx6	nd	digestive system, rhombomere, sensory system	[86]
Superoxide Dismutase	sod1	cytoplasm	whole organism	[86,100]
	sod2	mitochondria	blood, brain, digestive system, gill, kidney, muscle, sensory system	[90,101]
	sod3b	cytoplasm	whole organism	[102]
Thioredoxin	txn	nd	digestive system, Hypophysis, spinal cord, sensory system, tegmentum	[86,91]
	txn2	mitochondria	whole organism	[86]
Thioredoxin Reductase	txnrd3	mitochondria	blood, CNS, digestive system, muscle, pharyngeal arch, spinal cord, sensory system	[86,90]

No results have been reported for gpx2, gpx3, gpx9, nox4, sod3a, txnrd2-1, and txnrd2-2. CNS: central nervous system.

The use of the HyPer probe in a live animal revealed a surprisingly high level of oxidation during zebrafish development, and H_2O_2 levels proved to be heterogeneous and dynamic in space and time [61] (Figure 2). H_2O_2 levels occasionally exhibit a graded distribution with clear functional outcome. This distribution occurs in the embryonic tectum where the organization of the retinotectal projections is impaired by Pan NOX inhibition [61]. Nox2 invalidation using the CRISPR/Cas9 strategy induced the same phenotype [103], confirming Nox2 involvement in axon pathfinding during zebrafish development. Very recently, the group of E. Amaya proved that Duox activity was necessary for the development of zebrafish thyroid [104], and that ROS play a role in Xenopus mesoderm formation [59]. A role for Nox enzymes was also evidenced ex vivo for the differentiation of chondrocytes [81] or endometrial cells [105], the establishment of rat hippocampal neuron polarity in culture [106,107], the in vitro maturation of rat cerebellar granule neurons [108], and growth cone dynamics in *Aplysia* neuron [109–111]. Nox were also shown to be involved in epithelial-to-mesenchymal transition in normal or tumoral epithelial cell lines [112,113].

4.1.2. Catalase, Superoxide Dismutases and Glutathione Systems in Embryonic Development

Catalase, Sod1 and GPx expression and activities have been analysed during mice development from 8 days of gestation to adulthood [114]. mRNAs of these 3 proteins increase during somitogenesis. Then, catalase activity increases after birth, whereas Sod1 and GPx activities reach a plateau. In zebrafish, SOD activity is globally constant throughout morphogenesis until 7 dpf, whereas an increase in catalase activity is observed from 48 hpf onward when morphogenesis is almost completed [61], (Figure 2). In *Drosophila*, catalase protein is minimally detectable during embryogenesis and enhanced during the third instar larval stage and after the first day of pupal development [115]. It thus appears that low levels of catalase expression during development and high levels in mature tissues represent a general property that has been verified in mice, *Drosophila* and zebrafish.

GSH/GSSG balance is a crucial redox parameter during development [for review: [116]. Embryos from mutant mice invalidated for the enzymes responsible for GSH synthesis fail to gastrulate, do not form mesoderm, develop distal apoptosis, and die before day 8.5 [117,118]. In zebrafish, the total amount of glutathione doubles during embryonic development. The redox potential (E_{GSH}) is high in eggs and late larval stage but very low during morphogenesis [119]. This study defined 4 GSH contexts during embryonic development: (1) high E_{GSH} and low GSH in mature oocytes, (2) low E_{GSH} and low GSH from mid-blastula to 24 hpf, (3) high E_{GSH} and high GSH during organogenesis (30–48 hpf) and (4) high E_{GSH} and high GSH in mature larvae.

Grx patterns of expression during development have not been systematically investigated, but expression of Grx2 and its isoforms in vertebrate tissues at various stages (mostly adults) is described in [120–123]. Their role during development has been demonstrated in different models. One of the first hint (though outside metazoan) was that Grx1 affects cell fate decision in the culmination process in *Dictyostelium discoideum* [124]. It was subsequently shown that knocking down (KD) cytosolic Grx2 in zebrafish embryo impaired the development of the central nervous system. Grx2 is indeed essential for neuronal differentiation, survival and axon growth [87]. In mammals, Grx2 controls axon growth via a dithiol-disulfide switch affecting the conformation of CRMP2, a mediator of semaphorin-plexin signalling pathway [125]. The same Grx2 isoform is also required for correct wiring of embryonic vasculature in zebrafish by de-glutathionylation of the active site of the NAD-dependent deacetylase Sirtuin-1 [126]. Grx2 is also implicated in embryonic heart formation. Grx2 KD in zebrafish embryo prevented neural crest cell migration into the primary heart field, impairing heart looping [127]. The role of other members of the Grx family was also described in erythropoiesis. Grx5 is essential to zebrafish haeme synthesis through assembly of the Fe-S cluster, and this role is apparently conserved in humans [89]. Grx3 is also crucial to red blood cell formation in zebrafish as demonstrated by the reduced number of erythrocytes in embryos treated with Grx3 morpholinos and is required in human cells for the biogenesis of Fe-S proteins, as demonstrated by silencing Grx3 expression in HeLa cells [88]. Grx3 KO in mice induces a delay in development and eventually death approximately 12.5 days of gestation. Ex vivo analysis of $Grx3^{-/-}$ cells reveals impaired growth and cell cycle progression at the G2/M transition [128]. The loss of Grx3 also disturbs the development of mammary alveoli during pregnancy and lactation [129]. In addition to these roles during development proper, Grx proteins are also involved in various physio-pathological processes (for review [130]).

4.1.3. Thioredoxin System in Embryonic Development

Two distinct thioredoxin/thioredoxin reductase systems, Trx1 and Trx2, are present in mammalian cells in the cytosol and mitochondria. Trx1 and Trx2 are ubiquitously expressed, but Trx2 is expressed at higher levels in metabolically active tissues, such as heart, brain, and liver [131]. Interestingly, in mice, Trx1 (the cytosolic and nuclear Trx) KO is embryonic lethal early after implantation [132] and presents a dramatically reduced proliferation of inner mass cells. Trx2 (the mitochondrial form) KO is also embryonic lethal. Embryos die later during development (between 10.5 and 12.5 dpf), and lethality is associated with increased in apoptosis and exencephaly development [133]. Trx2 KD in zebrafish increases apoptosis and induces developmental defects in the liver [134]. In chick, Trx2 KD impairs post-mitotic neurons and induces massive cell death [135]. Thioredoxin reductases (TrxRs) KO reveals that TrxR1 (non-mitochondrial form) is embryonic lethal (E10.5) with multiple abnormalities in all organs except heart [136], whereas TrxR2 KO leads specifically to haematopoietic and heart defects [137]. These works demonstrate a clear dichotomy in the tissue specificity of TrxR1 and 2 actions during development: broad or restricted to heart and haematopoietic lineage development. In addition, Trxs play critical roles in immune response, cancer (for review [138]), and various pathologies in the nervous system (for reviews [72,74]).

4.1.4. Peroxiredoxin Systems in Embryonic Development

The peroxiredoxins (Prxs) family includes 6 members in vertebrates. These enzymes have been mostly studied in physio-pathological contexts or ex vivo where it was demonstrated that Prxs are major regulators of cell adhesion and migration [139]. In vivo, Prxs regulate cadherin expression during early *Drosophila* development [140]. Prxs are typically broadly expressed in embryos with mild specificity amongst isoforms. However, an exhaustive analysis in *Xenopus laevis* reveals maternal expression of prx1, 2, 3 and 6, which persists through all developmental stages. In contrast, Prx4 mRNA becomes detectable at gastrulation and increases afterwards, and Prx5 mRNA is always detected but at low levels. Additional specificities are revealed by in situ hybridization. Specifically, Prx1 is expressed in anterior structures, and Prx4, 5 and 6 expression is preferentially detected in somites [141]. This pattern of expression is not completely conserved between *Xenopus* and other vertebrates. In zebrafish, Prdx1 is expressed in developing vessels, and Prdx1 KD induces vascular defects [97]. In mice, proteomic analysis suggested that prx1 is involved in digit formation where it regulates interdigit apoptosis [142]. One of the best examples of Prxs' roles during normal differentiation is the involvement of Prx1 and Prx4 in the formation of motor neurons in the spinal cord of chicken and mouse [143,144].

Sexual reproduction and more precisely gamete formation is an unusual process regarding redox signalling. NOXs are involved in spermatogonia and germline stem cell renewal [145]. Moreover, a significant amount of H_2O_2 is needed to favour disulfide bridge formation during spermatozoa maturation and capacitation (for review [146]). Conversely, ROS insult should be neutralized to protect DNA and maintain genomic integrity from generation to generation. Several Gpx members are involved in these opposite aspects of redox signalling [147].

4.2. Adult Stem Cells and Tissue Homeostasis

During life, adult stem cells replenish damaged and lost tissues. In recent years, H_2O_2 appeared to be a major component of stem cell niche, stem cell renewal and recruitment for differentiation (for review [148,149]). In some studies, an increase in H_2O_2 is responsible for stem cell differentiation. In *Drosophila*, increased H_2O_2 induces the differentiation of haematopoietic progenitor cells, whereas a reduction delays the expression of differentiation markers [150]. In mammals, an increase in H_2O_2

induces vascular muscle cell [151] or blood stem cell [152] differentiation. In an apparent contradiction, high H_2O_2 levels are associated with stem cell renewal and proliferation. The renewal of intestinal stem cells in *Drosophila* is dependent on H_2O_2 [153], and self-renewal of neural stem cells is under the control of NOX activity in the mouse [154,155]. These opposite effects of H_2O_2 illustrate our limited understanding of cell fate regulation by redox signalling but both strengthen the relevance of ROS levels in the control of stem cell behaviour and the need for their tight regulation. Some clues come from the induced pluripotent stem cell (iPSC) field where H_2O_2 increase is essential during the early phase of iPSC generation. Reduction of H_2O_2 levels by NOX KD or antioxidant treatment suppresses nuclear reprogramming [156]. NOX2 is also involved in iPSc differentiation into endothelial cells [157].

4.3. Regeneration

Some species have the ability to regenerate damaged or removed body parts at adulthood [158–160]. Regeneration is probably the best paradigm of adult morphogenesis. The first step of regeneration is wound repair and the formation of a wound epidermis. Shortly after wound epidermis formation, progenitor cells are generated via stem cells that are recruited and the dedifferentiation of differentiated cells [159,161,162]. The newly formed progenitors (and stem cells in some systems) migrate to the wound epithelium to form a mass of undifferentiated cells called the blastema. The entire missing structure will be formed by differentiation and morphogenesis of blastema cells. The process of regeneration can be divided into 3 modules: (1) injury immediate response (wound healing), (2) regeneration induction (blastema formation), (3) formation of the missing structure (blastema growth, differentiation and patterning) [163] (Figure 3A).

It has been known for a long time that wounds generate ROS, specifically H_2O_2, in phylogenetically diverse organisms, such as *Drosophila melanogaster*, *C. elegans*, *Danio rerio*, [171–174], and plants [175–177]. In zebrafish, H_2O_2 production triggered by wounding is restricted to the first two hours following injury in larva [172] and adult [166]. After a lesion is generated, a gradient of H_2O_2 formed by NOX activity is required for leukocyte recruitment to the wound [172,178]. The role of H_2O_2 is not restricted to wounding and healing but extends to other steps of regeneration. When a body part is removed, induction of the regenerative programme correlates with a sustained production of H_2O_2 for several hours (Figure 3B) [166]. It has been demonstrated in different model organisms that after amputation, H_2O_2 signalling not only modulates the regeneration process but is indispensable for launching it (Table 2). Planarian and Hydra regenerate not only body parts but also their body axis [160,179]. In these animals, ROS are detected at the level of the amputation plane shortly after amputation, and reduction of H_2O_2 levels impairs regeneration [164,180] (Figure 3B). In *Drosophila*, regeneration can be observed in imaginal discs during larval development and adult gut regeneration [181–183]. In both systems, regeneration is redox-dependent [153,181,182,184–187]. In the *Xenopus* tadpole tail regeneration model, amputation-induced H_2O_2 production is necessary to activate Wnt/β-catenin, Fgf signalling and acetylation of lysine 9 of histone 3 (H3K9ac) [165,188] (Figure 3B). In the newt, ROS production is also necessary for neural stem cell proliferation, neurogenesis, and regeneration of dopamine neurons [189]. Moreover, recent data demonstrate that H_2O_2 activates voltage-gated Na^+ channels, linking redox signalling to bioelectric signalling during regeneration [190]. During gecko tail regeneration, H_2O_2 is produced by skeletal muscles and is also required for successful tail regeneration. In this case, H_2O_2 levels control autophagy in the skeletal muscles and consequently the length of the regenerated tail [191]. In adult zebrafish heart, caudal fin and superficial epithelial cells regeneration models, perturbation of ROS levels through the inhibition of NOX or overexpression of catalase impairs regeneration (Figure 3B) [166,167,192,193]. Different targets were identified at the cellular (neural cells), functional (apoptosis), and molecular levels (MAP kinases and Sonic Hedgehog). In mammals, regeneration at adulthood is very limited [194]. The comparison of regeneration of large circular defects through the ear pinna between regenerative mammals (*Acomys cahirinus*) and non-regenerative mammals (*Mus musculus*) revealed a strong correlation between H_2O_2 levels after injury and regenerative capacities (Figure 3B) [168]. In rats, it was further demonstrated that H_2O_2

participates in liver regeneration after partial hepatectomy. In this case, sustained and elevated H_2O_2 levels activate MAP kinase signalling that triggers the shift from quiescence to proliferation [195]. Recently, it has been demonstrated that H_2O_2 produced shortly after inguinal fat pad damage is responsible for its regeneration in MRL mice (Figure 3B). In the non-regenerative C57BL/6 strain, artificial enhancement of H_2O_2 leads to regeneration [169].

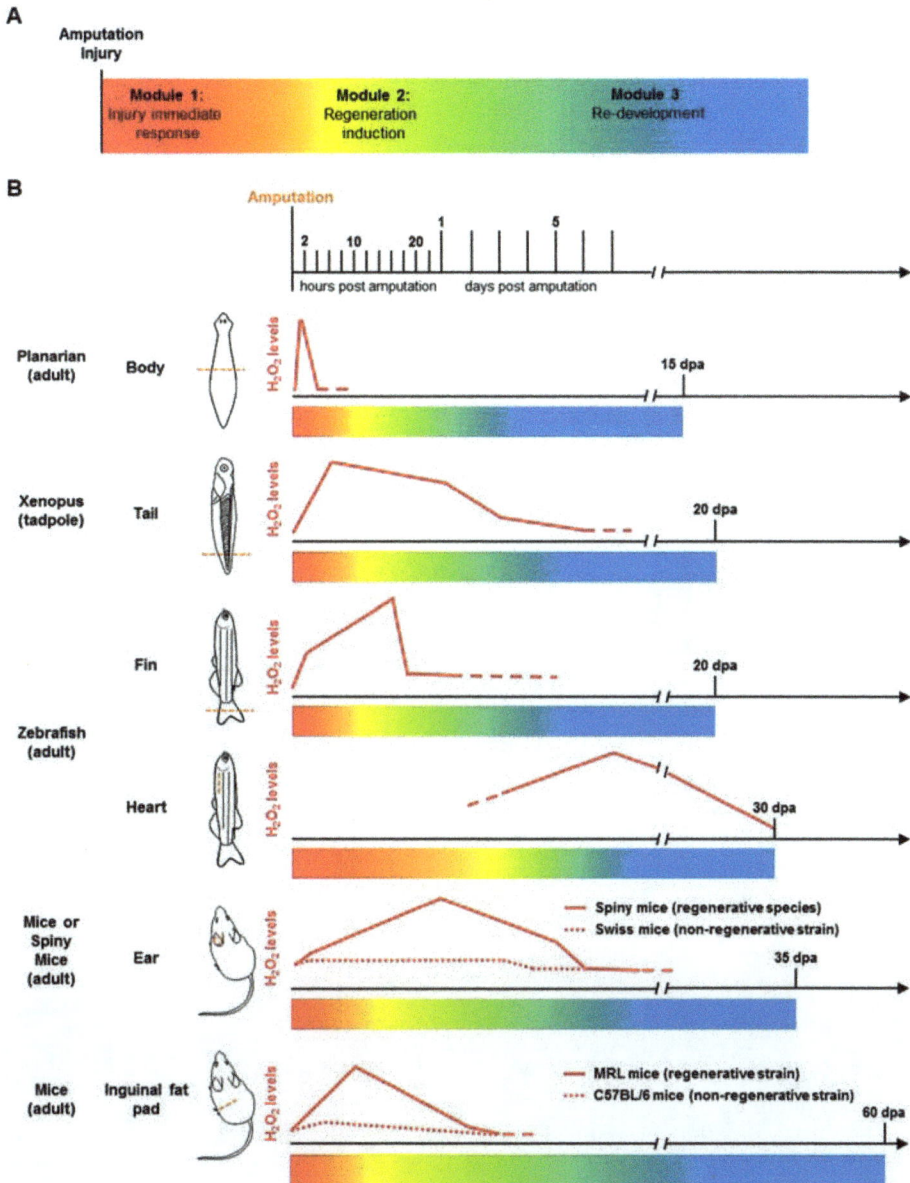

Figure 3. H_2O_2 detection during metazoan regeneration. (**A**): regeneration is divided in three modules. (**B**): H_2O_2 levels during regeneration in different models and organs. dpa: days post amputation. Adapted from [164–170].

Table 2. Redox regulation of regeneration among Phyla. APO: apocynin; DHE: dihydroethidium; dpa: days post-amputation; hpa: hours post amputation; n.s.: no significant.

Classification	Animals/Species	Stage	ROS Detection	Profil	ROS Modulation	Organ/Appendage	ROS Targets	Reference
Cnidaria	Hydra		DMPO	wound edge				[180]
Platyhelminthes	Planarian (*Schmidtea mediterranea*)		H_2DCFDA	burst at the wound site	DPI, APO	central nervous system	neuroregeneration	[164]
Arthropoda	Drosophila (*Drosophila melanogaster*)	Larvae	CellRox green	burst after apoptosis induction	NAC, vitamin C, Trolox, SOD, CAT misexpression of extracell.CAT	wing imaginal disc	p38 pathway JNK pathway	[181]
			DHE, H_2DCFDA	up to 24 h after apoptosis induction		eye and wing imaginal disc	macrophages	[182]
		Adult	H_2DCFDA	burst after oral admin. of HgCl2	Vitamin E	midgut		[183]
Amphibians	Xenopus (*Xenopus laevis or tropicalis*)	Tadpole	HyPer	production 6 h-4 dpa	DPI, APO, MCI-186	Tail	Wnt/b-catenin pathway FGF pathway	[165]
			DHE	nd	DPI, MCI, VAS, H_2O_2		bioelectric activity	[190]
			H_2DCFDA	nd	APO	Notochord in tail	acetylation of H3K9	[188]
Squamata	Gecko (*Gekko japonicus*)	Adult	H_2DCFDA	production (0-7 dpa), skeletal muscles	DPI, APO	tail	autophagy in skeletal muscles (ULK, MAPK)	[191]
Teleost fish	Zebrafish (*Danio rerio*)	Larvae	HyPer PfBS-F	nd	DPI	caudal fin	Src family kinase *ptch1, tcf7, raldh2, pea3, thbb* Apoptosis JNK pathway Hh pathway	[196] [91] [166]
			H_2DCFDA	production 0-16 hpa				[166]
		Adult	H_2DCFDA, HyPer	production 0-16 hpa	VAS2870, H_2O_2	nerve		[193]
			Myl7:HyPer, Redox sensor cc-1	production 3-14 dpa epicardium and adjacent myocardium, max 7 dpa	DPI, APO, CAT	heart	ERK pathway	[167]
			CellRox green	production 2 h-12 hpa	VAS2870	superficial epithelial cells (SECs) in caudal fin		[192]
Mammals	Rat (*Rattus norvegicus*)	Adult	H_2DCFDA, red H_2O_2 assay kit, amplex red H_2O_2 assay kit	production 1 h-3 dpa	GOX, CAT	liver	ERK pathway p38 pathway	[195]
	Mice (*Mus musculus*)	Adult	luminol	MRL mice (production 0-72 h pa, max 12 h), C57Bl6 (n.s. production)		inguinal fat pad		[169]
			lucigenin	n.s. production 3 h-10 dpa				
			luminol	production 3h-4 dpa				
	Spiny mice (*Acomys cahirinus*)	Adult	lucigenin	production 3 h-5 dpa		ear		[168]
			luminol	n.s. production 3h-10 dpa				

5. Conclusions: Towards the Redox Code

Redox signalling interacts directly or indirectly with most of the signalling pathways that control embryonic development. However, we are only starting to perceive the tip of the iceberg [197]. A comprehensive understanding of developmental redox biology will benefit from a better characterization of thiol targets. For this purpose, the optimization of redox proteomics and the in silico identification of reactive thiols susceptible to redox regulation based on 3D rather than 2D models are two promising strategies [198]. One can note that, unlike other posttranslational modifications (i.e., phosphorylation or ubiquitination), which behave as binary switches, thiol modifications are diverse, including the formation of sulfonic acids (–S–OH), S-nitro groups (S–NO) and disulfides bridges [199]. Each of these modifications could confer a specific status to the targeted protein, thus extending the spectrum of regulation provided by redox signalling [200]. Finally, because the different members of the redox machinery are interconnected, the modification of a specific thiol likely depends on the equilibrium of the entire redox machine. Modelling the entire process by integrating dynamic and quantitative information of the different redox machinery members would greatly help to decipher the physiological role of redox signalling.

Author Contributions: Writing-Review & Editing, C.R., M.V., A.J. and S.V.

Funding: This work has received support under the programme « Investissements d'Avenir » launched by the French Government and implemented by the ANR with the references: ANR-10-LABX-54 MEMO LIFE, ANR-11-IDEX-0001-02 PSL* Research University.

Acknowledgments: The authors thanks Eva-Maria Hanschmann, Carsten Berndt, Vsevolod Belousov and Helmut Sies for helpful discussions; France Maloumian for drawings; and Carole Gauron for providing zebrafish HyPer Images.

Conflicts of Interest: The authors declare no conflict of interest.

References

1. Manda-Handzlik, A.; Demkow, U. Neutrophils: The Role of Oxidative and Nitrosative Stress in Health and Disease. *Adv. Exp. Med. Biol.* **2015**, *857*, 51–60. [PubMed]
2. Liou, G.Y.; Storz, P. Reactive oxygen species in cancer. *Free Radic. Res* **2010**, *44*, 479–496. [CrossRef] [PubMed]
3. Li, J.; Li, W.; Jiang, Z.G.; Ghanbari, H.A. Oxidative stress and neurodegenerative disorders. *Int. J. Mol. Sci.* **2013**, *14*, 24438–24475. [CrossRef] [PubMed]
4. Chaturvedi, R.K.; Flint Beal, M. Mitochondrial diseases of the brain. *Free Radic. Biol. Med.* **2013**, *63*, 1–29. [CrossRef] [PubMed]
5. Suzuki, Y.J.; Forman, H.J.; Sevanian, A. Oxidants as stimulators of signal transduction. *Free Radic. Biol. Med.* **1997**, *22*, 269–285. [CrossRef]
6. D'Autreaux, B.; Toledano, M.B. ROS as signalling molecules: Mechanisms that generate specificity in ROS homeostasis. *Nat. Rev. Mol. Cell Biol.* **2007**, *8*, 813–824. [CrossRef] [PubMed]
7. Forman, H.J.; Maiorino, M.; Ursini, F. Signaling functions of reactive oxygen species. *Biochemistry* **2010**, *49*, 835–842. [CrossRef] [PubMed]
8. Forman, H.J.; Ursini, F.; Maiorino, M. An overview of mechanisms of redox signaling. *J. Mol. Cell Cardiol.* **2014**, *73*, 2–9. [CrossRef] [PubMed]
9. Covarrubias, L.; Hernandez-Garcia, D.; Schnabel, D.; Salas-Vidal, E.; Castro-Obregon, S. Function of reactive oxygen species during animal development: Passive or active? *Dev. Biol.* **2008**, *320*, 1–11. [CrossRef] [PubMed]
10. Hernandez-Garcia, D.; Wood, C.D.; Castro-Obregon, S.; Covarrubias, L. Reactive oxygen species: A radical role in development? *Free Radic. Biol. Med.* **2010**, *49*, 130–143. [CrossRef] [PubMed]
11. Sies, H. Hydrogen peroxide as a central redox signaling molecule in physiological oxidative stress: Oxidative eustress. *Redox Biol.* **2017**, *11*, 613–619. [CrossRef] [PubMed]
12. Nikolaidis, M.G.; Margaritelis, N.V. Same Redox Evidence But Different Physiological "Stories": The Rashomon Effect in Biology. *Bioessays* **2018**, *40*, e1800041. [CrossRef] [PubMed]

13. Di Meo, S.; Reed, T.T.; Venditti, P.; Victor, V.M. Role of ROS and RNS Sources in Physiological and Pathological Conditions. *Oxid Med. Cell Longev.* **2016**, *2016*, 1245049. [CrossRef] [PubMed]

14. Tyurina, Y.Y.; Shrivastava, I.; Tyurin, V.A.; Mao, G.; Dar, H.H.; Watkins, S.; Epperly, M.; Bahar, I.; Shvedova, A.A.; Pitt, B.; et al. Only a Life Lived for Others Is Worth Living: Redox Signaling by Oxygenated Phospholipids in Cell Fate Decisions. *Antioxid. Redox Signal.* **2018**, *29*, 1333–1358. [CrossRef] [PubMed]

15. Paulsen, C.E.; Carroll, K.S. Orchestrating redox signaling networks through regulatory cysteine switches. *ACS Chem. Biol.* **2010**, *5*, 47–62. [CrossRef] [PubMed]

16. Marinho, H.S.; Real, C.; Cyrne, L.; Soares, H.; Antunes, F. Hydrogen peroxide sensing, signaling and regulation of transcription factors. *Redox Biol.* **2014**, *2*, 535–562. [CrossRef] [PubMed]

17. Stone, J.R.; Yang, S. Hydrogen peroxide: A signaling messenger. *Antioxid. Redox Signal.* **2006**, *8*, 243–270. [CrossRef] [PubMed]

18. Holmstrom, K.M.; Finkel, T. Cellular mechanisms and physiological consequences of redox-dependent signalling. *Nat. Rev. Mol. Cell Biol.* **2014**, *15*, 411–421. [CrossRef] [PubMed]

19. Bedard, K.; Krause, K.H. The NOX family of ROS-generating NADPH oxidases: Physiology and pathophysiology. *Physiol. Rev.* **2007**, *87*, 245–313. [CrossRef] [PubMed]

20. Brandes, R.P.; Weissmann, N.; Schroder, K. Nox family NADPH oxidases: Molecular mechanisms of activation. *Free Radic. Biol. Med.* **2014**, *76*, 208–226. [CrossRef] [PubMed]

21. Miller, E.W.; Dickinson, B.C.; Chang, C.J. Aquaporin-3 mediates hydrogen peroxide uptake to regulate downstream intracellular signaling. *Proc. Natl. Acad. Sci. USA* **2010**, *107*, 15681–15686. [CrossRef] [PubMed]

22. Bienert, G.P.; Chaumont, F. Aquaporin-facilitated transmembrane diffusion of hydrogen peroxide. *Biochim. Biophys. Acta* **2014**, *1840*, 1596–1604. [CrossRef] [PubMed]

23. Bertolotti, M.; Farinelli, G.; Galli, M.; Aiuti, A.; Sitia, R. AQP8 transports NOX2-generated H_2O_2 across the plasma membrane to promote signaling in B cells. *J. Leukoc. Biol.* **2016**, *100*, 1071–1079. [CrossRef] [PubMed]

24. Hara-Chikuma, M.; Watanabe, S.; Satooka, H. Involvement of aquaporin-3 in epidermal growth factor receptor signaling via hydrogen peroxide transport in cancer cells. *Biochem. Biophys. Res. Commun.* **2016**, *471*, 603–609. [CrossRef] [PubMed]

25. Bechtel, W.; Bauer, G. Catalase protects tumor cells from apoptosis induction by intercellular ROS signaling. *Anticancer Res.* **2009**, *29*, 4541–4557. [PubMed]

26. Heinzelmann, S.; Bauer, G. Multiple protective functions of catalase against intercellular apoptosis-inducing ROS signaling of human tumor cells. *Biol. Chem.* **2010**, *391*, 675–693. [CrossRef] [PubMed]

27. Bohm, B.; Heinzelmann, S.; Motz, M.; Bauer, G. Extracellular localization of catalase is associated with the transformed state of malignant cells. *Biol. Chem.* **2015**, *396*, 1339–1356. [CrossRef] [PubMed]

28. Moran, E.C.; Kamiguti, A.S.; Cawley, J.C.; Pettitt, A.R. Cytoprotective antioxidant activity of serum albumin and autocrine catalase in chronic lymphocytic leukaemia. *Br. J. Haematol.* **2002**, *116*, 316–328. [CrossRef] [PubMed]

29. Poole, L.B. The basics of thiols and cysteines in redox biology and chemistry. *Free Radic. Biol. Med.* **2015**, *80*, 148–157. [CrossRef] [PubMed]

30. Delaunay, A.; Isnard, A.D.; Toledano, M.B. H_2O_2 sensing through oxidation of the Yap1 transcription factor. *EMBO J.* **2000**, *19*, 5157–5166. [CrossRef] [PubMed]

31. Delaunay, A.; Pflieger, D.; Barrault, M.B.; Vinh, J.; Toledano, M.B. A thiol peroxidase is an H_2O_2 receptor and redox-transducer in gene activation. *Cell* **2002**, *111*, 471–481. [CrossRef]

32. Veal, E.A.; Ross, S.J.; Malakasi, P.; Peacock, E.; Morgan, B.A. Ybp1 is required for the hydrogen peroxide-induced oxidation of the Yap1 transcription factor. *J. Biol. Chem.* **2003**, *278*, 30896–30904. [CrossRef] [PubMed]

33. Toledano, M.B.; Delaunay, A.; Monceau, L.; Tacnet, F. Microbial H_2O_2 sensors as archetypical redox signaling modules. *Trends Biochem. Sci* **2004**, *29*, 351–357. [CrossRef] [PubMed]

34. Gutscher, M.; Sobotta, M.C.; Wabnitz, G.H.; Ballikaya, S.; Meyer, A.J.; Samstag, Y.; Dick, T.P. Proximity-based protein thiol oxidation by H_2O_2-scavenging peroxidases. *J. Biol. Chem.* **2009**, *284*, 31532–31540. [CrossRef] [PubMed]

35. Calvo, I.A.; Boronat, S.; Domenech, A.; Garcia-Santamarina, S.; Ayte, J.; Hidalgo, E. Dissection of a redox relay: H_2O_2-dependent activation of the transcription factor Pap1 through the peroxidatic Tpx1-thioredoxin cycle. *Cell Rep.* **2013**, *5*, 1413–1424. [CrossRef] [PubMed]

36. Sobotta, M.C.; Liou, W.; Stocker, S.; Talwar, D.; Oehler, M.; Ruppert, T.; Scharf, A.N.; Dick, T.P. Peroxiredoxin-2 and STAT3 form a redox relay for H_2O_2 signaling. *Nat. Chem. Biol.* **2015**, *11*, 64–70. [CrossRef] [PubMed]

37. Veal, E.A.; Day, A.M.; Morgan, B.A. Hydrogen peroxide sensing and signaling. *Mol. Cell* **2007**, *26*, 1–14. [CrossRef] [PubMed]

38. Perkins, A.; Nelson, K.J.; Parsonage, D.; Poole, L.B.; Karplus, P.A. Peroxiredoxins: Guardians against oxidative stress and modulators of peroxide signaling. *Trends Biochem. Sci.* **2015**, *40*, 435–445. [CrossRef] [PubMed]

39. Netto, L.E.; Antunes, F. The Roles of Peroxiredoxin and Thioredoxin in Hydrogen Peroxide Sensing and in Signal Transduction. *Mol. Cells* **2016**, *39*, 65–71. [PubMed]

40. Stocker, S.; Van Laer, K.; Mijuskovic, A.; Dick, T.P. The Conundrum of Hydrogen Peroxide Signaling and the Emerging Role of Peroxiredoxins as Redox Relay Hubs. *Antioxid. Redox Signal.* **2018**, *28*, 558–573. [CrossRef] [PubMed]

41. Young, D.; Pedre, B.; Ezerina, D.; De Smet, B.; Lewandowska, A.; Tossounian, M.A.; Bodra, N.; Huang, J.; Astolfi Rosado, L.; Van Breusegem, F.; et al. Protein Promiscuity in H_2O_2 Signaling. *Antioxid. Redox Signal.* **2018**. [CrossRef] [PubMed]

42. Hanschmann, E.M.; Godoy, J.R.; Berndt, C.; Hudemann, C.; Lillig, C.H. Thioredoxins, glutaredoxins, and peroxiredoxins-molecular mechanisms and health significance: From cofactors to antioxidants to redox signaling. *Antioxid. Redox Signal.* **2013**, *19*, 1539–1605. [CrossRef] [PubMed]

43. Espinosa-Diez, C.; Miguel, V.; Mennerich, D.; Kietzmann, T.; Sanchez-Perez, P.; Cadenas, S.; Lamas, S. Antioxidant responses and cellular adjustments to oxidative stress. *Redox Biol.* **2015**, *6*, 183–197. [CrossRef] [PubMed]

44. Reczek, C.R.; Chandel, N.S. ROS-dependdent signal transduction. *Curr. Opin. Cell Boil.* **2015**, *33*, 8–13. [CrossRef] [PubMed]

45. Sies, H.; Berndt, C.; Jones, D.P. Oxidative Stress. *Ann. Rev Biochem.* **2017**, *86*, 715–748. [CrossRef] [PubMed]

46. Veal, E.A.; Underwood, Z.E.; Tomalin, L.E.; Morgan, B.A.; Pillay, C.S. Hyperoxidation of Peroxiredoxins: Gain or Loss of Function? *Antioxid. Redox Signal.* **2018**, *28*, 574–590. [CrossRef] [PubMed]

47. Veal, E.A.; Findlay, V.J.; Day, A.M.; Bozonet, S.M.; Evans, J.M.; Quinn, J.; Morgan, B.A. A 2-Cys peroxiredoxin regulates peroxide-induced oxidation and activation of a stress-activated MAP kinase. *Mol. Cell* **2004**, *15*, 129–139. [CrossRef] [PubMed]

48. Tachibana, T.; Okazaki, S.; Murayama, A.; Naganuma, A.; Nomoto, A.; Kuge, S. A major peroxiredoxin-induced activation of Yap1 transcription factor is mediated by reduction-sensitive disulfide bonds and reveals a low level of transcriptional activation. *J. Biol. Chem.* **2009**, *284*, 4464–4472. [CrossRef] [PubMed]

49. Iwai, K.; Naganuma, A.; Kuge, S. Peroxiredoxin Ahp1 acts as a receptor for alkylhydroperoxides to induce disulfide bond formation in the Cad1 transcription factor. *J. Biol. Chem.* **2010**, *285*, 10597–10604. [CrossRef] [PubMed]

50. Fomenko, D.E.; Koc, A.; Agisheva, N.; Jacobsen, M.; Kaya, A.; Malinouski, M.; Rutherford, J.C.; Siu, K.L.; Jin, D.Y.; Winge, D.R.; et al. Thiol peroxidases mediate specific genome-wide regulation of gene expression in response to hydrogen peroxide. *Proc. Natl. Acad. Sci. USA* **2011**, *108*, 2729–2734. [CrossRef] [PubMed]

51. Jarvis, R.M.; Hughes, S.M.; Ledgerwood, E.C. Peroxiredoxin 1 functions as a signal peroxidase to receive, transduce, and transmit peroxide signals in mammalian cells. *Free Radic. Biol. Med.* **2012**, *53*, 1522–1530. [CrossRef] [PubMed]

52. Stocker, S.; Maurer, M.; Ruppert, T.; Dick, T.P. A role for 2-Cys peroxiredoxins in facilitating cytosolic protein thiol oxidation. *Nat. Chem. Biol.* **2018**, *14*, 148–155. [CrossRef] [PubMed]

53. Gomes, A.; Fernandes, E.; Lima, J.L. Fluorescence probes used for detection of reactive oxygen species. *J. Biochem. Biophys. Methods* **2005**, *65*, 45–80. [CrossRef] [PubMed]

54. Rhee, S.G.; Chang, T.S.; Jeong, W.; Kang, D. Methods for detection and measurement of hydrogen peroxide inside and outside of cells. *Mol. Cells* **2010**, *29*, 539–549. [CrossRef] [PubMed]

55. Kalyanaraman, B.; Darley-Usmar, V.; Davies, K.J.; Dennery, P.A.; Forman, H.J.; Grisham, M.B.; Mann, G.E.; Moore, K.; Roberts, L.J., 2nd; Ischiropoulos, H. Measuring reactive oxygen and nitrogen species with fluorescent probes: Challenges and limitations. *Free Radic. Biol. Med.* **2012**, *52*, 1–6. [CrossRef] [PubMed]

56. Belousov, V.V.; Fradkov, A.F.; Lukyanov, K.A.; Staroverov, D.B.; Shakhbazov, K.S.; Terskikh, A.V.; Lukyanov, S. Genetically encoded fluorescent indicator for intracellular hydrogen peroxide. *Nat. Methods* **2006**, *3*, 281–286. [CrossRef] [PubMed]

57. Poburko, D.; Santo-Domingo, J.; Demaurex, N. Dynamic regulation of the mitochondrial proton gradient during cytosolic calcium elevations. *J. Biol. Chem.* **2011**, *286*, 11672–11684. [CrossRef] [PubMed]

58. Bilan, D.S.; Belousov, V.V. HyPer Family Probes: State of the Art. *Antioxid. Redox Signal.* **2016**, *24*, 731–751. [CrossRef] [PubMed]

59. Han, Y.; Ishibashi, S.; Iglesias-Gonzalez, J.; Chen, Y.; Love, N.R.; Amaya, E. Ca^{2+}-Induced Mitochondrial ROS Regulate the Early Embryonic Cell Cycle. *Cell Rep.* **2018**, *22*, 218–231. [CrossRef] [PubMed]

60. Knoefler, D.; Thamsen, M.; Koniczek, M.; Niemuth, N.J.; Diederich, A.K.; Jakob, U. Quantitative in vivo redox sensors uncover oxidative stress as an early event in life. *Mol. Cell* **2012**, *47*, 767–776. [CrossRef] [PubMed]

61. Gauron, C.; Meda, F.; Dupont, E.; Albadri, S.; Quenech'Du, N.; Ipendey, E.; Volovitch, M.; Del Bene, F.; Joliot, A.; Rampon, C.; et al. Hydrogen peroxide (H_2O_2) controls axon pathfinding during zebrafish development. *Dev. Biol.* **2016**, *414*, 133–141. [CrossRef] [PubMed]

62. Back, P.; De Vos, W.H.; Depuydt, G.G.; Matthijssens, F.; Vanfleteren, J.R.; Braeckman, B.P. Exploring real-time in vivo redox biology of developing and aging Caenorhabditis elegans. *Free Radic. Biol. Med.* **2012**, *52*, 850–859. [CrossRef] [PubMed]

63. Morgan, B.; Van Laer, K.; Owusu, T.N.; Ezerina, D.; Pastor-Flores, D.; Amponsah, P.S.; Tursch, A.; Dick, T.P. Real-time monitoring of basal H_2O_2 levels with peroxiredoxin-based probes. *Nat. Chem. Biol.* **2016**, *12*, 437–443. [CrossRef] [PubMed]

64. Albrecht, S.C.; Barata, A.G.; Grosshans, J.; Teleman, A.A.; Dick, T.P. In vivo mapping of hydrogen peroxide and oxidized glutathione reveals chemical and regional specificity of redox homeostasis. *Cell Metab.* **2011**, *14*, 819–829. [CrossRef] [PubMed]

65. Braeckman, B.P.; Smolders, A.; Back, P.; De Henau, S. In Vivo Detection of Reactive Oxygen Species and Redox Status in Caenorhabditis elegans. *Antioxid. Redox Signal.* **2016**, *25*, 577–592. [CrossRef] [PubMed]

66. Fujikawa, Y.; Roma, L.P.; Sobotta, M.C.; Rose, A.J.; Diaz, M.B.; Locatelli, G.; Breckwoldt, M.O.; Misgeld, T.; Kerschensteiner, M.; Herzig, S.; et al. Mouse redox histology using genetically encoded probes. *Sci. Signal* **2016**, *9*, rs1. [CrossRef] [PubMed]

67. Panieri, E.; Millia, C.; Santoro, M.M. Real-time quantification of subcellular H_2O_2 and glutathione redox potential in living cardiovascular tissues. *Free Radic. Biol. Med.* **2017**, *109*, 189–200. [CrossRef] [PubMed]

68. Gutscher, M.; Pauleau, A.L.; Marty, L.; Brach, T.; Wabnitz, G.H.; Samstag, Y.; Meyer, A.J.; Dick, T.P. Real-time imaging of the intracellular glutathione redox potential. *Nat. Methods* **2008**, *5*, 553–559. [CrossRef] [PubMed]

69. Tao, R.; Zhao, Y.; Chu, H.; Wang, A.; Zhu, J.; Chen, X.; Zou, Y.; Shi, M.; Liu, R.; Su, N.; et al. Genetically encoded fluorescent sensors reveal dynamic regulation of NADPH metabolism. *Nat. Methods* **2017**, *14*, 720–728. [CrossRef] [PubMed]

70. Ufer, C.; Wang, C.C.; Borchert, A.; Heydeck, D.; Kuhn, H. Redox control in mammalian embryo development. *Antioxid. Redox Signal.* **2010**, *13*, 833–875. [CrossRef] [PubMed]

71. Forman, H.J. Redox signaling: An evolution from free radicals to aging. *Free Radic. Biol. Med.* **2016**, *97*, 398–407. [CrossRef] [PubMed]

72. Ren, X.; Zou, L.; Zhang, X.; Branco, V.; Wang, J.; Carvalho, C.; Holmgren, A.; Lu, J. Redox Signaling Mediated by Thioredoxin and Glutathione Systems in the Central Nervous System. *Antioxid. Redox Signal.* **2017**, *27*, 989–1010. [CrossRef] [PubMed]

73. Wilson, C.; Munoz-Palma, E.; Gonzalez-Billault, C. From birth to death: A role for reactive oxygen species in neuronal development. *Semin. Cell Dev. Biol.* **2018**, *80*, 43–49. [CrossRef] [PubMed]

74. Olguin-Albuerne, M.; Moran, J. Redox Signaling Mechanisms in Nervous System Development. *Antioxid. Redox Signal.* **2018**, *28*, 1603–1625. [CrossRef] [PubMed]

75. Morishita, H.; Cabungcal, J.H.; Chen, Y.; Do, K.Q.; Hensch, T.K. Prolonged Period of Cortical Plasticity upon Redox Dysregulation in Fast-Spiking Interneurons. *Biol. Psychiatry* **2015**, *78*, 396–402. [CrossRef] [PubMed]

76. Do, K.Q.; Cabungcal, J.H.; Frank, A.; Steullet, P.; Cuenod, M. Redox dysregulation, neurodevelopment, and schizophrenia. *Curr. Opin. Neurobiol.* **2009**, *19*, 220–230. [CrossRef] [PubMed]

77. Le Belle, J.E.; Sperry, J.; Ngo, A.; Ghochani, Y.; Laks, D.R.; Lopez-Aranda, M.; Silva, A.J.; Kornblum, H.I. Maternal inflammation contributes to brain overgrowth and autism-associated behaviors through altered redox signaling in stem and progenitor cells. *Stem. Cell Rep.* **2014**, *3*, 725–734. [CrossRef] [PubMed]

78. Katsuyama, M. NOX/NADPH oxidase, the superoxide-generating enzyme: Its transcriptional regulation and physiological roles. *J. Pharmacol. Sci.* **2010**, *114*, 134–146. [CrossRef] [PubMed]

79. Weaver, C.J.; Leung, Y.F.; Suter, D.M. Expression dynamics of NADPH oxidases during early zebrafish development. *J. Comp. Neurol.* **2016**, *524*, 2130–2141. [CrossRef] [PubMed]

80. Milenkovic, M.; De Deken, X.; Jin, L.; De Felice, M.; Di Lauro, R.; Dumont, J.E.; Corvilain, B.; Miot, F. Duox expression and related H_2O_2 measurement in mouse thyroid: Onset in embryonic development and regulation by TSH in adult. *J. Endocrinol.* **2007**, *192*, 615–626. [CrossRef] [PubMed]

81. Kim, K.S.; Choi, H.W.; Yoon, H.E.; Kim, I.Y. Reactive oxygen species generated by NADPH oxidase 2 and 4 are required for chondrogenic differentiation. *J. Biol. Chem.* **2010**, *285*, 40294–40302. [CrossRef] [PubMed]

82. Li, J.; Stouffs, M.; Serrander, L.; Banfi, B.; Bettiol, E.; Charnay, Y.; Steger, K.; Krause, K.H.; Jaconi, M.E. The NADPH oxidase NOX4 drives cardiac differentiation: Role in regulating cardiac transcription factors and MAP kinase activation. *Mol. Biol. Cell* **2006**, *17*, 3978–3988. [CrossRef] [PubMed]

83. Buggisch, M.; Ateghang, B.; Ruhe, C.; Strobel, C.; Lange, S.; Wartenberg, M.; Sauer, H. Stimulation of ES-cell-derived cardiomyogenesis and neonatal cardiac cell proliferation by reactive oxygen species and NADPH oxidase. *J. Cell Sci.* **2007**, *120*, 885–894. [CrossRef] [PubMed]

84. Ren, F.; Wang, K.; Zhang, T.; Jiang, J.; Nice, E.C.; Huang, C. New insights into redox regulation of stem cell self-renewal and differentiation. *Biochim. Biophys. Acta* **2015**, *1850*, 1518–1526. [CrossRef] [PubMed]

85. Skonieczna, M.; Hejmo, T.; Poterala-Hejmo, A.; Cieslar-Pobuda, A.; Buldak, R.J. NADPH Oxidases: Insights into Selected Functions and Mechanisms of Action in Cancer and Stem Cells. *Oxid. Med. Cell Longev.* **2017**, *2017*, 9420539. [CrossRef] [PubMed]

86. Thisse, B.; Thisse, C. Fast Release Clones: A High Throughput Expression Analysis. *ZFIN Direct Data Submiss.* **2004**, in press.

87. Brautigam, L.; Schutte, L.D.; Godoy, J.R.; Prozorovski, T.; Gellert, M.; Hauptmann, G.; Holmgren, A.; Lillig, C.H.; Berndt, C. Vertebrate-specific glutaredoxin is essential for brain development. *Proc. Natl. Acad. Sci. USA* **2011**, *108*, 20532–20537. [CrossRef] [PubMed]

88. Haunhorst, P.; Hanschmann, E.M.; Brautigam, L.; Stehling, O.; Hoffmann, B.; Muhlenhoff, U.; Lill, R.; Berndt, C.; Lillig, C.H. Crucial function of vertebrate glutaredoxin 3 (PICOT) in iron homeostasis and hemoglobin maturation. *Mol. Biol. Cell* **2013**, *24*, 1895–1903. [CrossRef] [PubMed]

89. Wingert, R.A.; Galloway, J.L.; Barut, B.; Foott, H.; Fraenkel, P.; Axe, J.L.; Weber, G.J.; Dooley, K.; Davidson, A.J.; Schmid, B.; et al. Deficiency of glutaredoxin 5 reveals Fe-S clusters are required for vertebrate haem synthesis. *Nature* **2005**, *436*, 1035–1039. [CrossRef] [PubMed]

90. Thisse, B.; Pflumio, S.; Fürthauer, M.; Loppin, B.; Heyer, V. Expression of the zebrafish genome during embryogenesis (NIH R01 RR15402). *ZFIN Direct Data Submiss.* **2001**, in press.

91. Yang, L.; Kemadjou, J.R.; Zinsmeister, C.; Bauer, M.; Legradi, J.; Muller, F.; Pankratz, M.; Jakel, J.; Strahle, U. Transcriptional profiling reveals barcode-like toxicogenomic responses in the zebrafish embryo. *Genome Biol.* **2007**, *8*, R227. [CrossRef] [PubMed]

92. Seiler, C.; Davuluri, G.; Abrams, J.; Byfield, F.J.; Janmey, P.A.; Pack, M. Smooth muscle tension induces invasive remodeling of the zebrafish intestine. *PLoS Biol.* **2012**, *10*, e1001386. [CrossRef] [PubMed]

93. Mendieta-Serrano, M.A.; Schnabel, D.; Lomeli, H.; Salas-Vidal, E. Spatial and temporal expression of zebrafish glutathione peroxidase 4 a and b genes during early embryo development. *Gene Expr. Patterns* **2015**, *19*, 98–107. [CrossRef] [PubMed]

94. Rong, X.; Zhou, Y.; Liu, Y.; Zhao, B.; Wang, B.; Wang, C.; Gong, X.; Tang, P.; Lu, L.; Li, Y.; et al. Glutathione peroxidase 4 inhibits Wnt/beta-catenin signaling and regulates dorsal organizer formation in zebrafish embryos. *Development* **2017**, *144*, 1687–1697. [CrossRef] [PubMed]

95. Flores, M.V.; Crawford, K.C.; Pullin, L.M.; Hall, C.J.; Crosier, K.E.; Crosier, P.S. Dual oxidase in the intestinal epithelium of zebrafish larvae has anti-bacterial properties. *Biochem. Biophys. Res. Commun.* **2010**, *400*, 164–168. [CrossRef] [PubMed]

96. Opitz, R.; Maquet, E.; Zoenen, M.; Dadhich, R.; Costagliola, S. TSH receptor function is required for normal thyroid differentiation in zebrafish. *Mol. Endocrinol.* **2011**, *25*, 1579–1599. [CrossRef] [PubMed]

97. Huang, P.C.; Chiu, C.C.; Chang, H.W.; Wang, Y.S.; Syue, H.H.; Song, Y.C.; Weng, Z.H.; Tai, M.H.; Wu, C.Y. Prdx1-encoded peroxiredoxin is important for vascular development in zebrafish. *FEBS Lett.* **2017**, *591*, 889–902. [CrossRef] [PubMed]

98. Nakajima, H.; Nakajima-Takagi, Y.; Tsujita, T.; Akiyama, S.; Wakasa, T.; Mukaigasa, K.; Kaneko, H.; Tamaru, Y.; Yamamoto, M.; Kobayashi, M. Tissue-restricted expression of Nrf2 and its target genes in zebrafish with gene-specific variations in the induction profiles. *PLoS ONE* **2011**, *6*, e26884. [CrossRef] [PubMed]

99. Hegde, A.; Qiu, N.C.; Qiu, X.; Ho, S.H.; Tay, K.Q.; George, J.; Ng, F.S.; Govindarajan, K.R.; Gong, Z.; Mathavan, S.; et al. Genomewide expression analysis in zebrafish mind bomb alleles with pancreas defects of different severity identifies putative Notch responsive genes. *PLoS ONE* **2008**, *3*, e1479. [CrossRef] [PubMed]

100. Bazzini, A.A.; Lee, M.T.; Giraldez, A.J. Ribosome profiling shows that miR-430 reduces translation before causing mRNA decay in zebrafish. *Science* **2012**, *336*, 233–237. [CrossRef] [PubMed]

101. Peterman, E.M.; Sullivan, C.; Goody, M.F.; Rodriguez-Nunez, I.; Yoder, J.A.; Kim, C.H. Neutralization of mitochondrial superoxide by superoxide dismutase 2 promotes bacterial clearance and regulates phagocyte numbers in zebrafish. *Infect. Immun.* **2015**, *83*, 430–440. [CrossRef] [PubMed]

102. Priyadarshini, M.; Tuimala, J.; Chen, Y.C.; Panula, P. A zebrafish model of PINK1 deficiency reveals key pathway dysfunction including HIF signaling. *Neurobiol. Dis.* **2013**, *54*, 127–138. [CrossRef] [PubMed]

103. Weaver, C.J.; Terzi, A.; Roeder, H.; Gurol, T.; Deng, Q.; Leung, Y.F.; Suter, D.M. nox2/cybb Deficiency Affects Zebrafish Retinotectal Connectivity. *J. Neurosci.* **2018**, *38*, 5854–5871. [CrossRef] [PubMed]

104. Chopra, K.; Ishibashi, S.; Amaya, E. Zebrafish duox mutations provide a model for human congenital hypothyroidism. *bioRxiv* **2018**. [CrossRef]

105. Al-Sabbagh, M.; Fusi, L.; Higham, J.; Lee, Y.; Lei, K.; Hanyaloglu, A.C.; Lam, E.W.; Christian, M.; Brosens, J.J. NADPH oxidase-derived reactive oxygen species mediate decidualization of human endometrial stromal cells in response to cyclic AMP signaling. *Endocrinology* **2011**, *152*, 730–740. [CrossRef] [PubMed]

106. Wilson, C.; Nunez, M.T.; Gonzalez-Billault, C. Contribution of NADPH oxidase to the establishment of hippocampal neuronal polarity in culture. *J. Cell Sci.* **2015**, *128*, 2989–2995. [CrossRef] [PubMed]

107. Wilson, C.; Munoz-Palma, E.; Henriquez, D.R.; Palmisano, I.; Nunez, M.T.; Di Giovanni, S.; Gonzalez-Billault, C. A Feed-Forward Mechanism Involving the NOX Complex and RyR-Mediated Ca2+ Release During Axonal Specification. *J. Neurosci.* **2016**, *36*, 11107–11119. [CrossRef] [PubMed]

108. Olguin-Albuerne, M.; Moran, J. ROS produced by NOX2 control in vitro development of cerebellar granule neurons development. *ASN Neuro* **2015**, *7*. [CrossRef] [PubMed]

109. Munnamalai, V.; Suter, D.M. Reactive oxygen species regulate F-actin dynamics in neuronal growth cones and neurite outgrowth. *J. Neurochem.* **2009**, *108*, 644–661. [CrossRef] [PubMed]

110. Zhang, X.F.; Forscher, P. Rac1 modulates stimulus-evoked Ca^{2+} release in neuronal growth cones via parallel effects on microtubule/endoplasmic reticulum dynamics and reactive oxygen species production. *Mol. Biol. Cell* **2009**, *20*, 3700–3712. [CrossRef] [PubMed]

111. Munnamalai, V.; Weaver, C.J.; Weisheit, C.E.; Venkatraman, P.; Agim, Z.S.; Quinn, M.T.; Suter, D.M. Bidirectional interactions between NOX2-type NADPH oxidase and the F-actin cytoskeleton in neuronal growth cones. *J. Neurochem.* **2014**, *130*, 526–540. [CrossRef] [PubMed]

112. Boudreau, H.E.; Casterline, B.W.; Rada, B.; Korzeniowska, A.; Leto, T.L. Nox4 involvement in TGF-beta and SMAD3-driven induction of the epithelial-to-mesenchymal transition and migration of breast epithelial cells. *Free Radic. Biol. Med.* **2012**, *53*, 1489–1499. [CrossRef] [PubMed]

113. Wu, Y.H.; Lee, Y.H.; Shih, H.Y.; Chen, S.H.; Cheng, Y.C.; Tsun-Yee Chiu, D. Glucose-6-phosphate dehydrogenase is indispensable in embryonic development by modulation of epithelial-mesenchymal transition via the NOX/Smad3/miR-200b axis. *Cell Death Dis.* **2018**, *9*, 10. [CrossRef] [PubMed]

114. el-Hage, S.; Singh, S.M. Temporal expression of genes encoding free radical-metabolizing enzymes is associated with higher mRNA levels during in utero development in mice. *Dev. Genet.* **1990**, *11*, 149–159. [CrossRef] [PubMed]

115. Radyuk, S.N.; Klichko, V.I.; Orr, W.C. Catalase expression in Drosophila melanogaster is responsive to ecdysone and exhibits both transcriptional and post-transcriptional regulation. *Arch. Insect Biochem. Physiol.* **2000**, *45*, 79–93. [CrossRef]

116. Hansen, J.M.; Harris, C. Glutathione during embryonic development. *Biochim. Biophys. Acta* **2015**, *1850*, 1527–1542. [CrossRef] [PubMed]

117. Shi, Z.Z.; Osei-Frimpong, J.; Kala, G.; Kala, S.V.; Barrios, R.J.; Habib, G.M.; Lukin, D.J.; Danney, C.M.; Matzuk, M.M.; Lieberman, M.W. Glutathione synthesis is essential for mouse development but not for cell growth in culture. *Proc. Natl. Acad. Sci. USA* **2000**, *97*, 5101–5106. [CrossRef] [PubMed]

118. Winkler, A.; Njalsson, R.; Carlsson, K.; Elgadi, A.; Rozell, B.; Abraham, L.; Ercal, N.; Shi, Z.Z.; Lieberman, M.W.; Larsson, A.; et al. Glutathione is essential for early embryogenesis—Analysis of a glutathione synthetase knockout mouse. *Biochem. Biophys Res. Commun.* **2011**, *412*, 121–126. [CrossRef] [PubMed]

119. Timme-Laragy, A.R.; Goldstone, J.V.; Imhoff, B.R.; Stegeman, J.J.; Hahn, M.E.; Hansen, J.M. Glutathione redox dynamics and expression of glutathione-related genes in the developing embryo. *Free Radic. Biol. Med.* **2013**, *65*, 89–101. [CrossRef] [PubMed]

120. Jurado, J.; Prieto-Alamo, M.J.; Madrid-Risquez, J.; Pueyo, C. Absolute gene expression patterns of thioredoxin and glutaredoxin redox systems in mouse. *J. Biol. Chem.* **2003**, *278*, 45546–45554. [CrossRef] [PubMed]

121. Karunakaran, S.; Saeed, U.; Ramakrishnan, S.; Koumar, R.C.; Ravindranath, V. Constitutive expression and functional characterization of mitochondrial glutaredoxin (Grx2) in mouse and human brain. *Brain Res.* **2007**, *1185*, 8–17. [CrossRef] [PubMed]

122. Lonn, M.E.; Hudemann, C.; Berndt, C.; Cherkasov, V.; Capani, F.; Holmgren, A.; Lillig, C.H. Expression pattern of human glutaredoxin 2 isoforms: Identification and characterization of two testis/cancer cell-specific isoforms. *Antioxid. Redox Signal.* **2008**, *10*, 547–557. [CrossRef] [PubMed]

123. Hudemann, C.; Lonn, M.E.; Godoy, J.R.; Zahedi Avval, F.; Capani, F.; Holmgren, A.; Lillig, C.H. Identification, expression pattern, and characterization of mouse glutaredoxin 2 isoforms. *Antioxid. Redox Signal.* **2009**, *11*, 1–14. [CrossRef] [PubMed]

124. Choi, C.H.; Kim, B.J.; Jeong, S.Y.; Lee, C.H.; Kim, J.S.; Park, S.J.; Yim, H.S.; Kang, S.O. Reduced glutathione levels affect the culmination and cell fate decision in Dictyostelium discoideum. *Dev. Biol.* **2006**, *295*, 523–533. [CrossRef] [PubMed]

125. Gellert, M.; Venz, S.; Mitlohner, J.; Cott, C.; Hanschmann, E.M.; Lillig, C.H. Identification of a dithiol-disulfide switch in collapsin response mediator protein 2 (CRMP2) that is toggled in a model of neuronal differentiation. *J. Biol. Chem.* **2013**, *288*, 35117–35125. [CrossRef] [PubMed]

126. Brautigam, L.; Jensen, L.D.; Poschmann, G.; Nystrom, S.; Bannenberg, S.; Dreij, K.; Lepka, K.; Prozorovski, T.; Montano, S.J.; Aktas, O.; et al. Glutaredoxin regulates vascular development by reversible glutathionylation of sirtuin 1. *Proc. Natl. Acad. Sci. USA* **2013**, *110*, 20057–20062. [CrossRef] [PubMed]

127. Berndt, C.; Poschmann, G.; Stuhler, K.; Holmgren, A.; Brautigam, L. Zebrafish heart development is regulated via glutaredoxin 2 dependent migration and survival of neural crest cells. *Redox Biol.* **2014**, *2*, 673–678. [CrossRef] [PubMed]

128. Cheng, N.H.; Zhang, W.; Chen, W.Q.; Jin, J.; Cui, X.; Butte, N.F.; Chan, L.; Hirschi, K.D. A mammalian monothiol glutaredoxin, Grx3, is critical for cell cycle progression during embryogenesis. *FEBS J.* **2011**, *278*, 2525–2539. [CrossRef] [PubMed]

129. Pham, K.; Dong, J.; Jiang, X.; Qu, Y.; Yu, H.; Yang, Y.; Olea, W.; Marini, J.C.; Chan, L.; Wang, J.; et al. Loss of glutaredoxin 3 impedes mammary lobuloalveolar development during pregnancy and lactation. *Am. J. Physiol. Endocrinol. Metab.* **2017**, *312*, E136–E149. [CrossRef] [PubMed]

130. Lillig, C.H.; Berndt, C.; Holmgren, A. Glutaredoxin systems. *Biochim. Biophys. Acta* **2008**, *1780*, 1304–1317. [CrossRef] [PubMed]

131. Fujii, S.; Nanbu, Y.; Konishi, I.; Mori, T.; Masutani, H.; Yodoi, J. Immunohistochemical localization of adult T-cell leukaemia-derived factor, a human thioredoxin homologue, in human fetal tissues. *Virchows Arch. A Pathol. Anat. Histopathol.* **1991**, *419*, 317–326. [CrossRef] [PubMed]

132. Matsui, M.; Oshima, M.; Oshima, H.; Takaku, K.; Maruyama, T.; Yodoi, J.; Taketo, M.M. Early embryonic lethality caused by targeted disruption of the mouse thioredoxin gene. *Dev. Biol.* **1996**, *178*, 179–185. [CrossRef] [PubMed]

133. Nonn, L.; Williams, R.R.; Erickson, R.P.; Powis, G. The absence of mitochondrial thioredoxin 2 causes massive apoptosis, exencephaly, and early embryonic lethality in homozygous mice. *Mol. Cell. Biol.* **2003**, *23*, 916–922. [CrossRef] [PubMed]

134. Zhang, J.; Cui, X.; Wang, L.; Liu, F.; Jiang, T.; Li, C.; Li, D.; Huang, M.; Liao, S.; Wang, J.; et al. The mitochondrial thioredoxin is required for liver development in zebrafish. *Curr. Mol. Med.* **2014**, *14*, 772–782. [CrossRef] [PubMed]

135. Pirson, M.; Debrulle, S.; Clippe, A.; Clotman, F.; Knoops, B. Thioredoxin-2 Modulates Neuronal Programmed Cell Death in the Embryonic Chick Spinal Cord in Basal and Target-Deprived Conditions. *PLoS ONE* **2015**, *10*, e0142280. [CrossRef] [PubMed]

136. Jakupoglu, C.; Przemeck, G.K.; Schneider, M.; Moreno, S.G.; Mayr, N.; Hatzopoulos, A.K.; de Angelis, M.H.; Wurst, W.; Bornkamm, G.W.; Brielmeier, M.; et al. Cytoplasmic thioredoxin reductase is essential for embryogenesis but dispensable for cardiac development. *Mol. Cell Biol.* **2005**, *25*, 1980–1988. [CrossRef] [PubMed]

137. Conrad, M.; Jakupoglu, C.; Moreno, S.G.; Lippl, S.; Banjac, A.; Schneider, M.; Beck, H.; Hatzopoulos, A.K.; Just, U.; Sinowatz, F.; et al. Essential role for mitochondrial thioredoxin reductase in hematopoiesis, heart development, and heart function. *Mol. Cell Biol.* **2004**, *24*, 9414–9423. [CrossRef] [PubMed]

138. Lu, J.; Holmgren, A. The thioredoxin antioxidant system. *Free Radic. Biol. Med.* **2014**, *66*, 75–87. [CrossRef] [PubMed]

139. Hurd, T.R.; DeGennaro, M.; Lehmann, R. Redox regulation of cell migration and adhesion. *Trends Cell Biol.* **2012**, *22*, 107–115. [CrossRef] [PubMed]

140. DeGennaro, M.; Hurd, T.R.; Siekhaus, D.E.; Biteau, B.; Jasper, H.; Lehmann, R. Peroxiredoxin stabilization of DE-cadherin promotes primordial germ cell adhesion. *Dev. Cell* **2011**, *20*, 233–243. [CrossRef] [PubMed]

141. Shafer, M.E.; Willson, J.A.; Damjanovski, S. Expression analysis of the peroxiredoxin gene family during early development in Xenopus laevis. *Gene Expr. Patterns* **2011**, *11*, 511–516. [CrossRef] [PubMed]

142. Shan, S.W.; Tang, M.K.; Cai, D.Q.; Chui, Y.L.; Chow, P.H.; Grotewold, L.; Lee, K.K. Comparative proteomic analysis identifies protein disulfide isomerase and peroxiredoxin 1 as new players involved in embryonic interdigital cell death. *Dev. Dyn.* **2005**, *233*, 266–281. [CrossRef] [PubMed]

143. Yan, Y.; Sabharwal, P.; Rao, M.; Sockanathan, S. The antioxidant enzyme Prdx1 controls neuronal differentiation by thiol-redox-dependent activation of GDE2. *Cell* **2009**, *138*, 1209–1221. [CrossRef] [PubMed]

144. Yan, Y.; Wladyka, C.; Fujii, J.; Sockanathan, S. Prdx4 is a compartment-specific H_2O_2 sensor that regulates neurogenesis by controlling surface expression of GDE2. *Nat. Commun.* **2015**, *6*, 7006. [CrossRef] [PubMed]

145. Morimoto, H.; Iwata, K.; Ogonuki, N.; Inoue, K.; Atsuo, O.; Kanatsu-Shinohara, M.; Morimoto, T.; Yabe-Nishimura, C.; Shinohara, T. ROS are required for mouse spermatogonial stem cell self-renewal. *Cell Stem Cell* **2013**, *12*, 774–786. [CrossRef] [PubMed]

146. Conrad, M.; Ingold, I.; Buday, K.; Kobayashi, S.; Angeli, J.P. ROS, thiols and thiol-regulating systems in male gametogenesis. *Biochim. Biophys. Acta* **2015**, *1850*, 1566–1574. [CrossRef] [PubMed]

147. Noblanc, A.; Kocer, A.; Chabory, E.; Vernet, P.; Saez, F.; Cadet, R.; Conrad, M.; Drevet, J.R. Glutathione peroxidases at work on epididymal spermatozoa: An example of the dual effect of reactive oxygen species on mammalian male fertilizing ability. *J. Androl.* **2011**, *32*, 641–650. [CrossRef] [PubMed]

148. Bigarella, C.L.; Liang, R.; Ghaffari, S. Stem cells and the impact of ROS signaling. *Development* **2014**, *141*, 4206–4218. [CrossRef] [PubMed]

149. Tatapudy, S.; Aloisio, F.; Barber, D.; Nystul, T. Cell fate decisions: Emerging roles for metabolic signals and cell morphology. *EMBO Rep.* **2017**, *18*, 2105–2118. [CrossRef] [PubMed]

150. Owusu-Ansah, E.; Banerjee, U. Reactive oxygen species prime Drosophila haematopoietic progenitors for differentiation. *Nature* **2009**, *461*, 537–541. [CrossRef] [PubMed]

151. Su, B.; Mitra, S.; Gregg, H.; Flavahan, S.; Chotani, M.A.; Clark, K.R.; Goldschmidt-Clermont, P.J.; Flavahan, N.A. Redox regulation of vascular smooth muscle cell differentiation. *Circ. Res.* **2001**, *89*, 39–46. [CrossRef] [PubMed]

152. Jang, Y.Y.; Sharkis, S.J. A low level of reactive oxygen species selects for primitive hematopoietic stem cells that may reside in the low-oxygenic niche. *Blood* **2007**, *110*, 3056–3063. [CrossRef] [PubMed]

153. Hochmuth, C.E.; Biteau, B.; Bohmann, D.; Jasper, H. Redox regulation by Keap1 and Nrf2 controls intestinal stem cell proliferation in Drosophila. *Cell Stem. Cell* **2011**, *8*, 188–199. [CrossRef] [PubMed]

154. Le Belle, J.E.; Orozco, N.M.; Paucar, A.A.; Saxe, J.P.; Mottahedeh, J.; Pyle, A.D.; Wu, H.; Kornblum, H.I. Proliferative neural stem cells have high endogenous ROS levels that regulate self-renewal and neurogenesis in a PI3K/Akt-dependant manner. *Cell Stem. Cell* **2011**, *8*, 59–71. [CrossRef] [PubMed]

155. Dickinson, B.C.; Peltier, J.; Stone, D.; Schaffer, D.V.; Chang, C.J. Nox2 redox signaling maintains essential cell populations in the brain. *Nat. Chem. Biol.* **2011**, *7*, 106–112. [CrossRef] [PubMed]

156. Zhou, G.; Meng, S.; Li, Y.; Ghebre, Y.T.; Cooke, J.P. Optimal ROS Signaling Is Critical for Nuclear Reprogramming. *Cell Rep.* **2016**, *15*, 919–925. [CrossRef] [PubMed]

157. Kang, X.; Wei, X.; Wang, X.; Jiang, L.; Niu, C.; Zhang, J.; Chen, S.; Meng, D. Nox2 contributes to the arterial endothelial specification of mouse induced pluripotent stem cells by upregulating Notch signaling. *Sci. Rep.* **2016**, *6*, 33737. [CrossRef] [PubMed]

158. Sanchez Alvarado, A. Regeneration in the metazoans: Why does it happen? *Bioessays* **2000**, *22*, 578–590. [CrossRef]

159. Brockes, J.P.; Kumar, A. Comparative aspects of animal regeneration. *Ann. Rev. Cell Dev. Biol.* **2008**, *24*, 525–549. [CrossRef] [PubMed]

160. Galliot, B.; Ghila, L. Cell plasticity in homeostasis and regeneration. *Mol. Reprod. Dev.* **2010**, *77*, 837–855. [CrossRef] [PubMed]

161. Poss, K.D. Advances in understanding tissue regenerative capacity and mechanisms in animals. *Nat. Rev. Genet.* **2010**, *11*, 710–722. [CrossRef] [PubMed]

162. Sandoval-Guzman, T.; Wang, H.; Khattak, S.; Schuez, M.; Roensch, K.; Nacu, E.; Tazaki, A.; Joven, A.; Tanaka, E.M.; Simon, A. Fundamental differences in dedifferentiation and stem cell recruitment during skeletal muscle regeneration in two salamander species. *Cell Stem Cell* **2014**, *14*, 174–187. [CrossRef] [PubMed]

163. Meda, F.; Rampon, C.; Dupont, E.; Gauron, C.; Mourton, A.; Queguiner, I.; Thauvin, M.; Volovitch, M.; Joliot, A.; Vriz, S. Nerves, H_2O_2 and Shh: Three players in the game of regeneration. *Semin. Cell Dev. Biol.* **2018**, *80*, 65–73. [CrossRef] [PubMed]

164. Pirotte, N.; Stevens, A.S.; Fraguas, S.; Plusquin, M.; Van Roten, A.; Van Belleghem, F.; Paesen, R.; Ameloot, M.; Cebria, F.; Artois, T.; et al. Reactive Oxygen Species in Planarian Regeneration: An Upstream Necessity for Correct Patterning and Brain Formation. *Oxid Med. Cell Longev.* **2015**, *2015*, 392476. [CrossRef] [PubMed]

165. Love, N.R.; Chen, Y.; Ishibashi, S.; Kritsiligkou, P.; Lea, R.; Koh, Y.; Gallop, J.L.; Dorey, K.; Amaya, E. Amputation-induced reactive oxygen species are required for successful Xenopus tadpole tail regeneration. *Nat. Cell Biol.* **2013**, *15*, 222–228. [CrossRef] [PubMed]

166. Gauron, C.; Rampon, C.; Bouzaffour, M.; Ipendey, E.; Teillon, J.; Volovitch, M.; Vriz, S. Sustained production of ROS triggers compensatory proliferation and is required for regeneration to proceed. *Sci. Rep.* **2013**, *3*, 2084. [CrossRef] [PubMed]

167. Han, P.; Zhou, X.H.; Chang, N.; Xiao, C.L.; Yan, S.; Ren, H.; Yang, X.Z.; Zhang, M.L.; Wu, Q.; Tang, B.; et al. Hydrogen peroxide primes heart regeneration with a derepression mechanism. *Cell Res* **2014**, *29*, 1091–1107. [CrossRef] [PubMed]

168. Simkin, J.; Gawriluk, T.R.; Gensel, J.C.; Seifert, A.W. Macrophages are necessary for epimorphic regeneration in African spiny mice. *eLife* **2017**, *6*, e24623. [CrossRef] [PubMed]

169. Labit, E.; Rabiller, L.; Rampon, C.; Guissard, C.; Andre, M.; Barreau, C.; Cousin, B.; Carriere, A.; Eddine, M.A.; Pipy, B.; et al. Opioids prevent regeneration in adult mammals through inhibition of ROS production. *Sci. Rep.* **2018**, *8*, 12170. [CrossRef] [PubMed]

170. Chen, Y.; Love, N.R.; Amaya, E. Tadpole tail regeneration in Xenopus. *Biochem. Soc. Trans.* **2014**, *42*, 617–623. [CrossRef] [PubMed]

171. Roy, S.; Khanna, S.; Nallu, K.; Hunt, T.K.; Sen, C.K. Dermal wound healing is subject to redox control. *Mol. Ther.* **2006**, *13*, 211–220. [CrossRef] [PubMed]

172. Niethammer, P.; Grabher, C.; Look, A.T.; Mitchison, T.J. A tissue-scale gradient of hydrogen peroxide mediates rapid wound detection in zebrafish. *Nature* **2009**, *459*, 996–999. [CrossRef] [PubMed]

173. Moreira, S.; Stramer, B.; Evans, I.; Wood, W.; Martin, P. Prioritization of competing damage and developmental signals by migrating macrophages in the Drosophila embryo. *Curr. Biol.* **2010**, *20*, 464–470. [CrossRef] [PubMed]

174. Xu, S.; Chisholm, A.D. C. elegans epidermal wounding induces a mitochondrial ROS burst that promotes wound repair. *Dev. Cell.* **2014**, *31*, 48–60. [CrossRef] [PubMed]

175. Leon, J.; Rojo, E.; Sanchez-Serrano, J.J. Wound signalling in plants. *J. Exp. Bot.* **2001**, *52*, 1–9. [CrossRef] [PubMed]

176. Orozco-Cardenas, M.L.; Narvaez-Vasquez, J.; Ryan, C.A. Hydrogen peroxide acts as a second messenger for the induction of defense genes in tomato plants in response to wounding, systemin, and methyl jasmonate. *Plant Cell* **2001**, *13*, 179–191. [CrossRef] [PubMed]

177. Suzuki, N.; Mittler, R. Reactive oxygen species-dependent wound responses in animals and plants. *Free Radic. Biol. Med.* **2012**, *53*, 2269–2276. [CrossRef] [PubMed]

178. Yoo, S.K.; Starnes, T.W.; Deng, Q.; Huttenlocher, A. Lyn is a redox sensor that mediates leukocyte wound attraction in vivo. *Nature* **2011**, *480*, 109–112. [CrossRef] [PubMed]

179. Sanchez Alvarado, A.; Newmark, P.A. The use of planarians to dissect the molecular basis of metazoan regeneration. *Wound Rep. Regen.* **1998**, *6*, 413–420.

180. Vriz, S.; Reiter, S.; Galliot, B. Cell death: A program to regenerate. *Curr. Top. Dev. Biol.* **2014**, *108*, 121–151. [PubMed]

181. Santabarbara-Ruiz, P.; Lopez-Santillan, M.; Martinez-Rodriguez, I.; Binagui-Casas, A.; Perez, L.; Milan, M.; Corominas, M.; Serras, F. ROS-Induced JNK and p38 Signaling Is Required for Unpaired Cytokine Activation during Drosophila Regeneration. *PLoS Genet.* **2015**, *11*, e1005595. [CrossRef] [PubMed]

182. Fogarty, C.E.; Diwanji, N.; Lindblad, J.L.; Tare, M.; Amcheslavsky, A.; Makhijani, K.; Bruckner, K.; Fan, Y.; Bergmann, A. Extracellular Reactive Oxygen Species Drive Apoptosis-Induced Proliferation via Drosophila Macrophages. *Curr. Biol.* **2016**, *26*, 575–584. [CrossRef] [PubMed]

183. Chen, Z.; Wu, X.; Luo, H.; Zhao, L.; Ji, X.; Qiao, X.; Jin, Y.; Liu, W. Acute exposure of mercury chloride stimulates the tissue regeneration program and reactive oxygen species production in the Drosophila midgut. *Environ. Toxicol. Pharmacol.* **2016**, *41*, 32–38. [CrossRef] [PubMed]

184. Khan, S.J.; Abidi, S.N.F.; Skinner, A.; Tian, Y.; Smith-Bolton, R.K. The Drosophila Duox maturation factor is a key component of a positive feedback loop that sustains regeneration signaling. *PLoS Genet.* **2017**, *13*, e1006937. [CrossRef] [PubMed]

185. Brock, A.R.; Seto, M.; Smith-Bolton, R.K. Cap-n-Collar Promotes Tissue Regeneration by Regulating ROS and JNK Signaling in the Drosophila melanogaster Wing Imaginal Disc. *Genetics* **2017**, *206*, 1505–1520. [CrossRef] [PubMed]

186. Diwanji, N.; Bergmann, A. The beneficial role of extracellular reactive oxygen species in apoptosis-induced compensatory proliferation. *Fly (Austin)* **2017**, *11*, 46–52. [CrossRef] [PubMed]

187. Amcheslavsky, A.; Wang, S.; Fogarty, C.E.; Lindblad, J.L.; Fan, Y.; Bergmann, A. Plasma Membrane Localization of Apoptotic Caspases for Non-apoptotic Functions. *Dev. Cell.* **2018**, *45*, 450–464.e3. [CrossRef] [PubMed]

188. Suzuki, M.; Takagi, C.; Miura, S.; Sakane, Y.; Suzuki, M.; Sakuma, T.; Sakamoto, N.; Endo, T.; Kamei, Y.; Sato, Y.; et al. In vivo tracking of histone H3 lysine 9 acetylation in Xenopus laevis during tail regeneration. *Genes Cells* **2016**, *21*, 358–369. [CrossRef] [PubMed]

189. Hameed, L.S.; Berg, D.A.; Belnoue, L.; Jensen, L.D.; Cao, Y.; Simon, A. Environmental changes in oxygen tension reveal ROS-dependent neurogenesis and regeneration in the adult newt brain. *eLife* **2015**, *4*, e08422. [CrossRef] [PubMed]

190. Ferreira, F.; Luxardi, G.; Reid, B.; Zhao, M. Early bioelectric activities mediate redox-modulated regeneration. *Development* **2016**, *143*, 4582–4594. [CrossRef] [PubMed]

191. Zhang, Q.; Wang, Y.; Man, L.; Zhu, Z.; Bai, X.; Wei, S.; Liu, Y.; Liu, M.; Wang, X.; Gu, X.; et al. Reactive oxygen species generated from skeletal muscles are required for gecko tail regeneration. *Sci. Rep.* **2016**, *6*, 20752. [CrossRef] [PubMed]

192. Chen, C.H.; Puliafito, A.; Cox, B.D.; Primo, L.; Fang, Y.; Di Talia, S.; Poss, K.D. Multicolor Cell Barcoding Technology for Long-Term Surveillance of Epithelial Regeneration in Zebrafish. *Dev. Cell.* **2016**, *36*, 668–680. [CrossRef] [PubMed]

193. Meda, F.; Gauron, C.; Rampon, C.; Teillon, J.; Volovitch, M.; Vriz, S. Nerves Control Redox Levels in Mature Tissues Through Schwann Cells and Hedgehog Signaling. *Antioxid. Redox Signal.* **2016**, *24*, 299–311. [CrossRef] [PubMed]

194. Bely, A.E.; Nyberg, K.G. Evolution of animal regeneration: Re-emergence of a field. *Trends Ecol. Evol.* **2010**, *25*, 161–170. [CrossRef] [PubMed]

195. Bai, H.; Zhang, W.; Qin, X.J.; Zhang, T.; Wu, H.; Liu, J.Z.; Hai, C.X. Hydrogen peroxide modulates the proliferation/quiescence switch in the liver during embryonic development and posthepatectomy regeneration. *Antioxid. Redox Signal.* **2015**, *22*, 921–937. [CrossRef] [PubMed]

196. Yoo, S.K.; Freisinger, C.M.; Lebert, D.C.; Huttenlocher, A. Early redox, Src family kinase, and calcium signaling integrate wound responses and tissue regeneration in zebrafish. *J. Cell. Biol.* **2012**, *199*, 225–234. [CrossRef] [PubMed]

197. Vriz, S. Redox signalling in development and regeneration. *Semin. Cell. Dev. Biol.* **2018**, *80*, 1–2. [CrossRef] [PubMed]

198. Go, Y.M.; Chandler, J.D.; Jones, D.P. The cysteine proteome. *Free Radic. Biol. Med.* **2015**, *84*, 227–245. [CrossRef] [PubMed]
199. Herrmann, J.M.; Dick, T.P. Redox Biology on the rise. *Biol. Chem.* **2012**, *393*, 999–1004. [CrossRef] [PubMed]
200. Jones, D.P.; Sies, H. The Redox Code. *Antioxid. Redox Signal.* **2015**, *23*, 734–746. [CrossRef] [PubMed]

antioxidants

Review

Physiological Roles of Plant Methionine Sulfoxide Reductases in Redox Homeostasis and Signaling

Pascal Rey [1,*] and Lionel Tarrago [2]

1   Laboratoire d'Ecophysiologie Moléculaire des Plantes, Aix Marseille University, CEA, CNRS, BIAM,
    F-13108 Saint Paul-Lez-Durance, France
2   Laboratoire de Bioénergétique Cellulaire, Aix Marseille University, CEA, CNRS, BIAM,
    F-13108 Saint Paul-Lez-Durance, France; lioneltarrago@msn.com
*   Correspondence: pascal.rey@cea.fr; Tel.: +33-(0)4-42-25-47-76

Received: 6 July 2018; Accepted: 26 August 2018; Published: 29 August 2018

Abstract: Oxidation of methionine (Met) leads to the formation of two *S*- and *R*-diastereoisomers of Met sulfoxide (MetO) that are reduced back to Met by methionine sulfoxide reductases (MSRs), A and B, respectively. Here, we review the current knowledge about the physiological functions of plant MSRs in relation with subcellular and tissue distribution, expression patterns, mutant phenotypes, and possible targets. The data gained from modified lines of plant models and crop species indicate that MSRs play protective roles upon abiotic and biotic environmental constraints. They also participate in the control of the ageing process, as shown in seeds subjected to adverse conditions. Significant advances were achieved towards understanding how MSRs could fulfil these functions via the identification of partners among Met-rich or MetO-containing proteins, notably by using redox proteomic approaches. In addition to a global protective role against oxidative damage in proteins, plant MSRs could specifically preserve the activity of stress responsive effectors such as glutathione-*S*-transferases and chaperones. Moreover, several lines of evidence indicate that MSRs fulfil key signaling roles via interplays with Ca^{2+}- and phosphorylation-dependent cascades, thus transmitting ROS-related information in transduction pathways.

Keywords: methionine; methionine sulfoxide; methionine sulfoxide reductase; physiological function; protein; plant; repair; redox homeostasis; signaling; stress

1. Introduction

Post-translational modifications (PTMs) are covalent modifications that occur after protein synthesis, and change the chemical repertoire of amino acids by altering or introducing functional groups. Thereby, PTMs fulfill critical roles in the control of protein conformation, enzyme activity, or subcellular localization. Proteins are subject to numerous types of PTMs such as phosphorylation, glycosylation, acetylation or hydroxylation, which consist generally in enzymatic modifications of amino acid lateral chains. Special cases of PTMs are "redox PTMs", in which amino acids are modified by reductive or oxidative reactions. The most current redox PTMs are oxidative modifications due to reaction with reactive oxygen species (ROS). As these species are metabolically produced, redox PTMs are tightly linked to changes in the global cell redox homeostasis such as oxidative stress, but also to more subtle variations that are timely and spatially controlled at the subcellular level. All amino acids are subject to oxidative modifications in their lateral chain that are generally irreversible [1]. This is the case for example of carbonylation, the level of which is considered as a marker of oxidative damage [2]. Interestingly, sulfur-containing residues are highly prone to oxidative PTMs that are reversible in many cases. Cysteine (Cys) undergoes modifications such as disulfide bridge formation or *S*-glutathionylation that are reversed through the action of oxidoreductases belonging to the thioredoxin (TRX) and glutaredoxin (GRX) families [3]. TRXs are ubiquitous, small disulfide reductases carrying two catalytic cysteines in a Cys-Xxx-Xxx-Cys active site signature. They form complex families in plants, likely indicating functional specialization depending on subcellular localization and expression pattern [4]. TRXs get reducing power from two main sources: the reduced form of nicotinamide adenine dinucleotide phosphate (NADPH) and the photosynthetic electron transfer chain. GRXs, which are closely related to TRXs, generally use glutathione as an electron donor [5,6], and form large families in higher plants [7]. TRXs and GRXs control multiple metabolic reactions, developmental processes, and stress responses by modulating the redox status of Cys in a very broad range of proteins [4,8–10]. This is evidenced by the number and diversity of interacting partners in photosynthetic organisms [11–13]. Among them, methionine sulfoxide reductases (MSRs), which control the redox status of Met, are well characterized.

The oxidation of free and peptide-bound Met, the other sulfur-containing amino acid, results in the formation of Met sulfoxide (MetO) and Met sulfone (MetO$_2$), the latter modification being irreversible (Figure 1). Quantitative estimations indicate that MetO can range from low micromolar concentration in basal condition to low millimolar concentration during acute oxidative stress [14], corresponding to up to 40% of total Met [15]. MetO is reduced back to Met by MSRs [16]. Met oxidation leads to the formation of *S*- and *R*-MetO diastereoisomers, and two types of MSRs, A and B, specifically reduce the *S*- and *R*-MetO isomers, respectively. Although displaying a very close biochemical function, they do not share any sequence similarity, but have a mirrored active site designed to accommodate each MetO diastereoisomer [16–19]. Further, whereas MSRAs generally reduce both the peptide-bound MetO and the free form with similar efficiency, MSRBs are generally more efficient in reducing the peptide-bound form [20], except for some isoforms in plants [21]. Most MSRs harbor two redox-active Cys and function using a three-step catalytic mechanism involving the formation of a Cys sulfenic acid intermediate, the subsequent formation of a disulfide bond and the regeneration of activity, most often by a TRX reducing system [18,22–24] (Figure 1). Mechanisms involving GRXs, unusual TRX types or other thiol-compounds allow the regeneration of atypical MSR proteins carrying only one catalytic Cys [25–31]. Note that some plant MSRA isoforms do not harbor a catalytic Cys residue [32], raising the question of their biochemical function.

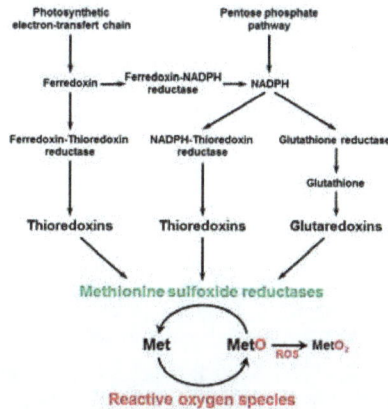

Figure 1. Reducing pathways of plant methionine sulfoxide reductases (MSRs). Methionine (Met) can be oxidized into Met sulfoxide and met sulfone, the latter modification being irreversible. Two main paths supply electrons to MSRs that allow Met regeneration from Met sulfoxide (MetO): on one hand, the path from photosynthetic electron chain to thioredoxins that takes place in chloroplast, and on the other hand, the one from reduced nicotinamide adenine dinucleotide phosphate (NADPH) to thioredoxin that is localized in cytosol. Other routes, involving notably glutaredoxins, participate in the reduction of atypical MSRs harboring for example one unique catalytic cysteine, in chloroplast. ROS: reactive oxygen species.

In bacterial, yeast, and mammal cells, MSRs fulfill essential functions in stress tolerance and during ageing. For instance, in yeast, deletion and overexpression of *MSRA* results in reduced and increased viability, respectively [33], and the abundance of MSRs decreases upon ageing and diseases in mammal cells [34,35]. Modifying the expression of *MSRA* genes revealed their participation in the responses to oxidative stress generated by hydrogen peroxide (H_2O_2) or methyl viologen (MV), which generates superoxide [36,37]. Based on these data, a direct antioxidant function was first attributed to these enzymes in elimination of ROS via cyclic oxidation of Met in proteins and reduction by MSRs [15,38]. Additionally, many evidence in various organisms revealed that the control of Met redox status is a key step in signaling pathways. For instance in *E. coli*, exposure to HOCl leads to Met oxidation in the hypochlorite-responsive transcription factor (HypT) and subsequent activation of the expression of genes participating in protective mechanisms against this toxic compound [39]. In animals, the reversible oxidation of Met by ROS in calcium regulatory proteins constitutes a switch modulating signaling and regulating apoptosis in mammal heart cells [40], and the redox status of two specific Met controls actin polymerization during development and immune response [41,42].

Regarding photosynthetic organisms, the report by Sanchez et al. [43] in 1983 provided the first evidence for MSR activity in extracts from various higher plants, the activity being largely localized in the chloroplastic fraction. The first molecular characterization of a plant *MSR* gene was performed in *Brassica napus* in 1996 [44], and the evidence for the participation of *MSR* genes in defense against oxidative stress was unveiled in *Arabidopsis thaliana* knockout mutants [45]. In the following years, numerous studies provided information about plant *MSR* genes and enzyme biochemical properties. In two previous reviews [46,47], we thoroughly described the organization of MSR families in photosynthetic organisms, which are more complex than in other organisms. Indeed, they include 14, 9, 8, and 8 members in Arabidopsis, poplar, maize and *Chlamydomonas reinhardtii*, respectively, against 2 and 4 in yeast and human, respectively [47,48]. These reviews also described the structure and sequence of *MSR* genes, the various catalytic mechanisms involved in the regeneration of the enzyme activity, and reported the first available data concerning the physiological functions of these reductases in the green lineage. At that time, plant MSRs were presumed to play mainly a direct antioxidant function to preserve cell structures via protein repair and avoidance of oxidative damage. Since then, as shown in other organisms, another key and exciting role in signaling pathways has emerged for plant MSRs. In this review, we present the current knowledge regarding MSRs in photosynthetic organisms. We focus more particularly on their subcellular localization and tissue distribution, on the stimuli and signaling actors controlling their expression, activity and substrate levels, Finally, we review the physiological functions and potential targets reported for plant MSRs in relation with antioxidant defense and signal transduction.

2. Subcellular Localization and Organ Distribution of Plant MSRs

2.1. Subcellular Localization

In Arabidopsis, the MSR family consists of 14 members, predicted to be localized in plastids, cytosol, and endoplasmic reticulum (Figure 2). The subcellular localization of some of them was proven experimentally, as reported for the three plastidial MSRA4, MSRB1, and MSRB2 isoforms [49,50], MSRB7 and MSRB8 in cytosol [51], and MSRB3 in endoplasmic reticulum [52]. Consistently, several MSRA and MSRB isoforms from other species (rice, tobacco, tomato, pepper, soybean, papaya, and banana) were found to localize in cytosol and plastids in Arabidopsis protoplasts using a bimolecular fluorescence complementation (BiFC) approach [53–62]. In other respects, data gained from BiFC experiments suggest that some cytosolic MSRA and MSRB isoforms from litchi and banana are present both in cytosol and in nucleus [61,62]. Taken collectively, these data indicate that MSRs are present in most subcellular compartments. Intriguingly, no MSR isoform is predicted to be addressed in mitochondria. Proteomic analyses performed on mitochondrial fractions found Arabidopsis MSRA4 and MSRB9 as potentially present in this compartment [63–65], but the double addressing of these isoforms remains to be experimentally validated. These data raise the question of how the Met redox status is preserved in plant mitochondria. Indeed, the maintenance of protein redox homeostasis is crucial in this organelle where ROS are produced as by-products in case of impairment of the electron transfer chain [66].

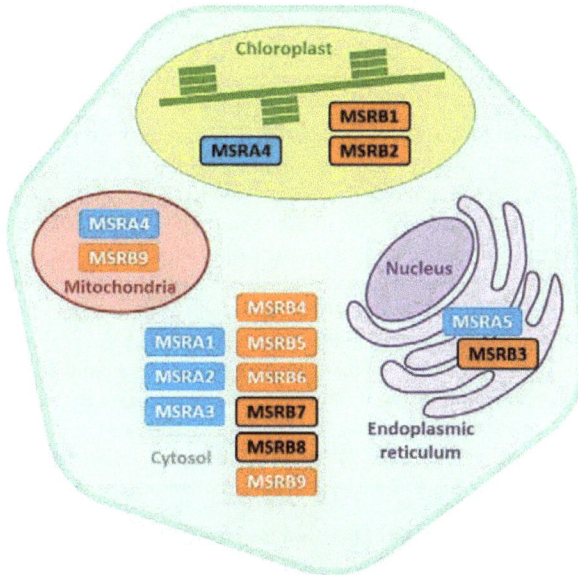

Figure 2. Subcellular localization of MSRs in *Arabidopsis thaliana*. Localization of MSRA4, MSRB1, and MSRB2 in plastid, of MSRB3 in endoplasmic reticulum and of MSRB7 and MSRB8 was proven experimentally [49–52]. These isoforms appear in black. For other isoforms (in white), the localization is based on predictions from sequence analysis or on proteomic analyses [63–65].

2.2. Organ Distribution

As inferred mainly from transcriptomic approaches in the plant model *A. thaliana* [46,47], *MSR* genes exhibit differential expression as a function of organ type. Briefly, *MSR4*, *MSRB1*, *MSRB2*, and *MSRB6* are specific of aerial photosynthetic organs while *MSRA2*, *MSRB5*, *MSRB7*, *MSRB8*, and *MSRB9* are preferentially expressed in root. Accordingly, *MSRB7* and *MSRB8* expression in root was confirmed by quantitative reverse transcription (qRT-PCR) analysis [51], and the high abundance of plastidial MSR proteins (A4, B1 and B2) in leaves was shown by Western analysis [50]. However, the three isoforms are also present in floral organs, and a relatively important amount of MSRA4 protein is detected in root [50], revealing the importance of investigating protein levels to gain an accurate overview about MSR distribution in plant organs and tissues. Regarding other species, data are still scarce at the protein level. Indeed, except the reports by In et al. [67] in rye leaves and by Châtelain et al. [68] in *Medicago truncatula* seeds, most studies are based on Northern, qRT-PCR or β-glucuronidase (GUS) expression analyses. Similarly to what reported in Arabidopsis, genes coding for plastidial isoforms are preferentially expressed in green organs. This is the case of pepper *MSRB2*, its transcript level being much higher in leaf, flower, and stem than in root [56]. In young *Glycine soja* seedlings, the messenger coding for plastidial MSRB5 is present in vascular tissues and two others coding for plastidial isoforms (B1 and B2a) are preferentially found in leaf and stem [60]. Another study in soybean indicated that four of the five *MSRB* genes are more expressed in leaf than in root or seed [69]. Similar differential patterns were reported in monocotyledons. For example, in rice, the *MSRB5* transcript level is higher in leaf than in other organs [70], and *MSRB1.1*, which codes for a plastidial isoform, is much more expressed in leaf and flower than in root or stem [55]. In contrast, the *MSRA4.1* transcript is detected at a similar level in all organs, although the encoded protein is localized in plastid [55]. Finally, an extensive qRT-PCR analysis in maize indicated that *MSRA2* and *A4* are mainly expressed in leaf and *MSRA5.1* and *A5.2* in seed [48]. Regarding *MSRB* genes, *MSRB1* and *MSRB2* are specifically expressed in leaf and *MSRB5.1* and *B5.2* in root [48]. In other respects, one strawberry *MSRA* gene is expressed only in the receptacle of red mature fruit [71]. Consistently with this finding, the expression of one *MSRA* gene is strongly up-regulated in the last stages of ripening and senescence in banana fruit [62]. In litchi fruit, the expression of *MSRA1, A2,* and *B1* genes decreases as senescence proceeds during storage [61].

These data indicate that MSRs are present in all plant organs, but display distinct expression patterns in many cases. They need to be deepened to determine whether transcript and protein abundances are correlated and to delineate expression in specific tissues and organs such as floral components, for which the knowledge is still scarce. By crossing the data related to subcellular and organ localization, it turns out that several cytosolic MSRs are specifically expressed in root while plastidial MSRs are preferentially found in aerial photosynthetic organs where photosynthesis takes place. This could mean that the function of the latter is linked to this metabolism, the activity of which can lead to the production of ROS altering redox homeostasis. On the other hand, and as mentioned earlier, MSRs fulfill key roles during ageing in bacterial, yeast, and animal cells. The expression patterns of plant genes reveal that plastidial MSRBs are more abundant in young leaves than in older ones [50], and that other isoforms are specifically expressed in fruit and seed at different stages of the ripening or maturation processes [61,62,68,71]. These findings are consistent with the participation of MSRs in the control of ageing and senescence processes in plant organs.

3. Regulation of the Expression of *MSR* Genes in Photosynthetic Organisms

3.1. Effect of Environmental Conditions

In agreement with the biochemical function of MSRs in the maintenance of Met redox status, the first microarray data gained in Arabidopsis revealed that environmental constraints leading to oxidative stress result in increased expression of most *MSR* genes [46]. In the last years, the expression patterns of these genes have been refined in various types of photosynthetic organisms in relation with abiotic, but also biotic constraints. The data gained in higher plants are summarized in Table 1.

Consistently with the data reported in Arabidopsis, oxidative stress conditions generated by manganese deficiency or copper excess in *C. reinhardtii* and *U. fasciata* algae lead to up-regulation of *MSRA* genes [72,73]. In *C. reinhardtii*, the expression of three *MSRB* genes (*1.1, 1.2* and *2.1*) is induced in very high light conditions and upon treatment with H_2O_2 [74]. Data that are more meaningful regarding the physiological factors regulating expression of *MSR* genes were gained in relation with the activity of the photosynthetic chain. In *U. fasciata*, the transcript levels of *MSRA* and *MSRB* genes peak following 1-h light exposure [75]. Most interestingly, the use of various inhibitors of the photosynthetic electron chain revealed that the expression of these genes differentially depends on the redox status of components belonging to the cytochrome b_6f complex or downstream complexes. This strongly supports the hypothesis that the photosynthetic activity level, which modulates plastidial redox homeostasis, plays an essential role in pathways regulating the expression of *MSR* genes. As these genes are nuclear-encoded, these pathways very likely involve retrograde signaling from plastid to nucleus.

Numerous studies report that oxidative stress conditions are associated with up-regulation of *MSR* gene expression in higher plants (Table 1). Thus, MV treatment leads to increased transcript or protein levels of tobacco MSRB3, tomato MSRA2, A4 and A5, rice MSRB1, rye MSRA, and Arabidopsis MSRA4, MSRB7, and MSRB8 [51,55,58,67,76,77]. More physiological constraints that impair the cell redox homeostasis enhance *MSR* expression. For instance, copper excess leads to *MSRB5* up-regulation in rice [70]. In Arabidopsis, exposure to cadmium triggers the antioxidant defense system, notably the expression of most *MSR* genes, but also provokes a decrease in the abundance of plastidial MSRBs [78,79]. In *Brassica juncea*, such a treatment results in a higher amount of cytosolic MSRA2 [80]. In other respects, increased amounts of plastidial MSRs (A4, B1 and B2) were observed in Arabidopsis plants exposed to photooxidative stress conditions generated by high light and low temperature conditions [50]. In rye, high light conditions induce the accumulation of a cytosolic MSRA protein [67].

Regarding other environmental constraints, many studies reported increased *MSR* expression in conditions leading to osmotic stress such as water shortage, high salt, and low temperature (Table 1). This was first established from microarray data in the Arabidopsis plant model [46,47]. In maize, most *MSR* genes are up-regulated in root, stem and leaf in the presence of polyethylene glycol (PEG) or NaCl with distinct kinetics depending on gene type and organ [48]. Consistently, in rice, *MSRA4.1* and *MSRB1.1* expression is enhanced by mannitol, high salt and low temperature [55]. In other respects, a higher MSRA protein abundance was observed in cold-hardened rye plants [67] and in maize seedlings, low temperature induces the expression of *MSRA5* in mesocotyl [81]. In soybean, Chu et al. [69] reported differential expression of the five *MSRB* genes in response to drought and high salt. Most importantly, they observed that three of them (*MSRB2, B3* and *B5*) show increased transcript levels in response to drought, but only in leaf and at distinct vegetative or reproductive stages. In tobacco, *MSRA4* expression is up-regulated by dehydration and cold, but not modified by high salt [59], whereas that of *MSRB3* is enhanced by cold and salt [58]. Finally, in tomato, *MSRA3* and *MSRA4* are substantially up-regulated by mannitol, high salt, and low temperature [82]. Altogether, these data give strong credence for essential functions of MSRs in plant responses to osmotic constraints. Accordingly, in the *Atriplex halimus* halophyte species, cultivation in the presence of 300 mM NaCl increased the abundance of plastidial MSRA concomitantly to a higher total MSR activity in a salt-tolerant genotype compared to a salt-sensitive one [83]. However, in barley, an increased protein amount of one MSR isoform was noticed using a proteomic approach in a salt-susceptible genotype compared to a salt-tolerant one [84], and no difference was noticed in the amount of plastidial MSRs in two cultivars exhibiting contrasted response to water deficit [85]. In other respects, exposure to carbonate, which induces alkaline stress in addition to osmotic and ionic stresses, leads to the expression of most *MSRB* genes in *Glycine soja* whether in leaf or in root [60].

In comparison, less is known regarding the expression of *MSR* genes in response to biotic stress. The first evidence was provided in Arabidopsis plants that display a strongly increased *MSRA4* transcript level following infection by the cauliflower mosaic virus, but no change in response to a virulent *Pseudomonas syringae* strain [49]. In papaya, infection by the ringspot virus leads to up-regulation of *MSRB1* expression in the late stages [54]. Most interestingly, some *MSR* genes could be involved in plant immune responses. Thus, in poplar leaves, the abundance of a plastidial MSRB is unchanged during infection by an incompatible rust *M. larici-populina* strain, whereas the protein level increases in the presence of a compatible strain. In contrast, the amount of another MSRB strongly decreases after infection either with compatible or incompatible fungi [50]. In pepper, the level of a transcript coding for a plastidial MSRB isoform first strongly decreases following infection both with compatible and incompatible *Xanthomonas axonopodis* strains, and then is restored to the initial level only in the case of the compatible reaction [56]. In *A. thaliana*, avirulent and virulent *P. syringae* strains lead to very distinct expression patterns for *MSRB7* and *MSRB8*, both being much more strongly up-regulated in the case of an incompatible reaction [86]. Moreover, an increased *MSRA2* transcript level was noticed early following infection of Arabidopsis seedlings by the parasite plant *Orobanche ramosa* [87]. Altogether, these data indicate that MSRs likely participate in immunity mechanisms and active defense against most types of biotic constraints, to which plants are exposed.

3.2. Signaling Actors Involved in the Control of MSR Gene Expression

3.2.1. Involvement of ROS and Reactive Nitrogen Species (RNS) in MSR Gene Expression

Most, if not all environmental conditions, leading to the expression of *MSR* genes reported in the previous section, involve the production of ROS due to metabolic impairment and subsequent changes in cell redox homeostasis that are associated with specific ROS signatures [88]. Reactive nitrogen species (RNS) constitute another type of oxidant molecules, tightly related to ROS, such as peroxynitrite produced by the reaction of nitric oxide (NO) with superoxide. Both ROS and RNS are deleterious at high level since they damage all macromolecules through oxidation, but fulfill critical signaling functions at basal level in developmental processes and responses to environmental constraints [89,90]. ROS and RNS are produced in very specific subcellular cell compartments, and very likely do not directly regulate gene expression at the transcriptional level due to their site of production within cell, diffusion properties and half-life time [2]. They are assumed to initiate or transfer signaling information through redox metabolic reactions with antioxidant molecules, lipids, and proteins [90,91]. Based on the increasing knowledge gained in bacterial, yeast, animal, and plant models, ROS/RNS are now considered as signals shifting cell redox homeostasis and major drivers in responses and adaptation to abiotic and biotic constraints [91,92]. For instance, the role of ROS is critical in the control of cell death upon incompatible reaction in plants [93].

Thereby, ROS and RNS species are good candidates to control *MSR* expression in photosynthetic organisms. Consistent with this hypothesis, Chang et al. [94] observed in Chlamydomonas distinct expression patterns for five *MSR* genes in high light conditions, four being up-regulated (*MSRA3, A5, B2.1* and *B2.2*) and the fifth down-regulated (*MSRA4*). Using various ROS scavengers and generators, they showed that these patterns are specifically linked to the ROS type. For example, hydrogen peroxide and superoxide differentially modulate *MSRB2.1* and *MSRB2.2* expression, respectively. Regarding RNS, a similar pharmaceutical approach in *U. fasciata* showed that NO induces acclimation to high light concomitantly to upregulation of *MSRA* and *MSRB* expression [95].

In higher plants, exposure to MV leads to up-regulation of various *MSR* genes in many species [51,55,59,67,76,82]. Treatment with H_2O_2 leads to more complex data since a decreased *MSRA4* transcript amount is noticed in tobacco seedlings [59] while the four tomato *MSRA* genes display contrasted responses [82]. In Arabidopsis, two *MSRA* genes, out of the four tested, display enhanced expression following H_2O_2 treatment [96]. Accumulation of singlet oxygen, a ROS produced when chlorophyll triplets excite O_2, induces *MSRB7* expression in Arabidopsis [97]. In other respects, priming with NO prevents up-regulation of *MSR* expression in cadmium-treated Arabidopsis plants, very likely due to reduction in ROS production and limitation of oxidative stress [78]. Altogether, these data reveal that ROS-related signals control the expression of *MSR* genes in higher plants. Although experimental evidence remains scarce, RNS also very likely regulate *MSR* expression as illustrated by the strong *MSRB7* up-regulation in Arabidopsis plants treated with *S*-nitrosoglutathione, a reservoir for NO [98].

3.2.2. Involvement of Phytohormones in MSR Gene Expression

Tight and complex interplays between ROS-dependent transduction pathways and other recognized signaling components like phytohormones, mitogen-activated protein kinases and calcium ions are established [99]. With regard to *MSR* genes, this is illustrated by the analysis of linolenic acid-responsive genes in Arabidopsis cell cultures subjected to osmotic stress [100]. Using a RNA-seq approach, this study reported that *MSRB7* and other antioxidant genes are strongly induced by linolenic acid, which is released from plastidial membrane galactolipids and is a precursor of jasmonic acid (JA). JA, together with derivatives such as methyljasmonate (MeJA), constitute a phytohormone family derived from oxidized lipids (oxylipins) and mediating responses to conditions modifying the redox homeostasis, notably biotic constraints [101]. In tomato seedlings treated with JA, the expression of the four *MSRA* genes is either down-regulated or unchanged [82]. However, in Arabidopsis, JA treatment increases the transcript level of two *MSRA* genes and decreases this level for two others [96]. In pepper, a strong decrease in *MSRB2* transcript level was observed following exposure to MeJA and to salicylic acid (SA) [56]. SA, another key hormone involved in plant immune responses to microbial pathogens [102], strongly triggers the expression of one *MSRA* gene in tomato, but decreases the expression of another [82].

Among phytohormones, abscisic acid (ABA) is central to plant responses to osmotic constraints such as drought or high salt partly through the transcriptional control of gene expression [103]. In many plant species, up-regulation of *MSR* expression was reported upon these constraints (cf. Section 3.1). Accordingly, exposure of plants to ABA triggers *MSRA4* expression in Arabidopsis [96], *MSRB2* in soybean [69], and *MSRB3* in tobacco [58]. Nevertheless, other *MSR* genes from these species and tomato are unresponsive to ABA or their expression is down-regulated by this phytohormone [59,69,82,96], revealing specific ABA-dependent patterns.

Table 1. Effects of various constraints and treatments on the expression of *methionine sulfoxide reductases* (*MSR*) genes in higher plants. The table was built from data gained on plants exposed to the mentioned constraints and treatments (in bold) either in vivo or in vitro. The cited reports aimed to investigate the expression of one or more *MSR* (*A* or *B*) genes at the transcript or protein levels. ↗, the expression of at least one *MSR* gene is up-regulated; ↘, the expression of at least one *MSR* gene is down-regulated; ↗ ↘, the expression of at least two *MSR* genes is modified (up or down).

Condition	Variation in *MSR* Expression. Species	References
Abiotic constraints		
High light	↗ *A. thaliana*	[76]
	↗ *S. cereale*	[67]
High light/low temperature	↗ *A. thaliana*	[50]
Low temperature	↗ *N. tabacum, S. lycopersicum*	[59,82]
	↗ *O. sativa, S. cereale, Z. mays*	[55,67,81]
Water deficit	↗ *G. max, N. tabacum*	[58,69]
High salt (NaCl)	↗ *A. halimus, A. thaliana, N. tabacum, S. lycopersicum*	[58,59,82,83,96]
	↗ *H. vulgare, O. sativa, Z. mays*	[48,55,84]
	↗ ↘ *G. max*	[69]
High carbonate	↗ *G. soja*	[60]
Cadmium	↗ ↘ *A. thaliana, B. juncea*	[78–80]
Biotic constraints		
Virus	↗ *A. thaliana, C. papaya*	[49,54]
Bacteria	↗ *A. thaliana*	[86]
	↘ *C. annuum*	[56]
Fungi	↗ ↘ *Populus × interamericana*	[50]
Parasite plants	↗ *A. thaliana*	[87]
Oxidative treatments		
Methyl viologen	↗ *A. thaliana, N. tabacum, S. lycopersicum*	[51,59,76,82]
	↗ *O. sativa, S. cereale*	[55,67]
Hydrogen peroxide	↗ ↘ *S. lycopersicum*	[82]
	↗ *A. thaliana*	[96]
Singlet oxygen	↗ *A. thaliana*	[97]
S-nitrosoglutathione	↗ *A. thaliana*	[98]
Copper excess	↗ *O. sativa*	[70]
Hormone treatments		
Abscisic acid	↗ ↘ *A. thaliana, G. max, N. tabacum, S. lycopersicum*	[58,59,69,82,96]
Jasmonic acid	↗ ↘ *A. thaliana, S. lycopersicum*	[82,96,100]
	↘ *C. annuum*	[56]
Salicylic acid	↗ ↘ *S. lycopersicum*	[82]
	↘ *C. annuum*	[56]
Ethylene	↗ *S. lycopersicum*	[82]
	↗ *M. acuminata*	[62]

Phytohormones are well known actors in the control of plant development. Most importantly, ROS and thiol-based mechanisms are also essential signaling players allowing proper development of plants in relation with varying environment and energy availability [10]. As mentioned in

Section 2.2, some *MSR* genes are specifically or highly expressed in fruits, such as banana and strawberry [61,62,71]. Climacteric fruits undergo ripening following harvest, a process during which the phytohormone ethylene plays a critical role. Interestingly, in the climacteric banana fruit, *MSRA7* expression is up-regulated during ripening and dramatically increased following ethylene treatment [62]. Consistently, the *MSRA2* transcript level is strongly up-regulated in another climacteric fruit, tomato, following treatment with ethephon, an ethylene-releasing compound [82]. In other respects, the transcript amounts of three *MSR* genes decrease in litchi, a non-climacteric species, along with the senescence process following harvest [61]. These findings suggest that the expression of *MSR* genes depends on fruit developmental stage and is controlled via the action of phytohormones, such as ethylene, which regulate maturation and senescence processes.

3.2.3. Conclusions

Based on all available data, we conclude that in higher plants ROS play a central role in the control of the expression of *MSR* genes at the transcriptional level upon environmental constraints. This is consistent with the role of MSR enzymes in the maintenance of protein redox status and the up-regulation of the expression of *MSR* genes generally observed upon these constraints (Table 1). ROS transfer signaling information through redox metabolic reactions with different compounds, and participate in transduction pathways involving other actors such as phytohormones [10,90]. Interestingly, phytohormones play more complex roles in regulating *MSR* gene expression, as mentioned above and shown in Table 1, probably in relation with both environmental condition and development stage.

4. MSR Activity and MetO Content in Higher Plants

4.1. MSR Activity in Plant Extracts

Another question emerges from the subcellular distribution of MSRs: do they display similar abundance and activity? The pioneer works by Sanchez et al., and Ferguson and Burke [43,104] indicated that in various plant species a large part of the leaf MSR activity (85%) is localized in the chloroplastic fraction, the remaining 15% being measured in the cytosol. Surprisingly, the number of MSR isoforms is low in this compartment: 3 out of 14 in Arabidopsis. Using a genetic approach i.e., mutants knockout for *MSRB1/MSRB2* genes and/or knockdown for *MSRA4* expression, we confirmed this finding by showing that the two plastidial MSRBs account for the greater part of leaf MSR capacity (75%), and that this capacity is further decreased in plants deficient in the three plastidial isoforms [105,106]. These data reveal that MSRAs and MSRBs do not fulfil equivalent roles in terms of activity and physiological function, at least in chloroplast, and suggest that the two MetO diastereoisomers are not generated in a racemic proportion *in planta* [106]. The predominant plastidial MSR capacity in leaf is very likely related to the fact that chloroplast is a major site of ROS production because of the photosynthetic electron transfer chain activity in light conditions [66]. Interestingly, the MSR activity is in the same range from 10 to 80 pmol $Met \cdot min^{-1} \cdot mg \cdot prot^{-1}$ in Arabidopsis root, stem, leaf, flower bud, flower, green silique, and seed extracts ([68,105]; our unpublished data). The contribution of each MSR isoform likely depends on organ type and physiological context. Accordingly, the MSR capacity in Arabidopsis lines lacking plastidial MSRBs is much less decreased in seed than in leaf [68,105] and the MSR activity in an Arabidopsis mutant lacking cytosolic *MSRA2* is specifically and strongly decreased by 50% in the second part of the dark period compared to wild-type (WT) [107]. Thorough measurements of MSR activity at the organ level in mutants knockout for the various *MSR* genes would greatly help to delineate the physiological functions of all isoforms.

Relatively few data are available concerning the effects of external factors on MSR activity in plants. Ferguson and Burke [108] measured the activity in various species subjected to high temperature or water deficit and observed species-dependent changes. For instance, water shortage leads to decreased and increased activity in cotton and pea, and in wheat, respectively. In Arabidopsis, photooxidative constraints result in increased MSR activity in chloroplastic or leaf fractions [76,105], and treatments with NaCl or ABA, but not exposure to low temperature, provoke substantial reductions in activity [96,105]. In wild soybean, moderately and strongly increased activities were noted upon high salt (NaCl) or carbonate alkaline stresses, respectively [60]. In *A. halimus*, high salinity provokes a noticeable increase in MSR activity in a tolerant genotype, but not in a sensitive one [85]. However, in barley, no difference in activity was recorded in two contrasted cultivars for water shortage tolerance [86]. In conditions of biotic stress, no significant change in MSR activity was observed in tomato plants challenged with Phytophtora [56].

4.2. MetO Content in Plants

Regarding protein-bound MetO, the substrate of MSR, its proportion in relation to the total quantity of Met and MetO is in the range from 5 to 20% in various species (pea, wheat, potato, Arabidopsis) grown in optimal conditions [45,76,108]. However, much lower levels (less than 1.5%) were also measured in Arabidopsis [52,105]. Interestingly, the MetO quantity varies during the day/night cycle, with a 4-fold higher content in the middle of the light period [109]. Genetic studies confirmed the biochemical function of MSRs via the determination of MetO content in extracts from modified plants. Arabidopsis mutants deficient in various types of MSRBs or MSRAs exhibit a higher MetO level compared to WT plants [45,52,76,105,109]. The increase in MetO proportion was recorded mainly in plants subjected to environmental constraints such as high light or low temperature [52,105]. Consistently, Arabidopsis plants overexpressing plastidial MSRBs or cytosolic MSRB3 display a lower protein-bound MetO level upon photooxidative constraints [52,76,105].

The consequences of environmental variations on MetO content in non-modified plants are less simple to interpret. In various species subjected to high temperature or water deficit, no great variation was noticed, except in pea plants exposed to high temperature where a strong decrease was noticed [108]. In 6-week old Arabidopsis WT plants, we reported a decreased peptide-bound MetO content upon high light and long photoperiod conditions, but no variation at low temperature [105]. However, other studies reported substantial increases in the peptide-bound MetO proportion in Arabidopsis plants subjected to severe photooxidative stress [45,76], or during cold acclimation [52].

4.3. Signals Involved in the Control of MSR Activity and MetO Level

Taken collectively, these findings reveal that in physiological conditions MSR activity and MetO content are regulated in a fine and complex manner in plant cells as a function of species type, environmental condition and stress intensity. Higher MSR activity and MetO level are generally recorded upon constraints leading to pronounced oxidative stress, consistent with the fact that Met oxidation occurs in the presence of ROS excess, and is a marker of protein damage [2]. Accordingly, a proteomic study of Arabidopsis *catalase 2* knockout plants exposed to very high light, identified more than 50 proteins displaying a higher MetO content compared to WT, due to deficiency in H_2O_2 scavenging [110]. However, another proteomic study on Arabidopsis cell cultures revealed that Met oxidation in proteins could result from the action of non-oxidative signaling molecules [111]. Thereby, it can be concluded that MetO formation in proteins even occurs in the absence of ROS excess during environmental changes and is a very finely controlled PTM mediated by interplaying transduction pathways and actors remaining to be unveiled.

5. Physiological Functions of Plant MSRs

5.1. Oxidative Treatments and Photooxidative Conditions

In animal, yeast and bacterial cells, many studies revealed the involvement of MSRs in the responses to MV or H_2O_2 [33,36,38]. In a first step, a similar physiological function has been searched in photosynthetic organisms. The use of knockdown and overexpression *C. reinhardtii* lines showed that two MSRB isoforms play specific protective roles against oxidative stress generated by H_2O_2 or high light [74]. Following exposure to MV, ozone or high light intensity, *MSRA4*-antisense or -overexpression Arabidopsis lines display poorly or better-preserved photosynthetic activity, respectively, compared to WT [76]. Moreover, these phenotypes are associated with increased or decreased MetO content, indicating that MSRA4 likely plays a role in the protection of photosynthetic structures against oxidative damage. Consistently, based on chlorophyll content, ion leakage and growth measurements, improved tolerance towards MV and H_2O_2 treatments was noticed in Arabidopsis seedlings overexpressing cytosolic MSRB7 or MSRB8 isoforms grown either on synthetic media or on soil, while a greater sensitivity to these compounds was observed in deficient lines [51]. Tomato plants ectopically expressing Arabidopsis MSRB7 also display increased tolerance to MV [51]. Interestingly, *MSRB7*- and *MSRB8*-overexpressing Arabidopsis plants exposed to MV exhibit reduced H_2O_2 level and increased glutathione-*S*-transferase (GST) activity, clearly showing the integration of MSRs in the antioxidant network. Based on the higher susceptibility to MV of Arabidopsis plants lacking MSRB3, this isoform located in endoplasmic reticulum is also presumed to preserve cell structures against oxidative damage [52]. In rice, decreased tolerance to MV and copper excess was reported in plants lacking the MSRB5 isoform, with both treatments leading to severe oxidative stress [70].

5.2. Abiotic Constraints

In natural environments, most abiotic constraints induce changes in cell redox homeostasis to a lesser extent than those described above following exposure to strong oxidizing agents, which are not relevant from a physiological point of view. The functions of MSRs in more meaningful conditions are thus somewhat harder to determine, and their expression patterns are very helpful to progress in this direction. Therefore, based on *MSRB3* induction in *A. thaliana* at low temperature, Kwon et al. [52] showed that a mutant deficient in this isoform loses the ability to tolerate freezing temperatures following cold acclimation. Overexpression of a mutated active version of the *MSRA4* gene, which is induced by NaCl treatment, results in enhanced tolerance to high salt in in vitro grown seedlings [112]. Regarding tolerance to heavy metals, ectopic expression of *Brassica rapa MSRA3* in Arabidopsis leads to better growth of in vitro plantlets in the presence of 50 μM cadmium [113]. Other, less severe abiotic constraints, but applied for a long time, have been reported to impair the growth of Arabidopsis lines lacking MSR isoforms. Thus, plants deficient in both plastidial MSRB1 and MSRB2, display a rosette weight reduced by ca. 25% compared to WT when grown at 10 °C for 18 days [105]. These plants also exhibit substantially reduced growth when continuously cultivated in long day/high light conditions compared to short day/moderate light [105]. In other respects, the growth of an Arabidopsis line lacking cytosolic MSRA2 is impaired in short-day and not in long-day conditions [107]. MSRA2 may limit the oxidative damage occurring in proteins at the end of a long dark period [105]. The long duration—several weeks in these experiments [105,107]—allowed to uncover physiologically relevant functions for plant MSRs, probably due to the fine and timely control of MetO formation in proteins as a function of environment and developmental stage.

In species other than Arabidopsis, several MSRs fulfil essential roles in responses to osmotic constraints. Thus, in rice, overexpression of plastidial MSRA4.1 is associated with improved tolerance to high salt (300 mM NaCl), as inferred from photosynthetic activity and oxidative damage measurements [55]. Furthermore, rice plants overexpressing pepper *MSRB2* exhibit better tolerance to water deficit than WT upon shortage and higher survival rate following re-watering [57,114]. In these plants, microarray analysis indicated that genes coding for photosystem components are much less down-regulated in drought conditions than in WT [57]. Regarding salt stress, few data are available regarding the protective effect of MSRs. Recently in *G. soja*, their involvement was reported in a specific constraint, i.e., high carbonate that leads to alkaline stress. Indeed plants overexpressing *MSRB5a.1* exhibit better tolerance to high carbonate either at the germination stage in vitro or during vegetative growth on soil [60].

5.3. Biotic Constraints

With regard to biotic constraints, two reports clearly provided physiological evidence for a role of plant MSRs upon attack of fungi or bacteria. Overexpression in tomato of pepper *MSRB2* results in enhanced resistance to two Phytophtora species that cause severe diseases in Solanaceae [56]. In parallel, these tomato lines display reduced H_2O_2 content following infection and are more tolerant to oxidative stress generated by high light or MV. This study also showed that pepper lines silenced for *MSRB2* expression exhibit increased ROS production, accelerated cell-death in the case of incompatible infection by a bacterial Xanthomonas race and increased susceptibility following infection by a virulent strain [56]. Data confirming the essential roles of MSRs in immune responses were also recently reported in Arabidopsis [86]. Compared to WT, knockout or overexpression lines for cytosolic *MSRB8* display increased sensitivity or tolerance, respectively, to an avirulent strain of *Pseudomonas syringae* while no modification in the responses of transgenic plants occur following infection by virulent strains [86]. Altogether, these data demonstrate that plant MSRs fulfill critical functions in plant immune mechanisms that could be explored for improving crop resistance.

It is worth mentioning that the plant pathogen *Erwinia chrysanthemi* when deficient for *msrA* displays reduced pathogenicity, revealing the requirement of this gene for the full virulence of the bacteria, likely via the participation in defense mechanisms against plant-produced ROS [115]. This finding and those reported on plant MSRs in compatible and incompatible reactions show that MSRs are essential actors in defense mechanisms in both partners during the infection process. Further, plant MSRs could even constitute targets of pathogens. Indeed, in papaya, the Nla-pro protein of the papaya ringspot virus interacts with the preprocessed MSRB1 protein and prevents its import in chloroplasts, thus possibly weakening antioxidant defenses of the host plant [54].

5.4. Involvement in Ageing Process

The participation of MSRs in ageing and lifespan control is well established in bacteria, yeasts, and animals [33,37]. The only evidence in plants concerns seed viability. Seeds are in a natural oxidative context leading to protein oxidation that at high level is deleterious and associated with ageing. In two *Medicago truncatula* genotypes contrasted for seed quality, a strong positive correlation was observed following controlled deterioration between the time to a 50% drop in viability and the MSR capacity of mature seeds. A similar correlation was recorded in seeds of *A. thaliana* lines, altered for *MSR* gene expression and capacity [68]. These data clearly reveal that the MSR repair system plays a critical role in the establishment and preservation of longevity in plant seeds.

6. Mode of Action and Substrates of Plant MSRs

Upon environmental constraints and development (Figure 3), plant MSRs likely fulfil an antioxidant function through MetO repair in proteins exhibiting accessible Met residues positioned on their surface [22,38]. Besides, the increasing knowledge gained recently gives strong credence for much more specific functions in the control of Met redox status in particular proteins involved in signaling pathways. Therefore, the identification of MSR partners is a prerequisite for giving accurate insight into the physiological roles of these reductases in plants. In recent years, in addition to targeted approaches, several global strategies based on biochemical, transcriptomic, and proteomic methods were set up to search for proteins interacting with MSRs, or displaying MetO residues, as well as genes exhibiting modified expression in plants up- or down-regulated for *MSR* genes.

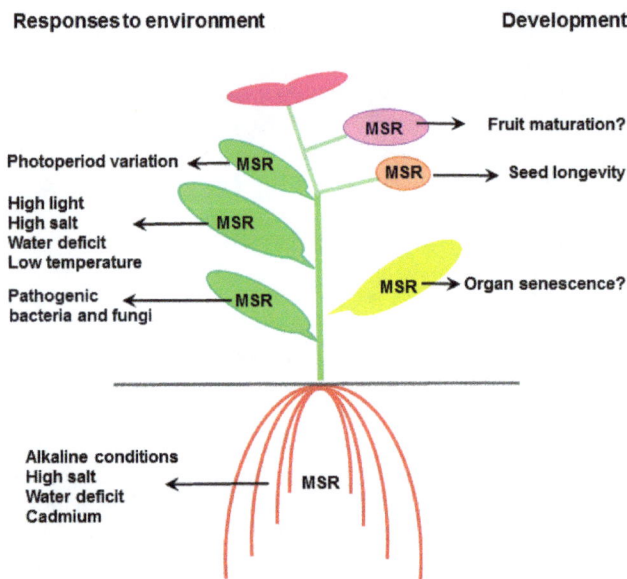

Figure 3. Physiological functions of plant MSRs. The functions of MSRs in the responses to high light, high salt, high carbonate, water deficit, low temperature, cadmium and in photoperiod adaptation was shown in Arabidopsis [50,52,76,105,107,112,113] rice, [55,57,114] or *Glycine soja* [60]. The participation in responses to biotic constraints was reported in Arabidopsis [86], pepper, or tomato [56] the involvement in seed ageing in Arabidopsis and *Medicago truncatula* [68]. The presumed role in organ senescence and fruit maturation is based on expression patterns and identification of possible partners in Aarabidopsis, strawberry, litchi, and banana [50,61,62,71].

6.1. Strategies for Searching Plant MSR Targets

6.1.1. Proteins Displaying High Met Content

One of the first strategies to identify MSR targets in plants is based on the hypothesis that proteins exhibiting a high Met percentage or Met-rich domains are good candidates. The Arabidopsis plastidial small heat shock protein of 21 kDa (HSP21) possesses a unique 19-residue domain carrying six Met in an amphipathic helix. In vitro, MSRA4 counteracts Met oxidation in HSP21, restores its oligomeric conformation and maintains its chaperone-like activity [116]. Similarly, based on the high Met content (4.6%) and on the fact that its homologue in *Escherichia coli* is a target of MSRs [117]. cpSRP54, the chloroplast signal recognition particle of 54 kDa that addresses light-harvesting complex (Lhc) proteins to thylakoids, was presumed to be a target of MSRs. Using recombinant forms, we showed that oxidized cpSRP54 is a substrate for plastidial MSRBs [105]. Interestingly, the oxidized form of the other component of the signal recognition particle, cpSRP43 that exhibits a lower Met content (1.9%) is also reduced by MSRBs [105]. More recently, a thorough survey of genomic data allowed identifying Met–rich proteins, MRPs, far more systematically in Arabidopsis and soybean. The search based on two criteria, peptide length of at least 95 residues and Met content higher than 6%, resulted in the isolation of 121 and 213 genes, respectively, coding for proteins meeting both conditions [118]. Of note, the function of 50% of encoded proteins is unknown. Such *in silico* approaches sound very promising and powerful to search proteins harboring Met-rich domains and identify physiological targets of MSRs.

6.1.2. Proteins Exhibiting Modified MetO Content in Response to Oxidative Treatments or Signaling Molecules

The proteins carrying MetO residues are also likely to be MSR substrates. Unfortunately, there is no efficient tool for rapidly isolating such proteins, notably due to the lack of highly reliable antibodies specific to MetO [119] to set up immunological-based analyses at the proteome scale. In the plant field, there is only one example reporting the use of these antibodies to search for MetO-containing proteins. The comparison of Western patterns from rice plants either WT or overexpressing pepper plastidial MSRB2 highlighted a higher signal level in the range of 40 kDa in WT. Mass spectrometry analysis identified 3 proteins located in plastid: porphobilinogen deaminase (PBGD), dihydrodipicolinate reductase I and ferredoxin-NADP reductase. Based on the MetO content of recombinant PBGD following H_2O_2 treatment and its ability to be reduced by pepper MSRB2, PBGD was proposed as a physiological substrate of MSRs [57]. To overcome the low specificity of MetO antibodies, other strategies based on redox proteomics were set up. Marondedze et al. [111] used titanium oxide treatment in combination with dihydroxybenzoic acid to enrich MetO containing peptides. They analyzed protein extracts of Arabidopsis cell suspension cultures treated with an analogue of cyclic guanosine monophosphate (cGMP), and identified by tandem mass spectrometry 94 and 224 unique proteins carrying MetO strongly enriched following 30 and 60 min of treatment, respectively. However, it is important to note that such a method of enrichment may provoke non-physiological oxidation of Met [120]. Another approach based on tandem-mass spectrometry was developed to isolate proteins differentially oxidized in WT and *MSRB7*-overexpressing Arabidopsis plants treated with MV [121]. It consisted to treat the protein extract with cyanogen bromide that hydrolyses the C-ter of Met, but not of MetO, before trypsin digestion and protein identification. This analysis identified more than 30 proteins that could be MSR substrates *in planta* [121]. Another strategy to identify proteins carrying MetO consists to take advantage of recombinant MSRs [110]. This combined fractional diagonal chromatography (COFRADIC) is made of three steps: (i) HPLC-fractionation of peptides; (ii) treatment of peptides with recombinant MSRs, which induces a hydrophobic shift in MetO containing peptide; and (iii) refractionation and identification of shifted peptides. Using this strategy, about 400 proteins carrying MetO were identified and differential oxidation was investigated in extracts from WT and from *catalase 2* knockout plants that over-produce H_2O_2 at ambient CO_2 concentration. Consistently, 51 proteins were significantly more oxidized at the level of Met in this genetic background [110]. These three studies clearly show the power of redox proteomic to identify potential physiological targets of plant MSRs at a large scale and indicate that these reductases very likely possess a broad range of substrates.

6.1.3. Proteins Interacting with MSRs

Proteins that interact with MSRs are also appropriate candidates to be reduced by these enzymes. Based on this hypothesis, we aimed at isolating plant MSR partners using affinity chromatography [122]. Using Arabidopsis recombinant MSRB1 and leaf extracts, we isolated 24 proteins, 13 being plastidial and potential physiological substrates of MSRB1. The other 11 could interact with non-plastidial MSRB isoforms. Several are actually substrates of MSRs when their recombinant forms are treated with H_2O_2 and others possess homologues in yeast or mammals known to interact with MSRs and/or to possess Met sensible to oxidation, arguing for the relevance of the affinity-based strategy to target physiological partners of MSRs in plants.

6.1.4. Genes Displaying Modified Expression in Lines Up- or Down-Regulated for *MSR* Expression

Proteins exhibiting differential abundance in plants down- or up-regulated for *MSR* expression are speculated to be MSR partners due for instance to decreased stability resulting from change in MetO content. Nonetheless, such proteins could participate in signaling pathways or metabolic processes altered by changes in the Met redox status. Comparative analysis of extracts from salt-treated WT or *MSRA4*-overexpressing Arabidopsis seedlings by two-dimensional electrophoresis coupled to mass spectrometry analysis identified five proteins with lower intensity in the modified line, among which two HSP70 isoforms [112]. A similar approach, pointed 9 proteins, including six located in plastid, that are more abundant upon cold-treatment in WT than in a *MSRB3*-deficient mutant [52]. Since MSRB3 is localized in reticulum endoplasmic, this isoform might participate in a transduction pathway regulating plastidial metabolism. In other respects, micro-array analysis indicated that overexpression of the pepper plastidial *MSRB2* in rice plants preserves the expression of numerous genes coding for photosynthetic proteins upon drought stress compared to WT [57].

Taken collectively, the data gained from different and complementary approaches for isolating MSR partners and targets in plants have led to the identification of numerous potential substrates. Nevertheless, most need to be validated from a biochemical point of view with regard to their capacity to be reduced by MSRs using reconstituted systems and recombinant proteins [105,122]. Further, investigations based on two-hybrid system in yeast and BiFC assays in plant cells will help to confirm their ability to interact in vivo with MSRs [61,62]. These approaches will have to be completed by redox proteome-scale analyses using plants modified for the expression of each *MSR* gene in relation with organ development and environmental conditions.

6.2. Identity and Functions of MSR Partners or Possible Targets

From the data acquired using the various strategies described in the previous sections, we can discuss how MSRs fulfill their functions in line with the identity of their proven or putative partners and the phenotype of modified plants.

6.2.1. Translation and Folding of Proteins

The plastidial elongation EFtu factor interacts with MSRB1 [122], and is less abundant in cold conditions in an *msrb3* mutant than in WT [52]. Further, recombinant EFtu is actually a substrate of MSRBs following treatment with H_2O_2 [122]. Consistently with these findings, proteins involved in translation are significantly enriched among the proteins prone to Met oxidation in *catalase 2*-deficient Arabidopsis plants [110]. Heat shock proteins (HSPs) or chaperones fulfill critical roles in proper folding of proteins, particularly in stress conditions. HSPs were among the first plant proteins presumed to be MSR substrates due to the presence of Met-rich domains in some like HSP21 [116]. Interestingly in Arabidopsis, other chaperone types are putative MSR targets, such as the chaperonin 60β and one heat shock cognate (HSC70-3) isolated using affinity chromatography [122], and one HSP70, the abundance of which is decreased in *MSRA4*-overexpressing plants [112]. Further, HSP70, HSP70B, and HSC70-2 are less oxidized at the level of Met in *MSRB7*-overexpressing plants [121]. As MetO is more hydrophilic than Met, MSRs would maintain the hydrophobic character of HSP regions binding unfolded proteins, thus preventing their aggregation [123].

From these reports, we conclude that plant MSRs are essential components preserving the activity of actors participating in proper elongation and folding of proteins. However, another hypothesis could be put forward in the sense that MSRs might associate to the complexes ensuring proper protein biogenesis and process, since yeast MSRs preferentially reduce MetO in unfolded proteins, protecting them from oxidative unfolding [20].

6.2.2. Chlorophyll Metabolism and Photosynthetic Activity

Numerous proteins involved in photosynthesis are potential MSR substrates. For instance, the two cpSRP43 and cpSRP54 components of the chloroplastic signal recognition particle, which targets Lhc proteins to thylakoids, are efficiently reduced by plastidial MSRBs [105]. The porphobilinogen deaminase catalyzes the polymerization of four monopyrrole units into a linear tetrapyrrole intermediate necessary for the formation of chlorophyll and heme. The recombinant form of this protein harbors two Met residues that are prone to oxidation upon H_2O_2 treatment, and is reduced by MSRs [57]. Non-targeted approaches confirmed the importance of MSRs regarding photosynthetic structures. Among the 24 proteins interacting with Arabidopsis plastidial MSRB1, six are involved in photosynthetic processes: ATPase subunits, RuBisCO, RubisCO activase, phosphoribulokinase, and glyceraldehyde-3-phosphate dehydrogenase B [122]. In other respects, in rice plants subjected to water deficit, a much more pronounced down-regulation of the expression of photosynthetic genes, such as those coding for photosystem I (PSI) subunits, was observed in WT than in *MSRB2*-overexpressing lines [57]. The authors concluded that MSRB2 maintains chloroplast function through the repair of Met upon water deficit and modulates retrograde signals involved in the regulation of gene expression in nucleus. Based on these data in rice [57] and those showing that an Arabidopsis mutant lacking plastidial MSRBs exhibits delayed growth and decreased photosynthesis in high light or low temperature conditions [105], these roles appear essential under environmental constraints impairing photosynthesis and plastidial redox homeostasis. From all these studies, we conclude that MSRs likely preserve photosynthetic structures along the whole process from pigment biogenesis and light capture to carbon assimilation.

In other respects, in cold conditions several photosynthetic proteins (RubisCO, sedoheptulose-1, 7-bisphosphatase, SBPase, and photosystem II oxygen-evolving complex subunits) are less abundant in an *msrb3* mutant than in WT [52]. These proteins are not targets of MSRB3 that is located in endoplasmic reticulum. Similarly, RubisCO, SBPase, RubisCO activase, and carbonic anhydrase are less oxidized at the Met level in Arabidopsis plants overexpressing cytosolic MSRB7 [121]. Thereby, we can infer from these studies the occurrence of signaling crosstalk between cell sub-compartments resulting from impaired Met redox status in endoplasmic reticulum or cytosol and controlling nuclear gene expression and/or plastidial redox balance. In line with this conclusion, chloroplastic proteins are more prone to Met oxidation compared to other compartments in an Arabidopsis mutant deficient in *catalase 2* displaying increased H_2O_2 production in peroxisome [110].

6.2.3. Antioxidant Mechanisms

As reviewed in Section 3, modifying expression of *MSR* genes is often associated with changes in cell redox homeostasis. Thus, higher MetO levels are observed in Arabidopsis *MSR*-deficient mutants particularly upon environmental constraints [52,105,109]. Moreover, the *msra2* mutant exhibits increased levels of protein nitration and glycation [109]. These findings demonstrate the integration of MSRs in the cell antioxidant network and indicate that impairment of one specific MSR-based repair system leads to general disturbance in the protein redox balance. This could originate from altered activity of other antioxidant proteins due to change in their Met redox status. This hypothesis is corroborated by the facts that two catalase isoforms interact with MSRB [122] and that Arabidopsis plants over-expressing cytosolic *MSRBs* display upon oxidative stress modified activity levels of peroxidase and catalase, and most importantly strongly increased GST activity [51]. This could indicate protection of GSTs by MSRs. Consistently, one GST isoform exhibits reduced abundance in a mutant lacking MSRB3 [52], and three GSTs are more subject to Met oxidation in a *catalase 2* mutant than in WT upon high light [110]. Further, three GSTFs (2, 3 and 8) exhibit less oxidized Met residues in an Arabidopsis line overexpressing *MSRB7* exposed to MV [121].

GSTs form a complex family of enzymes detoxifying a broad range of molecules such as secondary metabolites and exogenous substrates that are referred as xenobiotics, and include herbicides [124–126]. Depending on their catalytic residues, GSTs catalyze glutathione conjugation, perform deglutathionylation or bind non-substrate ligands. The higher GST activity in Arabidopsis *MSRB*-overexpressing plants prompted Chan's group [121] to investigate whether these enzymes are efficiently repaired by MSRs. Recombinant GSTFs 2, 3 and 8 interact with MSRB7 in BiFC, co-immunoprecipitation and yeast two-hybrid assays, and the activity of GSTFs 2 and 3 is restored in vitro by MSRB7 following oxidative treatment [121]. Moreover, plants overexpressing MSRB7 display a higher abundance of GSTFs 2 and 3 in oxidative stress conditions, revealing preserved protein stability possibly through the maintenance of Met redox status. In full agreement, MetO formation due to H_2O_2 treatment affects the activity of two other GSTs (GSTF9 and GSTT23) [110] and H_2O_2 leads to preferential oxidation of Met14 in GSTT23, that could alter GSH binding and/or catalytic activity of the enzyme [127]. Moreover, Met oxidation in GSTF9 results in increased flexibility in the H-site responsible of substrate binding and in lower enzyme activity towards hydrophobic substrates [128]. Taken collectively, these data give strong credence for a decisive role of plant MSRs in the redox maintenance of GSTs. Whether MSRs fulfil a similar function towards other antioxidant/detoxifying enzymes remains to be unveiled. Such a function seems very plausible, since in other organisms MSRs protect the activity of H_2O_2-scavenging enzymes [129].

6.2.4. Signaling in Relation with Calmodulin

Ca^{2+} is a major signal carrier and messenger in eukaryotic cells, and different sensors recognize specific calcium signatures caused by exogenous stimuli. Among them, the calmodulin (CAM) calcium receptor is a key actor in animal and plant cells. CAM proteins display Met-rich pockets binding partners harboring non-polar peptide sequences [130]. The Met redox status in these pockets is thus critical for partner recognition due to the hydrophobic character of Met compared to MetO [131]. Accordingly, oxidized CAM and MSR exhibit high affinity and cooperative interaction in in vitro assays [132]. Reversible Met oxidation finely tunes CAM-dependent signaling and modulates interaction with CAM-binding proteins [133–136]. Besides, Met oxidation can also directly regulate CAM-partners as inferred from the increased activity of CAM-dependent kinase II in *msrA*-deficient mice [40]. All these findings clearly provide evidence for crosstalk between redox modification of Met and calcium-dependent signaling pathways.

Compared to other organisms, plants possess much more CAM and CAM-like proteins (ca. 50) that regulate numerous binding partners involved in developmental processes and stress responses [137,138]. So far, the knowledge regarding their regulation via Met oxidation remains scarce. Several proteins participating in Ca^{2+}-dependent signaling are potential MSR substrates. Thus, among the 13 soybean drought-induced genes encoding Met-rich proteins, five code for CAM-related proteins [118]. In *G. soja*, the search of partners of a CAM-binding kinase led to the isolation of a MSRB isoform (B5a) [60]. The interaction was confirmed in BiFC assays and Arabidopsis lines overexpressing either MSRB5a or the CAM-binding kinase display enhanced tolerance to carbonate alkaline stress, suggesting that both fulfil a related physiological function. Intriguingly, the interaction takes place in the plasma membrane where the CAM-binding kinase is addressed, as well as MSRB5a when expressed without its plastidial transit peptide [60].

In plants, CAMs control many developmental processes including senescence. In litchi fruit pericarp, similar expression profiles were noticed for *CAM1* and three *MSR* genes during storage [61]. CAM1 physically interacts with two MSRA isoforms and can be repaired by MSRs after oxidation. Met oxidation in CAM1 does not alter its ability to bind two senescence-related transcription factors, but triggers their DNA-binding activity, revealing a possible role of MSRs in the control of the expression of senescence genes [61]. Very similar results were obtained in banana fruit [62]. These recent data reveal that like in animal cells the regulation of Met redox status in CAM-related proteins is likely a key step tuning their activity.

6.2.5. Proteins Responsive to Stress

Finally, most proteomic strategies highlight the importance of the control of Met redox status in stress responsive proteins. Indeed, among the 121 and 213 Met-rich proteins in Arabidopsis and soybean, respectively, many respond to drought or high salt and participate in regulation of transcription, modification of proteins and transport of metals [118]. Consistently, among the proteins exhibiting MetO following treatment of Arabidopsis cell cultures with a cGMP analogue, proteins responsive to various stress conditions such as tubulin or aconitase, are substantially enriched [111]. In *MSRB7*-overexpressing Arabidopsis plants, several proteins involved in stress responses and signaling processes (annexin D1, tubulins, transducins) carry less MetO residues than in WT [121]. Finally, in proteins subject to Met oxidation in an Arabidopsis *catalase 2*-deficient mutant, those belonging to the gene ontology (GO) biological process "response to stress" are enriched compared to WT [110]. Among them, not only executors like oxidoreductases, but also regulators involved in signal transduction like methylene-blue-sensitive 1 and mitogen-activated protein kinases are found.

7. Conclusions

The knowledge about the physiological functions of plant MSRs has considerably evolved in recent years, and their participation in defense mechanisms against abiotic and biotic constraints is now well established (Figure 3). They can act in the preservation of proteins upon oxidative stress, but could also be targets of pathogens as proposed by Gao et al. [54]. These data could open up avenues for improving crop responses to environmental stress conditions. Moreover, another function related to the control of longevity was unveiled in seeds subjected to adverse conditions [68], which is in agreement with the data reported in other organisms. The *MSR* expression patterns observed in leaf and fruit as a function of age or maturation stage, respectively, in various plant species prompt us to propose that MSRs participate in the control of senescence and ageing processes in plant organs (Figure 3).

Most interestingly, the search of partners provided accurate information how plant MSRs could fulfill their functions upon environmental constraints (Figure 4). Indeed, in addition to a global and direct protective role against oxidative damage in proteins, for instance in photosynthetic structures, MSRs on one hand could maintain the Met redox status preferentially in stress responsive effectors, thereby preserving their activity as shown for GSTs and chaperones. On the other hand, they very likely play key signaling roles in relation with Ca^{2+}-, hormone- and phosphorylation-dependent cascades as inferred from the identification of numerous MSR partners involved in these pathways [60–62,110,118]. The control of Met redox status by MSRs could be responsible for decoding ROS signatures and transmitting information in non-redox signaling pathways. Consistently, oxidation of Met538 in Arabidopsis nitrate reductase prevents in vivo phosphorylation of a nearby Ser residue, revealing the control of oxidative signals such as MetO formation on the capacity of kinase substrates to be phosphorylated due to modified recognition motif [139], MetO being more hydrophilic than Met. The hypothesis that Met oxidation participates in the control of protein phosphorylation is further supported by the fact that Met residues nearby phosphorylation sites are preferentially oxidized in vivo under stress conditions in human proteins [140]. These data indicate that MSRs are decisive components at the crosstalk of different transduction pathways within the complex signaling network (Figure 4).

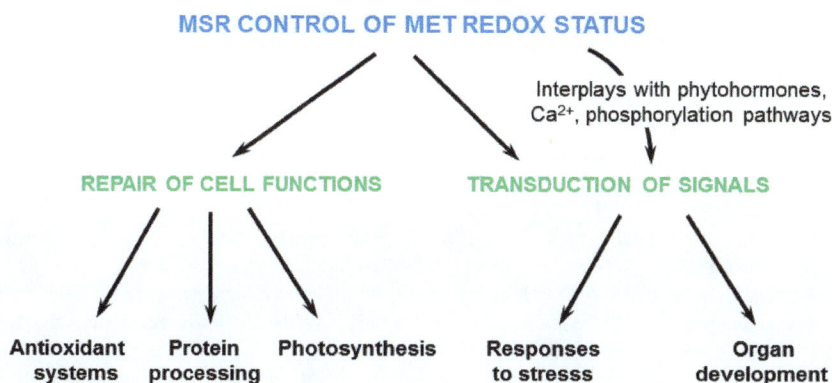

Figure 4. Presumed modes of action of plant MSRs. The proposed functions of MSRs in repair and preservation of antioxidant, protein processing, and photosynthetic systems and in signaling transduction pathways are based on the phenotype of plants modified for *MSR* expression and on the identity of their possible targets.

In addition to reductase activity, plant MSRs might fulfill other biochemical functions, since some do not harbor any catalytic cysteine [32], and bacterial and animal MSRAs exhibit in vitro methionine oxidase activity towards both free and peptide-bound Met [141]. Further investigation is thus needed to decipher more precisely *in planta* the functions of each MSR isoform, notably by identifying using redox proteomics the MetO-containing proteins in various genetic backgrounds and environmental conditions. Another promising approach consists of determining the Met redox status in vivo using redox fluorescent sensors like in bacterial and mammalian cells [14]. This will help with monitoring the MetO level in various subcellular compartments and genetic backgrounds as a function of environmental stimuli.

Author Contributions: Review of literature and writing of the paper: P.R. and L.T.

Funding: This research received no external funding.

Conflicts of Interest: The authors declare no conflict of interest.

References

1. Davies, M.J. The oxidative environment and protein damage. *Biochim. Biophys. Acta* **2005**, *1703*, 93–109. [CrossRef] [PubMed]
2. Møller, I.M.; Jensen, P.E.; Hansson, A. Oxidative modifications to cellular components in plants. *Annu. Rev. Plant Biol.* **2007**, *58*, 459–481. [CrossRef] [PubMed]
3. Couturier, J.; Chibani, K.; Jacquot, J.P.; Rouhier, N. Cysteine-based redox regulation and signaling in plants. *Front. Plant Sci.* **2013**, *4*, 105. [CrossRef] [PubMed]
4. Meyer, Y.; Reichheld, J.P.; Vignols, F. Thioredoxins in Arabidopsis and other plants. *Photosynth. Res.* **2005**, *86*, 419–433. [CrossRef] [PubMed]
5. Rouhier, N.; Couturier, J.; Jacquot, J.P. Genome-wide analysis of plant glutaredoxin systems. *J. Exp. Bot.* **2006**, *57*, 1685–1696. [CrossRef] [PubMed]
6. Rouhier, N.; Lemaire, S.D.; Jacquot, J.P. The role of glutathione in photosynthetic organisms: Emerging functions for glutaredoxins and glutathionylation. *Annu. Rev. Plant Biol.* **2008**, *59*, 143–166. [CrossRef] [PubMed]
7. Couturier, J.; Jacquot, J.P.; Rouhier, N. Evolution and diversity of glutaredoxins in photosynthetic organisms. *Cell. Mol. Life Sci.* **2009**, *66*, 2539–2557. [CrossRef] [PubMed]
8. Vieira Dos Santos, C.; Rey, P. Plant thioredoxins are key actors in the oxidative stress response. *Trends Plant Sci.* **2006**, *11*, 329–334. [CrossRef] [PubMed]
9. Meyer, Y.; Belin, C.; Delorme-Hinoux, V.; Reichheld, J.P.; Riondet, C. Thioredoxin and glutaredoxin systems in plants: Molecular mechanisms, crosstalks, and functional significance. *Antioxid. Redox Signal.* **2012**, *17*, 1124–1160. [CrossRef] [PubMed]
10. Rouhier, N.; Cerveau, D.; Couturier, J.; Reichheld, J.P.; Rey, P. Involvement of thiol-based mechanisms in plant development *Biochim. Biophys. Acta* **2015**, *1850*, 1479–1496. [CrossRef] [PubMed]
11. Rouhier, N.; Villarejo, A.; Srivastava, M.; Gelhaye, E.; Keech, O.; Droux, M.; Finkemeier, I.; Samuelsson, G.; Dietz, K.J.; Jacquot, J.P.; et al. Identification of plant glutaredoxin targets. *Antioxid. Redox Signal.* **2005**, *7*, 919–929. [CrossRef] [PubMed]
12. Montrichard, F.; Alkhalfioui, F.; Yano, H.; Vensel, W.H.; Hurkman, W.J.; Buchanan, B.B. Thioredoxin targets in plants: The first 30 years. *J. Proteom.* **2009**, *72*, 452–474. [CrossRef] [PubMed]
13. Pérez-Pérez, M.E.; Mauriès, A.; Maes, A.; Tourasse, N.J.; Hamon, M.; Lemaire, S.D.; Marchand, C.H. The deep thioredoxome in *Chlamydomonas reinhardtii*: New insights into redox regulation. *Mol. Plant* **2017**, *10*, 1107–1125. [CrossRef] [PubMed]
14. Tarrago, L.; Péterfi, Z.; Lee, B.C.; Michel, T.; Gladyshev, V.N. Monitoring methionine sulfoxide with stereospecific mechanism-based fluorescent sensors. *Nat. Chem. Biol.* **2015**, *11*, 332–338. [CrossRef]
15. Luo, S.; Levine, R.L. Methionine in proteins defends against oxidative stress. *FASEB J.* **2009**, *23*, 464–472. [CrossRef] [PubMed]
16. Brot, N.; Weissbach, L.; Werth, J.; Weissbach, H. Enzymatic reduction of protein-bound methionine sulfoxide. *Proc. Natl. Acad. Sci. USA* **1981**, *78*, 2155–2158. [CrossRef] [PubMed]

17. Grimaud, R.; Ezraty, B.; Mitchell, J.K.; Lafitte, D.; Briand, C.; Derrick, P.J.; Barras, F. Repair of oxidized proteins. Identification of a new methionine sulfoxide reductase. *J. Biol. Chem.* **2001**, *276*, 48915–48920. [CrossRef] [PubMed]

18. Lowther, W.T.; Brot, N.; Weissbach, H.; Honek, J.F.; Matthews, B.W. Thiol-disulfide exchange is involved in the catalytic mechanism of peptide methionine sulfoxide reductase. *Proc. Natl. Acad. Sci. USA* **2000**, *97*, 6463–6468. [CrossRef] [PubMed]

19. Weissbach, H.; Etienne, F.; Hoshi, T.; Heinemann, S.H.; Lowther, W.T.; Matthews, B.; St John, G.; Nathan, C.; Brot, N. Peptide methionine sulfoxide reductase: Structure, mechanism of action, and biological function. *Arch. Biochem. Biophys.* **2002**, *397*, 172–178. [CrossRef] [PubMed]

20. Tarrago, L.; Kaya, A.; Weerapana, E.; Marino, S.M.; Gladyshev, V.N. Methionine sulfoxide reductases preferentially reduce unfolded oxidized proteins and protect cells from oxidative protein unfolding. *J. Biol. Chem.* **2012**, *287*, 2448–2459. [CrossRef] [PubMed]

21. Le, D.T.; Tarrago, L.; Watanabe, Y.; Kaya, A.; Lee, B.C.; Tran, U.; Nishiyama, R.; Fomenko, D.E.; Gladyshev, V.N.; Tran, L.S. Diversity of plant methionine sulfoxide reductases B and evolution of a form specific for free methionine sulfoxide. *PLoS ONE* **2013**, *8*, e65637. [CrossRef] [PubMed]

22. Boschi-Muller, S.; Gand, A.; Branlant, G. The methionine sulfoxide reductases: Catalysis and substrate specificities. *Arch. Biochem. Biophys.* **2008**, *474*, 266–273. [CrossRef] [PubMed]

23. Rouhier, N.; Kauffmann, B.; Tete-Favier, F.; Palladino, P.; Gans, P.; Branlant, G.; Jacquot, J.P.; Boschi-Muller, S. Functional and structural aspects of poplar cytosolic and plastidial type a methionine sulfoxide reductases. *J. Biol. Chem.* **2007**, *282*, 3367–3378. [CrossRef] [PubMed]

24. Tarrago, L.; Laugier, E.; Zaffagnini, M.; Marchand, C.; Le Maréchal, P.; Rouhier, N.; Lemaire, S.D.; Rey, P. Regeneration mechanisms of *Arabidopsis thaliana* methionine sulfoxide reductases B by glutaredoxins and thioredoxins. *J. Biol. Chem.* **2009**, *284*, 18963–18971. [CrossRef] [PubMed]

25. Rey, P.; Cuine, S.; Eymery, F.; Garin, J.; Court, M.; Jacquot, J.P.; Rouhier, N.; Broin, M. Analysis of the proteins targeted by CDSP32, a plastidic thioredoxin participating in oxidative stress responses. *Plant J.* **2005**, *41*, 31–42. [CrossRef] [PubMed]

26. Sagher, D.; Brunell, D.; Hejtmancik, J.F.; Kantorow, M.; Brot, N.; Weissbach, H. Thionein can serve as a reducing agent for the methionine sulfoxide reductases. *Proc. Natl. Acad. Sci. USA* **2006**, *103*, 8656–8661. [CrossRef] [PubMed]

27. Vieira Dos Santos, C.; Laugier, E.; Tarrago, L.; Massot, V.; Issakidis-Bourguet, E.; Rouhier, N.; Rey, P. Specificity of thioredoxins and glutaredoxins as electron donors to two distinct classes of Arabidopsis plastidial methionine sulfoxide reductases B. *FEBS Lett.* **2007**, *581*, 4371–4376. [CrossRef] [PubMed]

28. Ding, D.; Sagher, D.; Laugier, E.; Rey, P.; Weissbach, H.; Zhang, X.H. Studies on the reducing systems for plant and animal methionine sulfoxide reductases B lacking the resolving cysteine. *Biochem. Biophys. Res. Commun.* **2007**, *361*, 629–633. [CrossRef] [PubMed]

29. Kim, H.Y.; Kim, J.R. Thioredoxin as a reducing agent for mammalian methionine sulfoxide reductases B lacking resolving cysteine. *Biochem. Biophys. Res. Commun.* **2008**, *371*, 490–494. [CrossRef] [PubMed]

30. Tarrago, L.; Laugier, E.; Zaffagnini, M.; Marchand, C.; Le Maréchal, P.; Lemaire, S.D.; Rey, P. Plant thioredoxin CDSP32 regenerates 1-Cys methionine sulfoxide reductase B activity through the direct reduction of sulfenic acid. *J. Biol. Chem.* **2010**, *285*, 14964–14972. [CrossRef] [PubMed]

31. Couturier, J.; Vignols, F.; Jacquot, J.P.; Rouhier, N. Glutathione- and glutaredoxin-dependent reduction of methionine sulfoxide reductase A. *FEBS Lett.* **2012**, *586*, 3894–3899. [CrossRef] [PubMed]

32. Le, D.T.; Nguyen, K.; Chu, H.D.; Vu, N.T.; Pham, T.T.L.; Tran, L.P. Function of the evolutionarily conserved plant methionine-S-sulfoxide reductase without the catalytic residue. *Protoplasma* **2018**, 1–10. [CrossRef] [PubMed]

33. Koc, A.; Gasch, A.P.; Rutherford, J.C.; Kim, H.Y.; Gladyshev, V.N. Methionine sulfoxide reductase regulation of yeast lifespan reveals reactive oxygen species-dependent and -independent components of aging. *Proc. Natl. Acad. Sci. USA* **2004**, *101*, 7999–8004. [CrossRef] [PubMed]

34. Gabbita, S.P.; Aksenov, M.Y.; Lovell, M.A.; Markesbery, W.R. Decrease in peptide methionine sulfoxide reductase in Alzheimer's disease brain. *J. Neurochem.* **1999**, *73*, 1660–1666. [CrossRef] [PubMed]

35. Petropoulos, I.; Mary, J.; Perichon, M.; Friguet, B. Rat peptide methionine sulphoxide reductase: Cloning of the cDNA, and down-regulation of gene expression and enzyme activity during aging. *Biochem. J.* **2001**, *355 Pt 3*, 819–825. [CrossRef]

36. Moskovitz, J.; Flescher, E.; Berlett, B.S.; Azare, J.; Poston, J.M.; Stadtman, E.R. Overexpression of peptide-methionine sulfoxide reductase in *Saccharomyces cerevisiae* and human T cells provides them with high resistance to oxidative stress. *Proc. Natl. Acad. Sci. USA* **1998**, *95*, 14071–14075. [CrossRef] [PubMed]

37. Ruan, H.; Tang, X.D.; Chen, M.L.; Joiner, M.L.; Sun, G.; Brot, N.; Weissbach, H.; Heinemann, S.H.; Iverson, L.; Wu, C.F.; et al. High-quality life extension by the enzyme peptide methionine sulfoxide reductase. *Proc. Natl. Acad. Sci. USA* **2002**, *99*, 2748–2753. [CrossRef] [PubMed]

38. Stadtman, E.R.; Moskovitz, J.; Berlett, B.S.; Levine, R.L. Cyclic oxidation and reduction of protein methionine residues is an important antioxidant mechanism. *Mol. Cell Biochem.* **2002**, *234–235*, 3–9. [CrossRef]

39. Drazic, A.; Miura, H.; Peschek, J.; Le, Y.; Bach, N.C.; Kriehuber, T.; Winter, J. Methionine oxidation activates a transcription factor in response to oxidative stress. *Proc. Natl. Acad. Sci. USA* **2013**, *110*, 9493–9498. [CrossRef] [PubMed]

40. Erickson, J.R.; Joiner, M.L.; Guan, X.; Kutschke, W.; Yang, J.; Oddis, C.V.; Bartlett, R.K.; Lowe, J.S.; O'Donnell, S.E.; Aykin-Burns, N.; et al. A dynamic pathway for calcium-independent activation of CaMKII by methionine oxidation. *Cell* **2008**, *133*, 462–474. [CrossRef] [PubMed]

41. Lee, B.C.; Péterfi, Z.; Hoffmann, F.W.; Moore, R.E.; Kaya, A.; Avanesov, A.; Tarrago, L.; Zhou, Y.; Weerapana, E.; Fomenko, D.E.; et al. MsrB1 and MICALs regulate actin assembly and macrophage function via reversible stereoselective methionine oxidation. *Mol. Cell* **2013**, *51*, 397–404. [CrossRef] [PubMed]

42. Hung, R.J.; Spaeth, C.S.; Yesilyurt, H.G.; Terman, J.R. SelR reverses Mical-mediated oxidation of actin to regulate F-actin dynamics. *Nat. Cell Biol.* **2013**, *15*, 1445–1454. [CrossRef] [PubMed]

43. Sanchez, J.; Nikolau, B.J.; Stumpf, P.K. Reduction of N-Acetyl methionine sulfoxide in plants. *Plant Physiol.* **1983**, *73*, 619–623. [CrossRef] [PubMed]

44. Sadanandom, A.; Piffanelli, P.; Knott, T.; Robinson, C.; Sharpe, A.; Lydiate, D.; Murphy, D.; Fairbairn, D.J. Identification of a peptide methionine sulphoxide reductase gene in an oleosin promoter from *Brassica napus*. *Plant J.* **1996**, *10*, 235–242. [CrossRef] [PubMed]

45. Rodrigo, M.J.; Moskovitz, J.; Salamini, F.; Bartels, D. Reverse genetic approaches in plants and yeast suggest a role for novel, evolutionarily conserved, selenoprotein-related genes in oxidative stress defense. *Mol. Genet. Genom.* **2002**, *267*, 613–621. [CrossRef] [PubMed]

46. Rouhier, N.; Vieira Dos Santos, C.; Tarrago, L.; Rey, P. Plant methionine sulfoxide reductase A and B multigenic families. *Photosynth. Res.* **2006**, *89*, 247–262. [CrossRef] [PubMed]

47. Tarrago, L.; Laugier, E.; Rey, P. Protein-repairing methionine sulfoxide reductases in photosynthetic organisms, gene organization, reduction mechanisms, and physiological roles. *Mol. Plant* **2009**, *2*, 202–217. [CrossRef] [PubMed]

48. Zhu, J.; Ding, P.; Li, Q.; Gao, Y.; Chen, F.; Xia, G. Molecular characterization and expression profile of methionine sulfoxide reductase gene family in maize (*Zea mays*) under abiotic stresses. *Gene* **2015**, *562*, 159–168. [CrossRef] [PubMed]

49. Sadanandom, A.; Poghosyan, Z.; Fairbairn, D.J.; Murphy, D.J. Differential regulation of plastidial and cytosolic isoforms of peptide methionine sulfoxide reductase in Arabidopsis. *Plant Physiol.* **2000**, *123*, 255–264. [CrossRef] [PubMed]

50. Vieira Dos Santos, C.; Cuine, S.; Rouhier, N.; Rey, P. The Arabidopsis plastidic methionine sulfoxide reductase B proteins. Sequence and activity characteristics, comparison of the expression with plastidic methionine sulfoxide reductase A, and induction by photooxidative stress. *Plant Physiol.* **2005**, *138*, 909–922. [CrossRef] [PubMed]

51. Li, C.W.; Lee, S.H.; Chieh, P.S.; Lin, C.S.; Wang, Y.C.; Chan, M.T. Arabidopsis root-abundant cytosolic methionine sulfoxide reductase B genes MsrB7 and MsrB8 are involved in tolerance to oxidative stress. *Plant Cell Physiol.* **2012**, *53*, 1707–1719. [CrossRef] [PubMed]

52. Kwon, S.J.; Kwon, S.I.; Bae, M.S.; Cho, E.J.; Park, O.K. Role of the methionine sulfoxide reductase MsrB3 in cold acclimation in Arabidopsis. *Plant Cell Physiol.* **2007**, *48*, 1713–1723. [CrossRef] [PubMed]

53. Dai, C.; Liu, L.; Wang, M.H. Characterization of a methionine sulfoxide reductase B from tomato (*Solanum lycopersicum*), and its protecting role in *Saccharomyces cerevisiae*. *Protein J.* **2013**, *32*, 39–47. [CrossRef] [PubMed]

54. Gao, L.; Shen, W.; Yan, P.; Tuo, D.; Li, X.; Zhou, P. NIa-pro of Papaya ringspot virus interacts with papaya methionine sulfoxide reductase B1. *Virology* **2012**, *434*, 78–87. [CrossRef] [PubMed]

55. Guo, X.; Wu, Y.; Wang, Y.; Chen, Y.; Chu, C. OsMSRA4.1 and OsMSRB1.1, two rice plastidial methionine sulfoxide isoforms, are involved in abiotic stress responses. *Planta* **2009**, *230*, 227–238. [CrossRef] [PubMed]

56. Oh, S.K.; Baek, K.H.; Seong, E.S.; Joung, Y.H.; Choi, G.J.; Park, J.M.; Cho, H.S.; Kim, E.A.; Lee, S.; Choi, D. MsrB2, pepper methionine sulfoxide reductase B2, is a novel defense regulator against oxidative stress and pathogen attack. *Plant Physiol.* **2010**, *154*, 245–261. [CrossRef] [PubMed]

57. Kim, J.S.; Park, H.M.; Chae, S.; Lee, T.H.; Hwang, D.J.; Oh, S.D.; Park, J.S.; Song, D.G.; Pan, C.H.; Choi, D.; et al. A pepper *MSRB2* gene confers drought tolerance in rice through the protection of chloroplast-targeted genes. *PLoS ONE* **2014**, *9*, e90588. [CrossRef] [PubMed]

58. Liu, L.; Wang, M.H. Cloning, expression, and characterization of a methionine sulfoxide reductase B gene from *Nicotiana tabacum. Protein J.* **2013**, *32*, 543–550. [CrossRef] [PubMed]

59. Liu, L.; Wang, M.H. Expression and biological properties of a novel methionine sulfoxide reductase A in tobacco (*Nicotiana tabacum*). *Protein J.* **2013**, *32*, 266–274. [CrossRef] [PubMed]

60. Sun, X.; Sun, M.; Jia, B.; Qin, Z.; Yang, K.; Chen, C.; Yu, Q.; Zhu, Y. A *Glycine soja* methionine sulfoxide reductase B5a interacts with the Ca(2+)/CAM-binding kinase GsCBRLK and activates ROS signaling under carbonate alkaline stress. *Plant J.* **2016**, *86*, 514–529. [CrossRef] [PubMed]

61. Jiang, G.; Xiao, L.; Yan, H.; Zhang, D.; Wu, F.; Liu, X.; Su, X.; Dong, X.; Wang, J.; Duan, X.; et al. Redox regulation of methionine in calmodulin affects the activity levels of senescence-related transcription factors in litchi. *Biochim. Biophys. Acta* **2017**, *1861*, 1140–1151. [CrossRef] [PubMed]

62. Jiang, G.; Wu, F.; Li, Z.; Li, T.; Gupta, V.K.; Duan, X.; Jiang, Y. Sulfoxidation regulation of Musa acuminata calmodulin (MaCaM) influences the functions of MaCaM-binding proteins. *Plant Cell Physiol.* **2018**, *59*, 1214–1224. [CrossRef] [PubMed]

63. Klodmann, J.; Senkler, M.; Rode, C.; Braun, H.P. Defining the protein complex proteome of plant mitochondria. *Plant Physiol.* **2011**, *157*, 587–598. [CrossRef] [PubMed]

64. Lee, J.; Lei, Z.; Watson, B.S.; Sumner, L.W. Sub-cellular proteomics of *Medicago truncatula. Front. Plant Sci.* **2013**, *4*, 112. [CrossRef] [PubMed]

65. Senkler, J.; Senkler, M.; Eubel, H.; Hildebrandt, T.; Lengwenus, C.; Schertl, P.; Schwarzländer, M.; Wagner, S.; Wittig, I.; Braun, H.P. The mitochondrial complexome of *Arabidopsis thaliana. Plant J.* **2017**, *89*, 1079–1092. [CrossRef] [PubMed]

66. Noctor, G.; Reichheld, J.P.; Foyer, C.H. ROS-related redox regulation and signaling in plants. *Semin. Cell Dev. Biol.* **2017**, *80*, 3–12. [CrossRef] [PubMed]

67. In, O.; Berberich, T.; Romdhane, S.; Feierabend, J. Changes in gene expression during dehardening of cold-hardened winter rye (*Secale cereale* L.) leaves and potential role of a peptide methionine sulfoxide reductase in cold-acclimation. *Planta* **2005**, *220*, 941–950. [CrossRef] [PubMed]

68. Châtelain, E.; Satour, P.; Laugier, E.; Ly Vu, B.; Payet, N.; Rey, P.; Montrichard, F. Evidence for the participation of the methionine sulfoxide reductase repair system in plant seed longevity. *Proc. Natl. Acad. Sci. USA* **2013**, *110*, 3633–3638. [CrossRef] [PubMed]

69. Chu, H.D.; Nguyen, K.L.; Watanabe, Y.; Le, D.T.; Tran, L.P. Expression analyses of soybean genes encoding methionine-R-sulfoxide reductase under various conditions suggest a possible role in the adaptation to stress. *Appl. Biol. Chem.* **2016**, *59*, 681–687. [CrossRef]

70. Xiao, T.; Memgmeng, M.; Wang, G.; Quian, M.; Chen, Y.; Zheng, L.; Zhang, H.; Hu, Z.; Shen, Z.; Xia, Y. A methionine-R-sulfoxide reductase, OsMSRB5, is required for rice defense against copper toxicity. *Environ. Exp. Bot.* **2018**, *153*, 45–53. [CrossRef]

71. Lopez, A.P.; Portales, R.; López-Ráez, J.; Medina-Escobar, N.; Blanco, J.; Franco, A. Characterization of a strawberry late-expressed and fruit-specific peptide methionine sulphoxide reductase. *Physiol. Plant* **2006**, *126*, 129–139. [CrossRef]

72. Allen, M.D.; Kropat, J.; Tottey, S.; Del Campo, J.A.; Merchant, S.S. Manganese deficiency in Chlamydomonas results in loss of photosystem II and MnSOD function, sensitivity to peroxides, and secondary phosphorus and iron deficiency. *Plant Physiol.* **2007**, *143*, 263–277. [CrossRef] [PubMed]

73. Wu, T.M.; Hsu, Y.T.; Sung, M.S.; Hsu, Y.T.; Lee, T.M. Expression of genes involved in redox homeostasis and antioxidant defense in a marine macroalga *Ulva fasciata* by excess copper. *Aquat. Toxicol.* **2009**, *94*, 275–285. [CrossRef] [PubMed]

74. Zhao, L.; Chen, M.; Cheng, D.; Yang, H.; Sun, Y.; Zhou, H.; Huang, F. Different B-type methionine sulfoxide reductases in Chlamydomonas may protect the alga against high-light, sulfur-depletion, or oxidative stress. *J. Integr. Plant Biol.* **2013**, *55*, 1054–1068. [CrossRef] [PubMed]

75. Hsu, Y.T.; Lee, T.M. Photosynthetic electron transport mediates the light-ciontrolled up-regulation of expression of methionine sulfoxide reductase A and B from marine macroalga *Ulva fasciata*. *J. Phycol.* **2010**, *46*, 112–122. [CrossRef]

76. Romero, H.M.; Berlett, B.S.; Jensen, P.J.; Pell, E.J.; Tien, M. Investigations into the role of the plastidial peptide methionine sulfoxide reductase in response to oxidative stress in Arabidopsis. *Plant Physiol.* **2004**, *136*, 3784–3794. [CrossRef] [PubMed]

77. Dai, C.; Singh, N.K.; Park, M. Characterization of a novel methionine sulfoxide reductase A from tomato (*Solanum lycopersicum*), and its protecting role in *Escherichia coli*. *BMB Rep.* **2011**, *44*, 805–810. [CrossRef] [PubMed]

78. Méndez, A.A.; Pena, L.B.; Benavides, M.P.; Gallego, S.M. Priming with NO controls redox state and prevents cadmium-induced general up-regulation of methionine sulfoxide reductase gene family in Arabidopsis. *Biochimie* **2016**, *131*, 128–136. [CrossRef] [PubMed]

79. Collin, V.C.; Eymery, F.; Genty, B.; Rey, P.; Havaux, M. Vitamin E is essential for the tolerance of Arabidopsis thaliana to metal-induced oxidative stress. *Plant Cell Environ.* **2008**, *31*, 244–257. [CrossRef] [PubMed]

80. Alvarez, S.; Berla, B.M.; Sheffield, J.; Cahoon, R.E.; Jez, J.M.; Hicks, L.M. Comprehensive analysis of the *Brassica juncea* root proteome in response to cadmium exposure by complementary proteomic approaches. *Proteomics* **2009**, *9*, 2419–2431. [CrossRef] [PubMed]

81. Simonović, A.D.; Anderson, M.D. Analysis of methionine oxides and nitrogen-transporting amino acids in chilled and acclimated maize seedlings. *Amino Acids* **2007**, *33*, 607–613. [CrossRef] [PubMed]

82. Dai, C.; Wang, M.H. Characterization and functional analysis of methionine sulfoxide reductase A gene family in tomato. *Mol. Biol. Rep.* **2012**, *39*, 6297–6308. [CrossRef] [PubMed]

83. Bouchenak, F.; Henri, P.; Benrebiha, F.Z.; Rey, P. Differential responses to salinity of two *Atriplex halimus* populations in relation to organic solutes and antioxidant systems involving thiol reductases. *J. Plant Physiol.* **2012**, *169*, 1445–1553. [CrossRef] [PubMed]

84. Fatehi, F.; Hosseinzadeh, A.; Alizadeh, H.; Brimavandi, T.; Struik, P.C. The proteome response of salt-resistant and salt-sensitive barley genotypes to long-term salinity stress. *Mol. Biol. Rep.* **2012**, *39*, 6387–6397. [CrossRef] [PubMed]

85. Marok, M.A.; Tarrago, L.; Ksas, B.; Henri, P.; Abrous-Belbachir, O.; Havaux, M.; Rey, P. A drought-sensitive barley variety displays oxidative stress and strongly increased contents in low-molecular weight antioxidant compounds during water deficit compared to a tolerant variety. *J. Plant Physiol.* **2013**, *170*, 633–645. [CrossRef] [PubMed]

86. Roy, S.; Nandi, A.K. *Arabidopsis thaliana* methionine sulfoxide reductase B8 influences stress-induced cell death and effector-triggered immunity. *Plant Mol. Biol.* **2017**, *93*, 109–120. [CrossRef] [PubMed]

87. Vieira Dos Santos, C.; Delavault, P.; Letousey, P.; Thalouarn, P. Identification by suppression subtractive hybridization and expression analysis of *Arabidopsis thaliana* putative defence genes during *Orobanche ramosa* infection. *Physiol. Mol. Plant Pathol.* **2003**, *62*, 297–303. [CrossRef]

88. Choudhury, F.K.; Rivero, R.M.; Blumwald, E.; Mittler, R. Reactive oxygen species, abiotic stress and stress combination. *Plant J.* **2017**, *90*, 856–867. [CrossRef] [PubMed]

89. Yu, M.; Lamattina, L.; Spoel, S.H.; Loake, G.J. Nitric oxide function in plant biology: A redox cue in deconvolution. *New Phytol.* **2014**, *202*, 1142–1156. [CrossRef] [PubMed]

90. Foyer, C.; Noctor, G. Stress-triggered redox signalling: What's in pROSpect? *Plant Cell Environ.* **2016**, *39*, 951–964. [CrossRef] [PubMed]

91. Umbreen, S.; Lubega, J.; Cui, B.; Pan, Q.; Jiang, J.; Loake, G.J. Specificity in nitric oxide signaling. *J. Exp. Bot.* **2018**, *69*, 3439–3448. [CrossRef] [PubMed]

92. D'Autréaux, B.; Toledano, M.B. ROS as signalling molecules: Mechanisms that generate specificity in ROS homeostasis. *Nat. Rev. Mol. Cell Biol.* **2007**, *10*, 813–824. [CrossRef] [PubMed]

93. Frederickson Matika, D.E.; Loake, G.J. Redox regulation in plant immune function. *Antioxid. Redox Signal.* **2014**, *21*, 1373–1388. [CrossRef] [PubMed]

94. Chang, H.L.; Tseng, Y.L.; Ho, K.L.; Shie, S.C.; Wu, P.S.; Hsu, Y.T.; Lee, T.M. Reactive oxygen species modulate the differential expression of methionine sulfoxide reductasegenes in *Chlamydomonas reinhardtii* under high light illumination. *Physiol. Plant* **2014**, *150*, 550–564. [CrossRef] [PubMed]

95. Hsu, Y.T.; Lee, T.M. Nitric oxide up-regulates the expression of methionine sulfoxide reductase genes in the intertidal macroalga *Ulva fasciata* for high light acclimation. *Plant Cell Physiol.* **2012**, *53*, 445–456. [CrossRef] [PubMed]

96. Oh, J.E.; Hong, S.W.; Lee, Y.; Koh, E.J.; Kim, K.; Seo, Y.W.; Chung, N.; Jeong, M.; Jang, C.S.; Lee, B.; et al. Modulation of gene expressions and enzyme activities of methionine sulfoxide reductases by cold, ABA or high salt treatments in Arabidopsis. *Plant Sci.* **2005**, *169*, 1030–1036. [CrossRef]

97. Danon, A.; Coll, N.S.; Apel, K. Cryptochrome-1-dependent execution of programmed cell death induced by singlet oxygen in *Arabidopsis thaliana*. *Proc. Natl. Acad. Sci. USA* **2006**, *103*, 17036–17041. [CrossRef] [PubMed]

98. Begara-Morales, J.C.; Sánchez-Calvo, B.; Luque, F.; Leyva-Pérez, M.O.; Leterrier, M.; Corpas, F.J.; Barroso, J.B. Differential transcriptomic analysis by RNA-Seq of GSNO-responsive genes between Arabidopsis roots and leaves. *Plant Cell Physiol.* **2014**, *55*, 1080–1095. [CrossRef] [PubMed]

99. Xia, X.J.; Zhou, Y.H.; Shi, K.; Zhou, J.; Foyer, C.H.; Yu, J.Q. Interplay between reactive oxygen species and hormones in the control of plant development and stress tolerance. *J. Exp. Bot.* **2015**, *66*, 2839–2856. [CrossRef] [PubMed]

100. Mata-Pérez, C.; Sánchez-Calvo, B.; Begara-Morales, J.C.; Luque, F.; Jiménez-Ruiz, J.; Padilla, M.N.; Fierro-Risco, J.; Valderrama, R.; Fernández-Ocaña, A.; Corpas, F.J.; et al. Yanscriptomic profiling of linolenic acid-responsive genes in ROS signaling from RNA-seq data in Arabidopsis. *Front. Plant Sci.* **2015**, *6*, 122. [CrossRef] [PubMed]

101. Wasternack, C.; Hause, B. Jasmonates: Biosynthesis, perception, signal transduction and action in plant stress response, growth and development. An update to the 2007 review in Annals of Botany. *Ann. Bot.* **2013**, *111*, 1021–1058. [CrossRef] [PubMed]

102. Yan, S.; Dong, X. Perception of the plant immune signal salicylic acid. *Curr. Opin. Plant Biol.* **2014**, *20*, 64–68. [CrossRef] [PubMed]

103. Zhu, J.K. Salt and drought stress signal transduction in plants. *Annu. Rev. Plant Biol.* **2002**, *53*, 247–273. [CrossRef] [PubMed]

104. Ferguson, D.L.; Burke, J.J. A new method of measuring protein-methionine-s-oxide reductase activity. *Plant Physiol.* **1992**, *100*, 529–532. [CrossRef] [PubMed]

105. Laugier, E.; Tarrago, L.; Vieira Dos Santos, C.; Eymery, F.; Havaux, M.; Rey, P. *Arabidopsis thaliana* plastidial methionine sulfoxide reductases B, MSRBs, account for most leaf peptide MSR activity and are essential for growth under environmental constraints through a role in the preservation of photosystem antennae. *Plant J.* **2010**, *61*, 271–282. [CrossRef] [PubMed]

106. Laugier, E.; Tarrago, L.; Courteille, A.; Innocenti, G.; Eymery, F.; Rumeau, D.; Issakidis-Bourguet, E.; Rey, P. Involvement of thioredoxin y2 in the preservation of leaf methionine sulfoxide reductase capacity and growth under high light. *Plant Cell Environ.* **2013**, *36*, 670–682. [CrossRef] [PubMed]

107. Bechtold, U.; Murphy, D.J.; Mullineaux, P.M. Arabidopsis peptide methionine sulfoxide reductase2 prevents cellular oxidative damage in long nights. *Plant Cell* **2004**, *16*, 908–919. [CrossRef] [PubMed]

108. Ferguson, D.L.; Burke, J.J. Methionyl sulfoxide content and protein-methionine-S-oxide reductase activity in response to water deficits or high temperature. *Physiol. Plant* **1994**, *90*, 253–258. [CrossRef]

109. Bechtold, U.; Rabbani, N.; Mullineaux, P.M.; Thornalley, P.J. Quantitative measurement of specific biomarkers for protein oxidation, nitration and glycation in Arabidopsis leaves. *Plant J.* **2009**, *59*, 661–671. [CrossRef] [PubMed]

110. Jacques, S.; Ghesquière, B.; De Bock, P.J.; Demol, H.; Wahni, K.; Willems, P.; Messens, J.; Van Breusegem, F.; Gevaert, K. Protein methionine sulfoxide dynamics in *Arabidopsis thaliana* under oxidative stress. *Mol. Cell Proteom.* **2015**, *14*, 1217–1229. [CrossRef] [PubMed]

111. Marondedze, C.; Turek, I.; Parrott, B.; Thomas, L.; Jankovic, B.; Lilley, K.S.; Gehring, C. Structural and functional characteristics of cGMP-dependent methionine oxidation in *Arabidopsis thaliana* proteins. *Cell Commun. Signal.* **2013**, *11*, 1. [CrossRef] [PubMed]

112. Oh, J.E.; Hossain Md, A.; Kim, J.H.; Chu, S.H.; Lee, E.H.; Hwang, K.Y.; Noh, H.; Hong, S.W.; Lee, H. The ectopic expression of methionine sulfoxide reductase 4 results in enhanced growth performance in arabidopsis. *Plant Sci.* **2010**, *178*, 265–270. [CrossRef]

113. Han, Y.; Du, Y.; Wang, J.; Wu, T. Overexpression of Chinese flowering cabbage BpPMSR3 enhances the tolerance of *Arabidopsis thaliana* to cadmium. *J. Plant Nutr. Soil Sci.* **2018**. [CrossRef]

114. Siddiqui, Z.S.; Cho, J.; Kwon, T.R.; Ahn, B.O.; Lee, S.B.; Jeong, M.J.; Ryu, T.H.; Lee, S.K.; Park, S.C.; Park, S.H. Physiological mechanism of drought tolerance in transgenic rice plants expressing *Capsicum annuum* methionine sulfoxide reductase B2 (*CaMsrB2*) gene. *Acta Physiol. Plant* **2014**, *36*, 1143–1153. [CrossRef]

115. El Hassouni, M.E.; Chambost, J.P.; Expert, D.; Van Gijsegem, F.; Barras, F. The minimal gene set member msrA, encoding peptide methionine sulfoxide reductase, is a virulence determinant of the plant pathogen *Erwinia chrysanthemi*. *Proc. Natl. Acad. Sci. USA* **1999**, *96*, 887–892. [CrossRef]

116. Gustavsson, N.; Kokke, B.; Härndahl, U.; Silow, M.; Bechtold, U.; Poghosyan, Z.; Murphy, D.; Boelens, W.; Sundby, C. A peptide methionine sulfoxide reductase highly expressed in photosynthetic tissue in *Arabidopsis thaliana* can protect the chaperone-like activity of a chloroplast-localized small heat shock protein. *Plant J.* **2002**, *29*, 545–553. [CrossRef] [PubMed]

117. Ezraty, B.; Grimaud, R.; El Hassouni, M.; Moinier, D.; Barras, F. Methionine sulfoxide reductases protect Ffh from oxidative damages in *Escherichia coli*. *EMBO J.* **2004**, *23*, 1868–1877. [CrossRef] [PubMed]

118. Chu, H.D.; Le, Q.N.; Nguyen, H.Q.; Le, D.T. Genome-wide analysis of genes encoding methionine-rich proteins in Arabidopsis and soybean suggesting their roles in the adaptation of plants to abiotic stress. *Int. J. Genom.* **2016**, *2016*, 5427062. [CrossRef]

119. Wehr, N.B.; Levine, R.L. Wanted and wanting: Antibody against methionine sulfoxide. *Free Radic. Biol. Med.* **2012**, *53*, 1222–1225. [CrossRef] [PubMed]

120. Ghesquière, B.; Gevaert, K. Proteomics methods to study methionine oxidation. *Mass Spectrom. Rev.* **2014**, *33*, 147–156. [CrossRef] [PubMed]

121. Lee, S.H.; Li, C.W.; Koh, K.W.; Chuang, H.Y.; Chen, Y.R.; Lin, C.S.; Chan, M.T. MSRB7 reverses oxidation of GSTF2/3 to confer tolerance of *Arabidopsis thaliana* to oxidative stress. *J. Exp. Bot.* **2014**, *65*, 5049–5062. [CrossRef] [PubMed]

122. Tarrago, L.; Kieffer-Jaquinod, S.; Lamant, T.; Marcellin, M.; Garin, J.; Rouhier, N.; Rey, P. Affinity chromatography: A valuable strategy to isolate substrates of methionine sulfoxide reductases? *Antioxid. Redox Signal.* **2012**, *16*, 79–84. [CrossRef] [PubMed]

123. Sundby, C.; Harndahl, U.; Gustavsson, N.; Ahrman, E.; Murphy, D.J. Conserved methionines in chloroplasts. *Biochim. Biophys. Acta* **2005**, *1703*, 191–202. [CrossRef] [PubMed]

124. Lallement, P.A.; Brouwer, B.; Keech, O.; Hecker, A.; Rouhier, N. The still mysterious roles of cysteine-containing glutathione transferases in plants. *Front. Pharmacol.* **2014**, *5*, 192. [CrossRef] [PubMed]

125. Martínez-Márquez, A.; Martínez-Esteso, M.J.; Vilella-Antón, M.T.; Sellés-Marchart, S.; Morante-Carriel, J.A.; Hurtado, E.; Palazon, J.; Bru-Martínez, R.A. Tau Class Glutathione-*S*-Transferase is Involved in *Trans*-Resveratrol Transport Out of Grapevine Cells. *Front. Plant Sci.* **2017**, *8*, 1457. [CrossRef] [PubMed]

126. Pégeot, H.; Mathiot, S.; Perrot, T.; Gense, F.; Hecker, A.; Didierjean, C.; Rouhier, N. Structural plasticity among glutathione transferase Phi members: Natural combination of catalytic residues confers dual biochemical activities. *FEBS J.* **2017**, *284*, 2442–2463. [CrossRef] [PubMed]

127. Tossounian, M.A.; Van Molle, I.; Wahni, K.; Jacques, S.; Gevaert, K.; Van Breusegem, F.; Vertommen, D.; Young, D.; Rosado, L.A.; Messens, J. Disulfide bond formation protects *Arabidopsis thaliana* glutathione transferase tau 23 from oxidative damage. *Biochim. Biophys. Acta* **2018**, *1862*, 775–789. [CrossRef] [PubMed]

128. Tossounian, M.A.; Wahni, K.; Van Molle, I.; Vertommen, D.; Astolfi Rosado, L.; Messens, J. Redox regulated methionine oxidation of *Arabidopsis thaliana* glutathione transferase Phi9 induces H-site flexibility. *Protein Sci.* **2018**. [CrossRef] [PubMed]

129. Benoit, S.L.; Bayyareddy, K.; Mahawar, M.; Sharp, J.S.; Maier, R.J. Alkyl hydroperoxide reductase repair by *Helicobacter pylori* methionine sulfoxide reductase. *J. Bacteriol.* **2013**, *195*, 5396–5401. [CrossRef] [PubMed]

130. O'Neil, K.T.; DeGrado, W.F. How calmodulin binds its targets: Sequence independent recognition of amphiphilic alpha-helices. *Trends Biochem. Sci.* **1990**, *15*, 59–64. [CrossRef]

131. Bernstein, H.D.; Poritz, M.A.; Strub, K.; Hoben, P.J.; Brenner, S.; Walter, P. Model for signal sequence recognition from amino-acid sequence of 54K subunit of signal recognition particle. *Nature* **1989**, *340*, 482–486. [CrossRef] [PubMed]

132. Xiong, Y.; Chen, B.; Smallwood, H.S.; Urbauer, R.J.; Markille, L.M.; Galeva, N.; Williams, T.D.; Squier, T.C. High-affinity and cooperative binding of oxidized calmodulin by methionine sulfoxide reductase. *Biochemistry* **2006**, *45*, 14642–14654. [CrossRef] [PubMed]

133. Bigelow, D.J.; Squier, T.C. Redox modulation of cellular signaling and metabolism through reversible oxidation of methionine sensors in calcium regulatory proteins. *Biochim. Biophys. Acta* **2005**, *1703*, 121–134. [CrossRef] [PubMed]

134. Carruthers, N.J.; Stemmer, P.M. Methionine oxidation in the calmodulin-binding domain of calcineurin disrupts calmodulin binding and calcineurin activation. *Biochemistry* **2008**, *47*, 3085–3095. [CrossRef] [PubMed]

135. Bigelow, D.J.; Squier, T.C. Thioredoxin-dependent redox regulation of cellular signaling and stress response through reversible oxidation of methionines. *Mol. Biosyst.* **2011**, *7*, 2101–2109. [CrossRef] [PubMed]

136. Snijder, J.; Rose, R.J.; Raijmakers, R.; Heck, A.J. Site-specific methionine oxidation in calmodulin affects structural integrity and interaction with Ca^{2+}/calmodulin-dependent protein kinase II. *J. Struct. Biol.* **2011**, *174*, 187–195. [CrossRef] [PubMed]

137. Bouché, N.; Yellin, A.; Snedden, W.A.; Fromm, H. Plant-specific calmodulin-binding proteins. *Annu. Rev. Plant Biol.* **2005**, *56*, 435–466. [CrossRef] [PubMed]

138. Yang, T.; Poovaiah, B.W. Calcium/calmodulin-mediated signal network in plants. *Trends Plant Sci.* **2003**, *8*, 505–512. [CrossRef] [PubMed]

139. Hardin, S.C.; Larue, C.T.; Oh, M.H.; Jain, V.; Huber, S.C. Coupling oxidative signals to protein phosphorylation via methionine oxidation in Arabidopsis. *Biochem. J.* **2009**, *422*, 305–312. [CrossRef] [PubMed]

140. Veredas, F.J.; Cantón, F.R.; Aledo, J.C. Methionine residues around phosphorylation sites are preferentially oxidized in vivo under stress conditions. *Sci. Rep.* **2017**, *7*, 40403. [CrossRef] [PubMed]

141. Lim, J.C.; You, Z.; Kim, G.; Levine, R.L. Methionine sulfoxide reductase A is a stereospecific methionine oxidase. *Proc. Natl. Acad. Sci. USA* **2011**, *108*, 10472–10477. [CrossRef] [PubMed]

antioxidants

MDPI

Review

The Oxidized Protein Repair Enzymes Methionine Sulfoxide Reductases and Their Roles in Protecting against Oxidative Stress, in Ageing and in Regulating Protein Function

Sofia Lourenço dos Santos, Isabelle Petropoulos * and Bertrand Friguet *

Sorbonne Université, CNRS, INSERM, Institut de Biologie Paris-Seine, Biological Adaptation and Ageing, B2A-IBPS, F-75005 Paris, France; sofia.lourenco.santos@gmail.com
* Correspondence: isabelle.petropoulos@sorbonne-universite.fr (I.P.); bertrand.friguet@sorbonne-universite.fr (B.F.)

Received: 15 November 2018; Accepted: 1 December 2018; Published: 12 December 2018

Abstract: Cysteine and methionine residues are the amino acids most sensitive to oxidation by reactive oxygen species. However, in contrast to other amino acids, certain cysteine and methionine oxidation products can be reduced within proteins by dedicated enzymatic repair systems. Oxidation of cysteine first results in either the formation of a disulfide bridge or a sulfenic acid. Sulfenic acid can be converted to disulfide or sulfenamide or further oxidized to sulfinic acid. Disulfide can be easily reversed by different enzymatic systems such as the thioredoxin/thioredoxin reductase and the glutaredoxin/glutathione/glutathione reductase systems. Methionine side chains can also be oxidized by reactive oxygen species. Methionine oxidation, by the addition of an extra oxygen atom, leads to the generation of methionine sulfoxide. Enzymatically catalyzed reduction of methionine sulfoxide is achieved by either methionine sulfoxide reductase A or methionine sulfoxide reductase B, also referred as to the methionine sulfoxide reductases system. This oxidized protein repair system is further described in this review article in terms of its discovery and biologically relevant characteristics, and its important physiological roles in protecting against oxidative stress, in ageing and in regulating protein function.

Keywords: protein oxidation; methionine oxidation; methionine sulfoxide reductases; oxidized protein repair; ageing

1. Introduction

Enzymatically repair of protein oxidative damage is only possible for certain oxidation products of the sulfur-containing amino acids, cysteine and methionine. In the case of cysteine, the major systems involved in reversing the oxidation of disulfide bridges and sulfenic acid include the reduced forms of small proteins such as thioredoxin and glutaredoxin. Methionine sulfoxide, on its turn, is reduced back to methionine by the methionine sulfoxide reductases enzymes that are then recycled by the thioredoxin/thioredoxin reductase system.

Thioredoxins (Trx) are small ubiquitous proteins with two catalytic redox active cysteines (Cys-XX-Cys), which catalyze the reversible reduction of protein disulfide bonds. Subsequently, oxidized thioredoxins are reduced back enzymatically by the NADPH-dependent thioredoxin reductase (TR) enzymes, which together with NADPH and Trx constitute the thioredoxin system [1] (Figure 1). Two Trx enzymes have been identified to date, Txr1, which is present in the cytosol and can be translocated into the nucleus in oxidative stress conditions and Trx2, which is present in the mitochondria. The antioxidant activity of these enzymes consists of providing electrons to thiol-dependent peroxidases (Prx), allowing the recycling of these Prx enzymes for the continuous removal of reactive oxygen species (ROS) and reactive nitrogen species (RNS). Furthermore, Trx are also involved in the protection against protein oxidative damages by reducing methionine sulfoxide reductases (Msrs), enzymes capable of repairing oxidized methionines. In mammals, Trx also regulates the activity of many redox-sensitive transcription factors, such as NF-κB, Nrf2 and p53 [1].

Glutaredoxins (Grx) are found in almost all living organisms and collaborate with thioredoxins for the reduction of protein disulfides and S-glutathionylated proteins. Four Grx isoenzymes (Grx1, Grx2, Grx3, and Grx5) exist in mammals. In terms of structure, they belong to the Trx superfamily having a dithiol or monothiol active motive, Cys-XX-Cys or Cys-XX-Ser, respectively [2,3]. In contrast to Trx, Grx are reduced back non-enzymatically by glutathione, which is recovered by the glutathione reductase enzyme in the presence of NADPH (Figure 1). An exception was observed with the Grx2 isoenzyme, which was shown to be reduced by Trx2 in mitochondria [4]. The reduction of disulfides and the participation on protein deglutathionylation state the importance of Grx enzymes in defense against oxidative stress as well as in redox regulation of signal transduction [5,6].

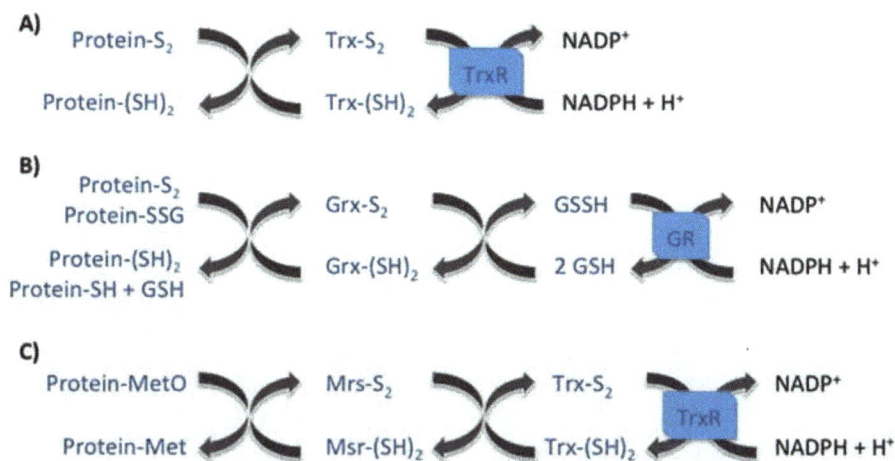

Figure 1. The three major protein repair systems. (**A**) Thioredoxin (Trx) system participates on the reduction of protein disulfide (Protein-S2); (**B**) Glutaredoxin (Grx) system reduces protein disulfide (Protein-S2) as well as protein glutathione mix disulfide (Protein-SSG); and (**C**) Methionine sulfoxide reductase (Msr) system reduces methionine sulfoxides (MetO).

Msrs in mammals constitute a group of four ubiquitous enzymes that catalyze the reduction of free and protein-derived methionine sulfoxides (MetO) to methionine (Met). Two diastereoisomers (rectus and sinister) of methionine sulfoxides can be formed upon protein methionine oxidation, methionine-R-sulfoxide (MetRO) and methionine-S-sulfoxide (MetSO). MsrA, which is present in the cytoplasm, in the nucleus and in the mitochondria reduces specifically the MetSO [7]. On the other hand, the three MsrB are responsible for the reduction of MetRO. In terms of their intracellular localization, MsrB1 is present in the cytoplasm as well as in nucleus, MsrB2 is only present in the mitochondria and MsrB3 is present in the mitochondria as well as in the endoplasmic reticulum of eukaryotic cells [7]. Msr enzymes possess one cysteine (a selenocysteine in the case of MsrB1), in their catalytic site, responsible for MetO reduction and whose recycling involves the formation of a disulfide bond with a second Msr cysteine. The disulfide bond can be further reduced restoring Msr activity by the Trx system (Figure 1). Msr have been intensely studied for their antioxidant roles as well as their protection against oxidative stress and apoptosis [8–10]. Furthermore, they have been suggested as being involved in longevity modulation of some models organisms such as *D. melanogaster* [11], *C. elegans* [12] and *S. cerevisiae* [13,14]. More recently, these enzymes have been considered as regulators of protein function [15] and as being involved in redox regulation of cellular signaling [16]. These proteins as well as their functions are further described and discussed in the following sections.

2. Methionine Sulfoxide Reductases Discovery

Due to the presence of a sulfur atom, methionine residues are very sensitive to oxidation leading to a modification or loss of protein function when oxidized within proteins. First evidences of the importance of keeping methionine in its reduced state for biological function appeared more than 70 years ago. Studying *L. arabinosus*, Waelsch and colleagues found that methionine oxidation of glutamine synthetase inhibited the conversion of glutamic acid into glutamine, an essential step for bacterial growth [17]. In addition, sporulation of *B. subtilis* was also described to be affected by methionine oxidation [18]. Few years after the identification of the first deleterious effects of MetO, a Msr activity, capable of reducing back MetO to Met was described, primarily in yeast [19], later in bacteria [20] and in higher organisms, such as plants [21] and animals [22]. Msr activity was evidenced in *E. coli* by their ability to grow in a culture medium with L-MetO as the only source of methionine, thus capable of catalyzing the reduction of MetO [20]. In 1981, Brot and colleagues partially purified one of the enzymes responsible for the reduction of MetO within proteins and showed it as essential for restoring the activity of the ribosomal protein L12 in *E. coli* [23]. The enzyme, later called MsrA, uses reduced Trx in vivo or dithiothreitol (DTT) in vitro as electron donor [24]. MsrA is a ubiquitous protein, differentially expressed in mammalian tissues and capable of reducing a variety of substrates such as free MetO and peptides or proteins containing MetO [25]. MsrA was found to be a stereospecific enzyme only capable of reducing the MetSO diastereoisomer of MetO, with an increased specificity for protein-bound MetO compared to free MetO [26,27].

Twenty years after the purification of MsrA, Grimaud et al. discovered that full reduction of oxidized calmodulin can be done by the combined action of MsrA and another enzyme called MsrB [28]. This new Msr is in fact responsible for the reduction of the MetRO diastereoisomer within proteins, which is not reduced by MsrA [29,30]. MsrB, later called MsrB1, SelX or SelR in mammals, was discovered in 1999 by Lescure and colleagues as a novel selenoprotein, but at this time, its function was unknown [31]. MsrB exclusively acts on peptidyl-MetO [28]. Intrigued by this, different authors have identified a novel Msr in *E. coli*, called fRMsr (for free MetRO reductase) or MsrC, which activity is specific for free MetO, being unable to reduce MetO present within proteins [30,32,33]. More recently, Gennaris and co-workers made another interesting discovery: they found a new Msr system, named MsrPQ (MsrP for periplasm and MsrQ for quinone), present in the envelope of *E. coli* bacteria, that, in contrast to the other known Msr, can reduce both MetO diastereoisomers using electrons directly from the respiratory chain, thus independently from Trx [34]. A similar system may exist in eukaryotic subcellular oxidizing compartments, such as the endoplasmic reticulum or lysosomes but, so far, fRMsr and MsrPQ systems were only found in prokaryotes or unicellular eukaryotes.

3. Methionine Sulfoxide Reductases Phylogenetic, Tissue and Cellular Distribution

The phylogenetic distribution of *Msr* genes was revealed by genomic analyses made in different organisms. These studies show the presence of *MsrA* and *MsrB* genes in all eukaryotes without exception. Bacteria can possess only *MsrA* genes, the two *Msr* genes or a bifunctional *MsrA/B* fusion gene [35]. This universal presence of *Msr* genes supports their essential roles for cell function, either in protecting them against oxidative damages as well as in regulating protein function. The greatest exception to the universal *Msr* representation among life domains is its absence in 12 archaea representative genomes [35]. Several hypothesis for this, although none of them was already been clearly proven, have been proposed: (i) the development of a functionally equivalent system; (ii) the low O_2 solubility at high temperatures which would avoid ROS production within hyperthermophiles; (iii) the observation of non-enzymatic MetO reduction at higher temperatures [36]; (iv) the existence of protein structures that protect Met residues; (v) the presence of *Msr*-containing plasmids in these archaea; (vi) the discovery of fRMsr in some archaea including some of those lacking *Msr* genes [37] or finally (vii) the existence of an efficient first line of defense against ROS production (catalase, superoxide dismutases and Prx enzymes), which would create a low-ROS environment that diminishes the frequency of protein oxidation.

The fact that no *MsrB*-containing organism exists without *MsrA* gene suggests to the authors that these two genes evolved independently and that in these organisms, a greater abundance of the S epimer of MetO may explain that MsrA protein was sufficient to perform this specific protein repair process [35]. According to this hypothesis, MsrB may have evolved to play other redox functions, increasing defenses against greater oxidative damage seen in more complex life forms such as animals and plants. In addition, the organization of the two *Msr* genes also differs, this can explain the massive *MsrB* gene duplication seen in the plant *Arabidopsis thaliana*: nine *MsrB* genes in contrast to five *MsrA* genes [38]. In contrast to plants and algae [38], animals and bacteria contain fewer *Msr* genes: mammals have one *MsrA* and three *MsrB* genes while *E. coli*, *S. cerevisiae*, *C. elegans* and *D. melanogaster* have only one *MsrA* and one *MsrB* gene [29,39]. Such gene diversity underlines the biological importance of this system and gene redundancy may be explained by the necessity for organisms to respond to the modification of environmental conditions such as oxidative [10] or thermal stress [40,41].

In mammals, Msr enzymes are ubiquitously expressed [23,25], with the only exception of leukemic cells that do not express MsrA [42]. Analyses of mouse, rat, and human tissues revealed a maximal expression level of MsrA in kidney and liver, followed by heart, lung, brain, skeletal muscle, retina, testis, bone marrow, and blood [25,42]. The highest Msr activities were found in rat kidney [43] and human neutrophils [44], which in the case of neutrophils was later shown to be due mainly to MsrB type [45]. In human skin, MsrA was shown to participate to tissue homeostasis and to be a sensitive target for UV [46,47]. MsrA and all three MsrB proteins are expressed in melanocytes [48] and keratinocytes [46]. A lower MsrA expression was found in dermal fibroblasts [46] while a greater MsrA expression was found in sebaceous glands [49]. MsrB1 and MsrB3 were both expressed within vascular endothelial cells [49].

Msr enzymes are differentially distributed in the mammal subcellular compartments (Figure 2), which indicates that each Msr may have an organelle-specific role. MsrA is present in mitochondrial matrix due to its N-terminal mitochondrial signal sequence [50] but was also found in rat liver cytosolic fractions [51] and in the nucleus of mouse cells [52]. If its N-terminal peptide sequence was sufficient for mitochondrial targeting, other structural and functional elements present in the MsrA sequence can determine its intra-cellular distribution. Correctly folded MsrA is retained in the cytosol, while partially misfolded MsrA appears to be targeted to the mitochondria [52]. An alternative first exon splicing generating an additional MsrA form lacking a mitochondrial signal, which resides in cytosol and nucleus, was also evidenced [53]. This protein produced by initiation at the second site has been shown to be myristoylated and localized in the late endosomes [54].

For the MsrB family, the situation is more complex than for MsrA, due to the existence of four proteins resulting from the transcription and consequent translation of three different genes. The selenoprotein protein MsrB1 is present both in the nucleus and the cytosol, while MsrB2, also known as CBS-1, is present only in the mitochondria due to the presence of a N-terminal signal peptide [39]. Interestingly, MsrB3A and MsrB3B result from alternative splicing of the first exon of the *MsrB3* gene with MsrB3A displaying an endoplasmic reticulum signal peptide while MsrB3B showed a mitochondrial signal peptide at the N-terminus, in addition to an endoplasmic reticulum retention signal peptide at their C-terminus. In mouse, however, there is no evidence for *MsrB3* alternative splicing and the only MsrB3 protein is present in the endoplasmic reticulum, even though it has both the endoplasmic reticulum and the mitochondrial signal peptides at its N-terminus [55].

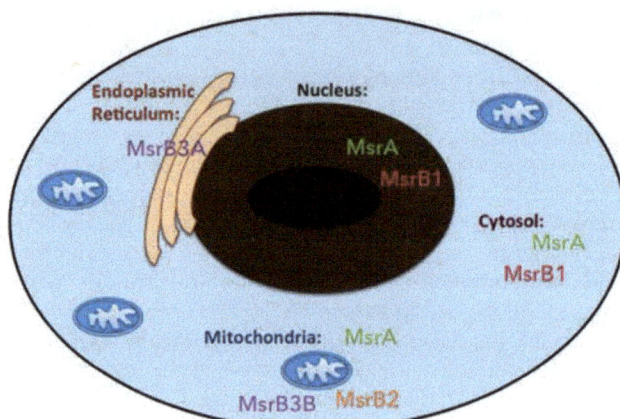

Figure 2. Subcellular distribution of methionine sulfoxide reductase enzymes in humans. MsrA and MsrB1 enzymes are present in the cytoplasm and in the nucleus of human cells. MsrA together with MsrB2 and MsrB3B enzymes can be found within mitochondria. The endoplasmic reticulum only contains a MsrB type enzyme called MsrB3A.

4. Methionine Sulfoxide Reductases Sequence, Structure and Catalytic Activity

Sequence alignment between the primary structures of MsrA protein from different organisms showed that there is a high homology among them, with *E. coli* and *B. taurus* MsrA having 67% and 88% sequence identity to human MsrA [56]. In addition, this alignment highlighted a conserved active-site sequence GCFWG in all organisms studied. The strictly conserved cysteine within this motif (Cys-51 in the case of *E. coli* MsrA, Cys-72 in the case of bovine MsrA and Cys-74 in the case of human MsrA) is essential for the MsrA reducing activity (Figure 3). Moreover, three other cysteine were shown to be conserved within 70% of all MsrA proteins: Cys-86, Cys-198 and Cys-206 in the case of *E. coli*; Cys-107, Cys-218 and Cys-227 for bovine MsrA; and Cys-109, Cys-220 and Cys-230 in the case of human MsrA [56]. Toward the C-terminus, the two last cysteines bracket a glycine-rich region on MsrA sequence and may serve as additional recycling cysteine for the catalytic mechanism.

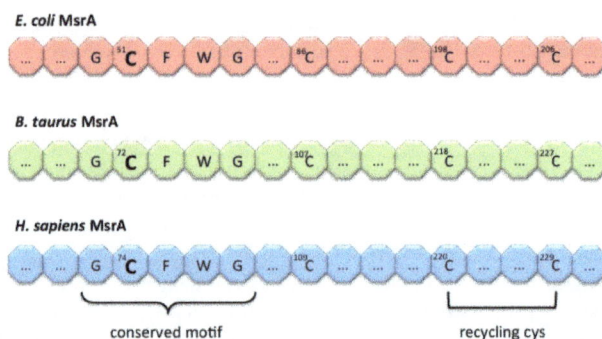

Figure 3. MsrA conserved GCFWG motif. Schematic representation of the conserved MsrA active-site sequence in *E. coli*, *B. Taurus* and *H. sapiens*, where the respective conserved cysteine is placed: Cys-51 in the case of bacterial MsrA, Cys-72 for bovine MsrA and Cys-74 in human MsrA. Three other cysteine residues were shown to be conserved: bacterial Cys-86, Cys-198 and Cys-206; bovine Cys-107, Cys-218 and Cys-227; and human Cys-109, Cys-220 and Cys-230; with the two last cysteines, toward the C-terminus of each MsrA, being responsible for recycling of the catalytic cysteine.

MsrA three-dimensional (3D)-structures obtained by X-ray crystallography from *E. coli* [57], *B. Taurus* [58], *M. tuberculosis* [59] and *P. trichocarpa* [60] were also essential to determine the mechanistic aspects of MsrA catalysis. MsrA folding belongs to an α/β class of proteins. The presence of the catalytic cysteine in the N-terminal α-helix of the protein allows it to face MetO residues present in other proteins, while the two recycling cysteines are buried in the C-terminal region of the protein. Analysis of the 3D structures of the MsrA with MetO showed that the oxygen of the methionine sulfoxide is strongly stabilized by a hydrophilic subsite composed of a network of hydrogen bonding interactions including Tyr-82, Glu-94 and Tyr-134 in MsrA (numbers based on the *E. coli* MsrA sequence) [61]. Using NMR technology, a high degree of flexibility of the C-terminal region of oxidized MsrA was evidenced, which favors the formation of an intramolecular disulfide bond between the two recycling cysteines [62].

The catalytic mechanism for MsrA has been described by Boschi-Muller and colleagues in the case of *E. coli* MsrA and by Lowther et al. for bovine MsrA [58,63]. Based on sulfenic acid chemistry, these two groups proposed a reaction mechanism for MsrA catalysis consisting in several steps (Figure 4). First, the catalytic Cys-51/72 will act as a nucleophilic agent attacking the sulfoxide moiety of the substrate leading to formation of a sulfenic acid on the catalytic cysteine with the concomitant release of 1 mol of methionine per mol of Msr. Subsequently, the recycling Cys-198/218 attack on the sulfenic intermediate will create an intramolecular disulfide bond between the catalytic and the recycling cysteine. In the case of another recycling cysteine, such as Cys-206 from *E. coli* and Cys-227 from *B. taurus*, there is subsequent nucleophilic attack of Cys-206/227 on Cys-198/218 creating a new intramolecular disulfide bond between these two recycling cysteines. The last step involves reduction of the disulfide bond by Trx in vivo or other reducing agents, such as DTT in vitro. Kinetic studies showed that the rate of formation of the sulfenic acid is high while the recycling process, which reduces back the oxidized catalytic cysteine, is overall rate-limiting [64]. Murine and human MsrA possess the same mechanism of catalysis but in the case of other bacteria, such as *N. meningitides* [65] and *M. tuberculosis* [59], MsrA proteins possess only one recycling cysteine equivalent to Cys-198 from *E. coli*. Recently, MsrA has been shown to also have an oxidase activity towards methionines, producing MetSO within proteins, including itself, or on free methionines [66], even in the presence of Trx [67].

The first protein showing a methionine-R-sulfoxide reductase activity was identified in *E. coli* and was named MsrB. It has no similarity with MsrA, presents 43% sequence homology with PilB from *N. gonorrhoeae* and contains a conserved signature sequence CGWP(S/A)F [28]. Indeed, the PilB protein has a MsrA- and a MsrB-like C-terminal domains that function with opposite substrate stereospecificity and a N-terminal thioredoxin-like domain that allows the regeneration of both Msr active sites [30,68,69]. The 3D-structures from the MsrB domain of PilB obtained by X-ray crystallography showed no similarity with the one from MsrA [68]. However, the active sites for both enzymes show an axial symmetry as if they were reflecting each other in a mirror, which can be explained by the stereospecificity of the two enzymes. This symmetry suggests a similar catalytic mechanism for both enzymes and indeed, the two catalytic cysteines of PilB, Cys-495 and Cys-440, function in a similar way to the *E. coli* MsrA Cys-51 and Cys-206: a nucleophilic attack by Cys-495 to MetO leads to the production of a trigonal intermediary compound that after ionic rearrangement and subsequent methionine release, will form a sulfenic acid on PilB Cys-495 [68]. As for MsrA, a series of proton exchanging events occurs leading to the formation of a disulfide intramolecular bond, which is consequently reduced by Trx [68].

MsrB1 is a mammalian MsrB enzyme that shows less homology with MsrBs from invertebrates, with only 29% of similarity between the sequences of human MsrB1 and the MsrB domain of *N. gonorrhoeae* PilB [31]. The presence of the selenium atom in its active site is critical for the catalytic function of this enzyme. In fact, the wild type selenoprotein MsrB1 is 800-fold more active than the corresponding cysteine-MsrB1 mutant form [39]. Similarly, a cysteine to selenocystein mutation in mammalian MsrB2 and MsrB3 resulted in 100-fold increase in the catalytic activity of the enzymes [70]. The incorporation of a selenocysteine into the primary sequence of MsrB1 protein is due to a SECIS

(SelenoCysteine Insertion Sequence) element localized in the 3'UTR of MsrB1 mRNA at a distance from the UGA codon [31]. The presence of a recycling cysteine, present in the N-terminal region of the MsrB1 protein is important for resolving the selenic acid intermediate formed during catalysis [70].

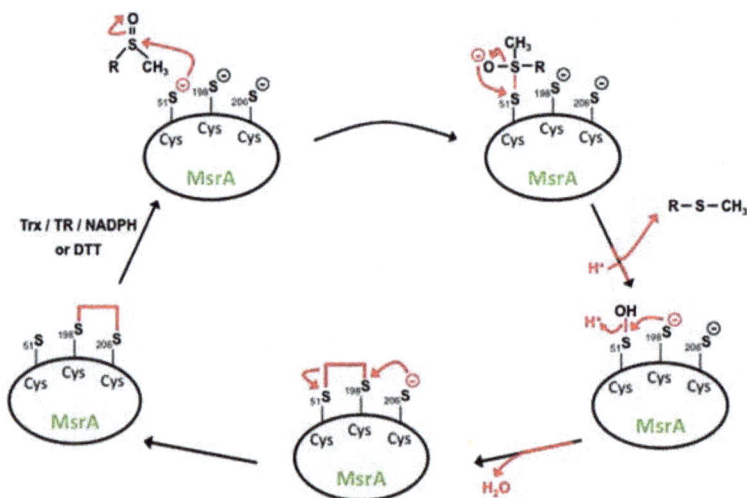

Figure 4. MsrA catalytic mechanism. The nucleophilic attack of the MsrA catalytic cystein (Cys)-51 on the sulfur atom of the methionine sulfoxide substrate leads to the formation of an unstable intermediate (enzyme bound to the substrate). Bacterial MsrA is used in this representation. Ionic rearrangement leads to the formation of a sulfenate ion with the concomitant release of the methionine molecule and protonation of the sulfenate ion to produce a sulfenic acid intermediate on MsrA. The nucleophilic attack of the recycling Cys-198 on the sulfur atom of the sulfenic acid intermediate leads to the formation of an intramolecular disulfide bond. MsrA full native state recovery is achieved after another nucleophilic attack of Cys-198 on Cys-206 generating a second intramolecular disulfide bond, which can be reduced either by the thioredoxin (Trx)/thioredoxin reductase (TR)/NADPH regenerating system or by dithiothreitol (DTT) [60,63].

MsrB2 was first identified in humans as a 21 kDa protein composed of 202 amino acids and carrying the conserved MsrB sequence GTGWP [71]. It presents 59% homology with *E. coli* MsrB and 42% with the C-terminal domain of *N. gonorrhoeae* PilB. MsrB2 is four times less efficient than MsrA for the reduction of a specific synthetic substrate. Similarly to MsrB1, MsrB2 possesses a CXXC motif responsible for zinc binding. In contrary to other MsrB, the mammalian MsrB2 proteins do not contain a recycling cysteine residue in the middle of the sequences and the sulfenic intermediate could be directly reduced by Trx [70].

Another gene, that seems to be only present in mammals' genome, encodes, by alternative splicing, two proteins of the MsrB family, MsrB3A and MsrB3B [39]. As for the other MsrB proteins, they contain a catalytic cysteine and the CXXC motif responsible for zinc binding. In terms of their catalytic activity, these enzymes act similarly to the MsrB2 enzymes (Figure 5). Despite of the presence of cysteine residues instead of selenocystein in their active sites, MsrB2 and MsrB3 exhibit good catalytic efficiencies but that are slightly lower than MsrB1 [39,70].

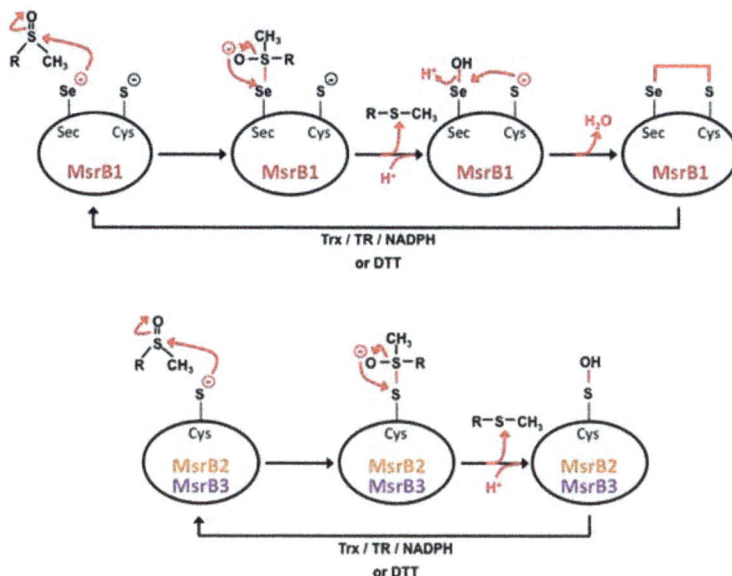

Figure 5. MsrB catalytic mechanism. In the case of MsrB1 (upper part of the figure), methionine sulfoxide reduction starts with the nucleophilic attack of selenocystein (Sec) on the sulfur atom of the substrate leading to the formation of an unstable intermediate. Ionic rearrangement leads to the formation of a selenic acid intermediate with the concomitant release of a methionine molecule. The nucleophilic attack of the recycling cysteine (Cys) on this selenic acid intermediate leads to the formation of an intramolecular selenenylsulfide bond, which is subsequently reduced by the thioredoxin (Trx)/thioredoxin reductase (TR)/NADPH regenerating system or by dithiothreitol (DTT). In contrast, the sulfenic intermediate of MsrB2 and MsrB3 (lower part of the figure), formed on Cys after methionine release, can be directly reduced to its fully active state by one of these mechanisms [68,70].

First studies on the biological reducing agents for MsrA revealed that reduced Trx, high levels of DTT or reduced lipoic acid can reduce oxidized MsrA in vitro. If *E. coli* MsrA and MsrB and bovine MsrA efficiently use either Trx or DTT as reducing agents, human MsrB2 and MsrB3 showed less than 10% of their activity with Trx as reducing agent when compared to the use of DTT [72]. This suggests that in animal cells, Trx may not be the only reducing power for these two enzymes. Thionein, the reduced metal-free form of metallothionein, and selenium compounds, such as selenocystamine, could also function as reducing agents for human MsrB3 and MsrB2 [72,73]. Furthermore, it was found that a few Msr such as *A. thaliana* MsrB1, *Clostridium* Sec-containing MsrA or the red alga *G. gracilis* MsrA can be regenerated by Grx/glutathione system [74–77]. In the particular case of *A. thaliana* MsrB1, the sulfenic acid is reduced by glutathione forming a glutathionylated intermediate that is attacked by glutaredoxins [76]. However, the rate of the recycling process by the Grx system is at least 10 to 100-fold lower compared to Trx acting in Msr with a recycling cysteine, suggesting that the Grx system would not be used in Msr in which a disulfide bond is formed. In agreement with this hypothesis is the fact that methionine auxotrophic *E. coli* is unable to grow in the presence of MetO when the *Trx1* gene is inactivated [78].

5. Methionine Sulfoxide Reductases in Protection Against Oxidative Stress

Several studies using different bacterial and eukaryotic models revealed that Msr enzymes and methionine amino acids work together in cellular protection against oxidative stress. Surface-exposed methionine residues in proteins are more easily oxidized by ROS, such as H_2O_2, chloramines or HOCl, due to the presence of sulfur atoms. However, they are believed to be more resistant to oxidative inactivation [79], thus keeping protein structure and catalytic function. Through this mechanism, methionines are proposed to act as a threshold barrier for other amino acid oxidation, which would lead to loss of the protein activity [79,80]. Another mechanism through which methionine can act as antioxidant amino acid, observed in *E. coli* [81], *S. cerevisiae* [82] and in mammalian cells [83], is its misacylation, i.e., the incorporation of methionines by non-methionyl-tRNAs during translation. To date, this translation infidelity was revealed to be specific to methionines and its frequency of occurrence was shown to increase upon innate immune or chemically induced oxidative stress [83], thus suggesting it plays a role in protecting against these kinds of stresses. In agreement, some Met-mistranslated forms were observed in the catalytic domain of the Ca^{2+}/calmodulin-dependent kinase II (CaMKII) under Ca^{2+} stress, resulting in/increased catalytic activity as well as alterations of proteins subcellular localization [84].

The antioxidant protection conferred by methionine residues is also due to their cyclic reduction by Msr enzymes, which by rendering methionines prone to new oxidation reactions, will lower ROS levels [85]. In fact, the absence of MsrA expression leads to reduced *E. coli* as well as *M. tuberculosis* viability when treated with H_2O_2, nitrite or S-nitrosoglutathione; effect that can be reversed by transformation of the mutant strain with a plasmid containing the wild-type MsrA gene from the respective bacteria species [86,87]. This MsrA protection against an oxidative stress state induced by different oxidants has been also verified in other bacteria such as *O. anthropi* [88] and *S. aureus* [89]. Moreover, the presence of MsrA was relevant for survival of *M. smegmatis* within macrophages producing high levels of ROS and RNS as a defensive response against those microorganisms [90]. Growth alteration and significant protein carbonyl accumulation were also observed in MsrA null yeast mutants when submitted to H_2O_2 treatment [91,92]. In agreement, overexpression of MsrA in this eukaryotic species reduced the levels of free and protein-bound MetO leading to increased resistance to toxic concentrations of H_2O_2 [93]. Overall the studies indicate that bacteria and yeast are dependent of MsrA to counteract the damaging effects of oxidative stress and consequently for their survival in these conditions. In the case of *S. cerevisiae* not only MsrA, but also MsrB was shown to protect against oxidative damages mediated by the toxic metal chromium [94].

Msr protection against oxidative stress is also notorious in higher organisms such as plants and animals. During the dark periods when *A. thaliana* produces more H_2O_2, it has been shown that the Msr system was important to prevent protein oxidative damage, thus minimizing protein turnover in these conditions of limited energy supply [95]. MsrB3 was identified in a proteomic study as a cold-responsive protein in Arabidopsis [96] as essential to reduce oxidized methionine and to lower H_2O_2 level that accumulates in the endoplasmic reticulum during plant cold acclimation [40]. In the case of invertebrate animals, overexpression of MsrA in the nervous system of *D. melanogaster* was shown to increase the resistance of these transgenic animals to paraquat-induced oxidative stress [97], while the suppression of this gene in *C. elegans* resulted in worms more sensitive to paraquat treatment, presenting chemotaxis and locomotor failure, partly due to muscle defects [12]. In line with these results, the induction of MsrA by ecdysone, was also shown to protect *Drosophila* against H_2O_2-induced oxidative stress [98].

In mammals, paraquat injection decreased the survival of $MsrA^{-/-}$ mice comparing to $MsrA^{+/+}$ or $MsrA^{+/-}$ mice [9]. MsrA null mutant mice present also higher level of protein carbonyls in the liver, kidney [99] and heart, correlated with mitochondria morphological changes in this last tissue [100]. Regarding the heart, our laboratory observed that the modulation of Msr activity in rat hearts along the course of cardiac ischemia/reperfusion may involve structural modification of the enzyme rather than a modification of MsrA protein level [101]. Indeed, later it was shown that the MsrA cytosolic form needs to be myristoylated in order to confer heart protection against ischemia/reperfusion damages, suggesting that it must interact with a hydrophobic domain [102]. This protective role of MsrA against ischemia/reperfusion injuries was also evidenced in mouse kidney, with increased oxidative stress markers, inflammation and fibrosis observed in kidneys of $MsrA^{-/-}$ mice after injury comparing to wild type [103,104]. In agreement to what have been found in $MsrA^{-/-}$ mice, increased GSH, protein and lipid oxidation were also found in the liver and kidney of the *MsrB1* K.O. mice [105]. MsrB1 is a selenoprotein, thus depending on selenium concentrations to be produced. Interestingly, Jacob Moskovitz has found that *MsrA* K.O. mice submitted to a selenium deficient diet presented decreased MsrB activity but also less Gpx and Trx activities in their brains [106]. MsrA or MsrB1 deficiency significantly accelerates acetaminophen-induced hepatic toxicity by aggravating GSH depletion and lipid peroxidation [107].

The protective role of Msr enzymes against oxidative damages was also suggested by various studies in different mammalian cell types in vitro, such as retinal pigment epithelial cells from human [108], monkey [109], rat [110], or human lens cells [111]. Overexpression of *MsrA* in human lens cells, for instance, gave them an increased resistance to H_2O_2-induced stress while *MsrA* gene silencing led to an their increased sensitivity towards oxidative treatment and to a loss of viability even in the absence of exogenously added stress [111]. On the other hand, silencing of all or individual *Msr* genes led to increased oxidative stress-induced cell death indicating that MsrB are also implicated in human lens epithelial cell viability in these conditions [112]. MsrB1 was the most studied MsrB in human lens epithelial cells. Silencing of this *Msr* gene resulted in increased ROS levels, lipid oxidation, ER stress, decreased mitochondrial potential and release of cytochrome c, ultimately leading to caspase-dependent apoptosis [113–115]. Furthermore, peroxynitrite treatment to MsrB1-deficient human lens epithelial cells will aggravate the oxidative damages and F-actin disruption, that normally occurs after this nitric stress [116], suggesting that MsrB1 protects lens cells from F-actin nitration. Using stable human embryonic kidney HEK293 clones with an altered Msr system due to silencing the expression of *MsrA*, *MsrB1*, or *MsrB2*, our laboratory performed a proteomic analysis on the Msr-silenced cells grown under basal conditions or submitted to oxidative stress, revealing that the disruption of the Msr system mainly affects proteins with redox, cytoskeletal or protein synthesis, and maintenance roles [117]. Interestingly, most of the proteins found altered in the Msr mutants were also identified as potential Msr substrates and have been associated with redox or ageing processes in previous studies. Furthermore, we and others have shown that human T lymphocyte cells presented an increase resistance to H_2O_2 or zinc treatments when transfected with *MsrA* and/or *MsrB2* genes by reducing the levels of intracellular ROS species and protein oxidative damages that would lead to cell death [8,93,118]. The role for MsrA in the prevention against the accumulation of protein and cellular oxidative damage provoked by H_2O_2-induced oxidative stress was also studied in fibroblast cells [119] and was associated in these cells to MsrA-dependent differentially expression proteins implicated in protection against oxidative stress, apoptosis, and premature ageing [120]. Even though overexpression of MsrA in the endoplasmic reticulum of mammalian cells increases their resistance to oxidative and endoplasmic reticulum stresses [121], the resistance to an endoplasmic reticulum stress is mainly conferred by MsrB3 [41,122]. Finally, in human skin cells, the behavior of the MsrA enzyme seems to be dependent on the type of ultraviolet (UV) exposure and the dose applied, suggesting a hormetic response to environmental stress. In fact, low doses of UVA stimulate MsrA expression, while UVB or high doses of UVA contribute to decrease MsrA expression and increase protein carbonyl [46,47], a profile that can be prevented by pre-treating the cells with MsrA [123]. In melanocytes, the absence of MsrA expression also increased sensibility to oxidative stresses and cell death even in the absence of exogenous stresses [124].

Together, these studies suggest that methionine amino acid residues along with the Msr system constitute a potent antioxidant ROS scavenging system, preserving macromolecules in their reduced state and thus contributing to protein homeostasis and function while protecting different cell types and organisms from different kinds of oxidative stresses.

6. Methionine Sulfoxide Reductases in Disease, Ageing and Longevity

Given that Msr system also protects proteins from irreversible oxidation as a result of a severe oxidative stress and that protein carbonyls levels are usually referred as a marker of oxidative stress in pathophysiological conditions and during ageing, it is expected that Msr would be implicated in diseases and in ageing process.

While it has been shown that oxidized proteins accumulate in tissues from patients exhibiting age-related diseases such as Alzheimer's, Parkinson's, and Huntington's diseases, and cataracts [125,126], reduced MsrA activity was found in the brains of Alzheimer's disease patients [127]. In agreement with this is the fact that *MsrA* K.O. mice demonstrated behavioral abnormality (tip-toe walking) consistent with cerebellar dysfunction [99], increased light scattering—a common cataract symptom [128] and enhanced neurodegeneration with characteristic features of neurodegenerative diseases [129]. Methionine oxidation in Met-35 of amyloid ß-peptide (Aβ peptide) is thought to be critical for aggregation and neurotoxicity [130] and it was shown that the absence of MsrA modifies Aβ solubility properties and causes mitochondrial dysfunction in a mouse model of Alzheimer's disease [131,132]. In the case of Parkinson's disease, oxidation of the methionine residues in α-synuclein is thought to be the main reason of protein fibrillation causing the pathology [133].

The importance of Msr in age-related diseases and the accumulation of modified proteins produced by the action of ROS as major hallmark of ageing, suggests that Msr would have also an important role in ageing phenotype. Indeed, the accumulation of oxidatively modified proteins during ageing has been largely attributed to declined efficacy of the systems involved in protein homeostasis such as protein degradation and protein repair [134]. Again, our laboratory has shown that MsrA is down-regulated in aged rats [135] and during replicative senescence of fibroblasts [136]. Both cytosolic and mitochondrial Msr activities were found to decline upon replicative senescence [137] and increased MetO levels were found in membrane proteins of senescent erythrocytes [138] as well as in senescent *E. coli* [139].

Several studies have been done to elucidate the implication of the Msr system in regulating lifespan but they are still controversial. The first two groups having tried to test this hypothesis used two different models: *MsrA* K.O. mice and *MsrA* overexpressing *Drosophila* [97,99]. Knockout of the *MsrA* gene in mice reduced its lifespan by 40% [99] while, in contrast, its overexpression in *Drosophila* accounted for a 70% extension in their healthy lifespan [97]. In both studies, MsrA-dependent lifespan modulation was related to its role in protection against oxidative stress but later, another study showed that while the lack of MsrA in mice increases sensitivity to oxidative stress, it does not diminish lifespan [9]. While the discussion about the effects of Msr system on the late survival of higher animals is still open, MsrA overexpression, however, was shown to increase lifespan of *S. cerevisiae* [13] whereas its inhibition in yeast or in *C. elegans* is accompanied by a shorten lifespan [12]. Moreover, ectopically expression of fRMsr in fruit flies, an enzyme lost during evolution, that reduces free MetO, increases stress resistance and extends lifespan of animals [140].

Until now, overall these studies agreed that the importance of Msr system in ageing and neurodegenerative diseases was dependent of its role as antioxidant enzyme protecting cells and organisms from the deleterious effects of oxidative stress. The discovery that alternation between methionine oxidation and reduction could serve as regulator of protein function, as reviewed below, raises the hypothesis that the Msr role on ageing and survival could also come from these intracellular signalling functions.

7. Methionine Sulfoxide Reductases as Regulators of Protein and Cellular Functions

Evidences that the cyclic interconversion between Met and MetO within proteins is implicated in the regulation of cell signaling functions are gaining space in the field of Msr studies. In general, it has been demonstrated that the oxidation of certain methionine residues in proteins induces mainly the loss of their biological activity, while their reduction by Msr is capable of reversing it. The types of proteins in which methionine oxidation has been involved in their function are very diverse: proteases or protease inhibitors, metabolic enzymes, cytoskeleton proteins, cytokines, heat shock proteins, hormones, heme proteins, proteins associated with neurodegenerative disorders, proteins involved in immunodefences, as well as different bacterial proteins and snake toxins (see [141] for review). Among the more than 50 proteins reported to have altered activity due to formation of MetO, only part of them was already described as being substrates of Msr enzymes either in vitro or in vivo experiments. In this last section, we will address some examples of these Msr substrates.

The *E. coli* ribosomal protein L12 was the first characterized substrate of MsrA [23]. Oxidation of three of its methionines by H_2O_2 decreases its ability to bind to ribosomes and to interact with other ribosomal proteins such as L10, impairing protein synthesis [142], while MsrA reduction of the oxidized methionines abled L12 ribosomal protein to regain its activity [23]. Another *E. coli* protein involved in protein machinery is the Ffh component of the ubiquitous signal recognition particle. This protein contains a methionine-rich domain whose oxidation compromises Ffh interaction with a small RNA. Oxidized Ffh is a substrate for MsrA and MsrB enzymes and reduction of Ffh MetO residues allows the recovery of its RNA-binding abilities [143].

Regarding serine protease inhibitors, methionine oxidation was also associated with a loss of function in α1-antitrypsin [144] and α2-antiplasmin [145]. In particular, α1-antitrypsin protein is important for lungs protection by preserving anti-neutrophil elastase activity, which is associated with the risk of developing emphysema. Oxidation of two methionines of α1-antitrypsin causes a loss of the anti-neutrophil elastase activity, which can be restored in part with the addition of MsrA in vitro [146]. Moreover, methionine oxidation of Human Immunodeficiency Virus 2 (HIV-2) protease, which cleaves proteins involved in the HIV-2 reproductive cycle, also inhibits its proteolytic activity while the addition of only MsrA partially restores it [147]. Another important protein participating in the immunological and oxidative stress response, the inhibitor of kappa B-alpha (IκBα), named for its inhibition activity of the transcription factor nuclear factor-κB (NF-κB), can also be oxidized on a methionine residue thereby increasing its resistance to proteasomal degradation [148]. When IκBα is oxidized by taurine chloride or chloramines, it cannot dissociate from NF-κB, thus preventing it from nucleus translocation and subsequent activation of its target genes [149]. Inhibition of NF-κB activation is prevented by MsrA [150]. Together, these findings could lead to new therapeutic strategies in order to fight against diseases such as pulmonary emphysema, AIDS or immunological diseases where NF-κB play an important role.

Methionine oxidation and its Msr-dependent reduction was also shown to be important for the regulation of cellular excitability, in particularly through the regulation of the voltage-gated and the calcium (Ca^{2+})-activated potassium channels [151,152]. Another protein involved in cellular excitability, and whose activity is regulated by Msr enzymes, is the Ca^{2+}-binding protein calmodulin (CaM) [153]. This protein detects the calcium signals in cells and coordinates energy metabolism, which in turn produce superoxide in the mitochondria. CaM loses its conformational stability upon Met oxidation [154] thus failing to activate the plasma membrane Ca^{2+}-ATPase [155]. In contrast, its full reduction by MsrA and MsrB leads to its binding to the inhibitory domain of the plasma membrane Ca^{2+}-ATPase, inducing helix formation within the CaM-binding sequence and releasing enzyme inhibition [28,156]. Full reduction of CaM by MsrA and MsrB was also found to restore its binding to *B. pertussis* adenylate cyclase [157]. Thus, one can think that upon oxidative stress, increasing levels of cytosolic Ca^{2+} are probably the consequence of the oxidation of specific methionines in CaM that is no longer able to activate the Ca^{2+}-ATPase in the plasma membrane. As a consequence, accumulation oxidized CaM will result in down-regulation of cellular metabolism thus, controlling the generation of ROS.

One of the CaM proteins targets, the Ca^{2+}/calmodulin-dependent protein kinase II (CaMKII), is also regulated by methionine oxidation. Indeed, apart from Ca^{2+}/CaM regulation or its autophosphorylation at Thr-287, oxidation of one of its methionine residues leads also to its Ca^{2+}/CaM-independent activity [158]. This is consistent with the notion that Met oxidation does not invariably induce enzyme inactivation. The same authors have shown that MsrA enzymes are essential for reversing CaMKII oxidation in myocardium in vivo [158] and this oxidation of CaMKII mediates the cardiotoxic effects of aldosterone, while MsrA overexpression reversed its effects [159].

If Met oxidations are involved in the regulation of biological functions, we could think that they cannot depend only on random oxidation by multiple forms of ROS. There should be also specialized oxidases that catalyze the oxidation of specific protein targets. Indeed, Hung and colleagues have found that a flavoprotein monooxidase (FMO) called Mical is able to bind to F-actin and to selectively oxidized 2 of its 16 methionine residues into the R stereoisomer of MetO, resulting on actin disassembly in vitro and in vivo [160]. This Mical-dependent redox regulation of actin that can be reversed by MsrB1/SelR, was involved in bristles formation in *Drosophila* [161], in membrane trafficking in bone marrow-derived macrophages [15] and in lens epithelial cells [116]. In addition to the Mical enzyme, other mammalian FMO mainly involved in oxidative xenobiotic metabolism in liver and kidney, have been shown to exhibit methionine oxidase catalytic activity. In *Aspergillus nidulans*, FMO-mediated oxidation of a specific methionine residue regulates the subcellular localisation of the transcription factor NirA [162].

The increasing work done in the reversibility of methionine oxidation in vitro as well as in vivo led to the accepted view that this sophisticated redox based mechanism must be considered as other post-translational modifications on proteins such as phosphorylation, acetylation or glutathionylation, implicated in the regulation of many important protein and cellular functions.

8. Conclusions

In this paper, we have tentatively provided a comprehensive review of the Msr system in terms of its discovery and biologically relevant characteristics as its important physiological roles in protecting against oxidative stress, in ageing and in regulating protein function. Indeed, since the specific reduction of L-methionine sulfoxide has been evidenced and the first Msr, now referred to as MsrA, partially purified and characterized in 1981 [23], much progress has been achieved in terms of both MsrA and MsrB, discovered 20 years later [28], gene organization and subcellular localization in different species from bacteria to mammals, as well as their enzymatic and structural characterization. Interestingly, the Msrs system is also connected to selenocysteine biology since mammalian MsrB1 is a selenoprotein [39].

Beside restoration of protein structure and function through the selective reduction of MetO within the polypeptide chain, protection against oxidative stress has been shown to also result from the cyclic reduction of protein surface exposed MetO by Msr enzymes, acting as a protein built-in antioxidant system [163].

Hence, methionine amino acid residues together with the Msr system constitute a potent ROS scavenging system, preserving proteins in their reduced state and thus protecting different cell types and organisms from oxidative stress, ultimately contributing to protein homeostasis and cell survival in endogenous or exogenous stress conditions.

Since accumulation of oxidized proteins is one of the hallmarks of ageing and many age-associated diseases, the fate of the Msr system has been investigated in these situations. The implication of the Msr system in regulating lifespan of model organisms has been studied with either overexpressing or knock-out strains for either *MsrA* and/or *MsrB* leading at first sight to somehow contradictory results that may be due to the multifunctionality of the Msr system and the complex relationship between longevity, protection against oxidative stress and redox regulation of signaling pathways.

Another important topic for which the Msr system has been implicated is the regulation of protein function and hence cellular signaling. Methionine oxidation of the first identified protein targets was mainly associated with a loss of function that was restored by the Msr system. However, the activation of the CaMKII by methionine oxidation in the absence of Ca^{2+}/CaM [158] has opened up the possibility that the Msr system could also play an important role in the regulation of protein function. The role of selective methionine oxidation of actin by monooxygenases that results in its disassembly which is reversed by MsrB1 [162] have further strengthened the concept that methionine oxidation and the Msr system indeed play a critical role in regulation protein function. Although, a number of protein targets for methionine oxidation and its reversion by the Msr system has already been identified, identification of new targets still needs to be achieved to appreciate the physiological relevance of this sophisticated redox based cellular regulatory system. Unfortunately, attempts to raise specific antibodies aimed at detecting MetO in proteins, which would be of great value for identifying protein targets, have been unsuccessful so far [164]. However, new alternative methods such as these aimed at characterizing in vivo methionine oxidation using mass spectrometry based and proteomics methods [165], and those aimed at determining the concentration of protein based MetO using fluorescent biosensor technology [166] are expected to be of valuable interest for further investigating the Msrs system and its role in redox biology.

Author Contributions: Review of literature and writing of the paper: S.L.d.S., I.P. and B.F.

Funding: This research was supported institutional funding from SU, CNRS and INSERM. S.L.d.S. was a recipient of a fellowship from a UPMC Emergence grant to I.P.

Conflicts of Interest: The authors declare no conflict of interest.

References

1. Lu, J.; Holmgren, A. The thioredoxin antioxidant system. *Free Radic. Biol. Med.* **2014**, *66*, 75–87. [CrossRef] [PubMed]
2. Lillig, C.H.; Berndt, C.; Holmgren, A. Glutaredoxin systems. *Biochim. Biophys. Acta* **2008**, *1780*, 1304–1317. [CrossRef] [PubMed]
3. Kalinina, E.V.; Chernov, N.N.; Novichkova, M.D. Role of glutathione, glutathione transferase, and glutaredoxin in regulation of redox-dependent processes. *Biochemistry* **2014**, *79*, 1562–1583. [CrossRef] [PubMed]
4. Johansson, C.; Lillig, C.H.; Holmgren, A. Human mitochondrial glutaredoxin reduces S-glutathionylated proteins with high affinity accepting electrons from either glutathione or thioredoxin reductase. *J. Biol. Chem.* **2004**, *279*, 7537–7543. [CrossRef] [PubMed]
5. Song, J.J.; Rhee, J.G.; Suntharalingam, M.; Walsh, S.A.; Spitz, D.R.; Lee, Y.J. Role of glutaredoxin in metabolic oxidative stress. Glutaredoxin as a sensor of oxidative stress mediated by H_2O_2. *J. Biol. Chem.* **2002**, *277*, 46566–46575. [CrossRef] [PubMed]
6. Starke, D.W.; Chock, P.B.; Mieyal, J.J. Glutathione-thiyl radical scavenging and transferase properties of human glutaredoxin (thioltransferase). Potential role in redox signal transduction. *J. Biol. Chem.* **2003**, *278*, 14607–14613. [CrossRef]
7. Lee, B.C.; Dikiy, A.; Kim, H.Y.; Gladyshev, V.N. Functions and evolution of selenoprotein methionine sulfoxide reductases. *Biochim. Biophys. Acta* **2009**, *1790*, 1471–1477. [CrossRef]
8. Cabreiro, F.; Picot, C.R.; Perichon, M.; Castel, J.; Friguet, B.; Petropoulos, I. Overexpression of mitochondrial methionine sulfoxide reductase B2 protects leukemia cells from oxidative stress-induced cell death and protein damage. *J. Biol. Chem.* **2008**, *283*, 16673–16681. [CrossRef]
9. Salmon, A.B.; Perez, V.I.; Bokov, A.; Jernigan, A.; Kim, G.; Zhao, H.; Levine, R.L.; Richardson, A. Lack of methionine sulfoxide reductase A in mice increases sensitivity to oxidative stress but does not diminish life span. *FASEB J.* **2009**, *23*, 3601–3608. [CrossRef]
10. Ugarte, N.; Petropoulos, I.; Friguet, B. Oxidized mitochondrial protein degradation and repair in aging and oxidative stress. *Antioxid. Redox Signal.* **2010**, *13*, 539–549. [CrossRef]
11. Chung, H.; Kim, A.K.; Jung, S.A.; Kim, S.W.; Yu, K.; Lee, J.H. The Drosophila homolog of methionine sulfoxide reductase A extends lifespan and increases nuclear localization of FOXO. *FEBS Lett.* **2010**, *584*, 3609–3614. [CrossRef] [PubMed]
12. Minniti, A.N.; Cataldo, R.; Trigo, C.; Vasquez, L.; Mujica, P.; Leighton, F.; Inestrosa, N.C.; Aldunate, R. Methionine sulfoxide reductase A expression is regulated by the DAF-16/FOXO pathway in Caenorhabditis elegans. *Aging Cell* **2009**, *8*, 690–705. [CrossRef] [PubMed]
13. Koc, A.; Gasch, A.P.; Rutherford, J.C.; Kim, H.Y.; Gladyshev, V.N. Methionine sulfoxide reductase regulation of yeast lifespan reveals reactive oxygen species-dependent and -independent components of aging. *Proc. Natl. Acad. Sci. USA* **2004**, *101*, 7999–8004. [CrossRef] [PubMed]
14. Kaya, A.; Koc, A.; Lee, B.C.; Fomenko, D.E.; Rederstorff, M.; Krol, A.; Lescure, A.; Gladyshev, V.N. Compartmentalization and regulation of mitochondrial function by methionine sulfoxide reductases in yeast. *Biochemistry* **2010**, *49*, 8618–8625. [CrossRef] [PubMed]

15. Lee, B.C.; Peterfi, Z.; Hoffmann, F.W.; Moore, R.E.; Kaya, A.; Avanesov, A.; Tarrago, L.; Zhou, Y.; Weerapana, E.; Fomenko, D.E.; et al. MsrB1 and MICALs regulate actin assembly and macrophage function via reversible stereoselective methionine oxidation. *Mol. Cell* **2013**, *51*, 397–404. [CrossRef] [PubMed]

16. Bigelow, D.J.; Squier, T.C. Thioredoxin-dependent redox regulation of cellular signaling and stress response through reversible oxidation of methionines. *Mol. Biosyst.* **2011**, *7*, 2101–2109. [CrossRef] [PubMed]

17. Waelsch, H.; Owades, P.; Miller, H.K.; Borek, E. Glutamic acid antimetabolites; the sulfoxide derived from methionine. *J. Biol. Chem.* **1946**, *166*, 273–281.

18. Krask, B.J. Methionine sulfoxide and specific inhibition of sporulation in Bacillus subtilis. *J. Bacteriol.* **1953**, *66*, 374.

19. Black, S.D.; Harte, E.M.; Hudson, B.; Wartofsky, L. A specific enzymatic reduction of L(-) methionine sulfoxide and a related non-specific reduction of disulfides. *J. Biol. Chem.* **1960**, *235*, 2910–2916.

20. Ejiri, S.I.; Weissbach, H.; Brot, N. Reduction of methionine sulfoxide to methionine by *Escherichia coli*. *J. Bacteriol.* **1979**, *139*, 161–164.

21. Doney, R.C.; Thompson, J.F. The reduction of S-methyl-L-cysteine sulfoxide and L-methionine sulfoxide in turnip and bean leaves. *Biochim. Biophys. Acta* **1966**, *124*, 39–49. [CrossRef]

22. Aymarda, C.; Seyera, L.; Cheftela, J.-C. Enzymatic Reduction of Methionine Sulfoxide. In Vitro Experiments with Rat Liver and Kidney. *Biol. Chem.* **1979**, *43*, 1869–1876. [CrossRef]

23. Brot, N.; Weissbach, L.; Werth, J.; Weissbach, H. Enzymatic reduction of protein-bound methionine sulfoxide. *Proc. Natl. Acad. Sci. USA* **1981**, *78*, 2155–2158. [CrossRef] [PubMed]

24. Brot, N.; Werth, J.; Koster, D.; Weissbach, H. Reduction of N-acetyl methionine sulfoxide: A simple assay for peptide methionine sulfoxide reductase. *Anal. Biochem.* **1982**, *122*, 291–294. [CrossRef]

25. Moskovitz, J.; Jenkins, N.A.; Gilbert, D.J.; Copeland, N.G.; Jursky, F.; Weissbach, H.; Brot, N. Chromosomal localization of the mammalian peptide-methionine sulfoxide reductase gene and its differential expression in various tissues. *Proc. Natl. Acad. Sci. USA* **1996**, *93*, 3205–3208. [CrossRef] [PubMed]

26. Sharov, V.S.; Ferrington, D.A.; Squier, T.C.; Schoneich, C. Diastereoselective reduction of protein-bound methionine sulfoxide by methionine sulfoxide reductase. *FEBS Lett.* **1999**, *455*, 247–250. [CrossRef]

27. Sharov, V.S.; Schoneich, C. Diastereoselective protein methionine oxidation by reactive oxygen species and diastereoselective repair by methionine sulfoxide reductase. *Free Radic. Biol. Med.* **2000**, *29*, 986–994. [CrossRef]

28. Grimaud, R.; Ezraty, B.; Mitchell, J.K.; Lafitte, D.; Briand, C.; Derrick, P.J.; Barras, F. Repair of oxidized proteins. Identification of a new methionine sulfoxide reductase. *J. Biol. Chem.* **2001**, *276*, 48915–48920. [CrossRef]

29. Kryukov, G.V.; Kumar, R.A.; Koc, A.; Sun, Z.; Gladyshev, V.N. Selenoprotein R is a zinc-containing stereo-specific methionine sulfoxide reductase. *Proc. Natl. Acad. Sci. USA* **2002**, *99*, 4245–4250. [CrossRef]

30. Olry, A.; Boschi-Muller, S.; Marraud, M.; Sanglier-Cianferani, S.; Van Dorsselear, A.; Branlant, G. Characterization of the methionine sulfoxide reductase activities of PILB, a probable virulence factor from Neisseria meningitidis. *J. Biol. Chem.* **2002**, *277*, 12016–12022. [CrossRef]

31. Lescure, A.; Gautheret, D.; Carbon, P.; Krol, A. Novel selenoproteins identified in silico and in vivo by using a conserved RNA structural motif. *J. Biol. Chem.* **1999**, *274*, 38147–38154. [CrossRef] [PubMed]

32. Etienne, F.; Spector, D.; Brot, N.; Weissbach, H. A methionine sulfoxide reductase in *Escherichia coli* that reduces the R enantiomer of methionine sulfoxide. *Biochem. Biophys. Res. Commun.* **2003**, *300*, 378–382. [CrossRef]

33. Lin, Z.; Johnson, L.C.; Weissbach, H.; Brot, N.; Lively, M.O.; Lowther, W.T. Free methionine-(R)-sulfoxide reductase from *Escherichia coli* reveals a new GAF domain function. *Proc. Natl. Acad. Sci. USA* **2007**, *104*, 9597–9602. [CrossRef] [PubMed]

34. Gennaris, A.; Ezraty, B.; Henry, C.; Agrebi, R.; Vergnes, A.; Oheix, E.; Bos, J.; Leverrier, P.; Espinosa, L.; Szewczyk, J.; et al. Repairing oxidized proteins in the bacterial envelope using respiratory chain electrons. *Nature* **2015**, *528*, 409–412. [CrossRef] [PubMed]

35. Zhang, X.H.; Weissbach, H. Origin and evolution of the protein-repairing enzymes methionine sulphoxide reductases. *Biol. Rev.* **2008**, *83*, 249–257. [CrossRef] [PubMed]

36. Fukushima, E.; Shinka, Y.; Fukui, T.; Atomi, H.; Imanaka, T. Methionine sulfoxide reductase from the hyperthermophilic archaeon Thermococcus kodakaraensis, an enzyme designed to function at suboptimal growth temperatures. *J. Bacteriol.* **2007**, *189*, 7134–7144. [CrossRef]

37. Kim, H.S.; Kwak, G.H.; Lee, K.; Jo, C.H.; Hwang, K.Y.; Kim, H.Y. Structural and biochemical analysis of a type II free methionine-R-sulfoxide reductase from Thermoplasma acidophilum. *Arch. Biochem. Biophys.* **2014**, *560*, 10–19. [CrossRef]

38. Rouhier, N.; Vieira Dos Santos, C.; Tarrago, L.; Rey, P. Plant methionine sulfoxide reductase A and B multigenic families. *Photosynth. Res.* **2006**, *89*, 247–262. [CrossRef]

39. Kim, H.Y.; Gladyshev, V.N. Methionine sulfoxide reduction in mammals: Characterization of methionine-R-sulfoxide reductases. *Mol. Biol. Cell* **2004**, *15*, 1055–1064. [CrossRef]

40. Kwon, S.J.; Kwon, S.I.; Bae, M.S.; Cho, E.J.; Park, O.K. Role of the methionine sulfoxide reductase MsrB3 in cold acclimation in Arabidopsis. *Plant Cell Physiol.* **2007**, *48*, 1713–1723. [CrossRef]

41. Lim, D.H.; Han, J.Y.; Kim, J.R.; Lee, Y.S.; Kim, H.Y. Methionine sulfoxide reductase B in the endoplasmic reticulum is critical for stress resistance and aging in Drosophila. *Biochem. Biophys. Res. Commun.* **2012**, *419*, 20–26. [CrossRef] [PubMed]

42. Kuschel, L.; Hansel, A.; Schonherr, R.; Weissbach, H.; Brot, N.; Hoshi, T.; Heinemann, S.H. Molecular cloning and functional expression of a human peptide methionine sulfoxide reductase (hMsrA). *FEBS Lett.* **1999**, *456*, 17–21. [CrossRef]

43. Moskovitz, J.; Weissbach, H.; Brot, N. Cloning the expression of a mammalian gene involved in the reduction of methionine sulfoxide residues in proteins. *Proc. Natl. Acad. Sci. USA* **1996**, *93*, 2095–2099. [CrossRef] [PubMed]

44. Brot, N.; Fliss, H.; Coleman, T.; Weissbach, H. Enzymatic reduction of methionine sulfoxide residues in proteins and peptides. *Methods Enzymol.* **1984**, *107*, 352–360. [PubMed]

45. Achilli, C.; Ciana, A.; Rossi, A.; Balduini, C.; Minetti, G. Neutrophil granulocytes uniquely express, among human blood cells, high levels of Methionine-sulfoxide-reductase enzymes. *J. Leukoc. Biol.* **2008**, *83*, 181–189. [CrossRef] [PubMed]

46. Ogawa, F.; Sander, C.S.; Hansel, A.; Oehrl, W.; Kasperczyk, H.; Elsner, P.; Shimizu, K.; Heinemann, S.H.; Thiele, J.J. The repair enzyme peptide methionine-S-sulfoxide reductase is expressed in human epidermis and upregulated by UVA radiation. *J. Investig. Dermatol.* **2006**, *126*, 1128–1134. [CrossRef] [PubMed]

47. Picot, C.R.; Moreau, M.; Juan, M.; Noblesse, E.; Nizard, C.; Petropoulos, I.; Friguet, B. Impairment of methionine sulfoxide reductase during UV irradiation and photoaging. *Exp. Gerontol.* **2007**, *42*, 859–863. [CrossRef]

48. Schallreuter, K.U.; Rubsam, K.; Chavan, B.; Zothner, C.; Gillbro, J.M.; Spencer, J.D.; Wood, J.M. Functioning methionine sulfoxide reductases A and B are present in human epidermal melanocytes in the cytosol and in the nucleus. *Biochem. Biophys. Res. Commun.* **2006**, *342*, 145–152. [CrossRef]

49. Taungjaruwinai, W.M.; Bhawan, J.; Keady, M.; Thiele, J.J. Differential expression of the antioxidant repair enzyme methionine sulfoxide reductase (MSRA and MSRB) in human skin. *Am. J. Dermatopathol.* **2009**, *31*, 427–431. [CrossRef]

50. Hansel, A.; Kuschel, L.; Hehl, S.; Lemke, C.; Agricola, H.J.; Hoshi, T.; Heinemann, S.H. Mitochondrial targeting of the human peptide methionine sulfoxide reductase (MSRA), an enzyme involved in the repair of oxidized proteins. *FASEB J.* **2002**, *16*, 911–913. [CrossRef]

51. Vougier, S.; Mary, J.; Friguet, B. Subcellular localization of methionine sulphoxide reductase A (MsrA): Evidence for mitochondrial and cytosolic isoforms in rat liver cells. *Biochem. J.* **2003**, *373*, 531–537. [CrossRef] [PubMed]

52. Kim, H.Y.; Gladyshev, V.N. Role of structural and functional elements of mouse methionine-S-sulfoxide reductase in its subcellular distribution. *Biochemistry* **2005**, *44*, 8059–8067. [CrossRef] [PubMed]

53. Kim, H.Y.; Gladyshev, V.N. Alternative first exon splicing regulates subcellular distribution of methionine sulfoxide reductases. *BMC Mol. Biol.* **2006**, *7*, 11. [CrossRef] [PubMed]

54. Lim, J.M.; Lim, J.C.; Kim, G.; Levine, R.L. Myristoylated methionine sulfoxide reductase A is a late endosomal protein. *J. Biol. Chem.* **2018**, *293*, 7355–7366. [CrossRef] [PubMed]

55. Kim, H.Y.; Gladyshev, V.N. Characterization of mouse endoplasmic reticulum methionine-R-sulfoxide reductase. *Biochem. Biophys. Res. Commun.* **2004**, *320*, 1277–1283. [CrossRef]

56. Lowther, W.T.; Brot, N.; Weissbach, H.; Honek, J.F.; Matthews, B.W. Thiol-disulfide exchange is involved in the catalytic mechanism of peptide methionine sulfoxide reductase. *Proc. Natl. Acad. Sci. USA* **2000**, *97*, 6463–6468. [CrossRef]

57. Tete-Favier, F.; Cobessi, D.; Leonard, G.A.; Azza, S.; Talfournier, F.; Boschi-Muller, S.; Branlant, G.; Aubry, A. Crystallization and preliminary X-ray diffraction studies of the peptide methionine sulfoxide reductase from *Escherichia coli*. *Acta Crystallogr. D Biol. Crystallogr.* **2000**, *56*, 1194–1197. [CrossRef]

58. Lowther, W.T.; Brot, N.; Weissbach, H.; Matthews, B.W. Structure and mechanism of peptide methionine sulfoxide reductase, an "anti-oxidation" enzyme. *Biochemistry* **2000**, *39*, 13307–13312. [CrossRef]

59. Taylor, A.B.; Benglis, D.M., Jr.; Dhandayuthapani, S.; Hart, P.J. Structure of Mycobacterium tuberculosis methionine sulfoxide reductase A in complex with protein-bound methionine. *J. Bacteriol.* **2003**, *185*, 4119–4126. [CrossRef]

60. Rouhier, N.; Kauffmann, B.; Tete-Favier, F.; Palladino, P.; Gans, P.; Branlant, G.; Jacquot, J.P.; Boschi-Muller, S. Functional and structural aspects of poplar cytosolic and plastidial type a methionine sulfoxide reductases. *J. Biol. Chem.* **2007**, *282*, 3367–3378. [CrossRef]

61. Ranaivoson, F.M.; Antoine, M.; Kauffmann, B.; Boschi-Muller, S.; Aubry, A.; Branlant, G.; Favier, F. A structural analysis of the catalytic mechanism of methionine sulfoxide reductase A from Neisseria meningitidis. *J. Mol. Biol.* **2008**, *377*, 268–280. [CrossRef]

62. Coudevylle, N.; Antoine, M.; Bouguet-Bonnet, S.; Mutzenhardt, P.; Boschi-Muller, S.; Branlant, G.; Cung, M.T. Solution structure and backbone dynamics of the reduced form and an oxidized form of *E. coli* methionine sulfoxide reductase A (MsrA): Structural insight of the MsrA catalytic cycle. *J. Mol. Biol.* **2007**, *366*, 193–206. [CrossRef] [PubMed]

63. Boschi-Muller, S.; Azza, S.; Sanglier-Cianferani, S.; Talfournier, F.; Van Dorsselear, A.; Branlant, G. A sulfenic acid enzyme intermediate is involved in the catalytic mechanism of peptide methionine sulfoxide reductase from *Escherichia coli*. *J. Biol. Chem.* **2000**, *275*, 35908–35913. [CrossRef] [PubMed]

64. Olry, A.; Boschi-Muller, S.; Branlant, G. Kinetic characterization of the catalytic mechanism of methionine sulfoxide reductase B from Neisseria meningitidis. *Biochemistry* **2004**, *43*, 11616–11622. [CrossRef] [PubMed]

65. Antoine, M.; Boschi-Muller, S.; Branlant, G. Kinetic characterization of the chemical steps involved in the catalytic mechanism of methionine sulfoxide reductase A from Neisseria meningitidis. *J. Biol. Chem.* **2003**, *278*, 45352–45357. [CrossRef] [PubMed]

66. Lim, J.C.; You, Z.; Kim, G.; Levine, R.L. Methionine sulfoxide reductase A is a stereospecific methionine oxidase. *Proc. Natl. Acad. Sci. USA* **2011**, *108*, 10472–10477. [CrossRef] [PubMed]

67. Kriznik, A.; Boschi-Muller, S.; Branlant, G. Kinetic evidence that methionine sulfoxide reductase A can reveal its oxidase activity in the presence of thioredoxin. *Arch. Biochem. Biophys.* **2014**, *548*, 54–59. [CrossRef]

68. Lowther, W.T.; Weissbach, H.; Etienne, F.; Brot, N.; Matthews, B.W. The mirrored methionine sulfoxide reductases of Neisseria gonorrhoeae pilB. *Nat. Struct. Biol.* **2002**, *9*, 348–352. [CrossRef]

69. Wu, J.; Neiers, F.; Boschi-Muller, S.; Branlant, G. The N-terminal domain of PILB from Neisseria meningitidis is a disulfide reductase that can recycle methionine sulfoxide reductases. *J. Biol. Chem.* **2005**, *280*, 12344–12350. [CrossRef]

70. Kim, H.Y.; Gladyshev, V.N. Different catalytic mechanisms in mammalian selenocysteine- and cysteine-containing methionine-R-sulfoxide reductases. *PLoS Biol.* **2005**, *3*, E375. [CrossRef]

71. Jung, S.; Hansel, A.; Kasperczyk, H.; Hoshi, T.; Heinemann, S.H. Activity, tissue distribution and site-directed mutagenesis of a human peptide methionine sulfoxide reductase of type B: HCBS1. *FEBS Lett.* **2002**, *527*, 91–94. [CrossRef]

72. Sagher, D.; Brunell, D.; Hejtmancik, J.F.; Kantorow, M.; Brot, N.; Weissbach, H. Thionein can serve as a reducing agent for the methionine sulfoxide reductases. *Proc. Natl. Acad. Sci. USA* **2006**, *103*, 8656–8661. [CrossRef] [PubMed]

73. Sagher, D.; Brunell, D.; Brot, N.; Vallee, B.L.; Weissbach, H. Selenocompounds can serve as oxidoreductants with the methionine sulfoxide reductase enzymes. *J. Biol. Chem.* **2006**, *281*, 31184–31187. [CrossRef] [PubMed]

74. Couturier, J.; Vignols, F.; Jacquot, J.P.; Rouhier, N. Glutathione- and glutaredoxin-dependent reduction of methionine sulfoxide reductase A. *FEBS Lett.* **2012**, *586*, 3894–3899. [CrossRef] [PubMed]

75. Kim, M.J.; Lee, B.C.; Jeong, J.; Lee, K.J.; Hwang, K.Y.; Gladyshev, V.N.; Kim, H.Y. Tandem use of selenocysteine: Adaptation of a selenoprotein glutaredoxin for reduction of selenoprotein methionine sulfoxide reductase. *Mol. Microbiol.* **2011**, *79*, 1194–1203. [CrossRef] [PubMed]

76. Tarrago, L.; Laugier, E.; Zaffagnini, M.; Marchand, C.; Le Marechal, P.; Rouhier, N.; Lemaire, S.D.; Rey, P. Regeneration mechanisms of Arabidopsis thaliana methionine sulfoxide reductases B by glutaredoxins and thioredoxins. *J. Biol. Chem.* **2009**, *284*, 18963–18971. [CrossRef] [PubMed]

77. Vieira Dos Santos, C.; Laugier, E.; Tarrago, L.; Massot, V.; Issakidis-Bourguet, E.; Rouhier, N.; Rey, P. Specificity of thioredoxins and glutaredoxins as electron donors to two distinct classes of Arabidopsis plastidial methionine sulfoxide reductases B. *FEBS Lett.* **2007**, *581*, 4371–4376. [CrossRef] [PubMed]

78. Jacob, C.; Kriznik, A.; Boschi-Muller, S.; Branlant, G. Thioredoxin 2 from *Escherichia coli* is not involved in vivo in the recycling process of methionine sulfoxide reductase activities. *FEBS Lett.* **2011**, *585*, 1905–1909. [CrossRef]

79. Levine, R.L.; Mosoni, L.; Berlett, B.S.; Stadtman, E.R. Methionine residues as endogenous antioxidants in proteins. *Proc. Natl. Acad. Sci. USA* **1996**, *93*, 15036–15040. [CrossRef]

80. Hsu, Y.R.; Narhi, L.O.; Spahr, C.; Langley, K.E.; Lu, H.S. In vitro methionine oxidation of *Escherichia coli*-derived human stem cell factor: Effects on the molecular structure, biological activity, and dimerization. *Protein Sci.* **1996**, *5*, 1165–1173. [CrossRef]

81. Jones, T.E.; Alexander, R.W.; Pan, T. Misacylation of specific nonmethionyl tRNAs by a bacterial methionyl-tRNA synthetase. *Proc. Natl. Acad. Sci. USA* **2011**, *108*, 6933–6938. [CrossRef] [PubMed]

82. Wiltrout, E.; Goodenbour, J.M.; Frechin, M.; Pan, T. Misacylation of tRNA with methionine in *Saccharomyces cerevisiae*. *Nucleic Acids Res.* **2012**, *40*, 10494–10506. [CrossRef] [PubMed]

83. Netzer, N.; Goodenbour, J.M.; David, A.; Dittmar, K.A.; Jones, R.B.; Schneider, J.R.; Boone, D.; Eves, E.M.; Rosner, M.R.; Gibbs, J.S.; et al. Innate immune and chemically triggered oxidative stress modifies translational fidelity. *Nature* **2009**, *462*, 522–526. [CrossRef] [PubMed]

84. Wang, X.; Pan, T. Methionine Mistranslation Bypasses the Restraint of the Genetic Code to Generate Mutant Proteins with Distinct Activities. *PLOS Genet.* **2015**, *11*, e1005745. [CrossRef] [PubMed]

85. Luo, S.; Levine, R.L. Methionine in proteins defends against oxidative stress. *FASEB J.* **2009**, *23*, 464–472. [CrossRef] [PubMed]

86. St John, G.; Brot, N.; Ruan, J.; Erdjument-Bromage, H.; Tempst, P.; Weissbach, H.; Nathan, C. Peptide methionine sulfoxide reductase from *Escherichia coli* and Mycobacterium tuberculosis protects bacteria against oxidative damage from reactive nitrogen intermediates. *Proc. Natl. Acad. Sci. USA* **2001**, *98*, 9901–9906. [CrossRef] [PubMed]

87. Moskovitz, J.; Rahman, M.A.; Strassman, J.; Yancey, S.O.; Kushner, S.R.; Brot, N.; Weissbach, H. *Escherichia coli* peptide methionine sulfoxide reductase gene: Regulation of expression and role in protecting against oxidative damage. *J. Bacteriol.* **1995**, *177*, 502–507. [CrossRef] [PubMed]

88. Tamburro, A.; Robuffo, I.; Heipieper, H.J.; Allocati, N.; Rotilio, D.; Di Ilio, C.; Favaloro, B. Expression of glutathione S-transferase and peptide methionine sulphoxide reductase in Ochrobactrum anthropi is correlated to the production of reactive oxygen species caused by aromatic substrates. *FEMS Microbiol. Lett.* **2004**, *241*, 151–156. [CrossRef] [PubMed]

89. Singh, V.K.; Moskovitz, J.; Wilkinson, B.J.; Jayaswal, R.K. Molecular characterization of a chromosomal locus in Staphylococcus aureus that contributes to oxidative defence and is highly induced by the cell-wall-active antibiotic oxacillin. *Microbiology* **2001**, *147*, 3037–3045. [CrossRef] [PubMed]

90. Douglas, T.; Daniel, D.S.; Parida, B.K.; Jagannath, C.; Dhandayuthapani, S. Methionine sulfoxide reductase A (MsrA) deficiency affects the survival of Mycobacterium smegmatis within macrophages. *J. Bacteriol.* **2004**, *186*, 3590–3598. [CrossRef] [PubMed]

91. Oien, D.; Moskovitz, J. Protein-carbonyl accumulation in the non-replicative senescence of the methionine sulfoxide reductase A (msrA) knockout yeast strain. *Amino Acids* **2007**, *32*, 603–606. [CrossRef] [PubMed]

92. Moskovitz, J.; Berlett, B.S.; Poston, J.M.; Stadtman, E.R. The yeast peptide-methionine sulfoxide reductase functions as an antioxidant in vivo. *Proc. Natl. Acad. Sci. USA* **1997**, *94*, 9585–9589. [CrossRef] [PubMed]

93. Moskovitz, J.; Flescher, E.; Berlett, B.S.; Azare, J.; Poston, J.M.; Stadtman, E.R. Overexpression of peptide-methionine sulfoxide reductase in *Saccharomyces cerevisiae* and human T cells provides them with high resistance to oxidative stress. *Proc. Natl. Acad. Sci. USA* **1998**, *95*, 14071–14075. [CrossRef] [PubMed]

94. Sumner, E.R.; Shanmuganathan, A.; Sideri, T.C.; Willetts, S.A.; Houghton, J.E.; Avery, S.V. Oxidative protein damage causes chromium toxicity in yeast. *Microbiology* **2005**, *151*, 1939–1948. [CrossRef] [PubMed]

95. Bechtold, U.; Murphy, D.J.; Mullineaux, P.M. Arabidopsis peptide methionine sulfoxide reductase2 prevents cellular oxidative damage in long nights. *Plant Cell* **2004**, *16*, 908–919. [CrossRef] [PubMed]

96. Bae, M.S.; Cho, E.J.; Choi, E.Y.; Park, O.K. Analysis of the Arabidopsis nuclear proteome and its response to cold stress. *Plant J.* **2003**, *36*, 652–663. [CrossRef] [PubMed]

97. Ruan, H.; Tang, X.D.; Chen, M.L.; Joiner, M.L.; Sun, G.; Brot, N.; Weissbach, H.; Heinemann, S.H.; Iverson, L.; Wu, C.F.; et al. High-quality life extension by the enzyme peptide methionine sulfoxide reductase. *Proc. Natl. Acad. Sci. USA* **2002**, *99*, 2748–2753. [CrossRef] [PubMed]

98. Roesijadi, G.; Rezvankhah, S.; Binninger, D.M.; Weissbach, H. Ecdysone induction of MsrA protects against oxidative stress in Drosophila. *Biochem. Biophys. Res. Commun.* **2007**, *354*, 511–516. [CrossRef] [PubMed]

99. Moskovitz, J.; Bar-Noy, S.; Williams, W.M.; Requena, J.; Berlett, B.S.; Stadtman, E.R. Methionine sulfoxide reductase (MsrA) is a regulator of antioxidant defense and lifespan in mammals. *Proc. Natl. Acad. Sci. USA* **2001**, *98*, 12920–12925. [CrossRef]

100. Nan, C.; Li, Y.; Jean-Charles, P.Y.; Chen, G.; Kreymerman, A.; Prentice, H.; Weissbach, H.; Huang, X. Deficiency of methionine sulfoxide reductase A causes cellular dysfunction and mitochondrial damage in cardiac myocytes under physical and oxidative stresses. *Biochem. Biophys. Res. Commun.* **2010**, *402*, 608–613. [CrossRef]

101. Picot, C.R.; Perichon, M.; Lundberg, K.C.; Friguet, B.; Szweda, L.I.; Petropoulos, I. Alterations in mitochondrial and cytosolic methionine sulfoxide reductase activity during cardiac ischemia and reperfusion. *Exp. Gerontol.* **2006**, *41*, 663–667. [CrossRef] [PubMed]

102. Zhao, H.; Sun, J.; Deschamps, A.M.; Kim, G.; Liu, C.; Murphy, E.; Levine, R.L. Myristoylated methionine sulfoxide reductase A protects the heart from ischemia-reperfusion injury. *Am. J. Physiol. Heart Circ. Physiol.* **2011**, *301*, H1513–H1518. [CrossRef] [PubMed]

103. Kim, J.I.; Choi, S.H.; Jung, K.J.; Lee, E.; Kim, H.Y.; Park, K.M. Protective role of methionine sulfoxide reductase A against ischemia/reperfusion injury in mouse kidney and its involvement in the regulation of trans-sulfuration pathway. *Antioxid. Redox Signal.* **2013**, *18*, 2241–2250. [CrossRef] [PubMed]

104. Kim, J.I.; Noh, M.R.; Kim, K.Y.; Jang, H.S.; Kim, H.Y.; Park, K.M. Methionine sulfoxide reductase A deficiency exacerbates progression of kidney fibrosis induced by unilateral ureteral obstruction. *Free Radic. Biol. Med.* **2015**, *89*, 201–208. [CrossRef] [PubMed]

105. Fomenko, D.E.; Novoselov, S.V.; Natarajan, S.K.; Lee, B.C.; Koc, A.; Carlson, B.A.; Lee, T.H.; Kim, H.Y.; Hatfield, D.L.; Gladyshev, V.N. MsrB1 (methionine-R-sulfoxide reductase 1) knock-out mice: Roles of MsrB1 in redox regulation and identification of a novel selenoprotein form. *J. Biol. Chem.* **2009**, *284*, 5986–5993. [CrossRef] [PubMed]

106. Moskovitz, J. Prolonged selenium-deficient diet in MsrA knockout mice causes enhanced oxidative modification to proteins and affects the levels of antioxidant enzymes in a tissue-specific manner. *Free Radic. Res.* **2007**, *41*, 162–171. [CrossRef] [PubMed]

107. Kim, K.Y.; Kwak, G.H.; Singh, M.P.; Gladyshev, V.N.; Kim, H.Y. Selenoprotein MsrB1 deficiency exacerbates acetaminophen-induced hepatotoxicity via increased oxidative damage. *Arch. Biochem. Biophys.* **2017**, *634*, 69–75. [CrossRef]

108. Sreekumar, P.G.; Kannan, R.; Yaung, J.; Spee, C.K.; Ryan, S.J.; Hinton, D.R. Protection from oxidative stress by methionine sulfoxide reductases in RPE cells. *Biochem. Biophys. Res. Commun.* **2005**, *334*, 245–253. [CrossRef]

109. Lee, J.W.; Gordiyenko, N.V.; Marchetti, M.; Tserentsoodol, N.; Sagher, D.; Alam, S.; Weissbach, H.; Kantorow, M.; Rodriguez, I.R. Gene structure, localization and role in oxidative stress of methionine sulfoxide reductase A (MSRA) in the monkey retina. *Exp. Eye Res.* **2006**, *82*, 816–827. [CrossRef]

110. Dun, Y.; Vargas, J.; Brot, N.; Finnemann, S.C. Independent roles of methionine sulfoxide reductase A in mitochondrial ATP synthesis and as antioxidant in retinal pigment epithelial cells. *Free Radic. Biol. Med.* **2013**, *65*, 1340–1351. [CrossRef]

111. Kantorow, M.; Hawse, J.R.; Cowell, T.L.; Benhamed, S.; Pizarro, G.O.; Reddy, V.N.; Hejtmancik, J.F. Methionine sulfoxide reductase A is important for lens cell viability and resistance to oxidative stress. *Proc. Natl. Acad. Sci. USA* **2004**, *101*, 9654–9659. [CrossRef] [PubMed]

112. Marchetti, M.A.; Pizarro, G.O.; Sagher, D.; Deamicis, C.; Brot, N.; Hejtmancik, J.F.; Weissbach, H.; Kantorow, M. Methionine sulfoxide reductases B1, B2, and B3 are present in the human lens and confer oxidative stress resistance to lens cells. *Investig. Ophthalmol. Vis. Sci.* **2005**, *46*, 2107–2112. [CrossRef] [PubMed]

113. Dai, J.; Liu, H.; Zhou, J.; Huang, K. Selenoprotein R Protects Human Lens Epithelial Cells against d-Galactose-Induced Apoptosis by Regulating Oxidative Stress and Endoplasmic Reticulum Stress. *Int. J. Mol. Sci.* **2016**, *17*, 231. [CrossRef] [PubMed]

114. Jia, Y.; Li, Y.; Du, S.; Huang, K. Involvement of MsrB1 in the regulation of redox balance and inhibition of peroxynitrite-induced apoptosis in human lens epithelial cells. *Exp. Eye Res.* **2012**, *100*, 7–16. [CrossRef] [PubMed]

115. Li, Y.; Jia, Y.; Zhou, J.; Huang, K. Effect of methionine sulfoxide reductase B1 silencing on high-glucose-induced apoptosis of human lens epithelial cells. *Life Sci.* **2013**, *92*, 193–201. [CrossRef] [PubMed]

116. Jia, Y.; Zhou, J.; Liu, H.; Huang, K. Effect of methionine sulfoxide reductase B1 (SelR) gene silencing on peroxynitrite-induced F-actin disruption in human lens epithelial cells. *Biochem. Biophys. Res. Commun.* **2014**, *443*, 876–881. [CrossRef] [PubMed]

117. Ugarte, N.; Ladouce, R.; Radjei, S.; Gareil, M.; Friguet, B.; Petropoulos, I. Proteome alteration in oxidative stress-sensitive methionine sulfoxide reductase-silenced HEK293 cells. *Free Radic. Biol. Med.* **2013**, *65*, 1023–1036. [CrossRef] [PubMed]

118. Cabreiro, F.; Picot, C.R.; Perichon, M.; Friguet, B.; Petropoulos, I. Overexpression of methionine sulfoxide reductases A and B2 protects MOLT-4 cells against zinc-induced oxidative stress. *Antioxid. Redox Signal.* **2009**, *11*, 215–225. [CrossRef] [PubMed]

119. Picot, C.R.; Petropoulos, I.; Perichon, M.; Moreau, M.; Nizard, C.; Friguet, B. Overexpression of MsrA protects WI-38 SV40 human fibroblasts against H_2O_2-mediated oxidative stress. *Free Radic. Biol. Med.* **2005**, *39*, 1332–1341. [CrossRef] [PubMed]

120. Cabreiro, F.; Picot, C.R.; Perichon, M.; Mary, J.; Friguet, B.; Petropoulos, I. Identification of proteins undergoing expression level modifications in WI-38 SV40 fibroblasts overexpressing methionine sulfoxide reductase A. *Biochimie* **2007**, *89*, 1388–1395. [CrossRef] [PubMed]

121. Kim, J.Y.; Kim, Y.; Kwak, G.H.; Oh, S.Y.; Kim, H.Y. Over-expression of methionine sulfoxide reductase A in the endoplasmic reticulum increases resistance to oxidative and ER stresses. *Acta Biochim. Biophys. Sin. (Shanghai)* **2014**, *46*, 415–419. [CrossRef] [PubMed]

122. Kwak, G.H.; Lim, D.H.; Han, J.Y.; Lee, Y.S.; Kim, H.Y. Methionine sulfoxide reductase B3 protects from endoplasmic reticulum stress in Drosophila and in mammalian cells. *Biochem. Biophys. Res. Commun.* **2012**, *420*, 130–135. [CrossRef] [PubMed]

123. Pelle, E.; Maes, D.; Huang, X.; Frenkel, K.; Pernodet, N.; Yarosh, D.B.; Zhang, Q. Protection against UVB-induced oxidative stress in human skin cells and skin models by methionine sulfoxide reductase A. *J. Cosmet. Sci.* **2012**, *63*, 359–364. [PubMed]

124. Zhou, Z.; Li, C.Y.; Li, K.; Wang, T.; Zhang, B.; Gao, T.W. Decreased methionine sulphoxide reductase A expression renders melanocytes more sensitive to oxidative stress: A possible cause for melanocyte loss in vitiligo. *Br. J. Dermatol.* **2009**, *161*, 504–509. [CrossRef] [PubMed]

125. Gil-Mohapel, J.; Brocardo, P.S.; Christie, B.R. The role of oxidative stress in Huntington's disease: Are antioxidants good therapeutic candidates? *Curr. Drug Targets* **2014**, *15*, 454–468. [CrossRef] [PubMed]

126. Swomley, A.M.; Butterfield, D.A. Oxidative stress in Alzheimer disease and mild cognitive impairment: Evidence from human data provided by redox proteomics. *Arch. Toxicol.* **2015**, *89*, 1669–1680. [CrossRef] [PubMed]

127. Gabbita, S.P.; Aksenov, M.Y.; Lovell, M.A.; Markesbery, W.R. Decrease in peptide methionine sulfoxide reductase in Alzheimer's disease brain. *J. Neurochem.* **1999**, *73*, 1660–1666. [CrossRef]

128. Brennan, L.A.; Lee, W.; Cowell, T.; Giblin, F.; Kantorow, M. Deletion of mouse MsrA results in HBO-induced cataract: MsrA repairs mitochondrial cytochrome c. *Mol. Vis.* **2009**, *15*, 985–999.

129. Pal, R.; Oien, D.B.; Ersen, F.Y.; Moskovitz, J. Elevated levels of brain-pathologies associated with neurodegenerative diseases in the methionine sulfoxide reductase A knockout mouse. *Exp. Brain Res.* **2007**, *180*, 765–774. [CrossRef]

130. Hou, L.; Kang, I.; Marchant, R.E.; Zagorski, M.G. Methionine 35 oxidation reduces fibril assembly of the amyloid abeta-(1-42) peptide of Alzheimer's disease. *J. Biol. Chem.* **2002**, *277*, 40173–40176. [CrossRef]

131. Moskovitz, J.; Du, F.; Bowman, C.F.; Yan, S.S. Methionine sulfoxide reductase A affects beta-amyloid solubility and mitochondrial function in a mouse model of Alzheimer's disease. *Am. J. Physiol. Endocrinol. Metab.* **2016**, *310*, E388–E393. [CrossRef] [PubMed]

132. Moskovitz, J.; Maiti, P.; Lopes, D.H.; Oien, D.B.; Attar, A.; Liu, T.; Mittal, S.; Hayes, J.; Bitan, G. Induction of methionine-sulfoxide reductases protects neurons from amyloid beta-protein insults in vitro and in vivo. *Biochemistry* **2011**, *50*, 10687–10697. [CrossRef] [PubMed]

133. Glaser, C.B.; Yamin, G.; Uversky, V.N.; Fink, A.L. Methionine oxidation, α-synuclein and Parkinson's disease. *Biochim. Biophys. Acta* **2005**, *1703*, 157–169. [CrossRef] [PubMed]

134. Chondrogianni, N.; Petropoulos, I.; Grimm, S.; Georgila, K.; Catalgol, B.; Friguet, B.; Grune, T.; Gonos, E.S. Protein damage, repair and proteolysis. *Mol. Aspects Med.* **2014**, *35*, 1–71. [CrossRef] [PubMed]

135. Petropoulos, I.; Mary, J.; Perichon, M.; Friguet, B. Rat peptide methionine sulphoxide reductase: Cloning of the cDNA, and down-regulation of gene expression and enzyme activity during aging. *Biochem. J.* **2001**, *355*, 819–825. [CrossRef] [PubMed]

136. Picot, C.R.; Perichon, M.; Cintrat, J.-C.; Friguet, B.; Petropoulos, I. The peptide methionine sulfoxide reductases, MsrA and MsrB (hCBS-1), are downregulated during replicative senescence of human WI-38 fibroblasts. *FEBS Lett.* **2004**, *558*, 74–78. [CrossRef]

137. Ahmed, E.K.; Rogowska-Wrzesinska, A.; Roepstorff, P.; Bulteau, A.L.; Friguet, B. Protein modification and replicative senescence of WI-38 human embryonic fibroblasts. *Aging Cell* **2010**, *9*, 252–272. [CrossRef]

138. Brovelli, A.; Seppi, C.; Castellana, A.M.; De Renzis, M.R.; Blasina, A.; Balduini, C. Oxidative lesion to membrane proteins in senescent erythrocytes. *Biomed. Biochim. Acta* **1990**, *49*, S218–S223.

139. Weissbach, H.; Resnick, L.; Brot, N. Methionine sulfoxide reductases: History and cellular role in protecting against oxidative damage. *Biochim. Biophys. Acta* **2005**, *1703*, 203–212. [CrossRef]

140. Lee, B.C.; Lee, H.M.; Kim, S.; Avanesov, A.S.; Lee, A.; Chun, B.H.; Vorbruggen, G.; Gladyshev, V.N. Gladyshev Expression of the methionine sulfoxide reductase lost during evolution extends Drosophila lifespan in a methionine-dependent manner. *Sci. Rep.* **2018**, *8*, 1010. [CrossRef]

141. Oien, D.B.; Moskovitz, J. Substrates of the methionine sulfoxide reductase system their physiological relevance. *Curr. Top. Dev. Biol.* **2008**, *80*, 93–133. [PubMed]

142. Caldwell, P.; Luk, D.C.; Weissbach, H.; Brot, N. Oxidation of the methionine residues of *Escherichia coli* ribosomal protein L12 decreases the protein's biological activity. *Proc. Natl. Acad. Sci. USA* **1978**, *75*, 5349–5352. [CrossRef] [PubMed]

143. Ezraty, B.; Grimaud, R.; El Hassouni, M.; Moinier, D.; Barras, F. Methionine sulfoxide reductases protect Ffh from oxidative damages in *Escherichia coli*. *EMBO J* **2004**, *23*, 1868–1877. [CrossRef] [PubMed]

144. Abrams, W.R.; Weinbaum, G.; Weissbach, L.; Weissbach, H.; Brot, N. Enzymatic reduction of oxidized α-1-proteinase inhibitor restores biological activity. *Proc. Natl. Acad. Sci. USA* **1981**, *78*, 7483–7486. [CrossRef] [PubMed]

145. Stief, T.W.; Aab, A.; Heimburger, N. Oxidative inactivation of purified human α-2-antiplasmin, antithrombin III, and C1-inhibitor. *Thromb. Res.* **1988**, *49*, 581–589. [CrossRef]

146. Glaser, C.B.; Karic, L.; Parmelee, S.; Premachandra, B.R.; Hinkston, D.; Abrams, W.R. Studies on the turnover of methionine oxidized α-1-protease inhibitor in rats. *Am. Rev. Respir. Dis.* **1987**, *136*, 857–861. [CrossRef] [PubMed]

147. Davis, D.A.; Newcomb, F.M.; Moskovitz, J.; Wingfield, P.T.; Stahl, S.J.; Kaufman, J.; Fales, H.M.; Levine, R.L.; Yarchoan, R. HIV-2 protease is inactivated after oxidation at the dimer interface activity can be partly restored with methionine sulphoxide reductase. *Biochem. J.* **2000**, *2 Pt 346*, 305–311. [CrossRef]

148. Kanayama, A.; Inoue, J.; Sugita-Konishi, Y.; Shimizu, M.; Miyamoto, Y. Oxidation of Ikappa Bα at methionine 45 is one cause of taurine chloramine-induced inhibition of NF-kappa B activation. *J. Biol. Chem.* **2002**, *277*, 24049–24056. [CrossRef]

149. Midwinter, R.G.; Cheah, F.C.; Moskovitz, J.; Vissers, M.C.; Winterbourn, C.C. IkappaB is a sensitive target for oxidation by cell-permeable chloramines: Inhibition of NF-kappaB activity by glycine chloramine through methionine oxidation. *Biochem. J.* **2006**, *396*, 71–78. [CrossRef]

150. Mohri, M.; Reinach, P.S.; Kanayama, A.; Shimizu, M.; Moskovitz, J.; Hisatsune, T.; Miyamoto, Y. Suppression of the TNFα-induced increase in IL-1α expression by hypochlorite in human corneal epithelial cells. *Investig. Ophthalmol. Vis. Sci.* **2002**, *43*, 3190–3195.

151. Chen, J.; Avdonin, V.; Ciorba, M.A.; Heinemann, S.H.; Hoshi, T. Acceleration of P/C-type inactivation in voltage-gated K(+) channels by methionine oxidation. *Biophys. J.* **2000**, *78*, 174–187. [CrossRef]

152. Ciorba, M.A.; Heinemann, S.H.; Weissbach, H.; Brot, N.; Hoshi, T. Modulation of potassium channel function by methionine oxidation reduction. *Proc. Natl. Acad. Sci. USA* **1997**, *94*, 9932–9937. [CrossRef] [PubMed]

153. Bigelow, D.J.; Squier, T.C. Redox modulation of cellular signaling metabolism through reversible oxidation of methionine sensors in calcium regulatory proteins. *Biochim. Biophys. Acta* **2005**, *1703*, 121–134. [CrossRef] [PubMed]

154. Lim, J.C.; Kim, G.; Levine, R.L. Stereospecific oxidation of calmodulin by methionine sulfoxide reductase A. *Free Radic. Biol. Med.* **2013**, *61*, 257–264. [CrossRef] [PubMed]

155. Yao, Y.; Yin, D.; Jas, G.S.; Kuczer, K.; Williams, T.D.; Schoneich, C.; Squier, T.C. Oxidative modification of a carboxyl-terminal vicinal methionine in calmodulin by hydrogen peroxide inhibits calmodulin-dependent activation of the plasma membrane Ca-ATPase. *Biochemistry* **1996**, *35*, 2767–2787. [CrossRef] [PubMed]

156. Sun, H.; Gao, J.; Ferrington, D.A.; Biesiada, H.; Williams, T.D.; Squier, T.C. Repair of oxidized calmodulin by methionine sulfoxide reductase restores ability to activate the plasma membrane Ca-ATPase. *Biochemistry* **1999**, *38*, 105–112. [CrossRef] [PubMed]

157. Vougier, S.; Mary, J.; Dautin, N.; Vinh, J.; Friguet, B.; Ladant, D. Essential role of methionine residues in calmodulin binding to Bordetella pertussis adenylate cyclase, as probed by selective oxidation and repair by the peptide methionine sulfoxide reductases. *J. Biol. Chem.* **2004**, *279*, 30210–30218. [CrossRef]

158. Erickson, J.R.; Joiner, M.L.; Guan, X.; Kutschke, W.; Yang, J.; Oddis, C.V.; Bartlett, R.K.; Lowe, J.S.; O'Donnell, S.E.; Aykin-Burns, N.; et al. A dynamic pathway for calcium-independent activation of CaMKII by methionine oxidation. *Cell* **2008**, *133*, 462–474. [CrossRef]

159. He, B.J.; Joiner, M.L.; Singh, M.V.; Luczak, E.D.; Swaminathan, P.D.; Koval, O.M.; Kutschke, W.; Allamargot, C.; Yang, J.; Guan, X.; et al. Oxidation of CaMKII determines the cardiotoxic effects of aldosterone. *Nat. Med.* **2011**, *17*, 1610–1618. [CrossRef]

160. Hung, R.J.; Yazdani, U.; Yoon, J.; Wu, H.; Yang, T.; Gupta, N.; Huang, Z.; van Berkel, W.J.; Terman, J.R. Mical links semaphorins to F-actin disassembly. *Nature* **2010**, *463*, 823–827. [CrossRef]

161. Hung, R.J.; Spaeth, C.S.; Yesilyurt, H.G.; Terman, J.R. SelR reverses Mical-mediated oxidation of actin to regulate F-actin dynamics. *Nat. Cell Biol.* **2013**, *15*, 1445–1454. [CrossRef] [PubMed]

162. Manta, B.; Gladyshev, V.N. Regulated methionine oxidation by monooxygenases. *Free Radic. Biol. Med.* **2017**, *109*, 141–155. [CrossRef] [PubMed]

163. Lim, J.M.; Kim, G.; Levine, R.L. Methionine in Proteins: It's Not Just for Protein Initiation Anymore. *Neurochem. Res.* **2018**. [CrossRef] [PubMed]

164. Wehr, N.B.; Levine, R.L. Wanted and wanting: Antibody against methionine sulfoxide. *Free Radic. Biol. Med.* **2012**, *53*, 1222–1225. [CrossRef] [PubMed]

165. Ghesquière, B.; Gevaert, K. Proteomics methods to study methionine oxidation. *Mass Spectrom. Rev.* **2014**, *33*, 147–156. [CrossRef] [PubMed]

166. Tarrago, L.; Oheix, E.; Péterfi, Z.; Gladyshev, V.N. Monitoring of methionine sulfoxide content and methionine sulfoxide reductase activity. *Methods Mol. Biol.* **2018**, *1661*, 285–299. [PubMed]

antioxidants

MDPI

Review

Reactive Oxygen Species and the Redox-Regulatory Network in Cold Stress Acclimation

Anna Dreyer and **Karl-Josef Dietz** *

Department of Biochemistry and Physiology of Plants, Faculty of Biology, University of Bielefeld,
33615 Bielefeld, Germany; andreyer@techfak.uni-bielefeld.de
* Correspondence: karl-josef.dietz@uni-bielefeld.de; Tel.: +49-521-106-5589

Received: 31 October 2018; Accepted: 16 November 2018; Published: 21 November 2018

Abstract: Cold temperatures restrict plant growth, geographical extension of plant species, and agricultural practices. This review deals with cold stress above freezing temperatures often defined as chilling stress. It focuses on the redox regulatory network of the cell under cold temperature conditions. Reactive oxygen species (ROS) function as the final electron sink in this network which consists of redox input elements, transmitters, targets, and sensors. Following an introduction to the critical network components which include nicotinamide adenine dinucleotide phosphate (NADPH)-dependent thioredoxin reductases, thioredoxins, and peroxiredoxins, typical laboratory experiments for cold stress investigations will be described. Short term transcriptome and metabolome analyses allow for dissecting the early responses of network components and complement the vast data sets dealing with changes in the antioxidant system and ROS. This review gives examples of how such information may be integrated to advance our knowledge on the response and function of the redox regulatory network in cold stress acclimation. It will be exemplarily shown that targeting the redox network might be beneficial and supportive to improve cold stress acclimation and plant yield in cold climate.

Keywords: chilling stress; cold temperature; posttranslational modification; regulation; ROS; thiol redox network; thioredoxin

---

1. Plant Response to Cold Temperature

Biological systems are unavoidably affected by changes in ambient temperature. Such temperature effects particularly concern the temperature-dependent rates of spontaneous and enzyme-catalyzed chemical and physical reactions, the structural and molecular dynamics, and strength of molecular interactions. Each of these effects interferes with the state of metabolism and cellular signal processing. Many plants have evolved the capacity to adapt to low-temperature climates and develop locally distinct adaptive traits [1]. Thus, the response of photosynthetic metabolism to 4 °C varies not just between different species, but also between differently adapted populations. Oakley et al. [1] demonstrated this phenomenon with Arabidopsis from Italy and Sweden, and their crossing, by measuring the fast recovery of non-photochemical quenching after 2 min of darkening. The cold-acclimation response involves a profound reorganization of gene expression and posttranscriptional processes employing abscisic acid (ABA)-independent and dependent pathways [2]. The fraction of unsaturated fatty acid residues in cell membranes in particular linoleic acid (C18:2) increases at the expense of saturated lipids during chilling stress acclimation [3]. The fatty acid desaturation increases the fluidity of the membrane at a lower temperature and involves activation of fatty acid desaturases such as stearoyl-acyl carrier-protein desaturase (SAD) [4]. In addition, higher amounts and often different types of sugars and other osmolytes accumulate at low temperatures. Osmolyte accumulation is assumed to prepare for dehydration stress during freezing-induced water deficit [5]. In addition, the antioxidant defense is enhanced during cold stress acclimation [6,7]. In concert with reactive oxygen species (ROS) production, the antioxidant system controls the redox regulatory network of the cell. Our review focuses on this aspect of chilling stress acclimation and is intended to synthesize a broader perspective on redox network function and consequences.

2. Cold Stress Experiments in the Laboratory

Chilling stress occurs if non-primed plants or non-adapted plants, which are unable to be hardened, are exposed to cold temperatures significantly below growth temperature. Table 1 summarizes eight experiments of recent years. The genetic model plants *Arabidopsis thaliana*, *Oryza sativa*, and *Sorghum bicolor*, the genetic model *Capsella bursa-pastoris*, two *Jatropha* species with potential in biofuel production, and the medicinal plant *Calendula officinalis* were grown at 20–28 °C during the light phase and 18–28 °C during the night and then mostly transferred to 4 °C, or 10, 12, and 15 °C, the latter for *Capsella*, rice, and soybean. Thus, the chosen temperature regimes covered down-shifts between 14 and 24 °C and were scrutinized for 30 min [8] to 14 d [6]. It is noteworthy that recovery from cold stress attracts increasing attention in *Arabidopsis thaliana* [6]. These authors recognized a positive relationship between the speed of recovery from 14 d chilling treatment and the strength of the plastid antioxidant system [6]. Thus, the *Arabidopsis* accessions N14, N13, Ms-0, and Kas-1, which grow under quite low temperatures in nature, displayed a weaker expression of some plastid antioxidant genes during the phase of 14 d cold treatment and simultaneously maintained a primed state after transfer to normal temperature. Maintenance of a primed state has an advantage if periods of chilling stress will return, but is disadvantageous for growth. Thus, maintenance of primed state has a trade-off for fitness at elevated temperatures [6].

Table 1. Experimental parameters used to explore chilling stress acclimation. See text for further description of the experiments.

Species	Plant Age	Growth Condition	Cold Treatment		Other Comments	Reference
			duration	temperature		
Arabidopsis thaliana	16 days	16/8 h light/dark 24 °C	0; 0.5; 1; 3; 6; 12; 24 h	4 °C	Decreased light intensity during cold stress treatment	[8]
Arabidopsis thaliana	42 days	16/8 h light/dark 20 °C/18 °C	14 days	4 °C	Decreased light intensity during cold stress Additional 1,2, and 3 days of deacclimation	[6]
Capsella bursa pastoris L.	30 days	16/8 h light/dark 25 °C	24; 48; 72; 96; 120 h	10 °C		[9]
Calendula officinalis	14 days	16/8 h light/dark 25 ± 2 °C	24; 48; 72; 96; 120 h	4 °C		[10]
Jatropha curcas	45 days	14/10 h light/dark 28 °C	48 h	4 °C	Partially pretreated at 15 °C for five days Cold sensitive *Jatropha*	[11]
Jatropha macrocarpa	45 days	14/10 h light/dark 28 °C	48 h	4 °C	Partially pretreated at 15 °C for 5 days, cold tolerant *Jatropha*	[11]
Oryza sativa	14 days	14/10 h light/dark 28 °C/22 °C	6 days	12 °C	2 days pretreatment with melatonin	[12]
Sorghum bicolor	30 days	16/8 h light/dark 28 °C/24 °C	6 days	15 °C	5 days recovery	[13]

3. Central Role of the Redox Regulatory Network in Stress Acclimation

All cells express a regulatory network of thiol-containing proteins which integrates information from reductive metabolism and electron drainage from the network by ROS [14]. Targets of regulation by thiol oxidation are metabolic enzymes of various pathways, signal transduction elements, transcription and translation factors, and, thus, essentially all functional levels of the cell [15]. These target proteins are reduced by the concerted action of redox input elements and redox transmitters.

Redox input elements transfer electrons from metabolism to the redox transmitters. The nicotinamide adenine dinucleotide phosphate (NADPH)-dependent thioredoxin reductase A (NTRA) functions as such redox input element in the cytosol by transferring electrons from NADPH to thioredoxin (TRX). The same role is played by NTRB in the mitochondrion [16] and by NTRC in the chloroplast and other plastids [17]. NTRC is a variant which carries a TRX domain in addition to the NTR domain. An additional input element is the ferredoxin-dependent thioredoxin reductase (FTR) which transfers electrons from the photosynthetic electron transport chain (ETC) to TRX [18].

Decisive elements in redox regulation are TRX which constitutes large families in plants comprising proper TRX and TRX-like proteins. They function as redox transmitters. A genome-wide association study (GWAS) with soybean accessions grown at 28 °C and transferred to 15 °C for cold stress (Table 1) [13] discovered 143 genomic sites considered as promising for improving cold acclimation of soybean. Soybean is a C4 plant which is chilling sensitive. The study focused on photosynthetic performance during cold treatment and recovery. Among the identified genes of interest were two TRX genes, *Sb03g004670* and *Sb06g029490*, which may contribute to the cold acclimation variation among the accessions [13]. These TRX-genes link cold stress to the regulation of the Calvin cycle [13,19,20].

Thiol peroxidases have a very high affinity to H_2O_2, alkylhydroperoxides and peroxynitrite (ONOO$^-$). This group of enzymes comprises peroxiredoxins (PRX) and glutathione peroxidase-like proteins (GPX) and was suggested to control the spreading of peroxide signals in the cell. They act as redox sensors [21,22]. Their affinity and abundance support their function as primary reactants with peroxides. Recently, it was shown that the chloroplast 2-cysteine peroxiredoxin functions as thioredoxin oxidase and thereby co-controls the activation state of target proteins of redox regulation, such as malate dehydrogenase, phosphoribulokinase, and fructose-1,6-bisphosphatase [23]. Based on this study on chloroplast 2-cysteine peroxiredoxin it may be hypothesized that other thiol peroxidases play the same role in the redox regulatory network of plastids and other cell compartments.

The final electron acceptors of the network are peroxides which oxidize the redox sensors. Generator systems are the photosynthetic and respiratory ETC, substrate oxidases in peroxisomes and apoplast, and the NADPH oxidases in the plasma membrane [24]. Antioxidants counteract the accumulation of ROS and, thus, lower the electron drainage from the network (Figure 1). Oxidation of TRX by O_2, direct oxidation of target proteins by peroxides or other mechanisms likely contribute to the oxidation of network components. But these mechanisms are poorly understood and may lack the specificity which is needed for tailored responses.

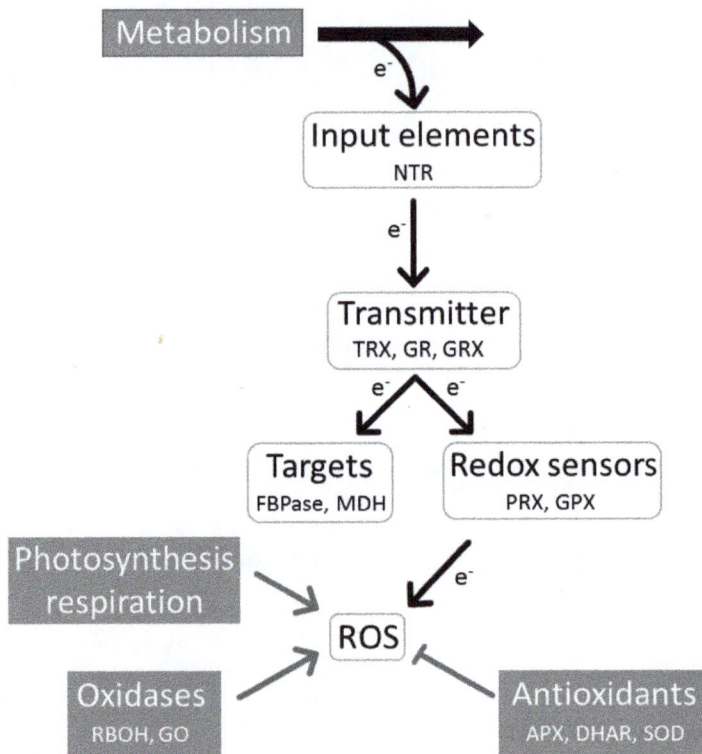

Figure 1. Basic structure of the redox regulatory network of the plant cell. Dependent on its state, metabolism feeds regulatory electrons via input elements (NTR: Nicotinamide adenine dinucleotide phosphate (NADPH)-dependent thioredoxin reductase) into the network. Redox transmitters (TRX: thioredoxin; GR: glutathione reductase, GRX: glutaredoxin) transfer electrons to regulated target proteins (FBPase: Fructose-1,6-bisphosphatase, MDH: malate dehydrogenase). Redox sensors (GPX: glutathione peroxidase (like), PRX: peroxiredoxin) also drain electrons from the transmitters in dependence on the ROS amount. The ROS amount is controlled by the activity of generator systems such as photosynthetic and respiratory electron transport chains and oxidases (RBOH: NADPH oxidase; GO: glycolate oxidase) and the decomposition of reactive oxygen species (ROS) by the antioxidant system (APX: ascorbate peroxidase, DHAR: dehydroascorbate reductase, SOD: superoxide dismutase). The proteins mentioned in the figure are typical representatives.

4. Variability of Cold Response Between Species

If the ambient temperatures drop significantly below growth temperature, an imbalance between photosynthetic light and dark reactions is established, and the photosynthetic electron transport chain releases more superoxide (O_2^-) by electron transfer to oxygen as alternative electron acceptor. An O_2^- increase is observed during chilling stress, e.g., in *Cynodon dactylon* [25].

Several studies indicate that increased superoxide dismutase (SOD) activity likely is related to chilling tolerance. SOD catalyzes the formation of hydrogen peroxide (H_2O_2) and H_2O from two superoxide molecules and two protons [26,27]. The response of SOD varies among species. Plants contain several *SOD* genes which often show species- and stress-specific expression patterns [28]. However, activity measurements usually cannot easily discriminate among the isoforms of Mn-, CuZn- and Fe-SOD. The transcript amounts and activity of SOD increase during low-temperature treatment in *Capsella bursa pastoris L.* [9] and *Calendula officinalis* [10]. The SOD activity in *Calendula officinalis* rises less than the transcript amount pointing to posttranscriptional regulation of synthesis or higher protein degradation rates [10,29]. Likewise, the SOD activities of *Phaseolus vulgaris* [30], *Cynodon dactylon* [25], and *Withania somnifera* [31] only increase slightly upon exposure to low temperature. In the ecotype AGB025 of *W. somnifera* the SOD activity was even constant during the first three days of low-temperature treatment. In a converse manner, the ecotype AGB002, which is known to be more tolerant to chilling stress than AGB025, showed an increased enzyme activity [31].

Similar contrasting results were observed for two cell cultures derived from *O. sativa* subspecies *Japonica cv.* Nipponbare and *Indica cv.* 9311. The genotype *Japonica* known to be more adapted to chilling temperatures showed induced SOD activity after 24 h of low-temperature treatment, whereas the activity in *Indica* was constant during the 72 h of stress treatment [32,33]. The significant role of SOD is plausible if considering the role of ROS in addressing the thiol network as discussed above. H_2O_2 acts as an efficient electron sink while O_2^- does not, but rather may cause damage, e.g., by peroxidation of lipids.

Unlike SOD, the catalase (CAT) activity in *O. sativa Indica* increased during the first 24 h of cold treatment but later on, returned to control level. CATs are localized in the peroxisomes and decompose H_2O_2 released from peroxisomal oxidases like glycolate oxidase with its role in photorespiration. CAT activity is important to minimize leakage of H_2O_2 from the peroxisome to the cytosol. The constitutively high SOD activity together with the transiently induced activity of CAT, could tune the accumulation of H_2O_2. In contrast to *Indica*, *Japonica* rice showed a relatively high, but constant, activity of CAT [32]. A similar increase of SOD and CAT activity was observed *C. bursa-pastoris* [9], *Citrullus lanatus* [34], and *C. dactylon* [25]. Therefore, SOD and CAT activity could play an important part during chilling stress, although some species like *C. officinalis* [10] and *P. vulgaris* [30] fail to increase the activity of these enzymes. The increased CAT is likely important to avoid accumulation of H_2O_2 if its production increases under cold due to imbalances and changes in metabolism.

The ascorbate-dependent Foyer–Halliwell–Asada cycle reduces H_2O_2 to water and is located in several subcellular compartments. This pathway relies on ascorbate peroxidases (APX) [27]. APX catalyzes the reduction of H_2O_2 to water by using ascorbate as electron donor, which results in the formation of dehydroascorbate (DHA). The produced DHA is recycled to ascorbate by dehydroascorbate reductase (DHAR) using glutathione as reductant (GSH). Liberated oxidized glutathione (GSSG) is regenerated by glutathione reductase (GR) using NADPH [26]. An ascorbate-independent water-water cycle for H_2O_2 reduction employs GPX and PRX which are found in plastids, mitochondria, and cytosol. The thiol peroxidases are part of the thiol network as described above. The oxidized thiol peroxidase reacts with an electron donor like TRX or glutathione with or without glutaredoxin (GRX) [35,36].

The APX activity decreased in *C. bursa pastoris* during cold treatment of 1 to 5 days duration. This is quite unexpected in the light of high transcript levels observed during the first 24 h of low-temperature treatment [9] and could be due to posttranscriptional regulation [37,38] or APX inactivation by reaction with H_2O_2 in the absence of ascorbate [39]. APX inactivation could be a means to enable H_2O_2 accumulation for signaling. Quantification of APX protein levels would help to explain the weak correlation between transcript level and enzyme activity. To compensate for the instability of enzymes and to increase the antioxidant capacity, a constant supply of *de novo* synthesized proteins is needed [40]. This may explain why the transcript levels of the corresponding antioxidant enzymes usually increase under stress, e.g., in *C. officinalis* where both the transcript level and the activity of APX increased with prolonged time of low-temperature treatment [10]. APX activity also increased in cold-tolerant *J. macrocarpa*, whereas the APX activity decreased >6-fold in the cold-sensitive *J. curcas* [11]. The ascorbate-dependent water-water cycle is supported by stimulated ascorbate biosynthesis, strengthened glutathione homeostasis, enhanced sulphate reduction, and TRX pathway [41].

Apparently, different species reveal partly controversial responses of components of the antioxidant system. Considering the central role of redox homeostasis in cell function, this variation is surprising. Thus, the dependency of priming and resilience of the primed state on the strength of the chloroplast antioxidant system offers an important explanation for this variation [6].

5. The Compartment-Specific Response of the Components of the Redox-Network to Cold in *A. thaliana*

The enzymes of the redox-regulatory network are localized in different subcellular compartments of the cell (Figure 2A,C). Frequently more than one isoform of the redox elements is present in the different compartments. The three schemes assemble published transcript changes for the three subcellular compartments cytosol, chloroplast, and mitochondrion during a time course of 24 h cold treatment [8].

As outlined above, NADPH-dependent thioredoxin reductases (NTR) function as redox input elements and isoforms are present in each compartment. The relative transcript amounts of the cytosolic *NTRA* (Figure 2A), chloroplast *NTRC* (Figure 2B) and mitochondrial *NTRB* (Figure 2C) slightly decreased during the 24 h of low-temperature treatment. NTRC, which regenerates oxidized 2-Cys-PRX in an NADPH dependent manner [17,73], showed a similar decrease in expression as *2-Cys-PRXB* (Figure 2B). The redox transmitters TRX and GRX transfer reducing equivalents to target proteins such as thiol peroxidases [74]. The downregulation of the h-type TRXs in the cytosol occurred in a peculiar time-dependent manner. *TRX-h3* was downregulated during the first 30 min, *TRX-h9* after 1 h, followed by *TRX-h4*, *TRX-h2* and *TRX-h1* which only changed after 24 h of low-temperature treatment (Figure 2A).

Transcript levels of most of the redox sensors *GPX* and *PRX* were also downregulated during the low-temperature treatment. Small upregulation of cytosolic and mitochondrial GPX6 was an exception. In comparison to the downregulated transcript level of the PRXs after short-term cold stress, the transcript level of *2-Cys-PRXA*, *PRXQ*, and *PRXIIE* increased after 14 days of cold stress treatment [6]. Transcripts for other enzymatic antioxidants like *APX*, *monodehydroascorbate reductase (MDAR)*, *dehydroascorbate reductase (DHAR)* and *SOD* slightly decreased during the low-temperature treatment, whereas an increase of transcript level was observed for *sAPX* and *MDAR* after 14 days of chilling stress [6]. Although the *APX* and *SOD* transcript levels decreased during the first 24 h of cold treatment, the activity of both enzymes increased [75]. Later on after 4–8 days, APX and SOD activity reached the activity of untreated plants. This could be due to posttranscriptional regulation [76]. It would be interesting to scrutinize the weak relationship of transcript levels and enzyme activity by quantification of APX and SOD protein levels.

ROS are continuously generated in the photosynthetic and respiratory electron transport chain and during photorespiration [77,78]. Previously, it was estimated that about 30% of the electrons of the photosynthetic electron transport chain flow into ROS metabolism [79,80]. These figures were recently challenged by Driever and Baker [81] who only detected low amounts of ROS generated by the photosynthetic ETC. The authors proposed that this ROS releases feature signalling functions rather than serving as a major alternative electron acceptor. Furthermore, ROS are generated under normal conditions at low rates by NADPH oxidases (RBOH) [82]. The transcript amounts of *RBOH* were downregulated, tentatively suggesting that the formation of ROS decreases at low temperatures. However, the activity and the protein level of these enzymes will have to be analyzed, since transcript levels poorly reflect the abundance or activity of their gene product [77].

In summary, it is striking that the main changes in transcript level only occurred after 24 h of low-temperature exposure. There were some exceptions: *APX2* is considered as a sensitive marker of oxidative stress and is upregulated under conditions of excess excitation of the photosynthetic apparatus [83]. Here, *APX2* was upregulated 2.4-fold after 30 min and maximally 5.6-fold after 6 h of cold treatment. Juszczak et al. [80] detected profound transcript changes after 14 d of cold treatment showing the persistence of metabolic disturbance at low temperatures. Thus, a global transcriptome analysis at later time points would be interesting to extend our analysis.

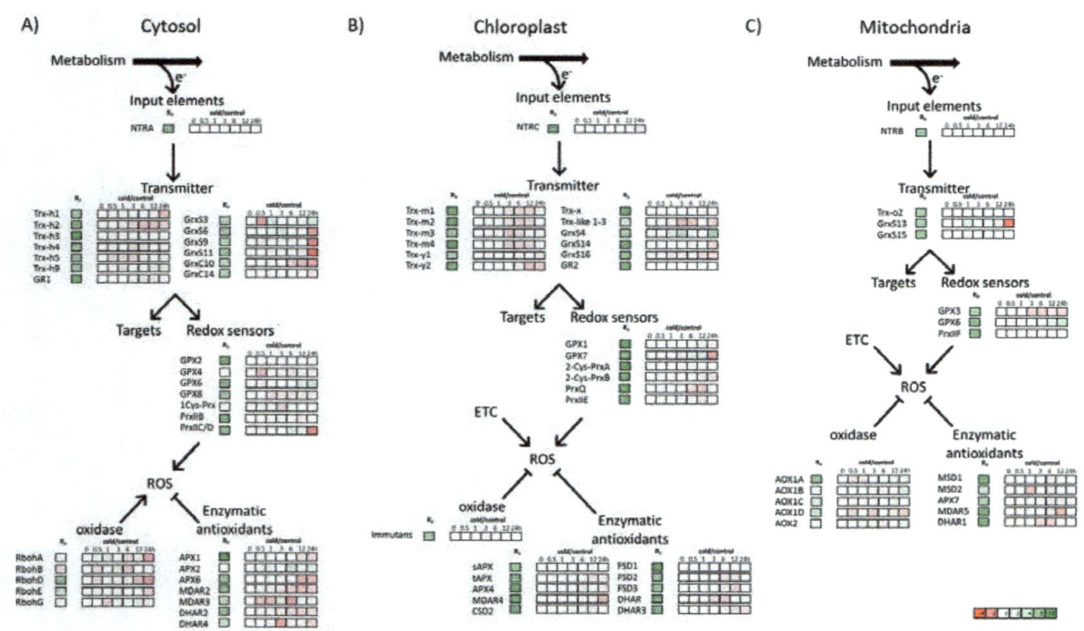

Figure 2. Time-dependent change in transcript levels of the redox regulatory network in the cytosol (**A**), chloroplast (**B**), and mitochondrion (**C**) during chilling stress. Relative transcript levels in *A. thaliana* were obtained from the Gene Expression Omnibus (GEO) Database of National Center for Biotechnology Information (NCBI) (Accession GSE5620 for control conditions and GSE5621 for low temperature treated plants) [8]. R0 shows the transcription level normalized by the Affymetrix system. The other boxes indicate the transcript levels in cold treated plants compared to the control conditions at the depicted time point. A complete list with the precise log2-fold changes is provided in the supplement. AOX: Alternative oxidase; APX: Ascorbate peroxidase; CSD: Cu/Zn-superoxide dismutase; DHAR: Dehydroascorbate reductase; ETC: electron transport chain; FSD: Fe-superoxide dismutase; GR: Glutathione reductase; GRX: Glutaredoxin; GPX: Glutathione peroxidase; MDAR: Monodehydroascorbate reductase; NTR: NADPH-dependent thioredoxin reductase; PRX: Peroxiredoxin; Rboh: Respiratory burst oxidase homolog protein; TRX: Thioredoxin. The assignment of subcellular localization of the proteins listed in this figure was based on the references [42–72], see also Table S1.

Furthermore, the cytosolic redox-regulatory network was stronger regulated than that of the chloroplast and mitochondria. The relatively small changes in transcript levels for chloroplast proteins of the network could be due to the setup in this particular experiment, where the light intensity was decreased from 150 µmol photons m^{-2} s^{-1} during growth to just 60 µmol photons m^{-2} s^{-1} during the low-temperature treatment [8]. The lowering of the incident photosynthetic active radiation probably minimizes the development of photoinhibition [84].

A. thaliana Col0 is sensitive to chilling stress and displays stunted growth, reduced metabolism and low biomass accumulation at low temperature [75]. Therefore, the transcriptional downregulation of the redox-regulatory network could participate in the chilling sensitivity of *A. thaliana* Col0.

6. Improvement of Cold Tolerance by Modulating the Redox-Network

Manipulation of the plant redox-network can enhance cold-stress tolerance [75,85–87]. Moon et al. [87] showed that overexpression of NTRC under control of the cauliflower mosaic 35S (CaMV35) promoter in *A. thaliana* enhances the tolerance to cold stress and freezing. These authors discussed NTRC and that it might affect the formation of secondary structures of mRNA at low temperatures to stimulate translation or protect nucleic acids against oxidative stress [87]. Since NTRC functions as a redox input element, it would be interesting to study the impact of NTRC overexpression on the redox network of the plastid including 2-Cys-PRX [17,82]. *2-Cys-PRXB*, and *NTRC* showed similar decreases in transcript levels during short-term chilling stress (Figure 2B).

Improved cold tolerance was observed in tomato (*Lycopersicon esculentum*) overexpressing tAPX [85]. The tAPX overexpressing lines also revealed disturbed GSH/GSSG ratios which could be due to the impact of chloroplast redox state on GSH biosynthesis [88]. Furthermore, the high tAPX level in the transgenic tomatoes resulted in a decreased photoinhibition of photosystem I and II under chilling stress [85]. Increased NTRC and tAPX activity will shift the redox state of the chloroplast to a more reduced state, either by stimulated feeding of electrons into the network (NTRC) or decreasing the electron drainage to ROS.

Shafi et al. [75] transferred both Cu/Zn-SOD from *Potentilla atrosanguinea* and APX from *Rheum australe* (with high similarity to the peroxisomal *APX* gene of *A. thaliana*) under control of the 35SCaMV promoter into *A. thaliana*. Since *P. atrosanguinea* and *R. australe* grow in alpine climate, the authors hypothesized that their enzymatic system might be better adapted to stress than *A. thaliana*. Total soluble sugars and proline content increased in the transgenic line, which could help to maintain the membrane integrity. As a result, the electrolyte leakage was decreased under cold treatment at 4 °C compared to the wildtype (WT) [75]. The accumulation of cytosolic $O_2^{\bullet-}$ and H_2O_2 should be suppressed in these plants and, indeed, less ROS accumulated in the transgenic lines.

The overexpression of the cytosolic redox transmitter AtGRXS17 in tomato resulted in an enhanced chilling tolerance [86]. The activities of SOD, CAT, and heme peroxidases were higher in the AtGRXS17 overexpressing lines compared to the WT, while the transcript levels were almost identical. These results indicate that the enhanced chilling tolerance mediated by AtGRXS17 is associated with the activity and/or stability, rather than the transcript level, of the enzymes. Other reasons for the enhanced chilling tolerance could be due to the increase in total soluble sugars, the reduced accumulation of H_2O_2, and the reduced electrolyte leakage [85]. This particular GRX was previously shown to link redox and ROS homeostasis with auxin signaling and development [89].

These findings demonstrate that improved cold tolerance can be engineered in plants by manipulating elements of the redox-regulatory network and underline the significance of the network in chilling stress acclimation.

7. Conclusions

The state of the thiol redox regulatory network is under the control of electron input by metabolism and electron drainage to ROS. ROS amounts, in turn, are controlled by the activity of ROS generator systems and antioxidant defense systems. Conclusive evidence shows that redox network elements control important features of the cold acclimation and recovery program. This review focused on the thiol-disulfide network and its interaction with ROS and the antioxidant defense system. However, other players, such as classical hormones including abscisic acid, salicylic acid, and oxylipins, interfere with acclimation to chilling temperatures. More recently novel players were discovered to affect cold acclimation. They include γ-amino butyric acid and melatonin [90]. Keeping this in mind, a few conclusions can be drawn for future directions.

(1)   Cold stress acclimation experiments often focus on leaves and photosynthetic metabolism. Response heterogeneity of different cell types has scarcely been addressed. Cell type-specific transcriptome, proteome, and metabolome analyses should reveal how other cell types respond to cold stress. But these approaches remain challenging and laborious.

(2)   Only a few methods allow researchers to address subcellular compartments. Transcriptome data provide easy access due to the predicted and often proven subcellular localization of the encoded gene products. This approach is straightforward and was applied here to the redox regulatory network. It would be interesting to see this type of data processing more frequently. However, the transcript amount is poorly linked to protein amount and activity. For a full understanding, we need compartment-specific proteomics and enzyme activity tests.

(3)   Metabolite-profiling of non-aqueous tissue fractions is another method which provides access to the major subcellular compartments. Non-aqueous fractions reflect the metabolic state of the compartments *in vivo* and are obtained from previously frozen and freeze-dried plant material like leaves [91]. This method was recently applied to cold-stress *A. thaliana* [92]. The latter study did not include metabolites with direct significance in the redox regulatory network.

(4)   Subcellular and cellular specificity can be addressed by imaging technologies detecting specific physicochemical properties such as Ca^{2+}-activity, specific compounds or the redox state of the glutathione system by using roGFP coupled to GRX [93]. The roGFP:GRX sensor can be targeted to different cell compartments and should be used to explore the glutathione redox state in dependence on cold stress intensity and duration.

(5)   To describe the state of the redox network in subcellular compartments, mathematical modeling and simulation combined with redox-proteomics for validation will be required. A pioneering modeling study presented a simulation of the fluxes through the ascorbate-dependent water-water cycle [94] and most recently, the thioredoxin oxidase-dependent inactivation of chloroplast enzymes was simulated [23]. Conceptually, cold stress appears to be an interesting target for this kind of simulation and prediction.

(6)   The question of acclimation and damage during the cold period is certainly of significant interest. However, the costs of priming and the speed of recovery likely play a major role when it comes to fitness and competitiveness. Thus, the report by Juszczak et al. [6] deserves attention as it provides clues on the advantages and disadvantages of expressing a strong antioxidant system.

Supplementary Materials: The following are available online at http://www.mdpi.com/2076-3921/7/11/169/s1, Table S1: Time-dependent changes in transcript levels of the redox network components.

Author Contributions: A.D. conceptualization, formal analysis, writing—original draft preparation, K.-J.D.; conceptualization, writing—Original draft preparation, project administration, funding acquisition.

Funding: The own research was funded by the Deutsche Foschungsgemeinschaft (Di 346, SPP1710).

Conflicts of Interest: The authors declare no conflict of interest.

References

1. Oakley, C.G.; Savage, L.; Lotz, S.; Larson, G.R.; Thomashow, M.F.; Kramer, D.M.; Schemske, D.W. Genetic basis of photosynthetic responses to cold in two locally adapted populations of Arabidopsis thaliana. *J. Exp. Bot.* **2018**, *69*, 699–709. [CrossRef] [PubMed]
2. Thomashow, M.F. PLANT COLD ACCLIMATION: Freezing Tolerance Genes and Regulatory Mechanisms. *Ann. Rev. Plant Physiol. Plant Mol. Biol.* **1999**, *50*, 571–599. [CrossRef] [PubMed]
3. Hayward, S.A.L.; Murray, P.A.; Gracey, A.Y.; Cossins, A.R. Beyond the Lipid Hypothesis. In *Molecular Aspects of the Stress Response: Chaperones, Membranes and Networks*; Csermely, P., Vígh, L., Eds.; Springer: New York, NY, USA, 2007; pp. 132–142.
4. Li, F.; Bian, C.S.; Xu, J.F.; Pang, W.F.; Liu, J.; Duan, S.G.; Lei, Z.-G.; Jiwan, P.; Jin, L.-P. Cloning and functional characterization of SAD genes in potato. *PloS ONE* **2015**, *10*, e0122036. [CrossRef] [PubMed]
5. Li, Z.-G.; Yuan, L.-X.; Wang, Q.-L.; Ding, Z.-L.; Dong, C.-Y. Combined action of antioxidant defense system and osmolytes in chilling shock-induced chilling tolerance in Jatropha curcas seedlings. *Acta Physiol. Plantarum* **2013**, *35*, 2127–2136. [CrossRef]
6. Juszczak, I.; Cvetkovic, J.; Zuther, E.; Hincha, D.K.; Baier, M. Natural Variation of Cold Deacclimation Correlates with Variation of Cold-Acclimation of the Plastid Antioxidant System in Arabidopsis thaliana Accessions. *Front. Plant Sci.* **2016**, *7*, 305. [CrossRef] [PubMed]
7. Li, X.; Cai, J.; Liu, F.; Dai, T.; Cao, W.; Jiang, D. Cold priming drives the sub-cellular antioxidant systems to protect photosynthetic electron transport against subsequent low temperature stress in winter wheat. *Plant Physiol. Biochem. PPB* **2014**, *82*, 34–43. [CrossRef] [PubMed]
8. Kilian, J.; Whitehead, D.; Horak, J.; Wanke, D.; Weinl, S.; Batistic, O.; D'Angelo, C.; Bornberg-Bauer, E.; Kudla, J.; Harter, K. The AtGenExpress global stress expression data set: Protocols, evaluation and model data analysis of UV-B light, drought and cold stress responses. *Plant J.* **2007**, *50*, 347–363. [CrossRef] [PubMed]
9. Wani, M.A.; Jan, N.; Qazi, H.A.; Andrabi, K.I.; John, R. Cold stress induces biochemical changes, fatty acid profile, antioxidant system and gene expression in Capsella bursa pastoris L. *Acta Physiol. Plantarum* **2018**, *40*, 167. [CrossRef]
10. Jan, N.; Majeed, U.; Andrabi, K.I.; John, R. Cold stress modulates osmolytes and antioxidant system in Calendula officinalis. *Acta Physiol. Plantarum* **2018**, *40*, 73. [CrossRef]
11. Spanò, C.; Bottega, S.; Ruffini Castiglione, M.; Pedranzani, H.E. Antioxidant response to cold stress in two oil plants of the genus Jatropha. *Plant Soil Environ.* **2017**, *63*, 271–276. [CrossRef]
12. Han, Q.-H.; Huang, B.; Ding, C.-B.; Zhang, Z.-W.; Chen, Y.-E.; Hu, C.; Zhou, L.-J.; Huang, Y.; Liao, J.-Q.; Yuan, S.; et al. Effects of Melatonin on Anti-oxidative Systems and Photosystem II in Cold-Stressed Rice Seedlings. *Front. Plant Sci.* **2017**, *8*, 785. [CrossRef] [PubMed]
13. Ortiz, D.; Hu, J.; Salas Fernandez, M.G. Genetic architecture of photosynthesis in Sorghum bicolor under non-stress and cold stress conditions. *J. Exp. Bot.* **2017**, *68*, 4545–4557. [CrossRef] [PubMed]
14. Dietz, K.-J. Redox signal integration: From stimulus to networks and genes. *Physiol. Plant* **2008**, *133*, 459–468. [CrossRef] [PubMed]
15. Buchanan, B.B. The Path to Thioredoxin and Redox Regulation in Chloroplasts. *Ann. Rev. Plant Biol.* **2016**, *67*, 1–24. [CrossRef] [PubMed]
16. Reichheld, J.-P.; Khafif, M.; Riondet, C.; Droux, M.; Bonnard, G.; Meyer, Y. Inactivation of thioredoxin reductases reveals a complex interplay between thioredoxin and glutathione pathways in Arabidopsis development. *Plant Cell* **2007**, *19*, 1851–1865. [CrossRef] [PubMed]
17. Serrato, A.J.; Pérez-Ruiz, J.M.; Spínola, M.C.; Cejudo, F.J. A novel NADPH thioredoxin reductase, localized in the chloroplast, which deficiency causes hypersensitivity to abiotic stress in Arabidopsis thaliana. *J. Biol. Chem.* **2004**, *279*, 43821–43827. [CrossRef] [PubMed]
18. Keryer, E.; Collin, V.; Lavergne, D.; Lemaire, S.; Issakidis-Bourguet, E. Characterization of Arabidopsis Mutants for the Variable Subunit of Ferredoxin:thioredoxin Reductase. *Photosynthes. Res.* **2004**, *79*, 265–274. [CrossRef] [PubMed]
19. Okegawa, Y.; Motohashi, K. Chloroplastic thioredoxin m functions as a major regulator of Calvin cycle enzymes during photosynthesis in vivo. *Plant J.* **2015**, *84*, 900–913. [CrossRef] [PubMed]
20. Raines, C.A. The Calvin cycle revisited. *Photosynthes. Res.* **2003**, *75*, 1–10. [CrossRef] [PubMed]

21. Flohe, L. The impact of thiol peroxidases on redox regulation. *Free Rad. Res.* **2016**, *50*, 126–142. [CrossRef] [PubMed]

22. Dietz, K.-J. Thiol-Based Peroxidases and Ascorbate Peroxidases: Why Plants Rely on Multiple Peroxidase Systems in the Photosynthesizing Chloroplast? *Mol. Cells* **2016**, *39*, 20–25. [CrossRef] [PubMed]

23. Vaseghi, M.-J.; Chibani, K.; Telman, W.; Liebthal, M.F.; Gerken, M.; Schnitzer, H.; Mueller, S.M.; Dietz, K.-J. The chloroplast 2-cysteine peroxiredoxin functions as thioredoxin oxidase in redox regulation of chloroplast metabolism. *eLife* **2018**, *7*. [CrossRef] [PubMed]

24. Hossain, M.S.; Dietz, K.-J. Tuning of Redox Regulatory Mechanisms, Reactive Oxygen Species and Redox Homeostasis under Salinity Stress. *Front. Plant Sci.* **2016**, *7*, 548. [CrossRef] [PubMed]

25. Shi, H.; Ye, T.; Zhong, B.; Liu, X.; Chan, Z. Comparative proteomic and metabolomic analyses reveal mechanisms of improved cold stress tolerance in bermudagrass (*Cynodon dactylon* (L.) Pers.) by exogenous calcium. *J. Integrat. Plant Biol.* **2014**, *56*, 1064–1079. [CrossRef] [PubMed]

26. Asada, K. THE WATER-WATER CYCLE IN CHLOROPLASTS: Scavenging of Active Oxygens and Dissipation of Excess Photons. *Ann. Rev. Plant Physiol. Plant Mol. Biol.* **1999**, *50*, 601–639. [CrossRef] [PubMed]

27. Foyer, C.H.; Shigeoka, S. Understanding oxidative stress and antioxidant functions to enhance photosynthesis. *Plant Physiol.* **2011**, *155*, 93–100. [CrossRef] [PubMed]

28. Hossain, M.S.; ElSayed, A.I.; Moore, M.; Dietz, K.-J. Redox and Reactive Oxygen Species Network in Acclimation for Salinity Tolerance in Sugar Beet. *J. Exp. Bot.* **2017**, *68*, 1283–1298. [CrossRef] [PubMed]

29. Yan, S.-P.; Zhang, Q.-Y.; Tang, Z.-C.; Su, W.-A.; Sun, W.-N. Comparative proteomic analysis provides new insights into chilling stress responses in rice. *Mol. Cell. Proteom. MCP* **2006**, *5*, 484–496. [CrossRef] [PubMed]

30. Soliman, M.H.; Alayafi, A.A.M.; El Kelish, A.A.; Abu-Elsaoud, A.M. Acetylsalicylic acid enhance tolerance of Phaseolus vulgaris L. to chilling stress, improving photosynthesis, antioxidants and expression of cold stress responsive genes. *Botan. Stud.* **2018**, *59*, 6. [CrossRef] [PubMed]

31. Mir, B.A.; Mir, S.A.; Khazir, J.; Tonfack, L.B.; Cowan, D.A.; Vyas, D.; Koul, S. Cold stress affects antioxidative response and accumulation of medicinally important withanolides in *Withania somnifera* (L.) Dunal. *Ind. Crops Prod.* **2015**, *74*, 1008–1016. [CrossRef]

32. Wang, X.; Fang, G.; Li, Y.; Ding, M.; Gong, H.; Li, Y. Differential antioxidant responses to cold stress in cell suspension cultures of two subspecies of rice. *Plant Cell Tiss Organ Cult* **2013**, *113*, 353–361. [CrossRef]

33. Glaszmann, J.C.; Kaw, R.N.; Khush, G.S. Genetic divergence among cold tolerant rices (*Oryza sativa* L.). *Euphytica* **1990**, *45*, 95–104. [CrossRef]

34. Cheng, F.; Lu, J.; Gao, M.; Shi, K.; Kong, Q.; Huang, Y.; Bie, Z. Redox Signaling and CBF-Responsive Pathway Are Involved in Salicylic Acid-Improved Photosynthesis and Growth under Chilling Stress in Watermelon. *Front. Plant Sci.* **2016**, *7*, 1519. [CrossRef] [PubMed]

35. Dietz, K.J.; Horling, F.; Konig, J.; Baier, M. The function of the chloroplast 2-cysteine peroxiredoxin in peroxide detoxification and its regulation. *J. Exp. Bot.* **2002**, *53*, 1321–1329.

36. Dietz, K.-J. Peroxiredoxins in plants and cyanobacteria. *Antioxid. Redox Signal.* **2011**, *15*, 1129–1159. [CrossRef] [PubMed]

37. Dat, J.; Vandenabeele, S.; Vranova, E.; van Montagu, M.; Inze, D.; van Breusegem, F. Dual action of the active oxygen species during plant stress responses. *Cell. Mol. Life Sci.* **2000**, *57*, 779–795. [CrossRef] [PubMed]

38. Cavalcanti, F.R.; Oliveira, J.T.A.; Martins-Miranda, A.S.; Viégas, R.A.; Silveira, J.A.G. Superoxide dismutase, catalase and peroxidase activities do not confer protection against oxidative damage in salt-stressed cowpea leaves. *New Phytol.* **2004**, *163*, 563–571. [CrossRef]

39. Baier, M.; Pitsch, N.T.; Mellenthin, M.; Guo, W. Regulation of Genes Encoding Chloroplast Antioxidant Enzymes in Comparison to Regulation of the Extra-plastidic Antioxidant Defense System. In *Ascorbate-Glutathione Pathway and Stress Tolerance in Plants*; Anjum, N.A., Chan, M.-T., Umar, S., Eds.; Springer Netherlands: Dordrecht, The Netherland, 2010.

40. Muthuramalingam, M.; Matros, A.; Scheibe, R.; Mock, H.-P.; Dietz, K.-J. The hydrogen peroxide-sensitive proteome of the chloroplast in vitro and in vivo. *Front Plant Sci.* **2013**, *4*, 54. [CrossRef] [PubMed]

41. Koç, I.; Yuksel, I.; Caetano-Anollés, G. Metabolite-Centric Reporter Pathway and Tripartite Network Analysis of Arabidopsis Under Cold Stress. *Front Bioeng. Biotechnol.* **2018**, *6*, 121. [CrossRef] [PubMed]

42. Kliebenstein, D.J.; Monde, R.-A.; Last, R.L. Superoxide Dismutase in Arabidopsis: An Eclectic Enzyme Family with Disparate Regulation and Protein Localization. *Plant Physiol.* **1998**, *118*, 637–650. [CrossRef] [PubMed]

43. Huang, C.-H.; Kuo, W.-Y.; Weiss, C.; Jinn, T.-L. Copper Chaperone-Dependent and -Independent Activation of Three Copper-Zinc Superoxide Dismutase Homologs Localized in Different Cellular Compartments in Arabidopsis. *Plant Physiol.* **2012**, *158*, 737–746. [CrossRef] [PubMed]

44. Panchuk, I.I.; Volkov, R.A.; Schöffl, F. Heat Stress- and Heat Shock Transcription Factor-Dependent Expression and Activity of Ascorbate Peroxidase in Arabidopsis. *Plant Physiol.* **2002**, *129*, 838–853. [CrossRef] [PubMed]

45. Mittler, R.; Poulos, T.L. Ascorbate Peroxidase. In *Antioxidants and Reactive Oxygen Species in Plants*; Wiley-Blackwell: Hoboken, NJ, USA, 2007; pp. 87–100.

46. Lisenbee, C.S.; Lingard, M.J.; Trelease, R.N. Arabidopsis peroxisomes possess functionally redundant membrane and matrix isoforms of monodehydroascorbate reductase. *Plant J.* **2005**, *43*, 900–914. [CrossRef] [PubMed]

47. Obara, K.; Sumi, K.; Fukuda, H. The Use of Multiple Transcription Starts Causes the Dual Targeting of Arabidopsis Putative Monodehydroascorbate Reductase to Both Mitochondria and Chloroplasts. *Plant Cell Physiol.* **2002**, *43*, 697–705. [CrossRef] [PubMed]

48. Chew, O.; Whelan, J.; Millar, A.H. Molecular definition of the ascorbate-glutathione cycle in Arabidopsis mitochondria reveals dual targeting of antioxidant defenses in plants. *J. Biol. Chem.* **2003**, *278*, 46869–46877. [CrossRef] [PubMed]

49. Reumann, S.; Quan, S.; Aung, K.; Yang, P.; Manandhar-Shrestha, K.; Holbrook, D.; Linka, N.; Switzenberg, R.; Wilkerson, C.G.; Weber, A.P.M.; et al. In-depth proteome analysis of Arabidopsis leaf peroxisomes combined with in vivo subcellular targeting verification indicates novel metabolic and regulatory functions of peroxisomes. *Plant Physiol.* **2009**, *150*, 125–143. [CrossRef] [PubMed]

50. Dixon, D.P.; Davis, B.G.; Edwards, R. Functional divergence in the glutathione transferase superfamily in plants. Identification of two classes with putative functions in redox homeostasis in Arabidopsis thaliana. *J. biol. chem.* **2002**, *277*, 30859–30869. [CrossRef] [PubMed]

51. Trivedi, D.K.; Gill, S.S.; Yadav, S.; Tuteja, N. Genome-wide analysis of glutathione reductase (GR) genes from rice and Arabidopsis. *Plant Signal. Behav.* **2012**, *8*, e23021. [CrossRef] [PubMed]

52. Mhamdi, A.; Queval, G.; Chaouch, S.; Vanderauwera, S.; van Breusegem, F.; Noctor, G. Catalase function in plants: A focus on Arabidopsis mutants as stress-mimic models. *J. Exp. Bot.* **2010**, *61*, 4197–4220. [CrossRef] [PubMed]

53. Milla, M.A.R.; Maurer, A.; Huete, A.R.; Gustafson, J.P. Glutathione peroxidase genes in Arabidopsis are ubiquitous and regulated by abiotic stresses through diverse signaling pathways. *Plant J.* **2003**, *36*, 602–615. [CrossRef]

54. Sagi, M.; Fluhr, R. Superoxide Production by Plant Homologues of the gp91phox NADPH Oxidase. Modulation of Activity by Calcium and by Tobacco Mosaic Virus Infection. *Plant Physiol.* **2001**, *126*, 1281–1290. [CrossRef] [PubMed]

55. Heazlewood, J.L.; Tonti-Filippini, J.S.; Gout, A.M.; Day, D.A.; Whelan, J.; Millar, A.H. Experimental analysis of the Arabidopsis mitochondrial proteome highlights signaling and regulatory components, provides assessment of targeting prediction programs, and indicates plant-specific mitochondrial proteins. *Plant Cell Online* **2004**, *16*, 241–256. [CrossRef] [PubMed]

56. Konert, G.; Trotta, A.; Kouvonen, P.; Rahikainen, M.; Durian, G.; Blokhina, O.; Fagerstedt, K.; Muth, D.; Corthals, G.L.; Kangasjärvi, S. Protein phosphatase 2A (PP2A) regulatory subunit B′γ interacts with cytoplasmic ACONITASE 3 and modulates the abundance of AOX1A and AOX1D in Arabidopsis thaliana. *New Phytol.* **2014**, *205*, 1250–1263. [CrossRef] [PubMed]

57. Saisho, D.; Nakazono, M.; Lee, K.-H.; Tsutsumi, N.; Akita, S.; Hirai, A. The gene for alternative oxidase-2 (AOX2) from Arabidopsis thaliana consists of five exons unlike other AOX genes and is transcribed at an early stage during germination. *Genes Genet. Syst.* **2001**, *76*, 89–97. [CrossRef]

58. Lennon, A.M.; Prommeenate, P.; Nixon, P.J. Location, expression and orientation of the putative chlororespiratory enzymes, Ndh and IMMUTANS, in higher-plant plastids. *Planta* **2003**, *218*, 254–260. [CrossRef] [PubMed]

59. Haslekås, C.; Viken, M.K.; Grini, P.E.; Nygaard, V.; Nordgard, S.H.; Meza, T.J.; Aalen, R.B. Seed 1-Cysteine Peroxiredoxin Antioxidants Are Not Involved in Dormancy, But Contribute to Inhibition of Germination during Stress. *Plant Physiol.* **2003**, *133*, 1148–1157. [CrossRef] [PubMed]

60. König, J.; Baier, M.; Horling, F.; Kahmann, U.; Harris, G.; Schürmann, P.; Dietz, K.-J. The plant-specific function of 2-Cys peroxiredoxin-mediated detoxification of peroxides in the redox-hierarchy of photosynthetic electron flux. *Proc. Natl. Acad. Sci. USA* **2002**, *99*, 5738–5743. [CrossRef] [PubMed]

61. Lamkemeyer, P.; Laxa, M.; Collin, V.; Li, W.; Finkemeier, I.; Schöttler, M.A.; Holtkamp, V.; Tognetti, V.B.; Issakidis-Bourguet, E.; Kandlbinder, A.; et al. Peroxiredoxin Q of Arabidopsis thaliana is attached to the thylakoids and functions in context of photosynthesis[†]. *Plant J.* **2006**, *45*, 968–981. [CrossRef] [PubMed]

62. Bréhélin, C.; Meyer, E.H.; de Souris, J.-P.; Bonnard, G.; Meyer, Y. Resemblance and Dissemblance of Arabidopsis Type II Peroxiredoxins: Similar Sequences for Divergent Gene Expression, Protein Localization, and Activity. *Plant Physiol.* **2003**, *132*, 2045–2057. [CrossRef] [PubMed]

63. Finkemeier, I.; Goodman, M.; Lamkemeyer, P.; Kandlbinder, A.; Sweetlove, L.J.; Dietz, K.-J. The mitochondrial type II peroxiredoxin F is essential for redox homeostasis and root growth of Arabidopsis thaliana under stress. *J. Biol. Chem.* **2005**, *280*, 12168–12180. [CrossRef] [PubMed]

64. Meyer, Y.; Siala, W.; Bashandy, T.; Riondet, C.; Vignols, F.; Reichheld, J.P. Glutaredoxins and thioredoxins in plants. *Biochim. Biophys. Acta* **2008**, *1783*, 589–600. [CrossRef] [PubMed]

65. Reichheld, J.-P.; Meyer, E.; Khafif, M.; Bonnard, G.; Meyer, Y. AtNTRB is the major mitochondrial thioredoxin reductase in Arabidopsis thaliana. *FEBS Lett.* **2005**, *579*, 337–342. [CrossRef] [PubMed]

66. Dangoor, I.; Peled-Zehavi, H.; Levitan, A.; Pasand, O.; Danon, A. A small family of chloroplast atypical thioredoxins. *Plant Physiol.* **2009**, *149*, 1240–1250. [CrossRef] [PubMed]

67. Lemaire, S.D. The Glutaredoxin Family in Oxygenic Photosynthetic Organisms. *Photosynthes. Res.* **2004**, *79*, 305–318. [CrossRef] [PubMed]

68. Cheng, N.-H. AtGRX4, an Arabidopsis chloroplastic monothiol glutaredoxin, is able to suppress yeast grx5 mutant phenotypes and respond to oxidative stress. *FEBS Lett.* **2008**, *582*, 848–854. [CrossRef] [PubMed]

69. Cheng, N.-H.; Liu, J.-Z.; Brock, A.; Nelson, R.S.; Hirschi, K.D. AtGRXcp, an Arabidopsis chloroplastic glutaredoxin, is critical for protection against protein oxidative damage. *J. Biol. Chem.* **2006**, *281*, 26280–26288. [CrossRef] [PubMed]

70. Liu, X.; Liu, S.; Feng, Y.; Liu, J.-Z.; Chen, Y.; Pham, K.; Deng, H.; Hirschi, K.D.; Wang, X.; Cheng, N. Structural insights into the N-terminal GIY-YIG endonuclease activity of Arabidopsis glutaredoxin AtGRXS16 in chloroplasts. *Proc. Natl. Acad. Sci. USA* **2013**, *110*, 9565–9570. [CrossRef] [PubMed]

71. Li, S.; Lauri, A.; Ziemann, M.; Busch, A.; Bhave, M.; Zachgo, S. Nuclear activity of ROXY1, a glutaredoxin interacting with TGA factors, is required for petal development in Arabidopsis thaliana. *Plant Cell Online* **2009**, *21*, 429–441. [CrossRef] [PubMed]

72. Murmu, J.; Bush, M.J.; DeLong, C.; Li, S.; Xu, M.; Khan, M.; Malcolmson, C.; Fobert, P.R.; Zachgo, S.; Hepworth, S.R. Arabidopsis basic leucine-zipper transcription factors TGA9 and TGA10 interact with floral glutaredoxins ROXY1 and ROXY2 and are redundantly required for anther development. *Plant Physiol.* **2010**, *154*, 1492–1504. [CrossRef] [PubMed]

73. König, J.; Muthuramalingam, M.; Dietz, K.-J. Mechanisms and dynamics in the thiol/disulfide redox regulatory network: Transmitters, sensors and targets. *Curr. Opin. Plant Biol.* **2012**, *15*, 261–268. [CrossRef] [PubMed]

74. Meyer, Y.; Buchanan, B.B.; Vignols, F.; Reichheld, J.-P. Thioredoxins and Glutaredoxins: Unifying Elements in Redox Biology. *Ann. Rev. Genet.* **2009**, *43*, 335–367. [CrossRef] [PubMed]

75. Shafi, A.; Dogra, V.; Gill, T.; Ahuja, P.S.; Sreenivasulu, Y. Simultaneous over-expression of PaSOD and RaAPX in transgenic Arabidopsis thaliana confers cold stress tolerance through increase in vascular lignifications. *PloS one* **2014**, *9*, e110302. [CrossRef] [PubMed]

76. Schwanhäusser, B.; Busse, D.; Li, N.; Dittmar, G.; Schuchardt, J.; Wolf, J.; Chen, W.; Selbach, M. Global quantification of mammalian gene expression control. *Nature* **2011**, *473*, 337. [CrossRef] [PubMed]

77. Mittler, R.; Vanderauwera, S.; Gollery, M.; van Breusegem, F. Reactive oxygen gene network of plants. *Trends Plant Sci.* **2004**, *9*, 490–498. [CrossRef] [PubMed]

78. Apel, K.; Hirt, H. REACTIVE OXYGEN SPECIES: Metabolism, Oxidative Stress, and Signal Transduction. *Ann. Rev. Plant Biol.* **2004**, *55*, 373–399. [CrossRef] [PubMed]

79. Miyake, C.; Michihata, F.; Asada, K. Scavenging of Hydrogen Peroxide in Prokaryotic and Eukaryotic Algae: Acquisition of Ascorbate Peroxidase during the Evolution of Cyanobacteria. *Plant Cell Physiol.* **1991**, *32*, 33–43. [CrossRef]

80. Juszczak, I.; Rudnik, R.; Pietzenuk, B.; Baier, M. Natural genetic variation in the expression regulation of the chloroplast antioxidant system among Arabidopsis thaliana accessions. *Physiol. Plantarum* **2012**, *146*, 53–70. [CrossRef] [PubMed]

81. Driever, S.M.; Baker, N.R. The water-water cycle in leaves is not a major alternative electron sink for dissipation of excess excitation energy when CO(2) assimilation is restricted. *Plant Cell Environ.* **2011**, *34*, 837–846. [CrossRef] [PubMed]

82. Liebthal, M.; Maynard, D.; Dietz, K.-J. Peroxiredoxins and Redox Signaling in Plants. *Antioxid. Redox Signal.* **2018**, *28*, 609–624. [CrossRef] [PubMed]

83. Karpinski, S.; Reynolds, H.; Karpinska, B.; Wingsle, G.; Creissen, G.; Mullineaux, P. Systemic Signaling and Acclimation in Response to Excess Excitation Energy in Arabidopsis. *Science* **1999**, *284*, 654–657. [CrossRef] [PubMed]

84. Krause, G.H.; Somersalo, S.; Osmond, C.B.; Briantais, J.-M.; Schreiber, U. Fluorescence as a Tool in Photosynthesis Research: Application in Studies of Photoinhibition, Cold Acclimation and Freezing Stress [and Discussion]. *Philosoph. Trans. Royal Soc. London B: Biol. Sci.* **1989**, *323*, 281–293. [CrossRef]

85. Duan, M.; Feng, H.-L.; Wang, L.-Y.; Li, D.; Meng, Q.-W. Overexpression of thylakoidal ascorbate peroxidase shows enhanced resistance to chilling stress in tomato. *J. Plant Physiol.* **2012**, *169*, 867–877. [CrossRef] [PubMed]

86. Hu, Y.; Wu, Q.; Sprague, S.A.; Park, J.; Oh, M.; Rajashekar, C.B.; Koiwa, H.; Nakata, P.A.; Cheng, N.; Hirschi, K.D.; et al. Tomato expressing Arabidopsis glutaredoxin gene AtGRXS17 confers tolerance to chilling stress via modulating cold responsive components. *Horticul. Res.* **2015**, *2*, 15051. [CrossRef] [PubMed]

87. Moon, J.C.; Lee, S.; Shin, S.Y.; Chae, H.B.; Jung, Y.J.; Jung, H.S.; Lee, K.O.; Lee, J.R.; Lee, S.Y. Overexpression of Arabidopsis NADPH-dependent thioredoxin reductase C (AtNTRC) confers freezing and cold shock tolerance to plants. *Biochem. Biophys. Res. Commun.* **2015**, *463*, 1225–1229. [CrossRef] [PubMed]

88. Maruta, T.; Tanouchi, A.; Tamoi, M.; Yabuta, Y.; Yoshimura, K.; Ishikawa, T.; Shigeoka, S. Arabidopsis Chloroplastic Ascorbate Peroxidase Isoenzymes Play a Dual Role in Photoprotection and Gene Regulation under Photooxidative Stress. *Plant Cell Physiol.* **2010**, *51*, 190–200. [CrossRef] [PubMed]

89. Cheng, N.-H.; Liu, J.-Z.; Liu, X.; Wu, Q.; Thompson, S.M.; Lin, J.; Chang, J.; Whitham, S.A.; Park, S.; Cohen, J.D.; et al. Arabidopsis monothiol glutaredoxin, AtGRXS17, is critical for temperature-dependent postembryonic growth and development via modulating auxin response. *J. Biol. Chem.* **2011**, *286*, 20398–20406. [CrossRef] [PubMed]

90. Baier, M.; Bittner, A.; Prescher, A.; van Buer, J. Preparing plants for improved cold tolerance by priming. *Plant Cell Environ.* **2018**. [CrossRef] [PubMed]

91. Dietz, K.-J. Subcellular metabolomics: The choice of method depends on the aim of the study. *J. Exp. Bot.* **2017**, *68*, 5695–5698. [CrossRef] [PubMed]

92. Hoermiller, I.I.; Naegele, T.; Augustin, H.; Stutz, S.; Weckwerth, W.; Heyer, A.G. Subcellular reprogramming of metabolism during cold acclimation in Arabidopsis thaliana. *Plant Cell Environ.* **2017**, *40*, 602–610. [CrossRef] [PubMed]

93. Meyer, A.J.; Brach, T.; Marty, L.; Kreye, S.; Rouhier, N.; Jacquot, J.-P.; Hell, R. Redox-sensitive GFP in Arabidopsis thaliana is a quantitative biosensor for the redox potential of the cellular glutathione redox buffer. *Plant J.* **2007**, *52*, 973–986. [CrossRef] [PubMed]

94. Polle, A. Dissecting the Superoxide Dismutase-Ascorbate-Glutathione-Pathway in Chloroplasts by Metabolic Modeling. Computer Simulations as a Step towards Flux Analysis. *Plant Physiol.* **2001**, *126*, 445–462. [CrossRef] [PubMed]

antioxidants

MDPI

Article

Crystal Structure of the Apo-Form of NADPH-Dependent Thioredoxin Reductase from a Methane-Producing Archaeon

Rubén M. Buey [1], Ruth A. Schmitz [2], Bob B. Buchanan [3] and Monica Balsera [4,*]

[1]   Metabolic Engineering Group. Dpto. Microbiología y Genética. Universidad de Salamanca, 37007 Salamanca, Spain; ruben.martinez@usal.es
[2]   Institut für Allgemeine Mikrobiologie, Christian-Albrechts-Universität Kiel, 24118 Kiel, Germany; rschmitz@ifam.uni-kiel.de
[3]   Department of Plant & Microbial Biology, University of California, 94720 Berkeley CA, USA; view@berkeley.edu
[4]   Instituto de Recursos Naturales y Agrobiología de Salamanca (IRNASA-CSIC), 37008 Salamanca, Spain
*   Correspondence: monica.balsera@csic.es

Received: 10 October 2018; Accepted: 14 November 2018; Published: 17 November 2018

Abstract: The redox regulation of proteins via reversible dithiol/disulfide exchange reactions involves the thioredoxin system, which is composed of a reductant, a thioredoxin reductase (TR), and thioredoxin (Trx). In the pyridine nucleotide-dependent Trx reduction pathway, reducing equivalents, typically from reduced nicotinamide adenine dinucleotide phosphate (NADPH), are transferred from NADPH-TR (NTR) to Trx and, in turn, to target proteins, thus resulting in the reversible modification of the structural and functional properties of the targets. NTR enzymes contain three functional sites: an NADPH binding pocket, a non-covalently bound flavin cofactor, and a redox-active disulfide in the form of CxxC. With the aim of increasing our knowledge of the thioredoxin system in archaea, we here report the high-resolution crystal structure of NTR from the methane-generating organism *Methanosarcina mazei* strain Gö1 (MmNTR) at 2.6 Å resolution. Based on the crystals presently described, MmNTR assumes an overall fold that is nearly identical to the archetypal fold of authentic NTRs; however, surprisingly, we observed no electron density for flavin adenine dinucleotide (FAD) despite the well-defined and conserved FAD-binding cavity in the folded module. Remarkably, the dimers of the apo-protein within the crystal were different from those observed by small angle X-ray scattering (SAXS) for the holo-protein, suggesting that the binding of the flavin cofactor does not require major protein structural rearrangements. Rather, binding results in the stabilization of essential parts of the structure, such as those involved in dimer stabilization. Altogether, this structure represents the example of an apo-form of an NTR that yields important insight into the effects of the cofactor on protein folding.

Keywords: redox active site; thioredoxin; disulfide; flavin; NADPH; X-ray crystallography; SAXS; methanoarchaea

---

1. Introduction

Thioredoxins (Trxs) participate in dithiol/disulfide exchange reactions that effect structural and functional changes in target proteins via reversible change in the redox state of selected cysteine (Cys) residues. Canonical members of the Trx family contain the common amino acid sequence motif WCGPC in the active site with two strictly conserved Cys that target proteins capable of reversibly forming a disulfide bridge. The Cys are reduced to the sulfhydryl form by a Trx reductase (TR) [1]. In many cases, TRs are dependent on reduced nicotinamide adenine dinucleotide phosphate

(NADPH) (NADPH-dependent TR or NTR). Based on phylogenetics, protein structure, and molecular mechanism, two groups of NTRs are recognized: (i) Bacteria, plants, lower eukaryotes, and archaea contain low-molecular weight (LMW)-NTRs of about 35 kDa with a non-covalently bound flavin adenine dinucleotide (FAD) cofactor. These enzymes accept reducing equivalents from NADPH and transfer them to a redox-active disulfide motif (CxxC) that subsequently reduces Trx to its sulfhydryl form in a dithiol–disulfide exchange reaction. These transfer reactions involve a large conformational rearrangement within the enzyme [2]; (ii) High-molecular-weight (HMW)-NTRs are enzymes of about 55 kDa present in eukaryotes that have developed an alternate mechanism for the transfer of reducing equivalents to Trx. Also homodimers with a non-covalently bound FAD, these enzymes are fitted with a third redox-active motif that contains either a Cys or selenocysteine at a mobile C-terminal tail. They have the capacity to transfer reducing equivalents from the CxxC motif to the disulfide in Trx [3].

The structural dynamics and functional mechanisms of NTRs have been extensively studied in bacteria and eukaryotes. In archaea, functional studies of the Trx system are available for several species, including *Aeropyrum pernix* [4], *Pyrococcus horikoshii* [5], *Methanosarcina acetivorans* [6,7], and *Sulfolobus solfataricus* [8]. The archaeal NADPH-linked Trx system is biochemically similar to that of bacteria. However, dissimilarities were detected in *Methanocaldococcus jannaschii* [9] and *Thermoplasma acidophilum* [10], where the thioredoxin system (TS) functions independently of NADPH. Here, the coenzyme F420 provides reducing equivalents for *M. jannaschii* TR (MjTR). *T. acidophilum* TR (TaTR), by contrast, is linked to NADH pyridine [10,11].

The present study was conducted (i) to apply in silico analysis for identifying members of the TS in the mesophilic *Methanosarcina mazei*, and (ii) to determine the X-ray crystal structure of its NTR. *M. mazei* is an anaerobic ecologically important methanoarchaeon with a broad substrate spectrum, able to generate methane using H_2 and CO_2, methanol, or methylamines as well as acetate as an energy and a carbon source [12,13]. It represents a model organism for the methanoarchaea.

2. Materials and Methods

2.1. MmNTR Protein Production and Purification

The full-length NTR of *M. mazei* strain Gö1 (MM2353) was cloned into pET28a, resulting in a cleavable N-terminal 6xHis-tagged fusion protein. Protein was produced in BL21pRIL *Escherichia coli* cells upon induction with 0.2 mM of isopropyl β-D-1-thiogalactopyranoside (IPTG) at 22 °C. Cells were disrupted by sonication in 20 mM Tris-HCl buffer, pH 7.6, containing 300 mM NaCl, 10% glycerol, and 1 mM phenylmethylsulfonyl fluoride. The crude extract was clarified by centrifugation at 40,000× *g* for 1 h. The enzyme was purified using nickel-nitriloacetic acid (Ni-NTA) chromatography (GE Healthcare) and digested with thrombin (EMD Chemicals, San Diego, CA, USA) in buffer, 20 mM Tris-HCl pH 8, 150 mM NaCl, and 2 mM CaCl₂, at room temperature overnight. The last step of purification involved gel filtration chromatography (Sephacryl S-300) using 10 mM Tris-HCl pH 7.6, 100 mM NaCl, and 2 mM 2-mercaptoethanol buffer. For protein molecular weight estimation by size exclusion chromatography, the column was calibrated with the Gel Filtration Standard mixture from Bio-Rad.

2.2. Protein Crystallization, Data Collection, and Structure Solution

One microliter of a 15 mg/mL protein solution in buffer, 10 mM Tris-HCl pH 8 and 50 mM NaCl, was mixed with 1 μL of a solution containing 0.1 M sodium acetate, pH 4.8, and 1 M ammonium chloride. Protein crystals appeared after 4 days of incubation at 25 °C on a sitting drop crystallization plate. Crystals were transferred to a solution containing Paratone Oil before flash freezing. Diffraction data were collected at a wavelength of 1 Å on the X06DA beamline of the Swiss Light Source synchrotron (SLS, Villigen, Switzerland). The diffraction data were processed and scaled using XDS and XSCALE programs [14]. The MmNTR structure was determined using an approach integrating molecular replacement using Phaser [15] and automated model building. Initially, a homology model of the NADPH-binding domain was used as search tool. The template was generated by FFAS03 server [16] based on the thioredoxin reductase structure of *Thermoplasma acidophilum* (Protein Data Bank (PDB) code 3CTY) [10], freed of non-conserved loops and side chains. Phenix.autobuild [17] was then used for automated model building of the FAD-binding domain. The final structure was refined using the Phenix crystallographic software suite [17], alternating with visual inspection of the electron density maps and manual modeling with Coot [18]. Data collection and refinement statistics are summarized in Table S1. The structures were rendered using Pymol [19]. Atomic coordinates and crystallographic structure factors have been deposited in the PDB under the accession code 4ZN0.

2.3. SAXS

SAXS data were collected at the B21 beamline at the Diamond Light Source using an online SEC-HPLC system coupled SAXS setup. In line SEC-SAXS was performed using an Agilent 1200 HPLC system equipped with a 2.4 mL Superdex 200 (GE Healthcare Life Sciences, Amersham, UK) column. Forty-five microliters of a protein sample at 5 mg/mL were loaded onto the size exclusion column equilibrated in 20 mM Tris-HCl pH 7.6, 150 mM NaCl, and 2 mM DTT. Data were collected all along the chromatogram at 3 s per frame, and recorded on a Pilatus 2M detector with a fixed camera length of 4.014 m, covering a scattering vector (q) range from 0.0032 to 0.38 Å^{-1}. The X-ray wavelength was set at 1.0 Å, equivalent to 12.4 keV. The scattering data were analyzed using the programs Chromixs [20] and Crysol [21] of the ATSAS package [22].

2.4. The Protein Sequence Analysis and Structure Modeling

Homolog protein sequences were retrieved from the National Center for Biotechnology Information's database using the Blastp program [23]. Protein multiple sequence alignments were performed with ClustalX [24]. Structural homology models were generated using Phyre2 [25] using SaNTR (PDB code 4GCM) as a template (100% confidence, 45% identity; 99% coverage).

3. Results

3.1. Description of the M. Mazei Trx System

The sequence analysis of the genome of *M. mazei* Gö1 revealed an open-reading frame (Locus tag MM_RS12200) that encoded a protein annotated as NADPH-dependent thioredoxin reductase (MmNTR). The amino acid sequence of the NTR-translated gene contained linear motifs for the binding of FAD (GGGPAG) and NADPH (GGGNSA and HRRDHLK) and a CxxC redox motif (Figure 1A), which are typically observed in NTRs [26]. A Blast search [23] of the non-redundant protein database using the MmNTR sequence as input revealed closest amino acid sequence similarity to low-molecular weight NTRs (Figure 1A). A comparison with the NTR enzyme from *M. acetivorans* C2A NTR (MaNTR) showed 83% identity and 92% similarity at the protein sequence level. Two other archaeal TR proteins have been included in the sequence analysis as they are of interest: the TR of *M. jannaschii* (MjTR) that employs F420 as an electron donor [9], and the TR of *T. acidophilum* (TaTR), which was initially described as a pyridine nucleotide-independent TR [10] and has recently been shown to display activity with NADH (but not NADPH) as an electron donor [11]. The MjTR and TaTR sequences lack the motifs necessary for NADPH coordination (the HRRxxxR motif in Figure 1A). Thereby, according to the protein sequence comparison, MmNTR is classified as a canonical LMW-NTR.

Thioredoxins are described as proteins of about 88 amino acids with the canonical motif WCGPC [27]. Using EcTrx as input in Blast homology searches [23], it was found that *M. mazei* contains two Trxs, herein named MmTrx1 (MM_RS02345) and MmTrx2 (MM_RS05160) (Table 1 and Figure 1B). MmTrx1 has a shorter sequence than canonical Trxs at its N-terminus, and its functional annotation awaits further investigations. The MmTrx2 amino acid sequence contains several insertions that provide the protein with additional structural elements compared to EcTrx; like the *M. acetivorans* homolog [6] (Table 1), the MmTrx2 gene is located upstream of the gene encoding the membranous CcdA oxidoreductase [28], suggesting a functional relation.

A

B

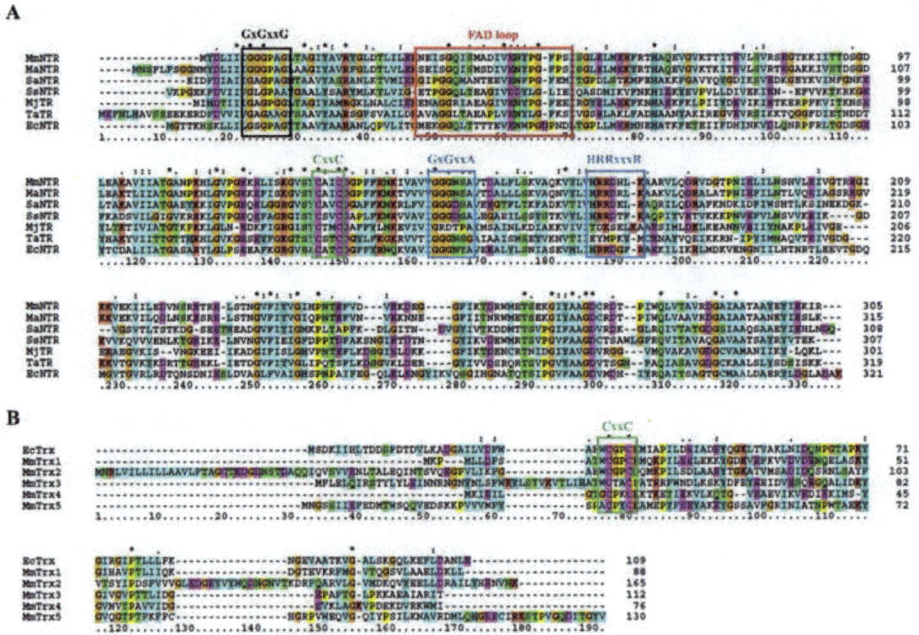

Figure 1. Protein multiple sequence alignment of selected NADPH-dependent Trx reductase (NTR) (**A**) and Trx (**B**). The positions of the conserved residues involved in FAD- and NADPH-binding are marked in panel (**A**) in black and blue, respectively; the region implicated in FAD coordination is depicted in red (see also Figure 3A below). The CxxC redox-active site motifs are shown in green in (**A**,**B**). Asterisks and column colors are assigned according to the ClustalX program's standard parameters. The ruler below indicates the amino acid position in the alignment. On the right, the number of amino acids for each Trx is stated. The figure includes the amino acid sequences from (**A**) *Methanosarcina mazei* NTR (MmNTR), *Methanosarcina acetivorans* NTR (MaNTR), *Staphylococcus aureus* NTR (SaNTR), *Sulfolobus solfataricus* NTR (SsNTR), *Methanocaldococcus jannaschii* TR (MjTR), *Thermoplasma acidophilum* TR (TaTR), *Escherichia coli* NTR (EcNTR); and, (**B**) *E. coli* Trx (EcTrx), and five Trx sequences from *M. mazei* (MmTrx1–MmTrx5).

Table 1. The thioredoxin system composition in *M. mazei* and a comparison with the *M. acetivorans* system.

M. Mazei Trx	Gene ID	Catalytic Motif	*M. Acetivorans* Homolog [6]	Reductant	Reduction of Insulin [7]
MmTrx1	MM_RS02345	WCGPC	MaTrx2	unknown	+
MmTrx2	MM_RS05160	WCGPC	MaTrx6	unknown	+
MmTrx3	MM_RS12205	WCTAC	MaTrx7	NTR [6]	+
MmTrx4	MM_RS11655	GCPKC	MaTrx5	unknown	-
MmTrx5	MM_RS10780	ACPYC	MaTrx1	FDR [29]	-

NTR, NADPH-thioredoxin reductase; FDR, ferredoxin:disulfide reductase.

A sequence analysis of the *M. mazei* Gö1 genome also identified atypical Trxs with varying active sites. NTR was placed adjacent to the gene MM_RS12205 that encoded a predicted Trx protein, herein called MmTrx3, with a WCTAC redox motif (Table 1 and Figure 1B); MmTrx3 is a homolog of Trx7 from *M. acetivorans* (MaTrx7) that has been biochemically demonstrated to be reduced by an NTR [6]. Other atypical Trxs include MM_RS11655 (MmTrx4) and MM_RS10780 (MmTrx5) with the non-canonical catalytic motifs GCPKC and ACPYC (Table 1 and Figure 1B). Their respective homologs in *M. acetivorans* (MaTrx5 and MaTrx1) failed to display insulin disulfide reductase activity and are of unknown function [7]. Recently, a ferredoxin:disulfide reductase (FDR) has been demonstrated to reduce MaTrx5 [29]. FDRs are Fe-S enzymes present in a number of archaea and bacteria that are related to the plant-type ferredoxin:thioredoxin reductase catalytic subunit (FTRc) [30]. Apparently, the NTR and FDR systems are both functional in *Methanosarcina* [29,31,32].

3.2. Purification and Structural Features of MmNTR

According to the protein sequence, MmNTR has a theoretical molecular weight of 35 kDa. A homogenous sample of MmNTR protein was prepared using the protocol described in the Materials and Methods section. The purified MmNTR migrated as a single band with an apparent molecular mass of 35 kDa in sodium dodecyl sulfate-polyacrilamide gel electrophoresis (SDS-PAGE) (Figure 2A). The purified MmNTR eluted as a single peak during gel filtration chromatography. The deduced molecular mass of 70 kDa indicates that MmNTR is a homodimer in solution (Figure 2B). The protein displayed a yellow color, and showed an absorption spectrum typical of a flavoprotein with peaks at about 274, 388, and 456 nm (Figure 2C).

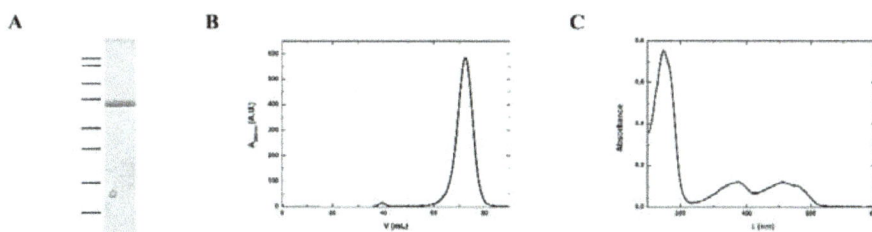

Figure 2. MmNTR was heterologously expressed in *E. coli*, and purified by affinity and size exclusion chromatography. (**A**) A Coomasie-stained SDS-PAGE image of purified MmNTR. The molecular weight markers are indicated on the left (100, 75, 50, 37, 25, 20, 15, and 10 kDa); (**B**) The size exclusion chromatography (Sephacryl S300) elution profile for MmNTR; (**C**) The UV-visible absorption spectrum of MmNTR in buffer, 10 mM Tris-HCl pH 7.6, 100 mM NaCl, and 2 mM 2-mercaptoethanol.

3.3. Apo-MmNTR is a Homodimer Composed of Two Domains

Protein crystals of MmNTR were obtained, and the X-ray structure was solved and refined to a resolution of 2.6 Å (Table S1). The X-ray structure shows that MmNTR is a homodimer, in agreement with the gel filtration data, with a structural organization that is analogous to other LMW-NTRs. Monomers consist of two Rossmann-fold domains that functionally correspond to FAD- and NADPH-binding domains [33] (Figure 3A). The FAD-binding domain is formed by two discontinuous segments of the polypeptide chain (residues 1–111 and 245–307 in MmNTR), and adopts a three-layer β/β/α-fold with a central four-stranded β-sheet flanked by a three-stranded β-sheet and three parallel α-helices. The NADPH-binding domain (residues 116–240) consists of a four-stranded β-sheet with three parallel α-helices and a three-stranded β-sheet at its flanks. The NADPH-binding domain is inserted into the FAD-binding domain via two antiparallel strands, β-7 (residues 112–114) and β-15 (residues 242–244) (Figure 3A). The redox motif CxxC (residues 132–136) is located on helix-α3 within the NADPH domain, with the two Cys forming a disulfide bridge in the crystal (Figure 3A).

Remarkably, no electron density of FAD or any other molecule was found at the presumed cofactor-binding pocket. The MmNTR structure thus represents the apo-form of the flavoprotein. Attempts to obtain crystals of the FAD-containing enzyme, including the addition of FAD to the crystallization drop, were unsuccessful. A search of structural homologues of the FAD domain of MmNTR using the Dali server [34] revealed that it is highly similar to the FAD-binding domains of other NTRs despite the absence of the cofactor. The FAD domain of MmNTR superimposes to the FAD domain of NTR from *Staphylococcus aureus* (SaNTR; PDB code 4GCM) with a root mean square deviation value of 1.7 Å (Figure 3B). The largest structural variations were found in a region next to the FAD-binding pocket, the so-called FAD-loop, which encompasses residues Ile37 through Pro55 (Figure 3B, in red), that lacks a defined electron density in the apo-structure and is, therefore, omitted from the final MmNTR model. The region constitutes a loop and a 3(10)-helix that covers the FAD isoalloxazine ring on its *si*-face, and a loop involved in dimerization [33]. Because the majority of FAD-contacting residues are strictly conserved in MmNTR (Figure 1A), a similar FAD binding mode is anticipated for this enzyme.

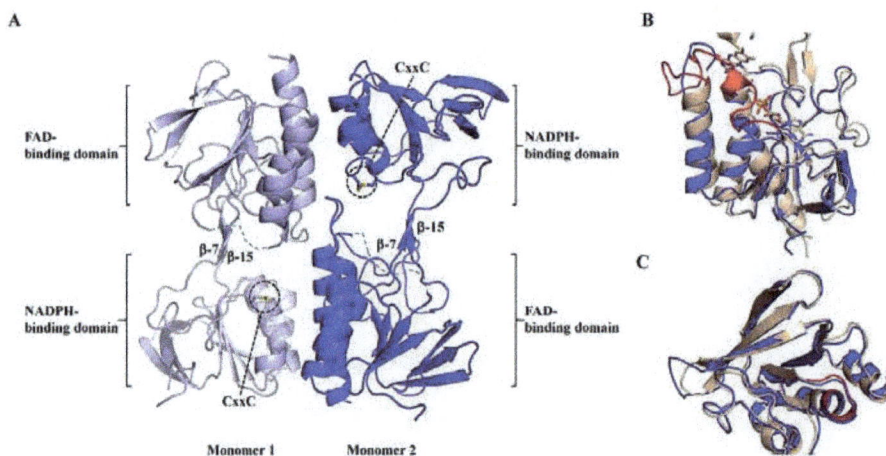

Figure 3. The crystal structure of MmNTR shown in a cartoon representation. (**A**) Monomers of MmNTR homodimers fold into two domains: The FAD-binding domain and the NADPH-binding domain. The Cys of the CxxC motif at the NADPH-binding domain are shown in sticks. The two domains are connected by two beta-strands (β-7 and β-15) that facilitate the conformational flexibility of the molecule. The redox-active disulfide faces inwards, where it would be accessible to reduction by the flavin, and the NADPH-binding site exposed to the solvent. The missing parts of the final model are indicated by dotted lines. (**B,C**) Fold comparison of FAD- and NADPH-binding domains, respectively, of MmNTR (in blue) and SaNTR (PDB code 4GCM, in brown). No electron density of FAD was found at the expected cofactor-binding pocket in MmNTR, despite the well-defined and conserved FAD-binding cavity in the folded module. The FAD from SaNTR is drawn as sticks. The FAD-loop and NADPH-binding motif are colored in red.

The NADPH-binding domain has no detectable pyridine nucleotide in the crystal structure, but displays essentially the same fold as the equivalent domain in SaNTR (Figure 3C). The binding site of the pyridine nucleotides in SaNTR includes a turn in a region containing conserved glycine amino acids (Gly152–154; Figure 3C, in red), and is dominated by electrostatic interactions between the cofactor phosphates and a cluster of arginine residues (motif HRRxxxR in Figure 1A). The lining of the residues is similar, implying that the cofactor-binding sites of both enzymes are comparable.

3.4. Apo-MmNTR Monomers Adopt a Flavin-Oxidizing Conformation

The NADPH-binding domain is positioned in an orientation with respect to the FAD-binding domain that exposes the NADPH-binding site to the bulk solvent and approaches the CxxC motif over the *re*-face of the isoalloxazine ring of the presumed flavin cofactor. This arrangement corresponds to the flavin-oxidizing (FO) conformation of NTRs that allows for the interaction of the disulfide bridge of the CxxC motif with the isoalloxazine ring of the flavin in a conformation competent for the reduction of the disulfide [35]. The two antiparallel beta-strands connecting the two domains are conserved (Figure 3A), and motion of the NADPH-binding domain relative to the FAD-binding domain is anticipated, as it required for the functional activity of the enzyme during the catalytic cycle.

3.5. The Absence of the Flavin Cofactor Compromises Dimer Formation

A structural comparison of MmNTR with *Escherichia coli* (EcNTR) homodimers reveals a significant difference at the dimer interface (Figure 4). In EcNTR, two-thirds of the dimer interface is formed by interactions between the FAD-binding domains that involve primarily three α-helices and two loops [36]. By contrast, the FAD-binding domains slide up to the dimer interface of MmNTR with the consequent loss of α-helix contacts. This results in new interactions between the NADPH-binding domain of one monomer with the FAD-binding domain of the opposite monomer (Figure 4 and Figure S1). Subsequently, the two domains in MmNTR are differently positioned each to the other with respect to canonical NTRs. This modification is likely a consequence of the loss of FAD, which results in the structural destabilization of the region comprising residues 36 to 68 that disrupts the dimer interface.

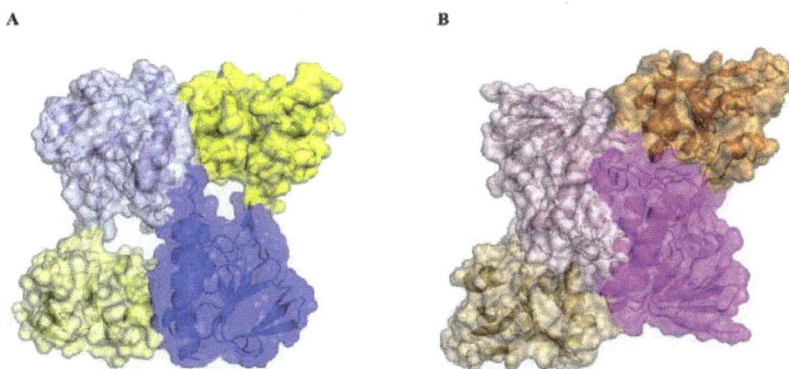

Figure 4. A structural comparison of the homodimer interface in (**A**) MmNTR, with the FAD- and NADPH-binding domains in blue and yellow, respectively; and (**B**) EcNTR (PDB code 1TRB), with the FAD- and NADPH-binding domains in magenta and orange, respectively. The interaction between two monomers within the crystal lattice buries 1696.9 Å^2 in MmNTR versus 2735.2 Å^2 in EcNTR of protein surface.

3.6. SAXS Reveals a Typical LMW-NTR Conformation of MmNTR

We further analyzed the structure of the enzyme using Size-Exclusion Chromatography coupled to with Small Angle X-ray Scattering (SEC-SAXS). As shown in Figure 5, the simulated scattering curve of the crystal structure fits the experimental SAXS data poorly (the curve was computed by Crysol [21]). These discrepancies indicate that the structure of MmNTR in solution differs significantly from the crystal model. Based on these results, a three-dimensional homology model for MmNTR was generated with the modeling server Phyre2 [25] using the crystal structure of SaNTR as a template. The theoretical scattering profiles calculated from the model fit the experimental SAXS curve quite well (Figure 5), and confirm that MmNTR, when bound to FAD, acquires an archetypal FO conformation in solution. These data confirm that MmNTR dimers in solution acquire the same structure as canonical LMW-NTRs.

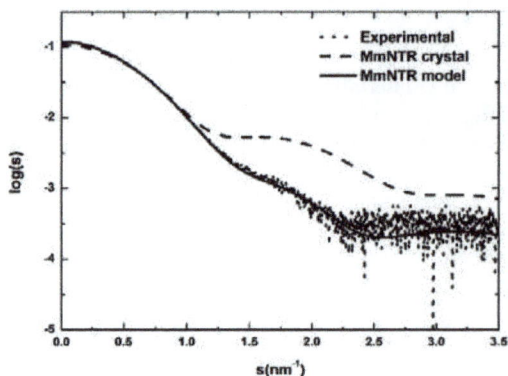

Figure 5. The SAXS study of MmNTR. Experimental SAXS data (dots) overlaid with the theoretical scattering curves that were computed from the crystal structure (dashed line) and the model (continuous line).

4. Discussion

A genome-wide sequence analysis showed that the composition of the Trx system of *M. mazei* Gö1 is similar to that of other *Methanosarcina* [7], with two Trxs having the canonical WCGPC redox-active motif as well as two Trx-related proteins of diverse features. Two different thiol reductases, NTR and FDR, have been shown to function in the reduction of Trxs in *M. acetivorans*, but not of canonical Trxs. This suggests the presence of other reduction pathways in this type of archaea. The elucidation of alternative reduction systems requires further work; possibilities might include the Fe-S proteins that have been detected in the genomes of *Methanosarcina*, including those that have been related to the catalytic subunit of the FTR protein that is functional in plants [30].

The high-resolution structural analyses of various bacterial NTRs show that LMW-NTRs are homodimers with two Rossmann-fold domains in each subunit [33]. One domain is responsible for FAD coordination (the FAD-binding domain); a second domain carries the CxxC redox-active motif and interacts with NADPH for transferring the reducing equivalents (the NADPH-binding domain). The two domains are joined by two antiparallel beta-strands that coordinate the spatial movement of one domain relative to the other during the catalytic cycle, in the so-called flavin-oxidizing (FO) and flavin-reducing (FR) conformations [35]. The crystal structure obtained in this work shows that MmNTR is a homodimer with two domains per monomer. Despite the lack of a cofactor, MmNTR has a canonical NTR fold. Further, the absence of the flavin results in a locally disordered conformation in a region of the protein that compromises dimer association, thus resulting in abnormal monomer–monomer interactions.

The protein was crystallized in the FO conformation, with the CxxC motif forming a disulfide bridge placed over the isoalloxazine ring of the assumed FAD. It is anticipated that MmNTR undergoes a general rearrangement that leads to interdomain motion to permit the transfer of reducing equivalents from NADPH to CxxC, and subsequently to Trx.

Our attempts to solve the holo-structure by X-ray crystallography have not been successful so far due to loss of the flavin cofactor during crystallization. Therefore, we performed SAXS studies to obtain structural information on the MmNTR holo-protein, concluding that the conformation of MmNTR in solution is similar to canonical LMW-NTR proteins. The differences between the crystal and the SAXS MmNTR structures might be attributed to the loss of the flavin cofactor, likely during the crystallization process.

5. Conclusions

We have presented a description of the thioredoxin system of *M. mazei* Gö1 and the structure of apo-MmNTR obtained by X-ray crystallography and SAXS. Our results show that the structural features of the LMW-NTR family of proteins are well-conserved in the MmNTR structure. The results suggest a primary important role of FAD. The cofactor, which may have been lost during crystallization, appears to function in protein folding since its absence impacts the ability of the enzyme to form the physiological dimer interface. In future work, the three-dimensional structure of the holo-form of the enzyme should provide more accurate data on the conformational organization of the redox-active sites.

Supplementary Materials: The following are available online at http://www.mdpi.com/2076-3921/7/11/166/s1, Table S1: Data collection and refinement statistics, Figure S1: Interaction between monomers in MmNTR. The interface is stabilized by hydrophobic interactions, mostly between the side chain of aromatic residues, including Tyr22, Tyr26, Phe161, Tyr301, and Trp264, among others. Hydrogen bonds have also formed between the side chain of the residue Arg142 and the main chain of the residue Arg25, the side chain of Try22 and the main chain of Ile135, as well as a salt bridge between the side chains of the residues Lys164 and Glu302.

Author Contributions: R.M.B., B.B.B., and M.B. conceived and designed the experiments; R.M.B. and M.B. performed the experiments and prepared the article; and R.M.B., R.A.S., B.B.B., and M.B. discussed and analyzed some of the data obtained during the experiments. All authors read and approved the final manuscript prior to submission.

Funding: This research was funded by Ministerio de Ciencia, Innovación y Universidades grant number BFU2016-80343-P.

Acknowledgments: We thank Ms. Elena Andrés Galván and Ms. M. Gloria González Holgado for excellent technical assistance; and the beamline staff at the Diamond Light Source and Swiss Light Synchrotron for their assistance in data collection. We thank Dr. José M. de Pereda for scientific discussions.

Conflicts of Interest: The authors declare no conflict of interest.

References

1. Jacquot, J.-P.; Eklund, H.; Rouhier, N.; Schürmann, P. Structural and evolutionary aspects of thioredoxin reductases in photosynthetic organisms. *Trends Plant Sci.* **2009**, *14*, 336–343. [CrossRef] [PubMed]
2. Lennon, B.W.; Williams, C.H. Reductive half-reaction of thioredoxin reductase from *Escherichia coli*. *Biochemistry* **1997**, *36*, 9464–9477. [CrossRef] [PubMed]
3. Fritz-Wolf, K.; Kehr, S.; Stumpf, M.; Rahlfs, S.; Becker, K. Crystal structure of the human thioredoxin reductase–thioredoxin complex. *Nat. Commun.* **2011**, *2*, 383. [CrossRef] [PubMed]
4. Jeon, S.-J.; Ishikawa, K. Identification and characterization of thioredoxin and thioredoxin reductase from *Aeropyrum pernix* K1. *Eur. J. Biochem.* **2002**, *269*, 5423–5430. [CrossRef] [PubMed]
5. Kashima, Y.; Ishikawa, K. A hyperthermostable novel protein-disulfide oxidoreductase is reduced by thioredoxin reductase from hyperthermophilic archaeon *Pyrococcus horikoshii*. *Arch. Biochem. Biophys.* **2003**, *418*, 179–185. [CrossRef] [PubMed]
6. McCarver, A.C.; Lessner, D.J. Molecular characterization of the thioredoxin system from *Methanosarcina acetivorans*. *FEBS J.* **2014**, *281*, 4598–4611. [CrossRef] [PubMed]
7. McCarver, A.C.; Lessner, F.H.; Soroeta, J.M.; Lessner, D.J. *Methanosarcina acetivorans* utilizes a single NADPH-dependent thioredoxin system and contains additional thioredoxin homologues with distinct functions. *Microbiology* **2017**, *163*, 62–74. [CrossRef] [PubMed]
8. Grimaldi, P.; Ruocco, M.R.; Lanzotti, M.A.; Ruggiero, A.; Ruggiero, I.; Arcari, P.; Vitagliano, L.; Masullo, M. Characterisation of the components of the thioredoxin system in the archaeon *Sulfolobus solfataricus*. *Extremophiles* **2008**, *12*, 553. [CrossRef] [PubMed]
9. Susanti, D.; Loganathan, U.; Mukhopadhyay, B. A novel F420-dependent Thioredoxin Reductase gated by low potential FAD: A tool for redox regulation in an anaerobe. *J. Biol. Chem.* **2016**, *291*, 23084–23100. [CrossRef] [PubMed]
10. Hernandez, H.; Jaquez, O.; Hamill, M.; Elliott, S.; Drennan, C. Thioredoxin reductase from *Thermoplasma acidophilum*: A new twist on redox regulation. *Biochemistry* **2008**, *47*, 9728–9737. [CrossRef] [PubMed]

11. Susanti, D.; Loganathan, U.; Compton, A.; Mukhopadhyay, B. A reexamination of thioredoxin reductase from *Thermoplasma acidophilum*, a thermoacidophilic euryarchaeon, identifies it as an NADH-dependent enzyme. *ACS Omega* **2017**, *2*, 4180–4187. [CrossRef] [PubMed]

12. Deppenmeier, U.; Johann, A.; Hartsch, T.; Merkl, R.; Schmitz, R.A.; Martinez-Arias, R.; Henne, A.; Wiezer, A.; Baumer, S.; Jacobi, C.; et al. The genome of *Methanosarcina mazei*: Evidence for lateral gene transfer between bacteria and archaea. *J. Mol. Microbiol. Biotechnol.* **2002**, *4*, 453–461. [PubMed]

13. Thauer, R.K.; Kaster, A.-K.; Seedorf, H.; Buckel, W.; Hedderich, R. Methanogenic archaea: Ecologically relevant differences in energy conservation. *Nat. Rev. Microbiol.* **2008**, *6*, 579. [CrossRef] [PubMed]

14. Kabsch, W. Integration, scaling, space-group assignment and post-refinement. *Acta Cryst. D* **2010**, *66*, 133–144. [CrossRef] [PubMed]

15. McCoy, A.J.; Grosse-Kunstleve, R.W.; Adams, P.D.; Winn, M.D.; Storoni, L.C.; Read, R.J. Phaser crystallographic software. *J. Appl. Cryst.* **2007**, *40*, 658–674. [CrossRef] [PubMed]

16. Rychlewski, L.; Jaroszewski, L.; Li, W.; Godzik, A. Comparison of sequence profiles. Strategies for structural predictions using sequence information. *Protein Sci.* **2000**, *9*, 232–241. [CrossRef] [PubMed]

17. Adams, P.D.; Afonine, P.V.; Bunkoczi, G.; Chen, V.B.; Davis, I.W.; Echols, N.; Headd, J.J.; Hung, L.W.; Kapral, G.J.; Grosse-Kunstleve, R.W.; et al. PHENIX: A comprehensive Python-based system for macromolecular structure solution. *Acta Cryst. D* **2010**, *66*, 213–221. [CrossRef] [PubMed]

18. Emsley, P.; Lohkamp, B.; Scott, W.G.; Cowtan, K. Features and development of Coot. *Acta Cryst. D* **2010**, *66*, 486–501. [CrossRef] [PubMed]

19. *The PyMOL Molecular Graphics System*, Version 1.8; Schrödinger, LLC: New York, NY, USA, 2015.

20. Panjkovich, A.; Svergun, D.I. CHROMIXS: Automatic and interactive analysis of chromatography-coupled small-angle X-ray scattering data. *Bioinformatics* **2018**, *34*, 1944–1946. [CrossRef] [PubMed]

21. Svergun, D.; Barberato, C.; Koch, M.H.J. CRYSOL—A program to evaluate X-ray solution scattering of biological macromolecules from atomic coordinates. *J. Appl. Cryst.* **1995**, *28*, 768–773. [CrossRef]

22. Konarev, P.V.; Petoukhov, M.V.; Volkov, V.V.; Svergun, D.I. ATSAS 2.1, a program package for small-angle scattering data analysis. *J. Appl. Cryst.* **2006**, *39*, 277–286. [CrossRef]

23. Altschul, S.F.; Madden, T.L.; Schaffer, A.A.; Zhang, J.; Zhang, Z.; Miller, W.; Lipman, D.J. Gapped BLAST and PSI-BLAST: A new generation of protein database search programs. *Nucleic Acids Res.* **1997**, *25*, 3389–3402. [CrossRef] [PubMed]

24. Thompson, J.D.; Gibson, T.J.; Plewniak, F.; Jeanmougin, F.; Higgins, D.G. The CLUSTAL_X windows interface: Flexible strategies for multiple sequence alignment aided by quality analysis tools. *Nucleic Acids Res.* **1997**, *25*, 4876–4882. [CrossRef] [PubMed]

25. Kelley, L.A.; Mezulis, S.; Yates, C.M.; Wass, M.N.; Sternberg, M.J.E. The Phyre2 web portal for protein modeling, prediction and analysis. *Nat. Protoc.* **2015**, *10*, 845. [CrossRef] [PubMed]

26. Gustafsson, T.; Sandalova, T.; Lu, J.; Holmgren, A.; Schneider, G. High-resolution structures of oxidized and reduced thioredoxin reductase from *Helicobacter pylori*. *Acta Cryst. D* **2007**, *63*, 833–843. [CrossRef] [PubMed]

27. Holmgren, A. Thioredoxin. *Annu. Rev. Biochem.* **1985**, *54*, 237–271. [CrossRef] [PubMed]

28. Cho, S.H.; Collet, J.F. Many roles of the bacterial envelope reducing pathways. *Antioxid. Redox Signal.* **2013**, *18*, 1690–1698. [CrossRef] [PubMed]

29. Prakash, D.; Walters, K.A.; Martinie, R.J.; McCarver, A.C.; Kumar, A.K.; Lessner, D.J.; Krebs, C.; Golbeck, J.H.; Ferry, J.G. Toward a mechanistic and physiological understanding of a ferredoxin:disulfide reductase from the domains Archaea and Bacteria. *J. Biol. Chem.* **2018**, *293*, 9198–9209. [CrossRef] [PubMed]

30. Balsera, M.; Uberegui, E.; Susanti, D.; Schmitz, R.; Mukhopadhyay, B.; Schürmann, P.; Buchanan, B. Ferredoxin:thioredoxin reductase (FTR) links the regulation of oxygenic photosynthesis to deeply rooted bacteria. *Planta* **2013**, *237*, 619–635. [CrossRef] [PubMed]

31. Kumar, A.K.; Kumar, R.S.; Yennawar, N.H.; Yennawar, H.P.; Ferry, J.G. Structural and biochemical characterization of a Ferredoxin:Thioredoxin reductase-like enzyme from *Methanosarcina acetivorans*. *Biochemistry* **2015**, *54*, 3122–3128. [CrossRef] [PubMed]

32. Wei, Y.; Li, B.; Prakash, D.; Ferry, J.G.; Elliott, S.J.; Stubbe, J. A ferredoxin disulfide reductase delivers electrons to the *Methanosarcina barkeri* class III Ribonucleotide Reductase. *Biochemistry* **2015**, *54*, 7019–7028. [CrossRef] [PubMed]

33. Kuriyan, J.; Krishna, T.S.R.; Wong, L.; Guenther, B.; Pahler, A.; Williams, C.H.; Model, P. Convergent evolution of similar function in two structurally divergent enzymes. *Nature* **1991**, *352*, 172–174. [CrossRef] [PubMed]

34. Holm, L.; Laakso, L.M. Dali server update. *Nucleic Acids Res.* **2016**, *44*, W351–W355. [CrossRef] [PubMed]
35. Lennon, B.W.; Williams, C.H., Jr.; Ludwig, M.L. Twists in catalysis: Alternating conformations of *Escherichia coli* thioredoxin reductase. *Science* **2000**, *289*, 1190–1194. [CrossRef] [PubMed]
36. Williams, C.H., Jr. Mechanism and structure of thioredoxin reductase from *Escherichia coli*. *FASEB J.* **1995**, *9*, 1267–1276. [CrossRef] [PubMed]

![antioxidants logo] *antioxidants*

MDPI

Review

Involvement of Glutaredoxin and Thioredoxin Systems in the Nitrogen-Fixing Symbiosis between Legumes and Rhizobia

Geneviève Alloing [1], Karine Mandon [1], Eric Boncompagni [1], Françoise Montrichard [2] and Pierre Frendo [1,*]

[1]   Université Côte d'Azur, INRA, CNRS, ISA, France; Genevieve.Alloing@unice.fr (G.A.); mandon@unice.fr (K.M.); Eric.BONCOMPAGNI@univ-cotedazur.fr (E.B.)
[2]   IRHS, INRA, AGROCAMPUS-Ouest, Université d'Angers, SFR 4207 QUASAV, 42 rue Georges Morel, 49071 Beaucouzé CEDEX, France; francoise.montrichard@univ-angers.fr
*   Correspondence: frendo@unice.fr

Received: 29 October 2018; Accepted: 1 December 2018; Published: 5 December 2018

Abstract: Leguminous plants can form a symbiotic relationship with Rhizobium bacteria, during which plants provide bacteria with carbohydrates and an environment appropriate to their metabolism, in return for fixed atmospheric nitrogen. The symbiotic interaction leads to the formation of a new organ, the root nodule, where a coordinated differentiation of plant cells and bacteria occurs. The establishment and functioning of nitrogen-fixing symbiosis involves a redox control important for both the plant-bacteria crosstalk and the regulation of nodule metabolism. In this review, we discuss the involvement of thioredoxin and glutaredoxin systems in the two symbiotic partners during symbiosis. The crucial role of glutathione in redox balance and S-metabolism is presented. We also highlight the specific role of some thioredoxin and glutaredoxin systems in bacterial differentiation. Transcriptomics data concerning genes encoding components and targets of thioredoxin and glutaredoxin systems in connection with the developmental step of the nodule are also considered in the model system *Medicago truncatula–Sinorhizobium meliloti*.

Keywords: thioredoxin; glutaredoxin; legume plant; symbiosis; redox homeostasis; stress

---

1. Introduction

Most terrestrial plants establish symbiotic relationships with fungi or bacteria that provide nutrients for their growth [1,2]. Nitrogen and phosphorous are critical determinants of plant growth and productivity. Amongst the plant families, leguminous plants can achieve a nitrogen-fixing symbiosis with soil bacteria of the family Rhizobiaceae to reduce atmospheric nitrogen (N_2) to ammonia [3]. The ability to reduce N_2 is restricted to bacteria and archaea which produce the enzyme nitrogenase. Legumes are an economically important plant family, for their contribution to animal and human nutrition on one hand, and for their ecosystemic services in cropping systems on the other hand, by participating to nitrogen enrichment of soils and thereby to a reduced use of nitrogen fertilizers. The study of these symbioses is therefore a major challenge to promote a more environmentally-friendly agriculture.

The nitrogen-fixing symbiosis (NFS) between rhizobia bacteria and legumes leads to the formation of new root organs, called nodules [4,5]. The development of the nodule requires many crucial steps to achieve the fixation of atmospheric nitrogen. The first step is the cross recognition between bacteria and the plant partner. This recognition involves the nodulation (Nod) factors produced by the bacteria that play a major role in the symbiotic specificity between the two partners. In parallel to bacterial recognition, Nod factors promote the development of a new meristem in the plant root that leads to the establishment of the root nodule. Subsequently, the formation of infection threads allows the transport of bacteria from the surface of the root to the plant cells that will host bacteria, in an endosymbiotic way. The accommodation of numerous bacteria inside plant cells and nitrogen fixation requirements involve modifications of the cellular structure and physiology of both partners for maintaining the symbiotic interaction. These modifications are achieved through differentiation of the plant cells which includes cell enlargement, DNA endoreduplication, and significant reprogramming of cellular structure and metabolism [6]. Cellular and biochemical changes are also observed during the differentiation of bacteria into N_2-fixing bacteroids. Amongst them, a high aerobic metabolism provides ATP and reductants necessary to sustain nitrogenase activity, whereas the nitrogen-fixing enzyme is irreversibly inactivated by oxygen. Thus nitrogen-fixation efficiency depends on oxygen protective mechanisms involving the formation of an oxygen barrier cell layer around the infected cells and the production of a symbiotic hemoglobin, called leghemoglobin. This later is involved in the protection of nitrogenase from denaturation, and in the supply of ample amount of oxygen to bacteria for respiration. The supply of energy from the plant to nitrogen-fixing bacteroids, and the export of ammonia from the bacteroids to the roots also require major metabolic adaptations in the nodules. In conclusion, the nodule functioning depends on a strict regulation of the development and the metabolism of plant and bacteria cells.

The nodules are considered as "indeterminate" or "determinate" according to their mode of development [7]. In the determinate nodules, such as those of soybean (*Glycine max*), the nodular meristems are transiently active. This results in spherical nodules, containing cells with a similar developmental state to each other. In the indeterminate nodules, such as those formed by pea (*Pisum sativum*), alfalfa (*Medicago sativa*) or barrel medic (*M. truncatula*), the meristems persist throughout the plant's life, giving an elongated nodule. Consequently, the functional nodule presents three zones: (I) the meristematic zone, (II) the infection zone, and (III) the N_2-fixing zone (Figure 1A,B). At later stage, there is a rupture in the symbiotic interaction, which occurs in the senescence zone (zone IV).

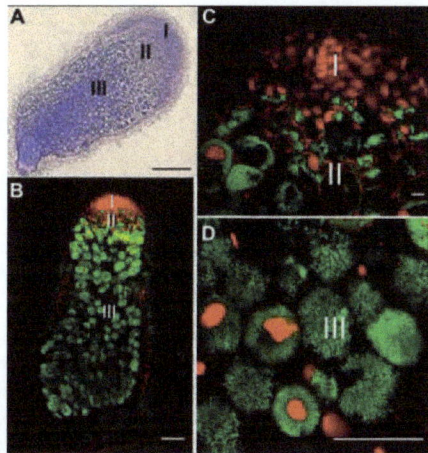

Figure 1. The indeterminate root nodule structure in *Medicago truncatula*. (**A**) Longitudinal section of indeterminate nodule 3 weeks post infection (wpi) with the apical meristem (I), the infection zone (II), and the nitrogen-fixing zone (III). (**B–D**) Longitudinal section of wild-type nodules 3 wpi analyzed by confocal microscopy with *S. meliloti* DNA stained with SYTO9 (green) and plant nuclei stained with propidium iodide (red) [8]. (**C**) The size of plant cells and of plant cell nuclei increases during cellular differentiation and intracellular bacterial infection occurs in zone II. (**D**) The nitrogen-fixing cells in zone III are fully packed with numerous elongated endosymbiotic bacteria called symbiosomes. Bars: (**A**) 200 μm; (**B**) 100 μm; (**C**) 10 μm; (**D**) 50 μm.

As mentioned above, the meristematic cells of the nodule destined to house the rhizobia undergo several DNA endoreduplication cycles. The endoreduplication (up to 64 C) is accompanied by an expansion (up to 80 times) of the infected cells (Figure 1C) [9,10]. These transformations are associated with metabolic changes allowing the bacteria reception and the assimilation of reduced nitrogen. Bacteroid differentiation depends on the host plant [11]. In some legumes such as soybean, the bacteroid morphology is little affected in comparison to the free-living bacteria. In contrast, in other legumes, such as faba bean (*Vicia faba*), pea or the *Medicago* genus, bacteroids present an extreme morphological change with an elongated phenotype (5 to 10 times longer than the free-living cells) (Figure 1D). This change is coupled with endoreduplication of the bacterial genome, and irreversible terminal differentiation, preventing subsequent bacterial multiplication. Transcriptomic analyses of the host plants, inducing (*M. truncatula*) or not (*Lotus japonicum*) the terminal bacterial differentiation, have allowed the identification of plant factors involved in this process. These factors, called Nodule Specific Cysteine Rich (NCR) peptides, are defensin-like peptides specifically expressed in the nodule [12]. This family of peptides has been extensively described in *M. truncatula*, and several homologs have been found in *M. sativa* and *P. sativum* [6]. The NCR peptides produced by the host plant are targeted to the bacteroids through the secretory pathway. In vitro treatment of *S. meliloti* culture with certain NCR peptides induces some aspects of terminal differentiation such as bacterial membrane permeabilization, cell division inhibition, genome endoreduplication, and bacterial elongation. In addition, mutants of *M. truncatula* deficient in two NCR peptides, NCR169 and NCR211, develop non-functional nodules [13,14].

Regulation of the cellular redox state represents a major regulatory component of the nitrogen-fixing symbiosis. During the last twenty years, analysis of numerous redox components of the nodules has shown their specific involvement in the functioning of the root nodule. Amongst them, some NADPH oxidases, which are involved in the production of reactive oxygen species (ROS), have been shown to regulate the symbiotic interaction throughout the lifetime of the nodule from its installation to its senescence [15]. Similarly, enzymes implicated in the steady state of nitric oxide (NO), a growth and metabolic regulator in plants, control nodule development and functioning [16]. Antioxidant components of the cells involved in the regulation of the cellular redox state also participate in the nodule development [17,18]. In this review, we will present an overview of the work performed on the glutaredoxin and thioredoxin systems, which regulate the redox state of the proteins, in the nitrogen-fixing symbiosis in both symbiotic partners.

2. The Glutaredoxin and Thioredoxin Systems of Plant Partner

2.1. The Glutaredoxin System

Glutaredoxins (Grxs) are small redox enzymes of approximately one hundred amino-acid residues that use glutathione (GSH) as a reducer, that is maintained in a reduced state by glutathione reductase (GR) and NADPH. The GSH synthesis has been extensively studied in leguminous plants [17]. In legumes, the structural homolog, homoglutathione (hGSH; γGlu-Cys-βAla), may partially or completely replace GSH [19–21]. Both compounds can be found at concentrations of 0.5–1.5 mM in nodules [22], similar to the estimated levels of 1–3 mM GSH and 0.4–0.8 mM hGSH in the chloroplast stroma [23] or in the cytosol [24]. However, the (h)GSH content is much higher in nodules than in roots due to the structural modifications of nodule cells with an increased cytosol volume compared to root cells (see Figure 1). (h)GSH synthesis derives from sulfur (S) metabolism which has been studied in N_2 fixing nodules of *L. japonicus* [25]. The high adenosine 5'-phosphosulfate reductase activity, the strong S-flux into cysteine and derivatives, and the up-regulation of the expression of several rhizobial and plant genes involved in S-assimilation showed the important function of nodules in S-assimilation [25]. Moreover, the higher thiol content observed in roots and leaves of N_2-fixing plants in comparison to uninoculated plants could not be attributed to local biosynthesis, showing that nodules are an important site for production of reduced S for the plants [25]. The S-metabolism of nodules is reduced in plants nodulated by mutant rhizobia unable to reduce N_2

indicating a strong interdependency between N_2-fixation and S-assimilation [25,26]. Sulfate transport is also modified in the nodule. Nodule-specific sulfate transporters have been identified [27]. Some of them are located on the peribacteroid membrane and allow the transport of inorganic sulfur to the bacteroid [28,29]. In soybean, this transport has been shown to be crucial for nitrogen-fixing efficiency. We have analyzed the transcriptome of *M. truncatula* using sulfate transporter as a key word in the symbimics website (https://iant.toulouse.inra.fr/symbimics/), which allows to compare the expression of genes in roots and nodules and to define the level of gene expression in the different nodule zones (Table 1). In *M. truncatula*, 22 putative sulfate transporter genes were identified. The expression of some of them (*Medtr3g087730*, *Medtr5g061860*, *Medtr6g086170*, *Mt0062_10115*) were significantly upregulated in nodules compared to roots. Transcriptomic analyses in the *M. truncatula* Gene Expression Atlas (https://mtgea.noble.org/v3/) showed that *Medtr5g061860* and *Medtr6g086170* expression were correlated with the nitrogen fixation efficiency as treatment of plants with nitrate, which reduces the nodule nitrogen fixation, led to a reduction of less than 20% of the expression of these genes as compared to control nodules.

The synthesis of GSH in plants and other organisms is accomplished in two sequential reactions catalyzed by γ-glutamylcysteine synthetase (γECS) and glutathione synthetase (GSHS), both showing a strict requirement for ATP and Mg^{2+} [23]. In legumes, the synthesis of hGSH is also carried out in two steps, involving the same γECS enzyme and a specific homoglutathione synthetase (hGSHS), which exhibits a much higher affinity for β-alanine than for glycine [19,20,30,31]. Site-directed mutagenesis of soybean and *M. truncatula* hGSHS has conclusively shown that two contiguous amino acid residues in the active site (Leu-487 and Pro-488, positions that are Ala in GSHS) mainly determine the substrate preference for β-alanine over glycine [20,32]. The *GSHS* and *hGSHS* genes share high homology (~70% amino acid identity) and are located in tandem on the same chromosome in the model plant legumes *M. truncatula* [20] and *L. japonicus* [21]. These findings are consistent with the hypothesis that the *hGSHS* gene derives from the *GSHS* gene by a duplication event occurring after the divergence between the Fabales, Solanales, and Brassicales [20]. Despite this close relationship, the two genes are differentially regulated in plant organs. This can be exemplified with studies performed on the two model legumes. Thus, *M. truncatula* produces exclusively GSH in the leaves and both GSH and hGSH in the roots and nodules, whereas *L. japonicus* produces almost exclusively hGSH in the roots and leaves, but more GSH than hGSH in the nodules. In legumes, the thiol contents are positively correlated with the GSHS and hGSHS activities and in general with their mRNA levels [21–33].

The concentration of (h)GSH and the N_2-fixing activity in nodules are positively correlated during nodule development [34]. The two parameters decline with aging [35,36] as well as during stress-induced senescence [37–41]. These findings suggest that (h)GSH is important for nodule activity, a hypothesis that was tested by modulating the nodule content of (h)GSH using pharmacological and genetic approaches. The application of buthionine sulfoximine (a specific inhibitor of γECS) or the expression of (h)GSHS in antisense orientation caused depletion of (h)GSH in *M. truncatula* roots [42]. The (h)GSH synthesis deficiency in roots decreased substantially the number of nascent nodules and the expression of some early nodulin genes [42]. These results, along with the proposed role of GSH in meristem formation in *Arabidopsis thaliana* [43–45], suggest that (h)GSH is required for the initiation and maintenance of the nodule meristem. The transcriptomic analysis of (h)GSH-depleted plants during early nodulation revealed downregulation of genes implicated in meristem formation and upregulation of salicylic acid-related genes after infection with *S. meliloti* [46]. The enhanced expression of defense-related genes provides a partial explanation for the negative effects of (h)GSH depletion on the symbiosis. The role of (h)GSH was also analyzed in the nitrogen-fixing zone. Downregulation of the γECS gene by RNA interference using the nodule nitrogen-fixing zone-specific NCR001 promoter resulted in significantly lower biological nitrogen fixation (BNF) associated with a significant reduction in the expression of nodule specific genes. This lower (h)GSH content was correlated with a reduction in the nodule size. Conversely, γECS overexpression using the same promoter resulted in an elevated GSH content associated with increased BNF and significantly higher expression of the sucrose synthase-1 and leghemoglobin genes. Taken together, these data show that the plant (h)GSH content of the nodule nitrogen-fixing zone modulates the efficiency of the BNF process, demonstrating their important role in the regulation of this process [47]. All these data show the importance of sulfur metabolism and more particularly of (h)GSH in the development and functioning of nodules.

Amongst the multiple roles of (h)GSH, these thiols serve as reducing power for Grxs. Numerous Grxs are present in plants. To date, no physiological analysis was performed to investigate the importance of Grxs in the nitrogen-fixing symbiosis. *M. truncatula* genome analysis using BLAST and publication data mining [48,49] allowed us to find thirty-six genes encoding putative Grxs of class I, II, and III (Table 2). Gene expression analysis in roots and nodules did not allow us to find class I and class II Grxs significantly upregulated in nodules compared to roots. In contrast, two Class III Grxs (*Medtr2g014760, Medtr1g088910*) are upregulated in nodules compared to roots suggesting that they play a significant role in nodule development or functioning. Nevertheless, three class III Grxs are also significantly downregulated in nodules compared to roots and multiple Class III Grxs are not expressed in roots and nodules. Taken together, these results showing the significant modification of the expression of multiple Grx genes suggest that redox regulation of nodule metabolism is extensively modified compared to roots.

Table 1. Expression of plant sulfate transporters in *M. truncatula* nodules. Gene accession numbers are indicated in the table. Gene annotation is based on candidate orthologs and interprodomain signature. The different columns correspond to root and nodule whole organ analysis (Root and Nodule) and to the nodule zones: meristematic zone (I), distal infection zone (IId), proximal infection zone (IIp), infection/fixation interzone (IZ II-III), and nitrogen-fixation zone (III). The numbers in the different columns correspond to Total Reads RiboMinus ™ rRNA depletion. All RNA-seq read values were normalized [50]. The total reads are reported from the symbimics bioinformatics website. The full organs are nitrogen starved Roots and 10-day old nodules. The red and blue colors correspond, respectively, to higher and lower significant differences between the organs (roots and nodules) and between the different nodule zones. The statistical differences are reported from the symbimics bioinformatics website.

Gene Name	Root	Nodule	I	IId	IIp	IZ II-III	III
Sulfate transporter							
Medtr7g095430	1307	1297	7	2	4	19	9
Medtr4g084620	298	176	13	9	43	44	29
Mt0062_10115	54	548	30	154	289	235	361
Medtr4g011970	633	1052	307	1721	3064	1094	1406
Medtr3g087730	4	876	118	530	1088	323	92
Medtr5g061860	87	1449	4	14	22	1313	1742
Medtr6g086170	45	48,527	14	291	566	6898	12,537
Medtr3g073730	220	20	2	4	1	1	0
Medtr5g061880	200	43	15	14	2	3	3
Medtr2g102243	322	193	19	18	1	1	0
Medtr2g008470	1406	253	1	1	1	7	8
Medtr3g087740	2150	854	19	17	16	4	13
Mt0050_00072	0	0	0	0	0	0	0
Medtr4g084640	0	0	0	0	0	0	0
Mt0008_01149	0	0	0	0	0	0	0
Mt0008_11083	0	0	0	0	0	0	0
Medtr4g063825	0	6	1	0	0	0	0
Mt0006_10002	1	0	2	3	0	0	0
Medtr2g082610	3	5	1	2	0	0	0
Medtr3g073780	9	1	0	0	0	0	1
Medtr1g071530	345	184	2	2	0	1	1
Medtr7g022870	464	449	48	52	17	20	39

Table 2. Expression of glutaredoxins in *M. truncatula* nodules. Gene accession numbers are indicated in the table. Gene annotation is based on candidate orthologues and interprodomain signature. The different columns correspond to root and nodule whole organ analysis (Root and Nodule) and to the nodule zones: meristematic zone (I), distal infection zone (IId), proximal infection zone (IIp), infection/fixation interzone (IZ II-III) and nitrogen-fixation zone (III). The numbers in the different columns correspond to Total Reads ribominus. All RNA-seq read values were normalized [50]. The total reads are reported from the symbimics bioinformatics website. The full organs are nitrogen starved Roots and 10 days old nodules. The red and blue colours correspond respectively to higher and lower significant differences between the organs (roots and nodules) and between the different nodule zones. The statistical differences are reported from the symbimics bioinformatics website.

Gene Name	Putative Redox Site	Root	Nodule	I	IId	IIp	IZ II-III	III
Glutaredoxins								
Class I								
Medtr7g035245	YCPFC	2612	2354	665	163	156	238	167
Medtr1g069255	WCSYC	121	153	54	41	135	109	92
Medtr3g077560	YCGYC	1314	366	2	1	1	2	0
Medtr3g077570	YCGYC	201	148	27	16	3	3	15
Medtr2g038560	YCPYC	1573	1200	425	682	837	509	297
Medtr5g021090	YCPYC	1284	1444	76	97	63	72	50
Class II								
Medtr2g103130	QCGFS	1332	1390	380	266	323	546	302
Medtr4g079110	GCCMS	968	1834	62	36	1	0	0
Medtr7g079520	QCGFS	771	640	170	143	103	107	62
Medtr4g088905	KCGFS	1444	1940	277	252	194	97	105
Medtr4g016930	LCGSF	124	218	57	66	70	65	71
Class III								
Medtr7g026770	TCCMC	13	13	21	5	0	0	0
Medtr3g104510	SCCMC	16	32	1	3	24	53	96
Medtr1g088910	SCYMC	62	260	2	0	0	0	0
Medtr1g015890	SCCMC	144	340	100	1437	562	164	258
Medtr2g090755	GCCMS	78	387	10	4	1	2	3
Medtr2g014760	GCCLC	71	467	33	19	8	4	10
Medtr1g088925	SCCLC	474	1	1	1	0	0	0
Medtr1g088920	LCCLC	460	3	1	0	0	0	0
Medtr7g108200	SCCLC	1650	308	2	1	0	0	0
Medtr4g119030	SCCMS	0	0	0	0	0	0	0
Medtr2g048970	SCCMS	0	0	0	0	0	0	0
Medtr2g019950	SCGMS	0	0	0	0	0	0	0
Medtr4g119050	SCCMS	0	0	0	0	0	0	0
Medtr7g108250	TCCLS	0	0	0	0	0	0	0
Medtr7g108220	SCYMC	0	0	0	0	0	0	0
Medtr7g108250	TCCLS	0	0	0	0	0	0	0
Medtr7g022690	SCCMC	0	0	0	0	0	0	0
Medtr5g077550	DCCFS	0	0	0	0	0	0	0
Medtr1g088905	TCCLS	0	0	1	0	0	0	0
Mt0001_10735	SCCMS	0	1	0	0	0	0	0
Medtr7g022710	SCCMC	0	0	1	0	0	0	0
Medtr7g022550	SCCMC	0	0	1	0	0	0	0
Medtr7g108260	TCPMS	2	4	0	0	0	0	0
Medtr2g019900	SCCMC	16	84	0	0	0	0	0
Medtr7g108210	SCYMC	30	30	1	0	0	0	0

2.2. The Thioredoxin System

The other biochemical system involved in the thiol-dependent redox regulation of enzyme activity is the thioredoxin system. Thioredoxins (Trxs) are small proteins similar to Grxs that reduce disulfide bounds. Oxidized Trxs are in turn re-reduced by NADP-dependent Trx reductases (TR) and NADPH or ferredoxin in plastids. Nonetheless, a few members of the Trx family use, as Grxs, glutathione as a reducer [51,52]. Trxs are able to reduce directly some of basic metabolic enzymes such as ribonucleotide reductase, and enzymes involved in the antioxidant systems such as peroxiredoxins (Prx), glutathione peroxidase (Gpx), and methionine sulfoxide reductase (MSR). In plant tissues, several groups of Trxs have been identified. The Trxs f, m, x, y, and z are localized in the plastids, Trxs o are addressed to the mitochondria and the Trxs h mainly accumulate in the cytoplasm [53]. Cytosolic Trxs h can also be transferred in nucleus in cells suffering oxidative stress [54]. Nucleoredoxins were also described as other redoxins located in the nucleus [55]. In legumes, the Trx family has been analyzed in detail in *M. truncatula* [52,56] and *L. japonicus* [57]. The analysis of *Trx* expression in *L. japonicus* showed that there is a differential expression pattern of the different isoforms in leaves, root, and nodules. However, no isoform seems to be significantly more expressed in nodules than in roots and leaves. In soybean, a Trx h expressed in infected cells of mature nodules is able to protect a yeast Trx mutant against hydrogen peroxide (H_2O_2) [58]. This Trx is crucial for nodule development and functioning as RNAi-mediated repression of the Trx gene severely impaired nodule development [58]. Nodulin-35, a subunit of uricase, was found to be a target of this thioredoxin suggesting a novel role of Trx in the regulation of enzyme activities involved in nodule nitrogen fixation [59]. In addition to all the classical types of Trxs found in plants, *M. truncatula* contains a novel type of Trxs, called Trxs s, comprising four isoforms which are associated with symbiosis [56,60]. No orthologs were found in *A. thaliana*, *L. japonicus* or soybean suggesting that the Trxs s isoforms could be unique to certain legume species. *Trx s1* and *s3*, are induced in the nodule infection zone where bacterial differentiation occurs. Trx s1 is targeted to the symbiosomes, the N_2-fixing organelles. Trx s1 interacted with NCR247 and NCR335 and increased the cytotoxic effect of NCR335 in *S. meliloti*. *Trx s1* silencing impairs bacteroid endoreduplication and enlargement, two features of terminal bacteroid differentiation, and the ectopic expression of *Trx s1* in *S. meliloti* partially complements the silencing phenotype. Thus, Trx s1 is targeted to the bacterial endosymbiont where it controls bacteroid terminal differentiation [60].

Gpxs and Prxs are also present in root nodules. In plants, most Gpxs reduce hydroperoxides using Trxs and TR, instead of GSH and GR, as a reducing system [61]. This is also true for Gpx1 in *M. truncatula* [62]. Based on genome analysis, six Gpxs were reported in *L. japonicus* [63]. Except the *LjGpx4*, the other isoforms were expressed in nodules with a higher level for *LjGpx1* and *LjGpx3* [63,64]. The two Gpx were Trx-dependent phospholipid hydroperoxidases and were upregulated in response to NO for LjGpx1 and in response to cytokinine and the ethylene precursor ACC for LjGpx3 [64]. Both genes were highly expressed in the nodule zone containing the bacteria, and the *LjGpx3* mRNA was also detected in the cortex and vascular bundles. Immunogold localization of Gpx allowed to localize LjGpx1 in plastids and nuclei and LjGpx3 in the cytosol and the endoplasmic reticulum [64]. Based on yeast complementation experiments, both enzymes protect against oxidative stress, salt stress, and membrane damage suggesting that both LjGpxs perform major antioxidative functions in nodules, preventing lipid peroxidation and other oxidative processes at different subcellular sites of vascular and infected cells [64].

There are four types of Prxs in plants (1-CysPrx, 2-CysPrx, PrxII, and PrxQ). Based on genome analysis, seven Prxs were reported in *L. japonicus* [63]. Eight transcripts were detected: *Lj1CPrx*, *LjPrxQ1a*, and *LjPrxQ1b* which derive from the gene *LjPrxQ1a* with an alternative splicing, and *Lj2CPrxA*, *Lj2CPrxB*, *LjPrxIIB*, *LjPrxIIE*, and *Lj1CPrxIIF* [63]. The expression profiles in the different plant tissues did not allow the detection of a Prx isoform which would be more expressed in the nodules. Nevertheless, reduction of *PrxIIB* and *PrxIIF* expression levels were associated to the nodule senescence process in bean nodules [65]. In contrast, whereas the level of PrxIIF protein remains constant in senescent nodules, the level of PrxIIB decreases in senescent nodules [65]. Similarly, the decrease of the putative PrxIIA content and a constant level of the mitochondrial PrxIIF protein were observed in senescent nodules compared to mature nodules [66]. Trx also serve as electron donors for MSRs that repair oxidized proteins. To our knowledge, no experiment was performed to analyze the roles of MSRs in root nodules. We have analyzed the transcriptome of *M. truncatula* searching for methionine sulfoxide in the symbimics website and BLAST sequence alignment program to validate the putative identity of the sequences. The comparison of gene expression in roots and nodules, and in the different nodule zones allowed the detection of a significant upregulation of *Medtr3g051460* in the infection zone and the nitrogen-fixing zone compared to the uninfected nodule zone I. Apart from this isoform, no clear difference in transcript level was observed for the seven other genes. Functional analysis of this enzymatic family awaits to be performed in root nodules.

3. The Glutaredoxin and Thioredoxin Systems of Bacterial Partner

Rhizobial genomes contain the genes of Grx and Trx systems (http://genome.annotation.jp/rhizobase). We summarize recent data on these systems emphasizing how they contribute to the efficiency of nitrogen-fixing symbiotic interaction.

3.1. The Gutaredoxin System

In most bacteria, the glutaredoxin system consists of GR, which catalyzes the NADPH-driven reduction of glutathione disulfide (GSSG) to GSH, which in turn reduces Grx. The two steps of GSH biosynthesis are catalyzed by γECS and GSHS, encoded by the *gshA* and *gshB* genes, respectively. The GSH recycling from GSSG is performed by a GR encoded by the *gor* gene. Studies of rhizobial mutants affected in GSH metabolism demonstrate the central role of GSH pool in free-living cells and in planta. In all cases, *gshB* inactivation alters the fitness of free-living bacteria, and *gshB* mutants develop poorly effective symbiosis with their plant partners. For example, the growth of a *S. meliloti gshB* mutant is altered in minimal medium whereas a *gshA* mutant does not grow under the same conditions, showing that GSH is essential and can be partially replaced by γ-glytamyl-cysteine [67]. The two mutants experience oxidative stress as both exhibit higher catalase activity, a biochemical marker of oxidative stress, when compared with the wild-type strain. *M. sativa* plants inoculated with the *gshA* mutant did not produce nodules, while *gshB* inactivation triggered a delayed nodulation phenotype and the development of abnormal, early senescing nodules associated with 75% reduction in the nitrogen-fixation capacity of bacteroids. A *gshB* mutant of *Rhizobium tropici* has a reduced ability to compete against the wild-type strain for nodule occupancy on common bean, while a *Rhizobium etli gshB* mutant has a delayed nodulation phenotype when inoculated onto bean [68,69]. Plants infected by either one of the other *gshB* mutant develop ineffective nitrogen-fixing nodules with obvious signs of early senescence. Nodule phenotype is associated with enhanced levels of superoxide anion in the case of *R. tropici* infection, showing that GSH-deficient bacteroids face an environmental oxidative stress [68]. In the same way, a *gshA* mutant of *Bradyrhizobium japonicum* gives rise to nodules with a strong nitrogen-fixation deficiency during interaction with soybean [70]. There are, however, variations in GSH requirement among rhizobial species since another *Bradyrhizobium*-legume interaction develops effective nodules independently of the bacterial GSH pool. The *gshA* mutant of *Bradyrhizobium* sp. SEMIA 6144 indeed induced functional nodules with peanut (*Arachis hypogaea* L.), even though GSH depletion affects nodule occupancy capacity and growth of the free-living bacterium in normal and stressful conditions [70]. Overall, the homeostasis (both level and redox status) of GSH in bacterial cells is important for nodule development, and this is also exemplified by the symbiotic deficiency of *S. meliloti gor* mutant [71]. The lack of GR in *gor* mutants causes a decrease in the GSH/GSSG ratio, triggering oxidative stress with an increased expression of catalase genes, and an enhanced sensitivity to oxidants. In planta, the *gor* mutant is affected in its ability to compete for nodule occupancy and displays a reduced nitrogen-fixing phenotype [71]. Altogether, these different studies highlight the major role of rhizobial GSH in regulating the intracellular redox environment and protecting cells against ROS.

Besides its role in redox balance, the GSH pool in nodules might also be crucial in regulating metabolic pathways. The *R. etli gshB* and *gor* mutants were shown to be affected in Gln uptake in free-living bacteria [69]. Similarly, a *gshB* mutant of *Rhizobium leguminosarum* is impaired in symbiosis with *P. sativum* and presents a defect in the uptake of several carbon source compounds in free-living bacteria [72].

Glutathione is involved in the maintenance of cellular redox homeostasis in particular as a reductant for Grxs. The function of bacterial Grxs during rhizobium–legume symbiosis has been investigated in *S. meliloti*. The genome of this bacterium encodes three Grxs, the dithiol SmGrx1

(CGYC redox active site), the monothiol SmGrx2 (CGFS redox active site), and the atypical SmGrx3 which carries two domains, an N-terminal Grx domain with a CPYG active site and a C-terminal domain with a methylamine utilization protein (MauE) motif. Both SmGrx1 and SmGrx2 orthologs are ubiquitously present in bacteria while SmGrx3 orthologs are found only in cyanobacteria and some proteobacteria [73]. Biochemical and genetic analyses established that the three proteins have distinct properties [73]. SmGrx1 was shown to play a key role in protein deglutathionylation: on one hand SmGrx1 recombinant protein displayed an efficient degluthationylation activity, on the other *Smgrx1* inactivation in free-living bacteria led to a higher level of glutathionylated proteins. The *Smgrx1* deficient mutant undergoes a severe growth defect under non-stress conditions and an increased sensitivity to H_2O_2 treatment. During the interaction with *M. truncatula* the *Smgrx1* mutant induces abortive nodules, containing bacteria unable to differentiate into bacteroids following release inside plant cells. This original symbiotic phenotype suggests that the control of protein and redox homeostasis by Grx1-mediated protein deglutathionylation is crucial for bacteroid differentiation.

Data obtained with SmGrx2 provide the first demonstration of Grx involvement in bacterial iron metabolism [73]. *Smgrx2* inactivation in free-living bacteria results in the decreased activity of Fe–S cluster containing enzymes, suggesting that SmGrx2 participates to Fe–S cluster assembly machinery. A deregulation of RirA (Rhizobial iron regulator)-dependent genes, and an increase of the total intracellular iron content, was also observed in the *Smgrx2* mutant. During the interaction between *S. meliloti* and *M. truncatula*, *Smgrx2* inactivation affects nodulation efficiency and the nitrogen-fixation capacity of bacteroids; *Smgrx2* bacteroids are fully differentiated, in contrast to those of *Smgrx1*. The nitrogen-fixation deficiency of mutant bacteroids could result from a direct effect on nitrogenase which contains many Fe–S clusters. Indeed, the nitrogen-fixing enzyme consists of two Fe–S cluster-containing proteins, the dimeric Fe protein that serves as the electron donor for N_2 reduction and as the site of ATP hydrolysis, and the heterotetrameric MoFe protein where substrates are reduced. The Fe protein contains a Fe–S cluster while the MoFe protein contains two unique metal clusters, the [8Fe:7S] P-cluster and the FeMo cofactor described as a [Mo:7Fe:9S]:C-homocitrate entity [74]. Consistently, a mutant in *sufT*, involved in Fe–S cluster metabolism, also has a lowered nitrogen fixation capacity [75].

Concerning SmGrx3, the same approaches used to analyze SmGrx1 and SmGrx2 function were performed. Whereas a SmGrx3 recombinant protein presents a low degluthationylation activity, a *Smgrx3* mutant did not display defective phenotype in the free-living and symbiotic states. The biological function of SmGrx3 still remains to be elucidated.

In conclusion, SmGrx1 and smGrx2 play distinct, critical roles in the control of *S. meliloti* physiology. The growth and symbiotic defects of *grx* mutants also indicate that Grx and Trx systems are not functionally redundant in *S. meliloti*, in contrary to the thiol-redox systems of *E. coli* [76]. The question arises as to whether these properties can be generalized to other rhizobial Grxs and deserves more studies.

3.2. The Thioredoxin System

The thioredoxin system consists of NADPH, the flavoprotein Trx reductase (TR), and Trxs. A very limited number of studies have investigated the role of thioredoxin system in rhizobia. Trx-like proteins were initially described as playing an important role in symbiosis. In *S. meliloti* CE52G, inactivation of a *trx*-like gene involved in melanine production increased the sensitivity of free-living bacteria to paraquat-induced stress and affected the nitrogen fixation capacity of bacteroids [77]. In *R. leguminosarum*, a mutant deficient in the Trx-like TlpA, involved in cytochrome c biogenesis, was unable to form nitrogen-fixing nodule [78]. TlpA was recently shown to act as a reductant for the copper metallochaperone ScoI and cytochrome oxidase subunit II CoxB [79].

In *S. meliloti* as in *E. coli*, the canonical Trx system contains two Trxs, TrxA and the product of *SMc03801* (TrxC in *E. coli*), and one TR (TrxB). Recent results showed that TrxB recombinant protein efficiently reduces Trx s1, a host-plant thioredoxin specifically addressed to the microsymbiont, which is able to reduce NCR and is involved in bacteroid differentiation [60]. These data suggest that TrxB is implicated in the redox regulation of differentiation by reactivating Trx s1 but further studies are required to characterize the physiological role of TrxB during symbiosis.

3.3. Transcriptional Regulation of Trx and Grx Systems in S. meliloti

Various gene expression studies underline the importance of rhizobial Trx and Grx systems during symbiosis; we will focus on *S. meliloti* for which most of the results were obtained.

A high expression level of Trx/Grx component genes was observed in bacteroids from different zones of the *M. truncatula* nodules ([50]; Table 3). Some of these genes belong to stress response regulons, markedly required for *S. meliloti* survival in host cells.

Regulation of the *S. meliloti* GSH metabolic pathway involves the activity of LsrB, a transcriptional regulator required for efficient alfalfa nodulation [80]. LsrB belongs to the LysR family of bacterial transcriptional regulators including the oxidative stress regulator OxyR [81]. An *lsrB* deletion mutant has a reduced pool of GSH, and LsrB inactivation accordingly results in the decreased expression of genes involved in GSH metabolism (*gshA*, *gshB*, *gor*) both in free-living and in planta [82,83]. The regulator was shown to directly activate the expression of *gshA*, and to respond to cellular redox changes via the three reactive cysteines in the substrate-binding domain [83]. LsrB also positively regulates the expression of genes involved in lipopolysaccharide biosynthesis [84]. Nodules induced by mutants defective in LsrB undergo premature senescence coupled to impaired bacteroid differentiation and ROS accumulation, which could be partly due to GSH deficiency [83]. Several genes of the Trx/Grx systems belong to the RpoH1 regulon. RpoH1 is one of the 14 alternative sigma factors encoded in the *S. meliloti* genome. The presence of multiple RpoHs in *S. meliloti* and other alpha proteobacteria is correlated with a diverse lifestyle. RpoH1 regulates gene expression in response to acidic pH stress [85,86], heat shock, and stationary phase [87], and was also involved in maintaining the redox status of the cell challenged with H_2O_2 [88]. A *rpoH1* mutant is capable of eliciting the formation of nodules on alfalfa plants, but shows poor survival after its release in plant cells and barely fixes N_2 [89]. In addition to environmental stresses encountered both in free-living state and in planta, the *S. meliloti* microsymbiont is challenged with hundreds of peptides secreted by the host-plant, and largely involved in controlling bacterial populations during nodule development and functioning [90]. Transcriptome analyses of cultures challenged with two cationic NCR peptides exhibiting antimicrobial activities, NCR247 and NCR335, showed upregulation of genes involved in stress adaptation such as *Smgrx1* and *trxA* [91], see Table 3. This effect might be mediated via RpoH1, as the *rpoH1* gene itself was induced by NCR treatment [91].

Other, still unknown transcription factors and signals are likely also to be involved in the regulation of Grx/Trx systems, and other regulatory mechanisms as well. For example, the expression of *gshB* and *Smgrx2* is very low in zone III whereas the activity of GSHS and SmGrx2 is required in this zone, indicating that post-transcriptional regulation mechanism(s) could play a significant role.

Some well-known target proteins of Trx or Grx have a high expression level inside the nodules and might contribute to their optimal development (Table 3). Ribonucleotide reductase (RNR) plays a central role in DNA replication and repair by catalyzing production of deoxyribonucleotides from the corresponding ribonucleotides. Both Trx and Grx were identified as being dithiol electron donors for the *E. coli* RNR [92]. There are three major classes of RNRs based on the metallocofactors necessary for nucleotide reduction. *S. meliloti* requires a cobalamin-dependent class II RNR for symbiosis with *M. sativa* [93]. This RNR is most likely involved in DNA synthesis during bacteroid differentiation, when cells undergo endoreduplication, and later in DNA repair within differentiated bacteroids. The *nrdJ* gene encoding RNR has a maximal expression in zone III, suggesting that the level of RNR synthesis and DNA repair mechanisms are tightly linked in bacteroids. Trxs are also involved in protein repair by providing the electrons to peptide methionine sulfoxide reductases (MsrA/MsrB), which catalyze the reduction of methionine sulfoxides (S- and R-MetSO diastereosisomers respectively) back to methionine [94]. Three *msrA* and three *msrB* genes are present in the genome of *S. meliloti*. The highest level of *msrA/msrB* expression in nodules is observed in the differentiation and nitrogen fixation zones, a feature probably correlated to an increased methionine oxidation once bacteroid differentiation has begun. Most *msrA/msrB* genes are controlled by RpoH1, which underlines the coordinated expression of genes encoding components of Trx system and their targets.

Table 3. Expression of *S. meliloti* genes from the Grx and Trx systems in *M. truncatula* nodules and regulation in free-living bacteria. Gene accession numbers are indicated in the table. Gene annotation is based on candidate orthologs and interprodomain signature. The values corresponding to gene expression in root nodules are, from left to right, total reads from laser-capture microdissection (LCM) and their distribution in each zone (%), as reported by Roux and colleagues [50]. All RNA-seq read values were normalized. The full organs were 10-day old nodules. IZ, interzone; ZIII, zone III; FI, fraction I; FIId, distal fraction II; FIIp, proximal fraction II.

S. meliloti Genes	Bacterial Gene Expression in M. truncatula Nodules						Transcription Factors	Inducing Conditions	References
	Total reads LCM	% FI	% FIIp	% FIId	% IZ	% FIII			
Grx system									
SMc00825 (gshA)	3819	16	22	30	16	16	LsrB	GSSG	[82,83]
SMc00419 (gshB)	6433	8	9	24	53	6	LsrB, RpoH1		[82,87]
SMc00154 (gor)	4477	19	12	28	24	17	LsrB, RpoH1		[82,87]
SMc02443 (Smgrx1)	9123	7	6	20	16	51	RpoH1	low pH NCR247, NCR335	[86] [91]
SMc00538 (Smgrx2)	10138	24	24	21	22	9			
SMa0280 (Smgrx3)	1571	17	18	22	22	21		HS	[87]
Trx system									
SMc02761 (trxA)	5519	13	10	19	21	37	RpoH1	HS NCR247, NCR335	[87] [91]
SMc03801	2780	20	18	31	14	17	RpoH1		
SMc01224 (trxB)	5394	18	17	28	13	24	RpoH1	low pH, HS	[86,87]
Grx/Trx targets									
SMc02885 (msrA1)	2016	11	8	17	32	33	RpoH1	low pH, HS NCR247, NCR335	[86,87] [91]
SMc02467 (msrA2)	1690	5	14	22	43	16			
SMa1896 (msrA3)	551	20	12	26	20	21	RpoH1	HS NCR247, NCR335 H$_2$O$_2$	[87] [91] [88]
SMc00117 (msrB1)	729	24	13	19	23	20	RpoH1	HS	[87]
SMa1894 (msrB2)	107	0	14	17	24	45	RpoH1	HS NCR247, NCR335 H$_2$O$_2$	[87] [91] [88]
SMc01724 (msrB3)	795	0	19	23	18	40			
SMc01237 (nrdJ)	33647	9	7	4	4	76			
SMb20964	1308	29	9	9	7	45	OxyR	H$_2$O$_2$	[88]

GSSG, glutathione disulfide; HS, heat shock.

S. meliloti encodes thiol-base peroxidases of distinct families, one typical 1-Cys peroxiredoxin (product of *SMb20964*), and two organic hydroperoxide resistance thiol peroxidase paralogs from the OsmC/Ohr family [95]. Whereas Ohr proteins used a lipoylated protein as reductant, the bacterial 1-Cys or 2-Cys peroxiredoxins can use Trx or GSH as reductants to support small alkyl peroxide or H$_2$O$_2$ reductase activity [96,97]. The *S. meliloti* 1-Cys peroxiredoxin is upregulated in cultures challenged with H$_2$O$_2$ via an OxyR-dependent mechanism [88]. The corresponding protein has been identified in nodules [98] and transcripts were detected in bacteroids mostly in zone III [50], suggesting a role in nodule functioning. In *R. etli*, the typical 2-Cys peroxiredoxin PrxS uses the thioredoxin system for H$_2$O$_2$ reductase activity. The *R. etli* double mutant *prxS-katG*, deficient for both peroxidoxin and catalase-dependent H$_2$O$_2$ reduction, induced nodules with reduced nitrogen-fixation capability [99].

In conclusion, these data suggest the existence of a complex oxidative stress response network involving Grx and Trx to control protein redox state, and which allow bacteria to adapt to the host cell environment.

4. Conclusions

During the last years, many advances have been made in the characterization of redox regulatory systems and their roles in the two partners of nitrogen-fixing symbiosis (Figure 2). The development of genomic and transcriptomic analyses allowed to better characterize the gene families involved in redox metabolism and to define their expression regulation. A new research track was also opened with the redox regulation of the cross-talk between plant and bacteria, exemplified by Trx s1. However, these advances are revealing the complexity of the regulatory mechanisms and an increased number of key regulatory actors, as illustrated by the amazing high number of Grxs and Trxs in the plant partner. Characterizing the most promising candidates represents an important task both at the scientific level and in terms of work amount. Moreover, many lines of research remain to be opened. One of them is to assemble the different pieces of the "redox puzzle", taking into account the redox post-transcriptional regulation signals including oxidation, nitrosylation, and glutathionylation. In this perspective, the development of redox proteomics and laser microdissection will allow the large-scale identification of proteins that are modified in response to specific stimuli in specific cell types. Another crucial task would be the use of this huge amount of knowledge to improve the resistance of nitrogen fixation efficiency to abiotic stresses. Intensive farming leads to a significant increase in surfaces affected by drought or salinity, which particularly impairs nitrogen fixation. In the context of sustainable agriculture, the use of redox components as markers for symbiotic efficiency or as genetic material to improve plant breeding or bacterial inoculum is of crucial importance.

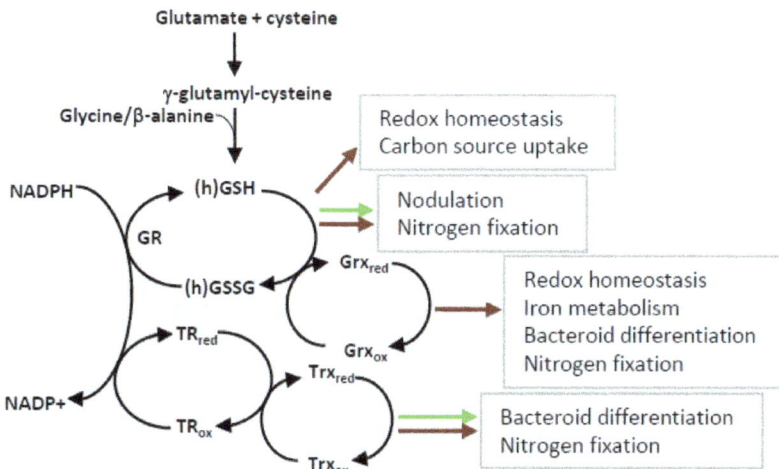

Figure 2. An overview of the physiological importance of Trx and Grx networks in rhizobium–legume symbiosis. Redox networks of glutaredoxin and thioredoxin systems in the two symbiotic partners is shown. The roles of (h)GSH, Grxs, and Trxs (grey squares) are indicated for bacteria (brown arrow) and for plants (green arrow). See text for details.

Funding: This work was supported by the "Institut National de la Recherche Agronomique", the "Centre National de la Recherche Scientifique", the University of Nice—Sophia Antipolis and the French Government (National Research Agency, ANR) through the "Investments for the Future" LABEX SIGNALIFE: program reference #ANR-11-LABX-0028-01.

Acknowledgments: We gratefully acknowledge Li Yang for providing the document for Figure 1. We thank the Microscopy and the Analytical Biochemistry Platforms—Sophia Agrobiotech Institut—INRA 1355—UNS—CNRS 7254—INRA PACA-Sophia Antipolis for access to instruments.

Conflicts of Interest: The authors declare no conflict of interest.

References

1. Wang, W.; Shi, J.; Xie, Q.; Jiang, Y.; Yu, N.; Wang, E. Nutrient exchange and regulation in arbuscular mycorrhizal symbiosis. *Mol. Plant* **2017**, *10*, 1147–1158. [CrossRef] [PubMed]

2. Martin, F.M.; Uroz, S.; Barker, D.G. Ancestral alliances: Plant mutualistic symbioses with fungi and bacteria. *Science* **2017**, *356*. [CrossRef] [PubMed]

3. Wang, Q.; Liu, J.; Zhu, H. Genetic and Molecular Mechanisms Underlying Symbiotic Specificity in Legume-Rhizobium Interactions. *Front. Plant Sci.* **2018**, *9*, 313. [CrossRef] [PubMed]

4. Oldroyd, G.E. Speak, friend, and enter: Signalling systems that promote beneficial symbiotic associations in plants. *Nat. Rev. Microbiol.* **2013**, *11*, 252–263. [CrossRef] [PubMed]

5. Oldroyd, G.E.; Murray, J.D.; Poole, P.S.; Downie, J.A. The rules of engagement in the legume-rhizobial symbiosis. *Annu. Rev. Genet.* **2011**, *45*, 119–144. [CrossRef] [PubMed]

6. Maroti, G.; Kondorosi, E. Nitrogen-fixing Rhizobium-legume symbiosis: Are polyploidy and host peptide-governed symbiont differentiation general principles of endosymbiosis? *Front. Microbiol.* **2014**, *5*, 326. [PubMed]

7. Hirsch, A.M. Developmental biology of legume nodulation. *New Phytol.* **1992**, *122*, 211–237. [CrossRef]

8. Pierre, O.; Hopkins, J.; Combier, M.; Baldacci, F.; Engler, G.; Brouquisse, R.; Herouart, D.; Boncompagni, E. Involvement of papain and legumain proteinase in the senescence process of *Medicago truncatula* nodules. *New Phytol.* **2014**, *202*, 849–863. [CrossRef]

9. Kondorosi, E.; Kondorosi, A. Endoreduplication and activation of the anaphase-promoting complex during symbiotic cell development. *FEBS Lett.* **2004**, *567*, 152–157. [CrossRef]

10. Vinardell, J.M.; Fedorova, E.; Cebolla, A.; Kevei, Z.; Horvath, G.; Kelemen, Z.; Tarayre, S.; Roudier, F.; Mergaert, P.; Kondorosi, A. Endoreduplication mediated by the anaphase-promoting complex activator CCS52A is required for symbiotic cell differentiation in *Medicago truncatula* nodules. *Plant Cell* **2003**, *15*, 2093–2105. [CrossRef]

11. Kondorosi, E.; Mergaert, P.; Kereszt, A. A paradigm for endosymbiotic life: Cell differentiation of Rhizobium bacteria provoked by host plant factors. *Annu. Rev. Microbiol.* **2013**, *67*, 611–628. [CrossRef]

12. Mergaert, P.; Nikovics, K.; Kelemen, Z.; Maunoury, N.; Vaubert, D.; Kondorosi, A.; Kondorosi, E. A novel family in *Medicago truncatula* consisting of more than 300 nodule-specific genes coding for small, secreted polypeptides with conserved cysteine motifs. *Plant Physiol.* **2003**, *132*, 161–173. [CrossRef] [PubMed]

13. Horvath, B.; Domonkos, A.; Kereszt, A.; Szucs, A.; Abraham, E.; Ayaydin, F.; Boka, K.; Chen, Y.; Chen, R.; Murray, J.D.; et al. Loss of the nodule-specific cysteine rich peptide, NCR169, abolishes symbiotic nitrogen fixation in the *Medicago truncatula dnf7* mutant. *Proc. Natl. Acad. Sci. USA* **2015**, *112*, 15232–15237. [CrossRef] [PubMed]

14. Kim, M.; Chen, Y.; Xi, J.; Waters, C.; Chen, R.; Wang, D. An antimicrobial peptide essential for bacterial survival in the nitrogen-fixing symbiosis. *Proc. Natl. Acad. Sci. USA* **2015**, *112*, 15238–15243. [CrossRef] [PubMed]

15. Puppo, A.; Pauly, N.; Boscari, A.; Mandon, K.; Brouquisse, R. Hydrogen peroxide and nitric oxide: Key regulators of the Legume-Rhizobium and mycorrhizal symbioses. *Antioxid. Redox Signal.* **2013**, *18*, 2202–2219. [CrossRef] [PubMed]

16. Hichri, I.; Boscari, A.; Castella, C.; Rovere, M.; Puppo, A.; Brouquisse, R. Nitric oxide: A multifaceted regulator of the nitrogen-fixing symbiosis. *J. Exp. Bot.* **2015**, *66*, 2877–2887. [CrossRef]

17. Frendo, P.; Matamoros, M.A.; Alloing, G.; Becana, M. Thiol-based redox signaling in the nitrogen-fixing symbiosis. *Front. Plant Sci.* **2013**, *4*, 376. [CrossRef] [PubMed]

18. Ribeiro, C.W.; Alloing, G.; Mandon, K.; Frendo, P. Redox regulation of differentiation in symbiotic nitrogen fixation. *Biochim. Biophys. Acta* **2015**, *1850*, 1469–1478. [CrossRef]

19. Klapheck, S. Homoglutathione: Isolation, quantification and occurrence in legumes. *Physiol. Plant.* **1988**, *74*, 727–732. [CrossRef]

20. Frendo, P.; Jimenez, M.J.; Mathieu, C.; Duret, L.; Gallesi, D.; Van de Sype, G.; Herouart, D.; Puppo, A. A *Medicago truncatula* homoglutathione synthetase is derived from glutathione synthetase by gene duplication. *Plant Physiol.* **2001**, *126*, 1706–1715. [CrossRef]

21. Matamoros, M.A.; Clemente, M.R.; Sato, S.; Asamizu, E.; Tabata, S.; Ramos, J.; Moran, J.F.; Stiller, J.; Gresshoff, P.M.; Becana, M. Molecular analysis of the pathway for the synthesis of thiol tripeptides in the model legume Lotus japonicus. *Mol. Plant Microbe Interact.* **2003**, *16*, 1039–1046. [CrossRef] [PubMed]

22. Matamoros, M.A.; Moran, J.F.; Iturbe-Ormaetxe, I.; Rubio, M.C.; Becana, M. Glutathione and homoglutathione synthesis in legume root nodules. *Plant Physiol.* **1999**, *121*, 879–888. [CrossRef] [PubMed]

23. Bergmann, L.R.H. Glutathione metabolism in plants. In *Sulphur Nutrition and Assimilation in Higher Plants*; DeKok, S.I., Rennenberg, H., Brunold, C., Rauser, W., Eds.; SPB Academic Publishing: The Hague, The Netherlands, 1993; pp. 109–124.

24. Meyer, A.J.; May, M.J.; Fricker, M. Quantitative in vivo measurement of glutathione in Arabidopsis cells. *Plant J.* **2001**, *27*, 67–78. [CrossRef] [PubMed]

25. Kalloniati, C.; Krompas, P.; Karalias, G.; Udvardi, M.K.; Rennenberg, H.; Herschbach, C.; Flemetakis, E. Nitrogen-fixing nodules are an important source of reduced sulfur, which triggers global changes in sulfur metabolism in *Lotus japonicus*. *Plant Cell* **2015**, *27*, 2384–2400. [CrossRef] [PubMed]

26. Thal, B.; Braun, H.P.; Eubel, H. Proteomic analysis dissects the impact of nodulation and biological nitrogen fixation on Vicia faba root nodule physiology. *Plant Mol. Biol.* **2018**, *97*, 233–251. [CrossRef] [PubMed]

27. Krusell, L.; Krause, K.; Ott, T.; Desbrosses, G.; Kramer, U.; Sato, S.; Nakamura, Y.; Tabata, S.; James, E.K.; Sandal, N.; et al. The sulfate transporter SST1 is crucial for symbiotic nitrogen fixation in *Lotus japonicus* root nodules. *Plant Cell* **2005**, *17*, 1625–1636. [CrossRef]

28. Wienkoop, S.; Saalbach, G. Proteome analysis. Novel proteins identified at the peribacteroid membrane from *Lotus japonicus* root nodules. *Plant Physiol.* **2003**, *131*, 1080–1090. [CrossRef]

29. Clarke, V.C.; Loughlin, P.C.; Gavrin, A.; Chen, C.; Brear, E.M.; Day, D.A.; Smith, P.M. Proteomic analysis of the soybean symbiosome identifies new symbiotic proteins. *Mol. Cell Proteom.* **2015**, *14*, 1301–1322. [CrossRef]

30. Macnicol, P.K.; Randall, P.J. Changes in the levels of major sulfur metabolites and free amino acids in pea cotyledons recovering from sulfur deficiency. *Plant Physiol.* **1987**, *83*, 354–359. [CrossRef]

31. Iturbe-Ormaetxe, I.; Heras, B.; Matamoros, M.A.; Ramos, J.; Moran, J.F.; Becana, M. Cloning and functional characterization of a homoglutathione synthetase from pea nodules. *Physiol. Plant* **2002**, *115*, 69–73. [CrossRef]

32. Galant, A.; Preuss, M.; Cameron, J.; Jez, J. Plant Glutathione Biosynthesis: Diversity in Biochemical Regulation and Reaction Products. *Front. Plant Sci.* **2011**, *2*, 45. [CrossRef]

33. Frendo, P.; Mathieu, C.; Van de Sype, G.; Herouart, D.; Puppo, A. Characterisation of a cDNA encoding gamma-glutamylcysteine synthetase in *Medicago truncatula*. *Free Radic. Res.* **1999**, *31*, S213–S218. [CrossRef]

34. Dalton, D.A.; Baird, L.M.; Langeberg, L.; Taugher, C.Y.; Anyan, W.R.; Vance, C.P.; Sarath, G. Subcellular localization of oxygen defense enzymes in Soybean (*Glycine max* [L.] Merr.) root nodules. *Plant Physiol.* **1993**, *102*, 481–489. [CrossRef]

35. Evans, P.J.; Gallesi, D.; Mathieu, C.; Hernandez, M.J.; de Felipe, M.; Halliwell, B.; Puppo, A. Oxidative stress occurs during soybean nodule senescence. *Planta* **1999**, *208*, 73–79. [CrossRef]

36. Groten, K.; Vanacker, H.; Dutilleul, C.; Bastian, F.; Bernard, S.; Carzaniga, R.; Foyer, C.H. The roles of redox processes in pea nodule development and senescence. *Plant Cell Environ.* **2005**, *28*, 1293–1304. [CrossRef]

37. Escuredo, P.R.; Minchin, F.R.; Gogorcena, Y.; Iturbe-Ormaetxe, I.; Klucas, R.V.; Becana, M. Involvement of activated oxygen in nitrate-Induced senescence of pea root nodules. *Plant Physiol.* **1996**, *110*, 1187–1195. [CrossRef]

38. Gogorcena, Y.; Gordon, A.J.; Escuredo, P.R.; Minchin, F.R.; Witty, J.F.; Moran, J.F.; Becana, M. N_2 fixation, carbon metabolism and oxidative damage in nodules of dark stressed common bean plants. *Plant Physiol.* **1997**, *113*, 1193–1201. [CrossRef]

39. Matamoros, M.A.; Baird, L.M.; Escuredo, P.R.; Dalton, D.A.; Minchin, F.R.; Iturbe-Ormaetxe, I.; Rubio, M.C.; Moran, J.F.; Gordon, A.J.; Becana, M. Stress induced legume root nodule senescence. Physiological, biochemical and structural alterations. *Plant Physiol.* **1999**, *121*, 97–112. [CrossRef]

40. Marino, D.; Frendo, P.; Ladrera, R.; Zabalza, A.; Puppo, A.; Arrese-Igor, C.; Gonzalez, E.M. Nitrogen fixation control under drought stress. Localized or systemic? *Plant Physiol.* **2007**, *143*, 1968–1974. [CrossRef]

41. Naya, L.; Ladrera, R.; Ramos, J.; Gonzalez, E.M.; Arrese-Igor, C.; Minchin, F.R.; Becana, M. The response of carbon metabolism and antioxidant defenses of alfalfa nodules to drought stress and to the subsequent recovery of plants. *Plant Physiol.* **2007**, *144*, 1104–1114. [CrossRef]

42. Frendo, P.; Harrison, J.; Norman, C.; Hernandez Jimenez, M.J.; Van de Sype, G.; Gilabert, A.; Puppo, A. Glutathione and homoglutathione play a critical role in the nodulation process of *Medicago truncatula*. *Mol. Plant Microbe Interact.* **2005**, *18*, 254–259. [CrossRef]

43. Vernoux, T.; Wilson, R.C.; Seeley, K.A.; Reichheld, J.P.; Muroy, S.; Brown, S.; Maughan, S.C.; Cobbett, C.S.; Van Montagu, M.; Inze, D.; et al. The ROOT MERISTEMLESS1/CADMIUM SENSITIVE2 gene defines a glutathione-dependent pathway involved in initiation and maintenance of cell division during postembryonic root development. *Plant Cell* **2000**, *12*, 97–110. [CrossRef]

44. Reichheld, J.P.; Khafif, M.; Riondet, C.; Droux, M.; Bonnard, G.; Meyer, Y. Inactivation of thioredoxin reductases reveals a complex interplay between thioredoxin and glutathione pathways in Arabidopsis development. *Plant Cell* **2007**, *19*, 1851–1865. [CrossRef]

45. Schippers, J.H.; Foyer, C.H.; van Dongen, J.T. Redox regulation in shoot growth, SAM maintenance and flowering. *Curr. Opin. Plant Biol.* **2016**, *29*, 121–128. [CrossRef]

46. Pucciariello, C.; Innocenti, G.; Van de Velde, W.; Lambert, A.; Hopkins, J.; Clement, M.; Ponchet, M.; Pauly, N.; Goormachtig, S.; Holsters, M.; et al. (Homo)glutathione depletion modulates host gene expression during the symbiotic interaction between *Medicago truncatula* and *Sinorhizobium meliloti*. *Plant Physiol.* **2009**, *151*, 1186–1196. [CrossRef]

47. El Msehli, S.; Lambert, A.; Baldacci-Cresp, F.; Hopkins, J.; Boncompagni, E.; Smiti, S.A.; Herouart, D.; Frendo, P. Crucial role of (homo)glutathione in nitrogen fixation in *Medicago truncatula* nodules. *New Phytol.* **2011**, *192*, 496–506. [CrossRef]

48. Rouhier, N.; Couturier, J.; Johnson, M.K.; Jacquot, J.P. Glutaredoxins: Roles in iron homeostasis. *Trends Biochem. Sci.* **2010**, *35*, 43–52. [CrossRef]

49. Meyer, Y.; Belin, C.; Delorme-Hinoux, V.; Reichheld, J.P.; Riondet, C. Thioredoxin and glutaredoxin systems in plants: Molecular mechanisms, crosstalks, and functional significance. *Antioxid. Redox Signal.* **2012**, *17*, 1124–1160. [CrossRef]

50. Roux, B.; Rodde, N.; Jardinaud, M.F.; Timmers, T.; Sauviac, L.; Cottret, L.; Carrère, S.; Sallet, E.; Courcelle, E.; Moreau, S.; et al. An integrated analysis of plant and bacterial gene expression in symbiotic root nodules using laser-capture microdissection coupled to RNA sequencing. *Plant J.* **2014**, *77*, 817–837. [CrossRef]

51. Gelhaye, E.; Rouhier, N.; Gerard, J.; Jolivet, Y.; Gualberto, J.; Navrot, N.; Ohlsson, P.I.; Wingsle, G.; Hirasawa, M.; Knaff, D.B.; et al. A specific form of thioredoxin h occurs in plant mitochondria and regulates the alternative oxidase. *Proc. Natl. Acad. Sci. USA* **2004**, *101*, 14545–14550. [CrossRef]

52. Renard, M.; Alkhalfioui, F.; Schmitt-Keichinger, C.; Ritzenthaler, C.; Montrichard, F. Identification and characterization of thioredoxin h isoforms differentially expressed in germinating seeds of the model legume *Medicago truncatula*. *Plant Physiol.* **2011**, *155*, 1113–1126. [CrossRef]

53. Meyer, Y.; Buchanan, B.B.; Vignols, F.; Reichheld, J.P. Thioredoxins and glutaredoxins: Unifying elements in redox biology. *Annu. Rev. Genet.* **2009**, *43*, 335–367. [CrossRef]

54. Pulido, P.; Cazalis, R.; Cejudo, F.J. An antioxidant redox system in the nucleus of wheat seed cells suffering oxidative stress. *Plant J.* **2009**, *57*, 132–145. [CrossRef]

55. Marchal, C.; Delorme-Hinoux, V.; Bariat, L.; Siala, W.; Belin, C.; Saez-Vasquez, J.; Riondet, C.; Reichheld, J.P. NTR/NRX define a new thioredoxin system in the nucleus of *Arabidopsis thaliana* cells. *Mol. Plant* **2014**, *7*, 30–44. [CrossRef]

56. Alkhalfioui, F.; Renard, M.; Frendo, P.; Keichinger, C.; Meyer, Y.; Gelhaye, E.; Hirasawa, M.; Knaff, D.B.; Ritzenthaler, C.; Montrichard, F. A novel type of thioredoxin dedicated to symbiosis in legumes. *Plant Physiol.* **2008**, *148*, 424–435. [CrossRef]

57. Tovar-Mendez, A.; Matamoros, M.A.; Bustos-Sanmamed, P.; Dietz, K.J.; Cejudo, F.J.; Rouhier, N.; Sato, S.; Tabata, S.; Becana, M. Peroxiredoxins and NADPH-dependent thioredoxin systems in the model legume *Lotus japonicus*. *Plant Physiol.* **2011**, *156*, 1535–1547. [CrossRef]

58. Lee, M.Y.; Shin, K.H.; Kim, Y.K.; Suh, J.Y.; Gu, Y.Y.; Kim, M.R.; Hur, Y.S.; Son, O.; Kim, J.S.; Song, E.; et al. Induction of thioredoxin is required for nodule development to reduce reactive oxygen species levels in soybean roots. *Plant Physiol.* **2005**, *139*, 1881–1889. [CrossRef]

59. Du, H.; Kim, S.; Nam, K.H.; Lee, M.S.; Son, O.; Lee, S.H.; Cheon, C.I. Identification of uricase as a potential target of plant thioredoxin: Implication in the regulation of nodule development. *Biochem. Biophys. Res. Commun.* **2010**, *397*, 22–26. [CrossRef]

60. Ribeiro, C.W.; Baldacci-Cresp, F.; Pierre, O.; Larousse, M.; Benyamina, S.; Lambert, A.; Hopkins, J.; Castella, C.; Cazareth, J.; Alloing, G.; et al. Regulation of differentiation of nitrogen-fixing bacteria by microsymbiont targeting of plant thioredoxin s1. *Curr. Biol.* **2017**, *27*, 250–256. [CrossRef]

61. Rouhier, N.; Jacquot, J.P. The plant multigenic family of thiol peroxidases. *Free Radic. Biol. Med.* **2005**, *38*, 1413–1421. [CrossRef]

62. Castella, C.; Mirtziou, I.; Seassau, A.; Boscari, A.; Montrichard, F.; Papadopoulou, K.; Rouhier, N.; Puppo, A.; Brouquisse, R. Post-translational modifications of *Medicago truncatula* glutathione peroxidase 1 induced by nitric oxide. *Nitric Oxide* **2017**, *68*, 125–136. [CrossRef]

63. Ramos, J.; Matamoros, M.A.; Naya, L.; James, E.K.; Rouhier, N.; Sato, S.; Tabata, S.; Becana, M. The glutathione peroxidase gene family of *Lotus japonicus*: Characterization of genomic clones, expression analyses and immunolocalization in legumes. *New Phytol.* **2009**, *181*, 103–114. [CrossRef]

64. Matamoros, M.A.; Saiz, A.; Penuelas, M.; Bustos-Sanmamed, P.; Mulet, J.M.; Barja, M.V.; Rouhier, N.; Moore, M.; James, E.K.; Dietz, K.J.; et al. Function of glutathione peroxidases in legume root nodules. *J. Exp. Bot.* **2015**, *66*, 2979–2990. [CrossRef]

65. Matamoros, M.A.; Fernandez-Garcia, N.; Wienkoop, S.; Loscos, J.; Saiz, A.; Becana, M. Mitochondria are an early target of oxidative modifications in senescing legume nodules. *New Phytol.* **2013**, *197*, 873–885. [CrossRef]

66. Groten, K.; Dutilleul, C.; van Heerden, P.D.; Vanacker, H.; Bernard, S.; Finkemeier, I.; Dietz, K.J.; Foyer, C.H. Redox regulation of peroxiredoxin and proteinases by ascorbate and thiols during pea root nodule senescence. *FEBS Lett.* **2006**, *580*, 1269–1276. [CrossRef]

67. Harrison, J.; Jamet, A.; Muglia, C.I.; Van de Sype, G.; Aguilar, O.M.; Puppo, A.; Frendo, P. Glutathione plays a fundamental role in growth and symbiotic capacity of *Sinorhizobium meliloti*. *J. Bacteriol.* **2005**, *187*, 168–174. [CrossRef]

68. Muglia, C.; Comai, G.; Spegazzini, E.; Riccillo, P.M.; Aguilar, O.M. Glutathione produced by *Rhizobium tropici* is important to prevent early senescence in common bean nodules. *FEMS Microbiol. Lett.* **2008**, *23*, 191–198. [CrossRef]

69. Tate, R.; Cermola, M.; Riccio, A.; Diez-Roux, G.; Patriarca, E.J. Glutathione is required by *Rhizobium etli* for glutamine utilization and symbiotic effectiveness. *Mol. Plant Microbe Interact.* **2012**, *25*, 331–340. [CrossRef]

70. Sobrevals, L.; Muller, P.; Fabra, A.; Castro, S. Role of glutathione in the growth of *Bradyrhizobium* sp. (peanut microsymbiont) under different environmental stresses and in symbiosis with the host plant. *Can. J. Microbiol.* **2006**, *52*, 609–616. [CrossRef]

71. Tang, G.; Li, N.; Liu, Y.; Yu, L.; Yan, J.; Luo, L. *Sinorhizobium meliloti* Glutathione Reductase Is Required for both Redox Homeostasis and Symbiosis. *Appl. Environ. Microbiol.* **2018**, *84*. [CrossRef]

72. Cheng, G.; Karunakaran, R.; East, A.K.; Munoz-Azcarate, O.; Poole, P.S. Glutathione affects the transport activity of *Rhizobium leguminosarum* 3841 and is essential for efficient nodulation. *FEMS Microbiol. Lett.* **2017**, *364*. [CrossRef]

73. Benyamina, S.M.; Baldacci-Cresp, F.; Couturier, J.; Chibani, K.; Hopkins, J.; Bekki, A.; de Lajudie, P.; Rouhier, N.; Jacquot, J.P.; Alloing, G.; et al. Two *Sinorhizobium meliloti* glutaredoxins regulate iron metabolism and symbiotic bacteroid differentiation. *Environ. Microbiol.* **2013**, *15*, 795–810. [CrossRef]

74. Spatzal, T.; Aksoyoglu, M.; Zhang, L.; Andrade, S.L.; Schleicher, E.; Weber, S.; Rees, D.C.; Einsle, O. Evidence for interstitial carbon in nitrogenase FeMo cofactor. *Science* **2011**, *334*, 940. [CrossRef]

75. Sasaki, S.; Minamisawa, K.; Mitsui, H. A *Sinorhizobium meliloti* RpoH-regulated gene is involved in iron-sulfur protein metabolism and effective plant symbiosis under intrinsic iron limitation. *J. Bacteriol.* **2016**, *198*, 2297–2306. [CrossRef]

76. Prinz, W.A.; Aslund, F.; Holmgren, A.; Beckwith, J. The role of the thioredoxin and glutaredoxin pathways in reducing protein disulfide bonds in the *Escherichia coli* cytoplasm. *J. Biol. Chem.* **1997**, *272*, 15661–15667. [CrossRef]

77. Castro-Sowinski, S.; Matan, O.; Bonafede, P.; Okon, Y. A thioredoxin of *Sinorhizobium meliloti* CE52G is required for melanin production and symbiotic nitrogen fixation. *Mol. Plant Microbe Interact.* **2007**, *20*, 986–993. [CrossRef]

78. Vargas, C.; Wu, G.; Davies, A.E.; Downie, J.A. Identification of a gene encoding a thioredoxin-like product necessary for cytochrome c biosynthesis and symbiotic nitrogen fixation in *Rhizobium leguminosarum*. *J. Bacteriol.* **1994**, *176*, 4117–4123. [CrossRef]

79. Abicht, H.K.; Scharer, M.A.; Quade, N.; Ledermann, R.; Mohorko, E.; Capitani, G.; Hennecke, H.; Glockshuber, R. How periplasmic thioredoxin TlpA reduces bacterial copper chaperone ScoI and cytochrome oxidase subunit II (CoxB) prior to metallation. *J. Biol. Chem.* **2014**, *289*, 32431–32444. [CrossRef]

80. Luo, L.; Yao, S.Y.; Becker, A.; Ruberg, S.; Yu, G.Q.; Zhu, J.B.; Cheng, H.P. Two new *Sinorhizobium meliloti* LysR-type transcriptional regulators required for nodulation. *J. Bacteriol.* **2005**, *187*, 4562–4572. [CrossRef]

81. Imlay, J.A. Transcription factors that defend bacteria against reactive oxygen species. *Annu. Rev. Microbiol.* **2015**, *69*, 93–108. [CrossRef]

82. Lu, D.; Tang, G.; Wang, D.; Luo, L. The *Sinorhizobium meliloti* LysR family transcriptional factor LsrB is involved in regulation of glutathione biosynthesis. *Acta Biochim. Biophys. Sin. (Shanghai)* **2013**, *45*, 882–888. [CrossRef]

83. Tang, G.; Xing, S.; Wang, S.; Yu, L.; Li, X.; Staehelin, C.; Yang, M.; Luo, L. Regulation of cysteine residues in LsrB proteins from *Sinorhizobium meliloti* under free-living and symbiotic oxidative stress. *Environ. Microbiol.* **2017**, *19*, 5130–5145. [CrossRef]

84. Tang, G.; Wang, Y.; Luo, L. Transcriptional regulator LsrB of *Sinorhizobium meliloti* positively regulates the expression of genes involved in lipopolysaccharide biosynthesis. *Appl. Environ. Microbiol.* **2014**, *80*, 5265–5273. [CrossRef]

85. Ono, Y.; Mitsui, H.; Sato, T.; Minamisawa, K. Two RpoH homologs responsible for the expression of heat shock protein genes in *Sinorhizobium meliloti*. *Mol. Gen. Genet.* **2001**, *264*, 902–912. [CrossRef]

86. De Lucena, D.K.; Puhler, A.; Weidner, S. The role of sigma factor RpoH1 in the pH stress response of *Sinorhizobium meliloti*. *BMC Microbiol.* **2010**, *10*, 265. [CrossRef]

87. Barnett, M.J.; Bittner, A.N.; Toman, C.J.; Oke, V.; Long, S.R. Dual RpoH sigma factors and transcriptional plasticity in a symbiotic bacterium. *J. Bacteriol.* **2012**, *194*, 4983–4994. [CrossRef]

88. Lehman, A.P.; Long, S.R. OxyR-dependent transcription response of *Sinorhizobium meliloti* to oxidative stress. *J. Bacteriol.* **2018**, *200*. [CrossRef]

89. Mitsui, H.; Sato, T.; Sato, Y.; Ito, N.; Minamisawa, K. *Sinorhizobium meliloti* RpoH1 is required for effective nitrogen-fixing symbiosis with alfalfa. *Mol. Genet. Genom.* **2004**, *271*, 416–425. [CrossRef]

90. Kereszt, A.; Mergaert, P.; Montiel, J.; Endre, G.; Kondorosi, E. Impact of plant peptides on symbiotic nodule development and functioning. *Front. Plant Sci.* **2018**, *9*, 1026. [CrossRef]

91. Tiricz, H.; Szucs, A.; Farkas, A.; Pap, B.; Lima, R.M.; Maroti, G.; Kondorosi, E.; Kereszt, A. Antimicrobial nodule-specific cysteine-rich peptides induce membrane depolarization-associated changes in the transcriptome of *Sinorhizobium meliloti*. *Appl. Environ. Microbiol.* **2013**, *79*, 6737–6746. [CrossRef]

92. Gon, S.; Faulkner, M.J.; Beckwith, J. In vivo requirement for glutaredoxins and thioredoxins in the reduction of the ribonucleotide reductases of *Escherichia coli*. *Antioxid. Redox Signal.* **2006**, *8*, 735–742. [CrossRef]

93. Taga, M.E.; Walker, G.C. *Sinorhizobium meliloti* requires a cobalamin-dependent ribonucleotide reductase for symbiosis with its plant host. *Mol. Plant Microbe Interact.* **2010**, *23*, 1643–1654. [CrossRef]

94. Ezraty, B.; Aussel, L.; Barras, F. Methionine sulfoxide reductases in prokaryotes. *Biochim. Biophys. Acta* **2005**, *1703*, 221–229. [CrossRef]

95. Fontenelle, C.; Blanco, C.; Arrieta, M.; Dufour, V.; Trautwetter, A. Resistance to organic hydroperoxides requires ohr and ohrR genes in *Sinorhizobium meliloti*. *BMC Microbiol.* **2011**, *11*, 100. [CrossRef]

96. Barranco-Medina, S.; Lazaro, J.J.; Dietz, K.J. The oligomeric conformation of peroxiredoxins links redox state to function. *FEBS Lett.* **2009**, *583*, 1809–1816. [CrossRef]

97. Cussiol, J.R.; Alegria, T.G.; Szweda, L.I.; Netto, L.E. Ohr (organic hydroperoxide resistance protein) possesses a previously undescribed activity, lipoyl-dependent peroxidase. *J. Biol. Chem.* **2010**, *285*, 21943–21950. [CrossRef]

98. Djordjevic, M.A.; Chen, H.C.; Natera, S.; Van Noorden, G.; Menzel, C.; Taylor, S.; Renard, C.; Geiger, O.; the Sinorhizobium DNA Sequencing Consortium; Weiller, G.F. A global analysis of protein expression profiles in *Sinorhizobium meliloti*: Discovery of new genes for nodule occupancy and stress adaptation. *Mol. Plant Microbe Interact.* **2003**, *16*, 508–524. [CrossRef]

99. Dombrecht, B.; Heusdens, C.; Beullens, S.; Verreth, C.; Mulkers, E.; Proost, P.; Vanderleyden, J.; Michiels, J. Defence of *Rhizobium etli* bacteroids against oxidative stress involves a complexly regulated atypical 2-Cys peroxiredoxin. *Mol. Microbiol.* **2005**, *55*, 1207–1221. [CrossRef]

antioxidants

MDPI

Article

Proteomic Analyses of Thioredoxins *f* and *m* *Arabidopsis thaliana* Mutants Indicate Specific Functions for These Proteins in Plants

Juan Fernández-Trijueque [1], Antonio-Jesús Serrato [2] and Mariam Sahrawy [2,*]

[1] Master Diagnóstica, Avenida del Conocimiento, 100. P.T. Ciencias de la Salud, 18016 Granada, Spain; trijuek@gmail.com

[2] Departamento de Bioquímica, Biología Molecular y Celular de Plantas, Estación Experimental del Zaidín, Consejo Superior de Investigaciones Científicas, C/Profesor Albareda 1, 18008 Granada, Spain; aserrato@eez.csic.es

* Correspondence: sahrawy@eez.csic.es; Tel.: +34-958-181-600; Fax: +34-958-129-600

Received: 11 January 2019; Accepted: 25 February 2019; Published: 2 March 2019

Abstract: A large number of plastidial thioredoxins (TRX) are present in chloroplast and the specificity versus the redundancy of their functions is currently under discussion. Several results have highlighted the fact that each TRX has a specific target protein and thus a specific function. In this study we have found that in vitro activation of the fructose-1,6-bisphosphatase (FBPase) enzyme is more efficient when *f1* and *f2* type thioredoxins (TRXs) are used, whilst the *m3* type TRX did not have any effect. In addition, we have carried out a two-dimensional electrophoresis-gel to obtain the protein profiling analyses of the *trxf1, f2, m1, m2, m3* and *m4* Arabidopsis mutants. The results revealed quantitative alteration of 86 proteins and demonstrated that the lack of both the *f* and *m* type thioredoxins have diverse effects on the proteome. Interestingly, 68% of the differentially expressed proteins in *trxf1* and *trxf2* mutants were downregulated, whilst 75% were upregulated in *trxm1, trxm2, trxm3* and *trxm4* lines. The lack of TRX *f1* provoked a higher number of down regulated proteins. The contrary occurred when TRX *m4* was absent. Most of the differentially expressed proteins fell into the categories of metabolic processes, the Calvin–Benson cycle, photosynthesis, response to stress, hormone signalling and protein turnover. Photosynthesis, the Calvin–Benson cycle and carbon metabolism are the most affected processes. Notably, a significant set of proteins related to the answer to stress situations and hormone signalling were affected. Despite some studies being necessary to find specific target proteins, these results show signs that are suggest that the *f* and *m* type plastidial TRXs most likely have some additional specific functions.

Keywords: thioredoxins; plastidial; specificity; function; proteomic; photosynthesis; Calvin cycle

---

1. Introduction

Thioredoxins (TRXs) are small proteins (12–14 kDa) present in every organelle with the canonical redox active site WC(G/P)PC and a conserved tertiary structure, which modify their target proteins through the post-translationally reduction of disulphide bonds [1,2]. In the chloroplast, the ferredoxin/thioredoxin system (FTS), composed of ferredoxin (Fdx), ferredoxin thioredoxin reductase (FTR), and TRX, is responsible for the reduction of target proteins involved in a wide range of processes [3]. TRXs have been classified into different groups depending on their primary structures, biochemical properties, and sub-cellular localizations. So far, about 20 TRX types have been identified in plants [4]. This diversity suggests a functional specificity for the different isoforms present in plants, rather than a redundancy. For many years, the best-known plastid TRXs have been of the *f* and *m* type [5–8]. The *Arabidopsis thaliana* genome contains two TRX *f* (TRX *f1* and *f2*) and four TRX *m* (TRX *m1, m2, m3* and *m4*). One of the most important biological processes in chloroplasts, the Calvin–Benson cycle, is controlled by TRX *f*, with the reduction mechanism of the fructose-1,6-bisphosphatase (cFBPase) being well known [9]. On the other hand, the *m* type TRXs, originally described as reducers of the malate dehydrogenase (MDH), are more related to photosynthesis [10–12]. However, in recent years, many other processes in the chloroplast, such as starch metabolism, photosynthetic electron-transport chain, oxidative-stress response, lipid biosynthesis, nitrogen metabolism, protein folding, and translation [13,14] have been highlighted to be regulated by plastid TRXs. Additionally, other studies have shown evidence of new roles for the plastidial TRXs in heterotrophic tissues, such as roots or flowers [15]. Therefore, it is evident that the importance of the redox regulation through thiol/disulphide interchanges mediated by the thioredoxins happens in almost all the processes of these organelles. For several years, numerous studies have focused on plastidial TRXs, however, from a functional point of view, the information is rather scarce and the debate regarding functional specificity versus redundancy is still open. Technical advances in the coming years will probably allow us to discover many other target proteins.

In order to identify specific functions, the extended use of mutated TRXs for the in vitro target search has proved to be a powerful method that has generated valuable knowledge [16]. However, as we were unable to preserve the in vivo conditions which avoid non-specific TRX-target interactions, the sequence similarity shown among TRX isoforms represents a clear disadvantage for the study of functional specificities. Therefore, to shed more light on possible functional specificities of plastid TRXs we have carried out a novel approach consisting of a wide protein-profiling analysis of Arabidopsis *trxf1, trxf2, trxm1, trxm2, trxm3* and *trxm4* knock-out/down mutants compared with the wild-type plants Columbia 0 (Col0) and Landsberg erecta (Ler). Despite this, more specific studies are necessary, our results suggest that the plastid TRXs we have analyzed are more functionally specialized than expected as we go on to describe in this paper.

2. Materials and Methods

2.1. Plant Material and Growth Conditions

Arabidopsis thaliana wild type plants (ecotype Columbia (Col0) and Landsberg erecta (Ler)), *trxf1* (SALK line SALK_128365), *trxf2* (Gabi-kat line GK_020E05), *trxm1* (SALK line SALK_087118), *trxm2* (SALK line SALK_130686), *trxm3* (ET_3878, background Ler), *trxm4* (SALK line SALK_032538) were grown in soil in a growth chamber at 22 °C under long-day conditions (16 h light/8 h dark) and with photosynthetically active radiation of 120 µmol photons m^{-2} s^{-1}. The observed phenotypes, described previously by [17], similar to the wild type lines, were caused by the disruption of the TRX genes. Rosettes from 25 day-old plants were immediately transferred to liquid nitrogen before storage at −80 °C. Expression level was performed by semiquantitative polymerase chain reaction (PCR) analysis using specific oligonucleotides (Table S1) following instructions of Barajas et al. [15].

2.2. Protein Extraction, Solubilisation

Protein extraction from a pool of a minimum of 6 rosette plants was performed by using trichloroacetic acid TCA–acetone–phenol protocol [18]. The final pellet was suspended in 600 µL of protein solubilization buffer (9 M urea, 4% 3-[(3-Cholamidopropyl)dimethylammonio]-1-propanesulfonate hydrate (CHAPS), 0.5% TritonX100, and 100 mM dithiothreitol (DTT)). Protein content was quantified by the method of Bradford [19], using bovine serum albumin (BSA) as standard. Three technical replicates of the quantified protein were performed per sample [20].

2.3. Isoelectrofocusing, 2-D Electrophoresis, Gel Staining, Image Capture and Analysis

IEF, 2-D electrophoresis, gel staining, image capture, protein spot digestion and MALDI (Matriz-Assisted Laser Desorption/Ionizacion)-TOF (Time of flight) were analyzed in the Universidad of Córdoba UCO-SCAI proteomics facility (Córdoba, Spain), a member of Carlos III Networked

Proteomics Platform, ProteoRed-ISCIII. The methodology of Soto et al. was followed [19] as below. Isoelectrofocusing (IEF) was carried out on Precast 17 cm IPG pH 5–8 linear gradient (Bio-Rad, Hercules, CA, USA) strips. The strips were allowed to rehydrate in a PROTEAN IEF Cell (Bio-Rad) for 14–16 h at 50 V and 20 °C with 315 µL of protein solubilization buffer containing 400 µg of proteins. The proteins were separated in the pH range 4–7 by using IEF in three-step procedure as follow, 15 min at 500 V, followed by 2 h at 10,000 V and a final step of 10,000 Vh to complete 60,000 Vh. The strips were immediately run after focusing. The strips were equilibrated after immersion for 20 min first in 375 mM Tris–HCl, pH 8.8, with 6 M urea, 2% sodium dodecyl sulfate (SDS), 20% glycerol, and 2% DTT, and then in the same solution containing 2.5% iodoacetamide as a substitute of DTT. After transferring the strips onto vertical slab 12% SDS-polyacrylamide gels (Bio-Rad PROTEAN Plus Dodeca Cell), the electrophoresis was run at 55 mA/gel until the dye front reached the bottom of the gel [20]. The gels were silver-stained as described by Yan et al. [21] or with coomassie brilliant blue G-250 (Sigma, (Sigma-Aldrich Chemical Co, St Louis, MO, USA). Gel images were captured, digitalized (Molecular Imager Pharos FX^{TM} Plus multi Imager System, Bio-Rad), and analyzed with PDQuest^{TM} 2-D analysis software (Bio-Rad laboratories, Hercules, CA, USA), and as a minimum criterion for presence/absence, ten-fold over background was used. Significant spots were excised automatically using *ProPic*) (Genomics Solutions Inc., Ann Arbor, USA), stored in milli-Q water until Matrix-Assited Laser Desorption/Ionization—Time-Of-Flight (MALDI-TOF/TOF) analysis. The protocols for digestion were performed as described previously [22]. For the MAILDI-TOF analysis, a combined Peptide Mass Fingerprinting (PMF) search Mass Spectrometry (MS plus MS/MS) was performed using GPS Explorer^{TM} software v 3.5 (Applied Biosystems, Waltham, Massachusetts, MA USA) over non-reductant NCBInr database using the MASCOT search engine (Matrix Science, London; http://www.matrixscience.com) following parameters reported previously [22].

3. Results

Differentially Regulated Proteins from Rosettes of trxf1, trxf2, trxm1, trxm2, trxm3 and trxm4 Mutants

The *trxm1* and *trxm4* mutants were described previously [23,24] as well as *trxm3* [25], *trxf1* [26,27] and *trxf2* [26]. Figure 1 shows that there is no detected TRX expression in *trxf1*, *trxm1*, *trxm3* and *trxm4*, with these lines being considered as knockout ones, whilst a slight level of transcripts can be observed for *trxf2* and *trxm2* as it was described. The level of expression of all the mutants was clearly sufficiently low to validate the results obtained. The proteomic approach was performed to study protein profiles in these plastidial TRX *f* and *m* mutants. For this, the total protein extracts from rosettes of 25 dpg plants of the mutant lines *trxf1*, *trxf2*, *trxm1*, *trxm2*, *trxm3* and *trxm4* and of the wild type lines Col0 and Ler were analyzed in the "Unidad de Proteómica of the Universidad de Córdoba" (Córdoba, Spain). An optimal concentration and purity degree of the protein extracts were reached for a high level of separation of the peptides using two dimensional (2-DE) electrophoresis. Figure 2 shows the spots corresponding to the peptides of up- and down-regulated proteins in rosettes from *trxf1* and *trxf2* (Figure 2A), *trxm1*, *trxm2* and *trxm4* (Figure 2B), and *trxm3* (Figure 2C), mutants in comparison with Columbia (Col0) and Landsberg erecta LE (Tables 1–3. The master gels from each three replicates gel mutant lines were obtained and normalized by using the software PDQuest$^{®}$ (Bio-rad). Three replicates gel of Col0 control was developed for each TRX type (*f* and *m*), as Ler for *trxm3* (Figure 2). Also, PDQuest$^{®}$ (Bio-rad) was used to select, in each gel of the mutant lines, those spots where the expression was down regulated or up regulated when compared to the same spot located in the gels of the control plants. The proteins contained in the spots were picked, digested and the peptides identified by MS (Table S2). The comparison with databases allowed us to identify the proteins that could contain the different peptides. Out of the 200 analyzed and resolved spots, a total of 86 differentially expressed proteins were identified (Table S2). The *trxf1* and *trxf2* mutants had a proportionally larger number of down regulated proteins (15 out of 20 and 13 out of 21, respectively), whilst the *trxm* mutants had a larger number of up regulated proteins

(16 out of 26 in *trxm1*, 18 out of 23 in *trxm2*, eight out of 10 in *trxm3*, and 21 out of 25 in *trxm4*) (Figure 3A). Despite the sequence similarities between the TRX *f1* and *f2*, in the *trxf1* and *trxf2* lines nine proteins showed a different regulation (Table 1): glutamate-glyoxylate aminotransferase 1 (GGAT1), aminomethyltransferase, 5-methyl tetrahydropteroyl triglutamate-homocysteine methyltransferase 1, ribulose-bisphosphate carboxylase oxygenase (RUBISCO) activase, β-D-glucopyranosyl abscisate β-glucosidase, glyceraldehyde-3-phosphate dehydrogenase (GAPC2), monodehydroascorbate reductase, V-type ATP synthase, and Chaperonin 60 subunit β1. Nevertheless, apart from these differences, 10 out of 24 (41.7%) of the analyzed proteins were down regulated in both mutant lines while four out of 24 (16.7%) were up regulated. Data analysis revealed that the largest number of down regulated biological processes corresponded to the *trxf* mutants (Figure 3B). The other *m* type TRXs mutants mostly showed up regulated processes, especially *trxm2*, *trxm3* and *trxm4* (Figure 3B).

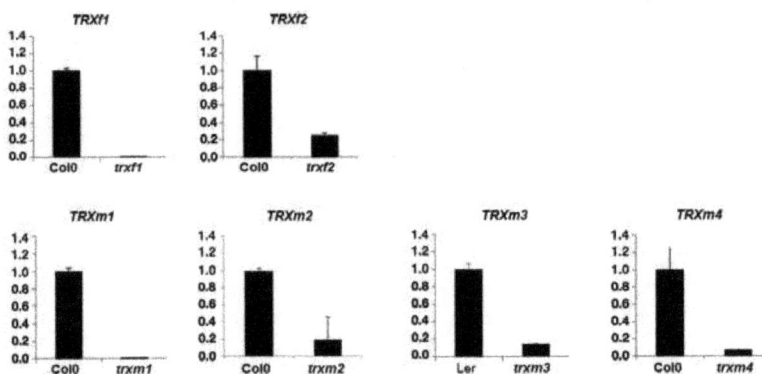

Figure 1. Thioredoxins (TRXs) *f* and *m* transcript levels in *trxf* and *trxm* mutants.

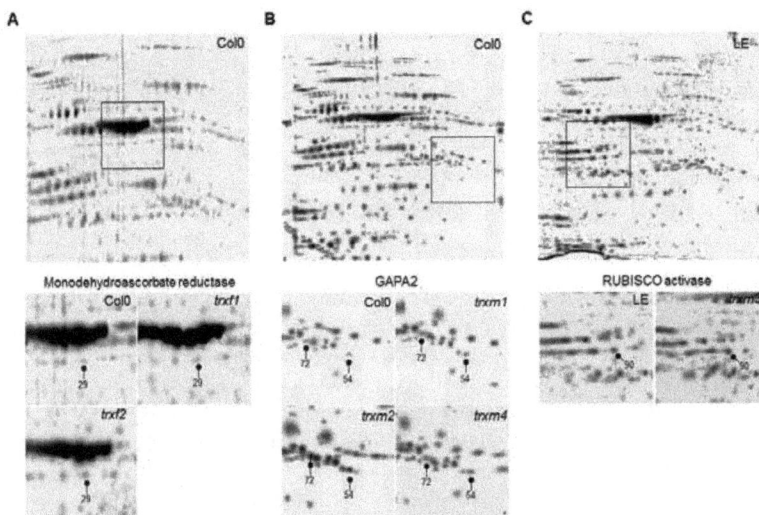

Figure 2. DE images from rosette of *trxf1, f2, m1, m2, m3* and *m4* mutants. 2-D images of total proteins from rosettes of *trx f1* and *f2* (**A**), *m1, m2* and *m4* (**B**), and *m3* (**C**) mutants in comparison with Col0 or Ler (LE). Numbers correspond to the protein spots identified by MALDI-TOF/TOF analysis (Table S2). The figure shows the representative experiments carried out with some examples of proteins identified in each gel.

Table 1. Differentially expressed proteins identified by MS in the *trxf1* and *trxf2* mutants, organized in the functional category, the gene code and the subcellular localization and whether the protein has been reported as a Thioredoxins (TRX) target. The color code indicates fold change in protein abundance.

Functional Category	Protein	Gene ID	Location	TRX Target	Spot	trxf1	trxf2
Amino acids metabolism	Glutamate-glyoxylate aminotransferase 1	At1g23310	Per.	-	16		
	Aminomethyltransferase	At1g11860	Mit.	-	8		
	Glutamate-glyoxylate aminotransferase 2	At1g70580	Per.	-	21		
	5-methyltetrahydropteroyltrigluta-mate-homocysteine methyltransferase 1	At5g17920	Per.	-	33		
	5-methyltetrahydropteroyltrigluta-mate-homocysteine methyltransferase 2	At3g03780	Cy.	-	12		
Calvin-Benson cycle	RUBISCO large subunit	AtCg00490	Ch.	Yes	5		
	Transketolase1	At3g60750	Ch.	Yes	20		
	RUBISCO activase	At2g39730	Ch.	Yes	27		
ATP synthesis	ATP synthase subunit ß	AtCg00480	Ch.	Yes	15		
	ATP synthase subunit α	AtCg00120	Ch.	Yes	4, 24		
	ATP synthase subunit 1	AtMg01190	Mit.	Yes	25		
Photosynthesis	PSII stability/assembly factor HCF136	At5g23120	Ch.	-	9		
	Ferredoxin-NADP reductase 1	At5g66190	Ch.	-	17		
ABA signalling	β-D-glucopyranosyl abscisate β-glucosidase	At1g52400	ER	-	3		
	Myrosinase 2	At5g25980	n.d.	Yes	22		
Glycolisis	Triosephosphate isomerase	At3g55440	Mit.	-	31		
	Glyceraldehyde-3-phosphate dehydrogenase C2 (GAPC2)	At1g13440	Cy.	-	26		
Stress response	Jacalin-Related lectin	At3g16470	n.d.	-	7		
	Monodehydroascorbate reductase	At1g63940	Ch., Mit.	Yes	29		
Protein biosynthesis	Elongation factor Tu	At4g20360	Ch.	Yes	6		
ATP hydrolysis	V-type ATP synthase	At1g78900	V.	-	2		
PSII biogenesis	PSII stability/assembly factor HCF136	At5g23120	Ch.	-	30		
Refolding activity	Chaperonin 60 subunit ß1	At1g55490	Ch.	-	13		
Tricarboxylic acid cycle	Malate dehydrogenase 1	At1g53240	Mit.	Yes	35		

Protein abundance change relative to the control (Col0).

<0.5 0.6 0.7 0.85 1 1.25 1.5 1.75 >2

Ch., chloroplast; Mit., mitochondria; Per., peroxisome; Cy., cytosol; V., vacuole; ER, endoplasmic reticulum; n.d., not determined. Proteins with a Confidence Interval C.I.% $\geq$ 95% are shown. According to Montrichard et al. (2009) [1], reported thioredoxin targets are shown. NADP: nicotinamide-adenine-dinucleotide phosphate; HCF: high chlorophyll fluorescence; ABA: absicic acid; PSII: photosystem II; RUBISCO: ribulose bisphosphate carboxylase/oxygenase.

The four *trxm* mutant lines only shared one differentially expressed protein (β carbonic anhydrase 2), while the *trxm1*, *trxm2*, and *trxm4* lines shared up to 13 differentially expressed proteins (39.4%, Tables 2 and 3). However, some of the differentially expressed proteins identified were up regulated in one mutant line but down regulated in another one, as in the case of the ferredoxin-NADP reductase 1 or the chlorophyll a–b binding protein 2. Interestingly, nine proteins underwent similar changes in these three *trxm* lines, eight up-regulated and only one down-regulated (ferredoxin-NADP reductase 2). Some interesting proteins up-regulated, affecting key pathways, were the two chloroplast isoforms of glyceraldehyde-3-phosphate dehydrogenase GAPB and GAPA2 and the mitochondrial enzymes serine hydroxymethyltransferase 1 and glycine dehydrogenase 1 (amino acid metabolism). Regarding carbon fixation, the RUBISCO large subunit was down regulated in *trxm1* and up regulated in *trxm4*, whereas its regulator RUBISCO activase was up-regulated in *trxm2*, *trxm3*, and *trxm4*. Interestingly, no photosynthetic genes were differentially expressed in the *trxm3* mutant; nevertheless, the up-regulation of photosystem II stability/assembly factor HCF136 in the *trxm1*, *trxm2* and *trxm4* mutant lines is significant.

Table 2. Differentially expressed proteins identified by MS in the *trxm1*, *trxm2*, and *trxm4* mutants, organized in the functional category, the gene code and the subcellular localization and whether the protein has been reported as a TRX target. The color code indicates fold change in protein abundance.

Functional Category	Protein	Gene ID	Location	TRX Target	Spot	*trxm1*	*trxm2*	*trxm4*
Calvin-Benson cycle	Transketolase	At3g60750	Ch.	Yes	38			
	RUBISCO large subunit	AtCg00490	Ch.	Yes	28			
	Fructose-bisphosphate aldolase 2	At4g38970	Ch.	Yes	59, 103			
	RUBISCO activase	At2g39730	Ch.	Yes	65			
	Glyceraldehyde-3-phosphate dehydrogenase B GAPB)	At1g42970	Ch.	Yes	68			
	Glyceraldehyde-3-phosphate dehydrogenase A2 (GAPA2)	At1g12900	Ch.	Yes	54, 72			
Amino acids metabolism	Serine hydroxymethyltransferase 1	At4g37930	Mit.	Yes	101			
	Glycine dehydrogenase (decarboxylating) 1	At4g33010	Mit.	-	100			
	Probable phosphoglycerate mutase 2	At3g08590	Mit.	-	47			
	Glutamate-glyoxylate aminotransferase 1	At1g23310	Per.	Yes	97			
	5-methyltetrahydropteroyltrigluta-mate-homocysteine methyltransferase 1	At5g17920	Cy.	Yes	45, 98			
Photosynthesis	Chlorophyll a-b binding protein 2	At1g29920	Ch.	Yes	42			
	Ferredoxin-NADP reductase 1	At5g66190	Ch.	-	43			
	Chlorophyll a-b binding protein	At2g34420	Ch.	-	56			
	Oxygen-evolving enhancer protein 2-1	At1g06680	Ch.	Yes	57			
	Oxygen-evolving enhancer protein 1-2	At3g50820	Ch.	Yes	41			
	Ferredoxin-NADP reductase 2	At1g20020	Ch.	-	104			
Stress response	Uncharacterized protein	At2g37660	Ch.	-	40			
	Heat shock 70 kDa protein 3	At3g09440	N.	-	46			
	Monodehydroascorbate reductase 1	At3g52880	Per.	Yes	70			
PSII stabilization/repair	Photosystem II stability/assembly factor HCF136	At5g23120	Ch.	-	67			
	Protease Do-like 1	At3g27925	Ch.	-	39			
Protein transport	Chaperone protein ClpC1	At5g50920	Ch.	-	37			
ATP synthesis	ATP synthase subunit beta	AtCg00480	Ch.	Yes	93			
Glycolysis	Glyceraldehyde-3-phosphate dehydrogenase GAPC2	At1g13440	Cy.	Yes	53			
Carbohydrate metabolism	Chloroplast stem-loop binding protein of 41 kDa b	At1g09340	Ch.	-	64			
Carbon utilization	β carbonic anhydrase 2	At5g14740	Ch.	-	55			
Lipid degradation	GDSL esterase/lipase ESM1	At3g14210	N.	-	105			
Protein refolding	Chaperonin 60 subunit beta 2	At3g13470	Ch.	Yes	96			
Unkown	Polyketide cyclase/dehydrase and lipid transport superfamily protein	At4g14500	Mit.	-	61			
	Uncharacterized protein	At2g37660	Ch.	-	40			
	Uncharacterized protein	At5g05113	Mit.	-	74			
	Disease resistance protein (NBS-LRR class) family	At5g40060	n.d.	-	75			

Protein abundance change relative to the control (Col0).

<0.5 0.6 0.7 0.85 1 1.25 1.5 1.75 >2

Ch., chloroplast; Mit., mitochondria; Per., peroxisome; Cy., cytosol; V., vacuole; N., nucleus; n.d., not determined. Proteins with Protein Scores C.I.% ≥ 95% are shown. According to Montrichard et al. (2009) [1], reported thioredoxin targets are shown. ESM1: epithiospecifier modifier 1.

Table 3. Differentially expressed proteins identified by MS in the *trxm3* mutant, organized in the functional category, the gene code and the subcellular localization and whether the protein has been reported as a TRX target. The color code indicates fold change in protein abundance.

Functional Category	Protein	Gene ID	Location	TRX Target	Spot	*trxm3*
Calvin-Benson cycle	RUBISCO activase	At2g39730	Ch.	Yes	90	
	Fructose-bisphosphate aldolase 2	At4g38970	Ch.	Yes	91	
ATP synthesis	ATP synthase subunit 1	AtMG01190	Mit.	Yes	82	
	ATP synthase γ chain 1	At4g04640	Ch.	Yes	85	
JA signalling /response	Epithiospecifier protein	At1g54040	N.	-	84	
	Lipoxygenase 2	At3g45140	Ch.	Yes	76	
ABA signalling	Myrosinase 2	At5g25980	n.d.	Yes	77	
Carbon utilization	β carbonic anhydrase 2	At5g14740	Ch.	Yes	92	
Refolding activity	Chaperonin 60 subunit β 2	At3g13470	Ch.	-	87	
Stress response	Monodehydroascorbate reductase	At1g63940	Ch., Mit.	Yes	83	

<0.5 0.6 0.7 0.85 1.25 1.5 1.75 >2

Protein abundance change relative to the control (Ler).

Ch., chloroplast; Mit., mitochondria; N., nucleus; n.d., not determined. Proteins with Protein Scores C.I.% ≥ 95% are shown. According to Montrichard et al. (2009) [1], reported thioredoxin targets are shown. JA: jamonic acid.

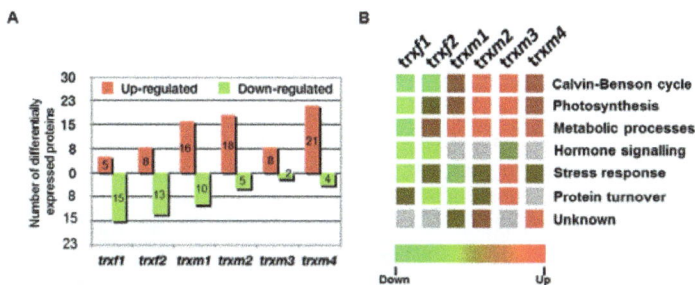

Figure 3. Processes down- or up-regulated in the *trx* mutants, number of coincident spots identified by proteomic analyses. (**A**) number of proteins down- or up-regulated in *trxf* or *trxm* mutants; (**B**) biological processes affected in each mutant line.

No coincident protein was observed in the *trxf* and *trxm* mutants. Nevertheless, the *trxf1* and *trxf2* lines shared one differentially expressed protein with the *trxm1*, *trxm2* and *trxm4* mutants: the ferredoxin-NADP reductase 1, though this protein was up-regulated in *trxf1*, *trxf2*, and *trxm4* and down-regulated in *trxm1* and *trxm2*. In addition, in the *trxf* and *trxm3* mutants only myrosinase 2, a ABA signalling protein, was differentially expressed (down-regulated).

Regarding hormonal processes, in the *trxm* mutants only *trxm3* displayed changes in ABA and JA signalling. Adjustments in the *trxf* mutants were limited to ABA signalling with a down-regulation of myrosinase 2 in both mutant lines and of β-D-glucopyranosyl abscisate β-glucosidase (activating the glucose-conjugated inactive ABA), but only in the *trxf1* mutant. In Tables 1–3, we can observe the metabolic alterations mostly corresponded to changes in the level of glycine, alanine, glutamine, and methionine synthesizing/hydrolyzing enzymes in chloroplasts, mitochondria, peroxisomes, and cytosol. However, we could not observe any change in the amino acid metabolism in the *trxm3* mutant.

All the identified spots corresponded to proteins with putative general functions related to the Calvin–Benson cycle, photosynthesis, stress response, hormone signalling, protein turnover, and to unknown processes (Figure 4, Table 1, Table 2, Table 3 and Table S2). The most represented functions fell into the metabolism category, containing 30–58% of the spots analysed, especially in the *trxf* mutants. The Calvin–Benson cycle was also well represented in all the mutants, ranging from 10.5% to

20%. Surprisingly, photosynthesis was affected in all the mutants except for *trxm3*, especially in the *trxm1*, *trxm2*, and *trxm4* lines (Figure 4).

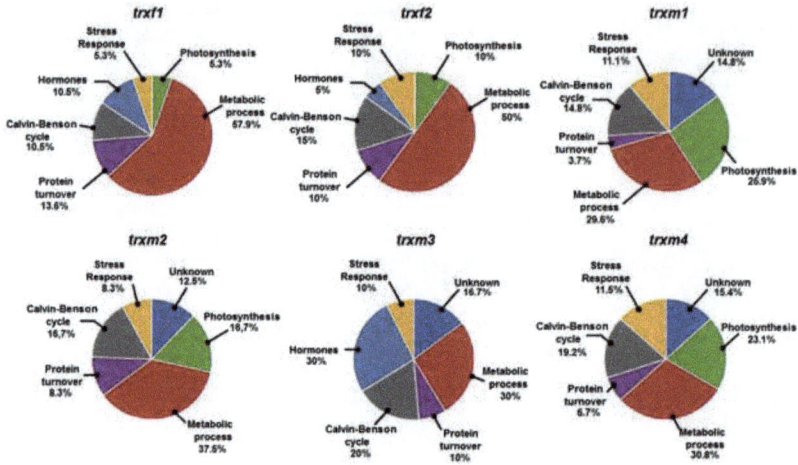

Figure 4. Degree of significance of the biological processes affected in the *trx* mutants in relation to the number peptides predicted to be involved in a process.

At a sub-cellular level, most of the differential expressed proteins were predicted to be chloroplastic, ranging from 47% to 59% of the total spots identified in each mutant (Figure 5). Differential expression in TRX mutants also affected non-chloroplast proteins as they were also predicted to be located in other sub-cellular compartments such as mitochondria, peroxisomes, cytosol, nucleus, Golgi/endoplasmic reticulum, and vacuole (enumerated according to the frequency of appearance in the proteomic analyses). Taking into account the affected processes, we applied a hierarchical clustering to our proteomic data (Figure 6). Two major clusters contained the *trxf* and the *trxm* mutants. Within the *trxm* cluster, two sub-clusters separated the *trxm3* line from the *trxm1*, *trxm2*, and *trxm4* mutants. Interestingly, according to our clustering analysis, *trxm1* might be functionally closer to *trxm4* than to *trxm2*. These data are according to recent results obtained in our laboratory (unpublished data).

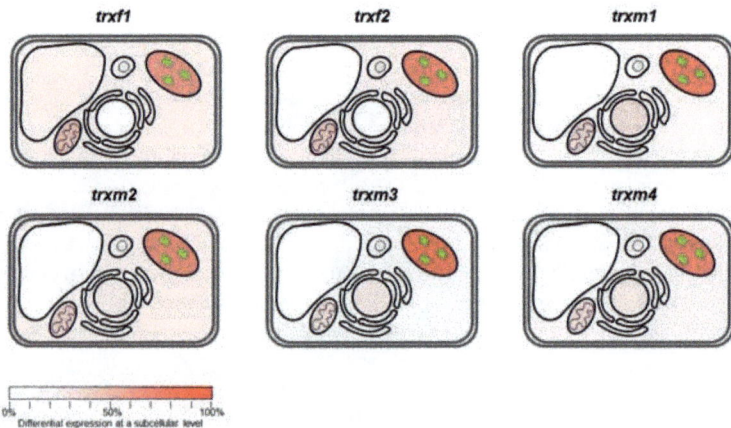

Figure 5. Differential expression shown at a sub-cellular level. The red color intensity is indicating if differential expressions are significant in a given organelle.

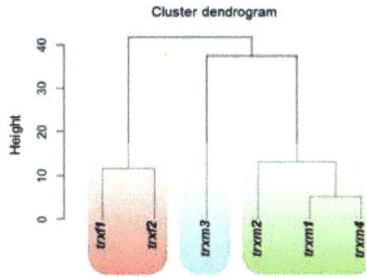

Figure 6. Hierarchical clustering segregates *trx* mutants according to the biological processes affected. For calculations, the R statistical environment (The R Foundation, Vienna, Austria) available at http://www.R-project.orgwasused.

4. Discussion

The existence of an elevated number of plastidial thioredoxins, amongst which are two *f* type TRXs and four *m* type TRXs in the chloroplast makes it difficult to identify and separate their different functions. A great deal of effort has been made to reach important conclusions and different authors have reported previously protein targets and suggested specific roles for several of the plastidial TRXs. Due to similarities with the *f* and *m* type TRXs, these isoforms might both carry out overlapping as specific functions in the redox regulation of the plant physiological processes. However, it has been difficult to find a clear border to delimit the role of each plastidial TRX described up to now. Beyond the study of the different reduction degree of the target proteins, our aim has been to know whether the pattern response of the Arabidopsis proteome is specific for each plastidial TRX mutant characterized. In this study we have carried out a proteomic profile of Arabidopsis mutant lines lacking *f1*, *f2*, *m1*, *m2*, *m3* and *m4* type TRXs.

The analysis of the 2-DE pattern revealed a total of 86 different protein-spot intensities out of the 200 that were resolved in all images (Figure 2 and Table S2). The differentially expressed proteins in the mutants showed particular patterns, supporting the hypothesis of a specific functionality for each TRX *f* or *m* isoform. A higher number of down regulated proteins were observed for *trxf* mutants, mainly *trxf1*; whilst more up regulated proteins occurred in the *trxm* lines, principally the *trxm4* line (Figure 3A), suggesting that their roles are separate and affected.

Almost 59% of the differentially expressed proteins are plastidial, but other affected proteins are localized in mitochondria, cytosol, peroxisomes and nucleus (Figure 5), indicating the connection between the processes occurring either in the chloroplast or in other localizations when one plastidial TRX is missing. Likely, redox imbalance provoked by the lack of TRXs *f*/*m* isoforms in chloroplasts is triggering retrograde signalling to readjust the biological functions affected as we noticed differentially expressed proteins in other organelles. Moreover, we cannot discard other factors regulating the protein levels as protein degradation or stabilization by proteases and chaperonins, respectively, as discussed later.

As expected for the chloroplast redox enzymes involved in photosynthesis-related processes, among the proteins found that predominate are those which are functionally related with electronic transport. The presence of several subunits of the chloroplastidic and mithocondrial ATP synthase complexes is noteworthy. The subunit α of the chloroplastidic ATP synthase was down regulated in the *trxf1* and *trxf2*, whilst the subunit β was up regulated in these lines. Interestingly, the subunits β and γ (the latter being up-regulated in *trxm3*) have been reported to be targets of plastidial TRXs [1,28]. However, it has been described that under low irradiance, NADPH thioredoxin reductase C (NTRC) is required for the redox modulation of the subunit γ of ATP synthase [29]. As the subunit 1 of the mitochondrial ATP synthase is also affected in *trxf1*, *trxf2* and *trxm3* mutants, these patterns of up/down regulation would be an indication of the importance and complexity of the regulation of ATP synthesis in plants suggesting the significance of balancing chloroplastic and mitochondrial ATP levels in response to redox changes. Remarkably, and similarly to the above-mentioned chloroplast subunits, the mitochondrial subunit 1 also seems to be redox regulated [1,30].

The link between chloroplast and mitochondrion is also evidenced with proteins responding to redox changes in the chloroplast involved in amino acid metabolism, such as serine hydroxyl methyl transferase or glycine dehydrogenase in *trxm1*, *trxm2* and *trxm4* and amino methyl transferase in *trxf1*; and even proteins with unknown functions such as the protein with identity number B3H6G1 in the *trxm1*, *trxm2* and *trxm4* mutants (Table S2).

These results do not necessarily imply dual localization chloroplast/mitochondria of the referred isoforms (though this cannot be ruled out), but rather they demonstrate the relationship between the metabolism of the chloroplast and the mitochondria. The observed differences at a proteomic level reveal, once again, a certain degree of functional specificity of each isoform of analyzed plastidial TRX.

Metabolism, the Calvin–Benson cycle, photosynthesis, response to stress, and hormone signaling are the functional categories most affected when one of the *f* or *m* type TRXs is absent. However, the lack of TRX *f1* or *f2* provokes a down-regulation in cell general processes, whilst the influence of the absence of TRXs *m* is the opposite (Figure 3). Most of the proteins differentially expressed in the *trxf* mutants relate to its function in metabolism in general, and to the synthesis of ATP specifically. In the same way, it seems that TRXs *m1*, *m2* and *m4* are involved in the redox regulation of the photosynthesis. TRXs *f* and *m3* are likely to be involved in the redox regulation during stress situations as monodehydroascorbate reductase (in the stress response category) and myrosinase 2 (in the ABA signaling category) have been reported as TRXs targets (Tables 1 and 3).

Taking into consideration the data obtained, it seems that the protein differentially expressed in the mutants fell into different functional categories which can be organized in function of the processes affected or their sub-cellular localization. Thus, as expected in a photosynthetic organ like the leaf, among the predominate proteins found are those which are functionally related with electronic transport with the only exception of *trxm3*.

Likewise, photosynthesis appears highly unstable in these mutants, amongst which a number of the electron transport or Calvin-Benson cycle proteins are up regulated or down regulated, as is described in the Results section. Interestingly, factor HCF136 was up regulated in the *trxm1*, *trxm2* and *trxm4* mutant lines, but down regulated in the *trxf1* and *trxf2* lines. It has been reported that factor HCF136 is involved in the assembly of an intermediary complex of PSII [31], suggesting that it is possible to detect putative alterations or increased instability of PSII in these mutants.

Previous studies have reported that TRXs are able to transfer reducing equivalents to the redox protein HCF164 in the thylakoid lumen [32], and that these proteins are necessary for the biogenesis of PSI. Additionally, evidence showed that HCF164 serves as transducer of reducing equivalents to proteins in the thylakoid lumen. Consequently, decreased levels of *m* type TRXs might lead to a redox imbalance in the thylakoid lumen and the instability of redox regulated proteins components of the PSII as PsbO subunits [14]. Interestingly, PSI subunit levels do not seem to be influenced in the TRX mutants analyzed, ruling out the possibility of different spot intensities (detection threshold) as both photosystems have a similar stoichiometry in the thylakoid membranes. Curiously, apart from proteins participating in primary metabolism or photosynthesis, other less represented proteins did not appear in our analyses, suggesting that these spots were not abundant enough to be detected with our experimental conditions. A clear example is the absence among the identified spots of the plastid TRXs that we were analyzing. However, this fact does not invalidate our results.

Some authors have attributed the *f* and *m* type TRXs to being part of the mechanisms involved in the control of the binding of the light harvesting chlorophyl (LHC) antenna complexes to the PSI and PSII through the activation of a serine/threonine kinase, SNT7 [33]. This is relevant because two components of binding factors to the chlorophyll *a/b* of the PSII are differentially expressed in *trxm1* (Lhcb1B2 is up regulated and Lhcb2C1 is down regulated), *trxm2* (Lhcb1B2 is up regulated) and *trxm4* (Lhcb2C1 is up regulated). Additionally, *trxm1* was the only one to have differentially expressed two subunits of the photolysis water complex (PsbO2 is up regulated and PsbP1 is down regulated), essential for generating the reducing power during the electron transport chain. TRX *m1* is probably closely related to the redox regulation of the photosynthesis process.

In relation to the carbon fixation, it was not surprising to detect the large subunit of the RUBISCO protein among the differentially expressed proteins, due to its relevant position in carbon fixation. The RUBISCO large subunit was found to be down regulated in all the *trxf* and *trxm1* lines. However, we cannot rule out the possibility that these changes might correspond to post-translationally modifications of the RUBISCO large subunit as we did not observe differences in the phenotypes with respect to wild-type plants due to a putative defect in carbon fixation. RUBISCO activase, involved in the regulation of the small subunit of RUBISCO appeared up regulated in the *trxm2*, *trxm3* and *trxm4* lines and down regulated in the *trxf2* mutant, indicating a specific action when a change occurs in CO_2 fixation during the loss of the redox homeostasis in plants. It is well known that several Calvin-Benson cycle enzymes are redox regulated, so it is not surprising to find them in the proteome of the mutants. However, the list is certainly lower due to the high plasticity of the plant to adapt under different adverse environments.

Due to their importance in key positions in the Calvin-Benson cycle the following proteins are particularly relevant: fructose-bisphosphate aldolase (up regulated in *trxm1*, *trxm2* and *trxm3*) and transketolase 1 (down regulated in *trxf1*, *trxf2* and *trxm1*); and the plastidial isoforms of GADPH, found to be up regulated in *trxm1*, *trxm2* and *trxm4* mutants; which were confirmed in the gene expression analysis (data not shown). It seems clear that redox activation/regulation by TRXs has a direct effect on the balance of carbohydrate synthesis and distribution as an optimal cell redox homeostasis environment is mandatory for the proper functioning of several processes, essentially carbon metabolism and the photosynthesis.

An interesting group was that of the proteins involved in redox homeostasis, with the most relevant being the subunits 1 and 2 of ferredoxin $NADP^+$ reductase (FNR) (not described so far as a target of plastidial TRXs) and chloroplastidic and peroxisomal monodehydroascorbate reductase (MDAR), the relationship between these proteins and the redox regulation through TRXs is well known [34]. One of the possible functions of plastidial TRXs could be to maintain homeostasis against abiotic stress conditions, such as salinity during germination [35]. It seems that *f* type TRXs possibly play a more relevant role in controlling the homeostasis conditions inside the cells than previously thought.

Enzymes were also found to be involved in the response to biotic stress, such as the myrosinase (down regulated in *trxf1*, *trxf2* and *trxm3*). In addition, the epithiospecifier protein, up regulated in *trxm3*, is also functionally related to myrosinase [36]. Similarly, GDSL esterase/lipase endothelial cell specific molecule1 (ESM1) was found to be up-regulated in *trxm1*, *trxm2* and *trxm4* mutants and it has been related to the response to insects [37].

Moreover, several interesting proteins related with the metabolism of amino acids, proteins and lipids have also been identified. Differentially expressed enzymes, mainly related with the synthesis and transfer of amino and methyl groups were found to be up regulated. Concerning the metabolism of proteins, proteases appeared, as well as factors involved in the translation process and proteins with chaperone activity. Particularly Relevant are the chaperonin 60 subunit β1 (up regulated in *trxf1* and down regulated in *trxf2*) and chaperonin 60 subunit β2 (up-regulated in *trxm2* and *trxm3*), suggesting that the protein folding and assembly are likely to be related to redox regulation. Previous studies have shown that the chaperonin 60 subunit β may be protecting the photosynthtetic components during stress situations [38].

Although the number of proteins with unknown functions was very low, a few examples without characterization were found (four), such as the protein with identity number B3H6G1 up regulated in *trxm1*, *trxm2* and *trxm4* (mentioned above). It is reasonable to think that a deep characterization would be necessary to identify new and specific functions of these plastidial thioredoxins.

5. Conclusions

In this study we have attempted to shed some light and to get closer to functionally defining the different isoforms of *f* and *m* type TRXs in Arabidopsis and to clarify the blurred frontier that exists, in most cases, between specificity and functional redundancy in multi-gene families in plants. Even though there are numerous studies on plastidial TRXs, the existing information is still scarce, from a functional point of view. Despite this, more studies are necessary, this is the first time a broad view is showing what processes are affected when one of the *f* or *m* type TRXs is lacking. We were aware of the complexity of the task. The results we have achieved from the approach followed in this study, which consisted of a comparative analysis of knockout/down mutants for the six isoforms of *f* and *m* type TRXs, should help to better understand the functional role of *f* and *m* type TRXs in Arabidopsis and to open up new research paths in the study of processes regulated by these enzymes.

Supplementary Materials: The following are available online at http://www.mdpi.com/2076-3921/8/3/54/s1, Table S1: gene-specific oligonucleotides used for quantitative PCR, Table S2: list of identified proteins in *trxf1*, *trxf2 trxm1*, *trxm2*, *txm3*, and *trxm4* mutants.

Author Contributions: Funding acquisition, Mariam Sahrawy (M.S.); investigation, Juan Fernández–Trijueque (J.F.-T.), Antonio-Jesús Serrato (A.-J.S.) and M.S.; methodology, J.F.-T. and A.-J.S.; project administration, M.S.; supervision, A.-J.S. and M.S.; visualization, A.-J.S. and M.S.; writing—original draft, A.-J.S. and M.S.; writing—review and editing, A.-J.S. and M.S.

Funding: This work has been funded by research project BIO2009-07297, BIO2015-65272 from the Spanish Ministry of Science and Innovation and the European Fund for Regional Development, project P07-CVI-2795 and BIO 154 from the Andalusian Regional Government, Spain, and project BIO2012-33292, from the Spanish Ministry of Economy and Competitiveness.

Acknowledgments: The authors thank Trinidad Moreno and Sabrina De Brasi for their technical support, and Angela Tate for helpful editorial feedback on the manuscript. Protein spot digestion and MALDI-TOF analysis were carried out in the Universidad of Cordoba UCO-SCAI proteomics facility, a member of Carlos III Networked Proteomics Platform, ProteoRed-ISCIII.

Conflicts of Interest: The authors declare no conflict of interest.

References

1. Montrichard, F.; Alkhalfioui, F.; Yano, H.; Vensel, W.H.; Hurkman, W.J.; Buchanan, B.B. Thioredoxin targets in plants: The first 30 years. *J. Proteom.* **2009**, *72*, 452–474. [CrossRef] [PubMed]

2. Buchanan, B.B.; Schurmann, P.; Wolosiuk, R.A.; Jacquot, J.P. The ferredoxin/thioredoxin system: From discovery to molecular structures and beyond. *Photosynth. Res.* **2002**, *73*, 215–222. [CrossRef] [PubMed]

3. Barajas-López, J.D.; Serrato, A.J.; Cazalis, R.; Meyer, Y.; Chueca, A.; Reichheld, J.P.; Sahrawy, M. Circadian regulation of chloroplastic *f* and *m* thioredoxins through control of the CCA1 transcription factor. *J. Exp. Bot.* **2011**, *62*, 2039–2051. [CrossRef] [PubMed]

4. Chibani, K.; Wingsle, G.; Jacquot, J.P.; Gelhaye, E.; Rouhier, N. Comparative genomic study of the thioredoxin family in photosynthetic organisms with emphasis on *Populus trichocarpa*. *Mol. Plant* **2009**, *2*, 308–322. [CrossRef] [PubMed]

5. Motohashi, K.; Kondoh, A.; Stumpp, M.T.; Hisabori, T. Comprehensive survey of proteins targeted by chloroplast thioredoxin. *Proc. Natl. Acad. Sci. USA* **2001**, *8*, 11224–11229. [CrossRef] [PubMed]

6. Collin, V.; Issakidis-Bourguet, E.; Marchand, C.; Hirasawa, M.; Lancelin, J.M.; Knaff, D.B.; Miginiac-Maslow, M. The *Arabidopsis* plastidial thioredoxins: New functions and new insights into specificity. *J. Biol. Chem.* **2003**, *278*, 23747–23752. [CrossRef] [PubMed]

7. Lemaire, S.D.; Michelet, L.; Zaffagnini, M.; Massot, V.; Issakidis-Bourguet, E. Thioredoxins in chloroplasts. *Curr. Genet.* **2007**, *51*, 343–365. [CrossRef] [PubMed]

8. Serrato, A.J.; Fernández-Trijueque, J.; Barajas-López, J.D.; Chueca, A.; Sahrawy, M. Plastid thioredoxins: A "one-for-all" redox-signaling system in plants. *Front. Plant Sci.* **2013**, *4*, 1–10. [CrossRef] [PubMed]

9. Chiadmi, M.; Navaza, A.; Miginiac-Maslow, M.; Jacquot, J.P.; Cherfils, J. Redox signalling in the chloroplast: Structure of oxidized pea fructose-1,6-bisphosphate phosphatase. *EMBO J.* **1999**, *18*, 6809–6815. [CrossRef] [PubMed]

10. Schürmann, P.; Wolosiuk, R.A.; Breazeale, V.D.; Buchanan, B.B. 2 Proteins function in regulation of photosynthetic CO$_2$ assimilation in chloroplasts. *Nature* **1976**, *263*, 257–258. [CrossRef]

11. Buchanan, B.B.; Crawford, N.A.; Wolosiuk, R.A. Ferredoxin-thioredoxin system functions with effectors in activation of NADP glyceraldehyde 3-phosphate dehydrogenase of barley seedlings. *Plant Sci. Let.* **1978**, *12*, 257–264. [CrossRef]

12. Jacquot, J.P.; Vidal, J.; Gadal, P.; Schürmann, P. Evidence for existence of several enzyme-specific thioredoxins in plants. *FEBS Let.* **1978**, *96*, 243–246. [CrossRef]

13. Buchanan, B.B.; Balmer, Y. Redox regulation: A broadening horizon. *Ann. Rev. Plant Biol.* **2005**, *56*, 187–220. [CrossRef] [PubMed]

14. Balmer, Y.; Vensel, W.H.; DuPont, F.M.; Buchanan, B.B.; Hurkman, W.J. Proteome of amyloplasts isolated from developing wheat endosperm presents evidence of broad metabolic capability. *J. Exp. Bot.* **2006**, *57*, 1591–1602. [CrossRef] [PubMed]

15. Barajas-López, J.D.; Serrato, A.J.; Olmedilla, A.; Chueca, A.; Sahrawy, M. Localization in roots and flowers of pea chloroplast thioredoxin *f* and *m* proteins reveals new roles in non-photosynthetic organs. *Plant Physiol.* **2007**, *145*, 1–15.

16. Hisabory, T. Thioredoxin affinity chromatography: A useful method for further understanding the thioredoxin network. *J. Exp. Bot.* **2005**, *56*, 1463–1468. [CrossRef] [PubMed]

17. Yoshida, K.; Hisabory, T. Two distinct redox cascades cooperatively regulate chloroplast functions and sustain plant viability. *Proc. Natl. Acad. Sci. USA* **2016**, *113*, E3967–E3976. [CrossRef] [PubMed]

18. Valot, B.; Negroni, L.; Zivy, M.; Gianinazzi, S.; Dumas-Gaudot, E. A mass spectrometric approach to identify arbuscular mycorrhiza-related proteins in root plasma membrane fractions. *Proteom. J.* **2006**, S145–S155. [CrossRef] [PubMed]

19. Soto-Suarez, M.; Serrato, A.J.; Rojas-González, J.A.; Bautista, R.; Sahrawy, M. Transcriptomic and proteomic approach to identify differentially expressed genes and proteins in Arabidopsis thaliana mutants lacking chloroplastic 1 and cytosolic FBPases reveals several levels of metabolic regulation. *BMC Plant Biol.* **2016**, *16*, 258–274. [CrossRef] [PubMed]

20. Ramagli, L.S.; Rodríguez, L.V. Quantitation of microgram amounts of protein in two-dimensional polyacrylamide-gel sample buffer electrophoresis electrophoresis. *Electrophoresis* **1985**, *6*, 559–563. [CrossRef]

21. Yan, J.X.; Wait, R.; Berkelman, T.; Harry, R.A.; Westbrook, J.A.; Wheeler, C.H.; Dunn, M.J. A modified silver staining protocol for visualization of proteins compatiblewith matrix-assisted laser desorption/ionization and electrospray ionization massspectrometry. *Electrophoresis Int. J.* **2000**, *21*, 3666–3672. [CrossRef]

22. Castillejo, M.; Fernández-Aparicio, M.; Rubiales, D. Proteomic analysis by twodimensional differential in gel electrophoresis (2D DIGE) of the early response of Pisum sativum to Orobanchecrenata. *J. Exp. Bot.* **2012**, *63*, 107–119. [CrossRef] [PubMed]

23. Courteille, A.; Vesa, S.; Sanz-Barrio, R.; Cazale, A.C.; Becuwe-Linka, N.; Farran, I.; Havaux, M.; Rey, P.; Rumeau, D. Thioredoxin m4 controls photosynthetic alternative electron pathways in Arabidopsis. *Plant Physiol.* **2013**, *161*, 508–520. [CrossRef] [PubMed]

24. Wang, P.; Liu, J.; Liu, B.; Feng, D.; Da, Q.; Shu, S.; Su, J.; Zhang, Y.; Wang, J.; Wang, H.-B. Evidence for a role of chloroplastic m-type thioredoxins in the biogenesis of photosystem II in Arabidopsis. *Plant Physiol.* **2013**, *163*, 1710–1728. [CrossRef] [PubMed]

25. Benitez-Alfonso, Y.; Cilia, M.; San Roman, A.; Thomas, C.; Maule, A.; Hearn, S.; Jackson, D. Control of Arabidopsis meristem development by thioredoxin-dependent regulation of intercelular transport. *Proc. Natl. Acad. Sci. USA* **2009**, *106*, 3615–3620. [CrossRef] [PubMed]

26. Yoshida, K.; Hara, S.; Hisabori, T. Thioredoxin Selectivity for Thiol-Based Redox Regulation of Target Proteins in Chloroplasts. *J. Biol. Chem.* **2015**, *115*, 14278–14288. [CrossRef] [PubMed]

27. Thormaehlen, I.; Ruber, J.; Von Roepenack-Lahaye, E.; Ehrlich, S.; Massot, V.; Huemmer, C.; Tezycka, J.; Issakidis-Bourguet, E.; Geigenberger, P. Inactivation of thioredoxin *f*1 leads to decreased light activation of ADP-glucose pyrophosphorylase and altered diurnal starch turnover in leaves of *Arabidopsis* plants. *Plant Cell Environ.* **2013**, *36*, 16–29. [CrossRef] [PubMed]

28. Stumpp, M.T.; Motohashi, K.; Hisabori, T. Chloroplast thioredoxin mutants without active-site cysteines facilitate the reduction of the regulatory disulphide bridge on the gamma-subunit of chloroplast ATP synthase. *Biochem. J.* **1999**, *341 Pt 1*, 157–163. [CrossRef]

29. Carrillo, L.R.; Froehlich, J.E.; Cruz, J.A.; Savage, L.J.; Kramer, D.M. Multi-level regulation of the chloroplast ATP synthase: The chloroplast NADPH thioredoxin reductase C (NTRC) is required for redox modulation specifically under low irradiance. *Plant J.* **2016**, *87*, 654–663. [CrossRef] [PubMed]

30. Balmer, Y.; Vensel, W.H.; Tanaka, C.K.; Hurkman, W.J.; Gelhay, E.; Rouhier, N.; Jacquot, J.P.; Manieri, W.; Schürmann, P.; Droux, M.; et al. Thioredoxin links redox to the regulation of fundamental processes of plant mitochondria. *Proc. Natl. Acad. Sci. USA* **2004**, *101*, 2642–2647. [CrossRef] [PubMed]

31. Plücken, H.; Müller, B.; Grohmann, D.; Westhoff, P.; Eichacker, L. The HCF136 protein is essential for assembly of the photosystem II reaction center in *Arabidopsis thaliana*. *FEBS Lett.* **2002**, *532*, 85–90. [CrossRef]

32. Motohashi, K.; Hisabori, T. HCF164 receives reducing equivalents from stromal thioredoxin across the thylakoid membrane and mediates reduction of target proteins in the thylakoid lumen. *J. Biol. Chem.* **2006**, *281*, 35039–35047. [CrossRef] [PubMed]

33. Rintamäki, E.; Martinsuo, P.; Pursisheimo, S.; Aro, E.M. Coorperative regulation of light-harvesting complex II phosphorylation via the plastoquinol and ferredoxin-thioredoxin system in chloroplasts. *Proc. Natl. Acad. Sci. USA* **2000**, *97*, 11644–11649. [CrossRef] [PubMed]

34. Balmer, Y.; Vensel, W.H.; Cai, N.; Manieri, W.; Schürmann, P.; Hurkman, W.J.; Buchanan, B.B. A complete ferredoxin/thioredoxin system regulates fundamental processes in amyloplasts. *Proc. Natl. Acad. Sci. USA* **2006**, *103*, 2988–2993. [CrossRef] [PubMed]

35. Fernández-Trijueque, J.; Barajas-López, J.D.; Chueca, A.; Cazalis, R.; Sahrawy, M.; Serrato, A.J. Plastid thioredoxins *f* and *m* are related to the developing and salinity response of post-germinating seeds of Pisum sativum. *Plant Sci.* **2012**, *188–189*, 82–88. [CrossRef] [PubMed]

36. Lambrix, V.; Reichelt, M.; Mitchell-Olds, T.; Kliebenstein, D.J.; Gershenzon, J. The Arabidopsis epithiospecifier protein promotes the hydrolysis of glucosinolates to nitriles and influences *Trichoplusiani herbivory*. *Plant Cell* **2001**, *13*, 2793–2807. [CrossRef] [PubMed]

37. Zhang, Z.-Y.; Ober, J.A.; Kliebenstein, D.J. The gene controlling the quantitative trait locus EPITHIOSPECIFIER MODIFIER1 alters glucosinolate hydrolysis and insect resistance in Arabidopsis. *Plant Cell* **2006**, *18*, 1524–1536. [CrossRef] [PubMed]

38. Salvucci, M.E. Association of Rubisco activase with chaperonin-60β: A possible mechanism for protecting photosynthesis during heat stress. *J. Exp. Bot.* **2008**, *59*, 1923–1933. [CrossRef] [PubMed]

antioxidants

MDPI

Article

Crystal Structure of Chloroplastic Thioredoxin f2 from *Chlamydomonas reinhardtii* Reveals Distinct Surface Properties

Stéphane D. Lemaire [1], Daniele Tedesco [2], Pierre Crozet [1], Laure Michelet [1], Simona Fermani [3], Mirko Zaffagnini [4,*] and Julien Henri [1,*]

[1] Laboratoire de Biologie Moléculaire et Cellulaire des Eucaryotes, Institut de Biologie Physico-Chimique, Unité Mixte de Recherche 8226 CNRS Sorbonne Université, 13 rue Pierre et Marie Curie, 75005 Paris, France; lemaire@ibpc.fr (S.D.L.); crozet@ibpc.fr (P.C.); michelet@ibpc.fr (L.M.)

[2] Bio-Pharmaceutical Analysis Section (Bio-PhASe), Department of Pharmacy and Biotechnology, University of Bologna, via Belmeloro 6, 40126 Bologna, Italy; daniele.tedesco@unibo.it

[3] Department of Chemistry "Giacomo Ciamician", University of Bologna, via Selmi 2, 40126 Bologna, Italy; simona.fermani@unibo.it

[4] Laboratory of Molecular Plant Physiology, Department of Pharmacy and Biotechnology, University of Bologna, via Irnerio 42, 40126 Bologna, Italy

* Correspondence: mirko.zaffagnini3@unibo.it (M.Z.); julien.henri@sorbonne-universite.fr (J.H.); Tel.: +39-051-209-1314 (M.Z.); +33-158-415-007 (J.H.)

Received: 31 October 2018; Accepted: 20 November 2018; Published: 23 November 2018

Abstract: Protein disulfide reduction by thioredoxins (TRXs) controls the conformation of enzyme active sites and their multimeric complex formation. TRXs are small oxidoreductases that are broadly conserved in all living organisms. In photosynthetic eukaryotes, TRXs form a large multigenic family, and they have been classified in different types: f, m, x, y, and z types are chloroplastic, while o and h types are located in mitochondria and cytosol. In the model unicellular alga *Chlamydomonas reinhardtii*, the TRX family contains seven types, with f- and h-types represented by two isozymes. Type-f TRXs interact specifically with targets in the chloroplast, controlling photosynthetic carbon fixation by the Calvin–Benson cycle. We solved the crystal structures of TRX f2 and TRX h1 from *C. reinhardtii*. The systematic comparison of their atomic features revealed a specific conserved electropositive crown around the active site of TRX f, complementary to the electronegative surface of their targets. We postulate that this surface provides specificity to each type of TRX.

Keywords: thioredoxin; Calvin-Benson cycle; photosynthesis; carbon fixation; chloroplast; macromolecular crystallography; protein-protein recognition; electrostatic surface; *Chlamydomonas reinhardtii*

1. Introduction

Thioredoxins (TRXs) are small oxidoreductases of 10–16 kDa exhibiting a characteristic three dimensional structure classified as TRX fold [1], composed of a single canonical globular domain comprising a mixed β-sheet surrounded by four α-helices [2–4]. These proteins play a key role in controlling the redox status of protein disulfide bonds in all non-parasitic organisms [5]. The redox activity of TRXs is guaranteed by the presence of a solvent-exposed motif (most commonly Trp-Cys-Gly-Pro-Cys) containing two cysteine (Cys) residues that catalyze protein disulfide reduction. TRXs are recognized as having diverse roles in numerous cellular processes and human diseases [6–9]. Non-photosynthetic organisms contain a limited number of TRXs (two in *Escherichia coli*, three in *Saccharomyces cerevisiae*, and two in *Homo sapiens*), which are localized in the cytosol and mitochondria, and are reduced by the nicotinamide adenine dinucleotide phosphate

(NADPH)-dependent flavoenzyme thioredoxin reductase (NTR). By contrast, in photosynthetic organisms, TRXs are part of a large multigenic family (four in *Synechocystis* sp. PCC6803, 21 in *Arabidopsis thaliana*, and nine in *Chlamydomonas reinhardtii*). Phylogenetic and sequence analyses led to the classification of plant TRXs in different types: TRXs f, m, x, y, and z are chloroplastic, while o-type and h-type are found in mitochondria and cytosol [10–13]. Cytosolic and mitochondrial TRXs are reduced by NTRs, while chloroplastic TRXs are specifically reduced by the iron-sulfur containing ferredoxin–thioredoxin reductase, which derives electrons from ferredoxin and the photosynthetic electron transfer chain [14–17].

In photoautotrophic eukaryotes, TRXs were originally identified for their ability to modulate the activity of chloroplastic enzymes involved in carbon metabolism, such as the Calvin–Benson cycle fructose-1,6-bisphosphatase (FBPase) [18], NADP malate dehydrogenase (NADP-MDH) [19], or glucose-6-phosphate dehydrogenase [20]. In the dark, chloroplast 2-cysteine peroxiredoxins (2-CysPRX) inactivate FBPase, phosphoribulokinase (PRK), NADP-MDH, and glyceraldehyde-3-phosphate dehydrogenase (GAPDH) by oxidation [21,22]. Subsequent activation by light proceeds through the TRX-dependent reduction of regulatory disulfide bonds. Proteomic studies revealed that TRXs potentially reduce more than 1000 targets [23]. The large number of putative targets highlights the crucial role of TRXs in the control of a myriad of metabolic pathways and processes. Nevertheless, the molecular mechanisms underlying the TRX-dependent regulation of numerous targets are not clearly established.

The TRX fold is one of the most conserved throughout evolution, as suggested by paleobiochemistry [24]. Conservation of the tridimensional fold and of the active site residues account for the functional redundancy of plant TRX family members, as exemplified by the functional compensation of yeast deletion mutants by plant orthologues [25,26]. Nevertheless, several studies provided evidence for a functional specialization of TRX types for the regulation of specific targets [16]. In particular, systematic evaluation of the specificity of the different TRX types for the activation of Calvin–Benson enzymes revealed that they are all specifically or preferentially activated by f-type TRX, including 3-phosphoglycerate kinase (PGK) [27], FBPase [28–30], sedoheptulose-1,7-bisphosphatase (SBPase) [31,32], GAPDH [33–35], and PRK [33,36] (for reviews see [10,14,16]). This specificity does not appear to be linked to the redox potential of the Cys couple, which ranges from −368 mV to −336 mV at pH 7.9 for the different TRX types [29,30,37,38].

Despite the wealth of data gathered from biochemical studies on the functionality of TRXs, the molecular rules for the selective reduction of a given target disulfide by a specific TRX remain open to speculation. Understanding the physico-chemistry and structural features of TRXs hence appears as a powerful entry point to estimate the physiological significance of TRX-dependent regulation, along with specificity towards target proteins [39].

Here, we describe the novel crystal structure of chloroplastic TRX f2 (CrTRXf2), and a crystal structure of TRX h1 (CrTRXh1) from *Chlamydomonas reinhardtii*. Extensive comparison with other known 3D structures of algal or land plants TRXs and their targets allowed for structural features likely responsible for target recognition to be distinguished.

2. Materials and Methods

2.1. Cloning, Expression, and Purification of CrTRXf2 and CrTRXh1

The gene at locus Cre05.g243050.t1.2 encodes chloroplastic TRX f2 from *Chlamydomonas reinhardtii* (CrTRXf2). The complementary DNA (cDNA)-encoding mature CrTRXf2 was amplified by polymerase chain reaction using 5′-AGCAAACCATGGGCGGCAGCGTTGACGGCCAG as a forward primer introducing a 5′-*NcoI* restriction site, and 5′-GGTGTGGGATCCTCAGTTCTTGGGCGGCTG as a reverse primer introducing a *BamHI* restriction site downstream of the stop codon. The cleavage site of the chloroplast transit peptide was predicted using multiple sequence alignments of plant TRX f sequences and the ChloroP prediction program [40]. Residues are numbered according to Uniprot reference sequences (ID: A0A2K3DSC9). CrTRXf2 was cloned in a modified pET-3d vector containing additional codons upstream of the *Nco*I site to express a His-tagged protein with six N-terminal histidines [41]. The expression vector was then used to transform *E. coli* BL21 Rosetta™ 2 (DE3) (Novagen). Bacterial transformants were grown at 37 °C in lysogeny broth (LB) medium supplemented with 100 µg mL^{-1} ampicillin, and the production was induced at an Abs$_{600}$ of 0.5 with 0.2 mM isopropyl-β-D-thiogalactopyranoside at 37 °C for 3 h. Cells were then harvested by centrifugation, re-suspended in 30 mM Tris-HCl (pH 7.9), and broken using a French press (6.9 × 10^7 Pa). Cell debris were removed by centrifugation at 20,000× *g* for 20 min at 4 °C, and the supernatant was then applied onto a Ni^{2+} Hi-Trap chelating resin (HIS-Select® nickel affinity gel, Sigma-Aldrich, St. Louis, MO, USA) pre-equilibrated with 30 mM Tris-HCl (pH 7.9) and 150 mM NaCl. The recombinant protein was purified according to the manufacturer's instructions. The molecular mass and purity of the protein were analyzed by denaturing gel electrophoresis (SDS-PAGE) after dialysis against 30 mM Tris-HCl (pH 7.9) and 1 mM ethylenediaminetetraacetic acid EDTA. The concentration of purified CrTRXf2 was determined spectrophotometrically using a molar extinction coefficient at 280 nm of 17,085 M^{-1} cm^{-1} [42]. CrTRXh1 was expressed and purified, as previously described [43]. Samples of recombinant proteins were stored at −20 °C. The recombinant CrTRXf2 contains 125 residues, starting at the N-terminus with the introduced MHHHHHHHM peptide, followed by the mature protein sequences (i.e., upon removal of the chloroplast targeting sequence), beginning with a glycine (Gly65, Figure 1). Throughout the paper, residues are numbered according to the mature protein sequence (Gly65 in the preprotein becomes Gly1 in the mature protein).

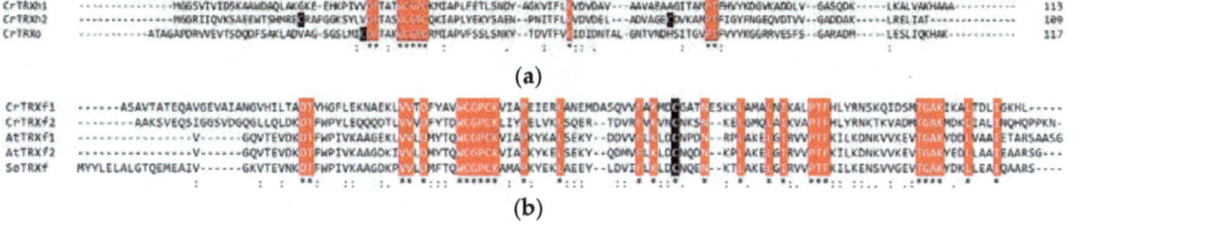

(a)

(b)

Figure 1. Multiple sequence alignment of plant thioredoxins (TRXs). (**a**) Sequence alignment of TRXs from *Chlamydomonas reinhardtii* (CrTRXs). CrTRXs were aligned using Clustal Omega [44,45]. The sequences used correspond to mature forms either known or predicted by the ChloroP prediction software [40]. Abbreviations and accession numbers: CrTRXf1, *Chlamydomonas reinhardtii* TRX f1 (UniProt ID: Q84XR8); CrTRXf2, *Chlamydomonas reinhardtii* TRX f2 (UniProt ID: A0A2K3DSC9); CrTRXm, *Chlamydomonas reinhardtii* TRX m (UniProt ID: P23400); CrTRXx, *Chlamydomonas reinhardtii* TRX x (UniProt ID: Q84XR9); CrTRXy, *Chlamydomonas reinhardtii* TRX y (UniProt ID: Q84XS2); CrTRXz, *Chlamydomonas reinhardtii* TRX z (UniProt ID: A8J0Q8); CrTRXh1, *Chlamydomonas reinhardtii* TRX h1 (UniProt ID: P80028); CrTRXh2, *Chlamydomonas reinhardtii* TRX h2 (UniProt ID: Q84XS1); CrTRXo, *Chlamydomonas reinhardtii* TRX o (UniProt ID: Q84XS0). An asterisk (*) indicates positions that have a single, fully conserved residue. These residues are highlighted in white on a red background. Catalytic Cys are marked with two arrows. A colon (:) indicates conservation between groups of strongly similar properties, scoring > 0.5 in the Gonnet point accepted mutation (PAM) 250 matrix. A period (.) indicates conservation between groups of weakly similar properties, scoring ≤ 0.5 in the Gonnet PAM 250 matrix. Cys residues other than the catalytic ones are highlighted in white on a black background. (**b**) Sequence alignment of *Chlamydomonas reinhardtii*, *Arabidopsis thaliana*, and *Spinacia oleracea* f-type TRXs. Mature proteins were aligned as described in panel (**a**). Abbreviations and accession numbers: CrTRXf1, *Chlamydomonas reinhardtii* TRX f1 (UniProt ID: Q84XR8); CrTRXf2,

Chlamydomonas reinhardtii TRX f2 (UniProt ID: A0A2K3DSC9); AtTRXf1, *Arabidopsis thaliana* TRX f1 (UniProt ID: Q9XFH8); AtTRXf2, *Arabidopsis thaliana* TRX f2 (UniProt ID: Q9XFH9); SoTRXf, *Spinacia oleracea* TRX f (UniProt ID: P09856). An asterisk (*) indicates positions that have a single, fully conserved residue. These residues are highlighted in white on a red background, except the conserved non-catalytic Cys that is highlighted in white on a black background. A colon (:) indicates conservation between groups of strongly similar properties, scoring > 0.5 in the Gonnet PAM 250 matrix. A period (.) indicates conservation between groups of weakly similar properties, scoring ≤ 0.5 in the Gonnet PAM 250 matrix.

2.2. Crystallization and Diffraction Data Collection

Sparse-matrix screening of candidate crystallization conditions was set up on iQ plates from TTP Labtech Ltd. (Melbourn, United Kingdom) with mixes of 100 nL protein and 100 nL commercial precipitant solutions (Qiagen) and incubated at 20 °C. Monocrystals of CrTRXf2 grown in condition JCSG II 26 (100 mM HEPES-NaOH, pH 7.5, 2% polyethylene glycol (PEG) 400, 2.0 M ammonium sulfate) were harvested and flash-frozen in liquid nitrogen. A complete, 2.01 Å resolution, diffraction dataset was collected on beamline ID29 at the European Synchrotron Radiation Facility (Grenoble, France). Monocrystals of CrTRXh1 grown in condition Classics 70 (200 mM ammonium sulfate, 100 mM sodium cacodylate, pH 6.5, 30% PEG 8000) were harvested and cryo-protected with an additional 25% ethylene glycol before flash-freezing in liquid nitrogen. A complete, 1.38 Å resolution, diffraction dataset was collected on beamline Proxima-1 at the SOLEIL synchrotron (Saint-Aubin Gif-sur-Yvette, France).

2.3. Structure Determination, Model Building, and Analysis

The native I222 dataset of CrTRXf2 crystal was used for molecular replacement by PHENIX.PHASER-MR [46], with an homology model of the protein calculated by PHYRE2 [47] and three copies of each per asymmetric unit. The top solution was refined by PHENIX.REFINE [48]; the resulting molecular model was manually adjusted into experimental electron density in COOT software [49] and further refined until reaching the final R-work = 0.2262 and R-free = 0.2638 with 98.34% favored Ramachandran dihedrals (Table 1). The native P3121 dataset of CrTRXh1 crystal was used for molecular replacement by PHENIX.PHASER-MR, with chain A from Protein Data Bank entry 1EP7.pdb [50] as a search model and two molecules per asymmetric unit. The top solution was refined by PHENIX.REFINE, the resulting molecular model manually adjusted into experimental electron density with COOT and further refined until reaching final R-work = 0.1812 and R-free = 0.2168 with 98.17% favored Ramachandran dihedrals. Reflection files and final models coordinates were deposited in the Protein Data Bank under accession codes 6I1C.pdb and 6I19.pdb, for CrTRXf2 and CrTRXh1, respectively. Protein models were analyzed with the webservers Structural Classification of Proteins (SCOPe), PDBeFold structure similarity, ConSurf server for the identification of functional regions in proteins, Pictorial database of tridimensional structures in the Protein Data Bank (PDBsum), and CATH Protein structure classification database. TRX surface electrostatic potentials were computed by the eF-surf algorithm [51] on Protein Data Bank Japan portal with the self-consistent boundary method, or by the Adaptive Poisson-Boltzmann Solver (APBS) Electrostatics plugin of PyMOL. Figures were drawn with PyMOL version 2.0.3 (The PyMOL Molecular Graphics System, Schrödinger, LLC).

Table 1. Crystallographic data collection and refinement statistics.

	CrTRXf2	CrTRXh1
Wavelength (Å)	0.9762	0.9677
Resolution range	42.85–2.01 (2.082–2.01)	36.42–1.378 (1.427–1.378)
Space group	I 2 2 2	P 31 2 1
Unit cell	65.383 97.475 139.545 90 90 90	48.76 48.76 143.97 90 90 120
Total reflections	57,413 (5796)	83,812 (8115)
Unique reflections	29,702 (2973)	41,955 (4085)
Multiplicity	1.9 (2.0)	2.0 (2.0)
Completeness (%)	98.74 (99.80)	99.80 (98.81)
Mean I/sigma (I)	6.69 (0.95)	25.45 (3.98)
Wilson B-factor	34.89	16.29
R-merge	0.0697 (0.7309)	0.01107 (0.1238)
R-meas	0.09857 (1.034)	0.01565 (0.1751)
R-pim	0.0697 (0.7309)	0.01107 (0.1238)
CC1/2	0.997 (0.485)	1 (0.968)
CC *	0.999 (0.808)	1 (0.992)
Reflections used in refinement	29,749 (2970)	41,938 (4085)
Reflections used for R-free	1998 (200)	2095 (204)
R-work	0.2262 (0.3131)	0.1812 (0.2281)
R-free	0.2638 (0.3606)	0.2168 (0.2690)
CC (work)	0.951 (0.709)	0.964 (0.914)
CC (free)	0.915 (0.649)	0.946 (0.940)
Number of non-hydrogen atoms	2764	2033
Macromolecules	2585	1638
Solvent	179	395
Protein residues	318	222
RMS (bonds)	0.008	0.005
RMS (angles)	0.97	0.77
Ramachandran favored (%)	98.34	98.17
Ramachandran allowed (%)	1.33	1.83
Ramachandran outliers (%)	0.33	0.00
Rotamer outliers (%)	0.34	0.60
Clashscore	5.18	4.24
Average B-factor	39.23	19.69
Macromolecules	39.04	17.65
Solvent	42.01	28.15

2.4. Circular Dichroism (CD) Spectroscopy

CD analysis was performed at room temperature on a J-810 spectropolarimeter (Jasco, Tokyo, Japan). Samples of CrTRXf2 were prepared at a nominal concentration of 10 μM in 30 mM Tris-HCl buffer (pH 7.9). Reduced CrTRXf2 was obtained following 30 min incubation in the presence of a 10-fold molar excess of tris(2-carboxyethyl)phosphine (TCEP). The exact concentration of samples was determined from the absorbance at 280 nm (1 cm path-length) based on the theoretical molar absorption coefficients of 16,960 and 17,085 M^{-1} cm^{-1} for reduced and oxidized CrTRXf2, respectively [42]. The solutions were then transferred into a QS quartz cell with a 0.5 mm path length (Hellma, Milan, Italy) for far-ultraviolet (UV) CD measurements in the 250–195 nm spectral range, using a 20 nm min^{-1} scanning speed, a 4 nm response, a 2 nm spectral bandwidth, and an accumulation cycle of 3. Solvent-corrected CD spectra of reduced and oxidized CrTRXf2 were converted to molar units per residue ($\Delta\varepsilon_{res}$) and analyzed using the BeStSel web server (http://bestsel.elte.hu [52,53]) to estimate the secondary structure contents.

3. Results

3.1. Sequence Analysis of Chlamydomonas TRX f2

To obtain insights on putative regions providing target specificity, *Chlamydomonas reinhardtii* TRX sequences were compared. Multiple sequence alignments revealed that Chlamydomonas TRXs exhibit low similarity ranging from ~21% to ~45% (Figure 1a). The highest homology was found between isozymes of chloroplastic f-type CrTRX (45.5%, f1 versus f2) and cytoplasmic h-type CrTRX (44.4%, h1 versus h2), whereas the lowest homology is observed between CrTRXh2 and CrTRXy (21.2%). The presence of fully conserved residues is restricted to the active site motif WCGPC containing the two catalytic Cys, and five amino acids (Figure 1a, highlighted in red). The latter strictly conserved residues were shown to be important for the function or the structure of diverse TRXs [54–58]. Comparison of CrTRXf2 sequence with plastidial and mitochondrial CrTRXs (CrTRX m, x, y, z, and o) revealed sequence identity in the ~22–29% range (Figure 1a), which is consistent with the low similarity shared by CrTRXs. By contrast, CrTRXf2 has a slightly higher homology when compared to h-type CrTRXs (35.2 and 30.5% with h1 and h2, respectively) (Figure 1a). When compared with other f-type TRXs from plants, the sequence identity increased to ~40%, AtTRXf1 having the highest homology (41.4%, Figure 1b). Moreover, 29 out of 125 amino acids in CrTRXf2 are fully conserved in all f-type TRXs (Figure 1b, highlighted in red), including the extra Cys located in position 65 of mature CrTRXf2. This homology analysis displayed a strong diversity, even for TRX from the same type. To gain further insights into the structural determinants of TRX specificity, we determined the crystal structure of two TRXs: CrTRXf2 and CrTRXh1. The latter is highly similar to previously described structures (RMSD = 0.287 Å to 1EP7.pdb [50]), and to structures described in a companion paper of this journal issue [59].

3.2. Chlamydomonas TRX f2 Folds as a Canonical TRX

The crystal structure of CrTRXf2 was solved at 2.01 Å resolution. The closest structural match revealed by tridimensional comparison with the PDB archive is SoTRXf (PDB identifier: 2PVO) [60]. The structural alignment of the two enzymes gave an RMSD = 0.859 Å, despite CrTRXf2 exhibiting only 39.7% sequence identity with its spinach ortholog. As shown in Figure 2A–C, the secondary structures are organized from the N-terminus to the C-terminus, as follows (residue boundaries indicated in parentheses): α-helix 1 (16–23), β-strand 1 (29–34), α-helix 2 (39–54), β-strand 2 (59–64), α-helix 3 (70–76), β-strand 3 (83–88), β-strand 4 (91–97), α-helix 4 (101–111). β-strands order in the mixed β-sheet is 4–3–1–2, with β-strand 3 being antiparallel to the others. The β-sheet is sandwiched between α helices 1 and 3 on one side, and α-helices 2 and 4 on the other side. The overall structure is a flattened spheroid of 49 Å equatorial diameter and 25 Å polar diameter, and conforms to the classical TRX fold.

Structural alignments with CrTRXm (PDB ID: 1DBY) [61], and CrTRXh1 (this study) yield RMSDs of 0.927 Å and 0.953 Å, respectively. These close alignment scores further confirm the high structural similarity irrespective of a low sequence homology (24.2% and 35.2% identity with CrTRXm and CrTRXh1, respectively; Figure 1a). Hence, the cell requirements for TRX function maintained strong selection towards the TRX fold, despite the specialization of diverse types.

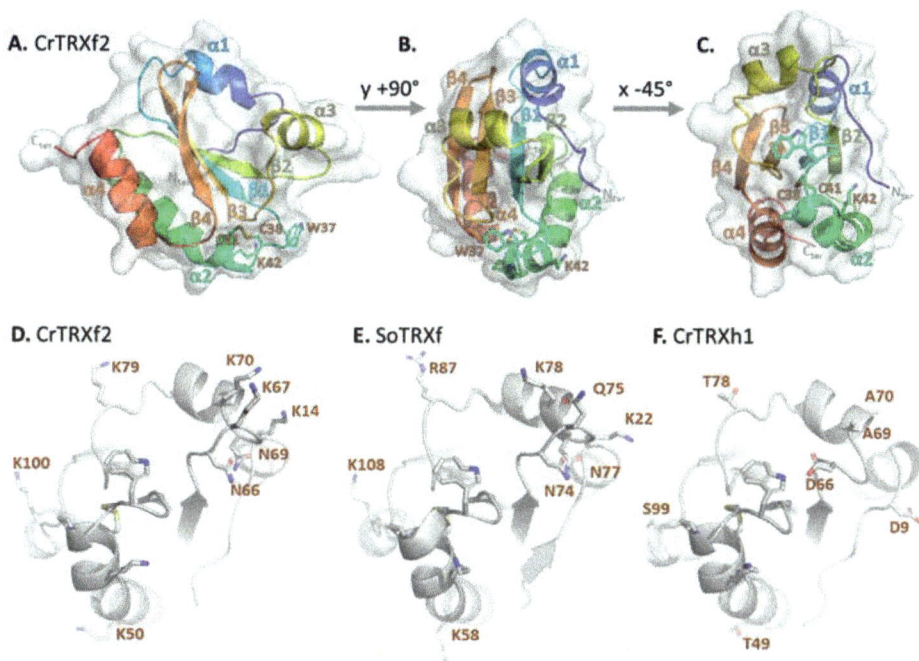

Figure 2. Crystal structure of *Chlamydomonas reinhardtii* thioredoxin f2 (CrTRXf2). (**A–C**) Three rotated projections of the crystal structure of CrTRXf2. The main chain is drawn as cartoon, colored from blue (amino-terminus) to red (carboxy-terminus). Side chains of the ^{37}WCGPCK^{42} motif are drawn as sticks. The protein surface is displayed in transparent light gray. (**D–F**) Side chains of the ^{37}WCGPCK^{42} motif and of electropositive residues conserved in (**D**) CrTRXf2 and (**E**) *Spinacia oleracea* SoTRXf, but not in (**F**) CrTRXh1, are shown as sticks on the cartoon-drawn main chain.

3.3. The Redox Site of CrTRXf2

In CrTRXf2, the first active site Cys (hereafter referred to as Cys$_N$) is located at position 38 at the N-terminal kinked tip of α-helix 2. The second catalytic Cys (hereafter referred to as Cys$_C$) is located at position 41, and it belongs to the same α-helix 2 (Figure 2A–C). In the CrTRXf2 structure, the catalytic Cys are covalently disulfide-bonded in accordance with the non-reducing conditions of the purification and crystallization procedures. Trp37, Gly39, Pro40, and Lys42 of the conserved ^{37}WCGPCK^{42} motif arch over the disulfide bond, leaving two gates to interact with the solvent. In the solved crystal structure, the deep pocket on Tyr45 side of the bond is filled with water oxygens 20, 80, 89, 122, and 131 while the shallow crevice on the Pro82 side of the bond is occupied by water oxygen 110. The redox activity of the Cys pair either requires target disulfide docking on these gates, or rearrangement of the ^{37}WCGPCK^{42} arch, to increase Cys$_N$ thiol exposure.

Structural alignment of CrTRXf2 with CrTRXh1 revealed that the oxygen of water 20 of CrTRXf2 localizes 0.4 Å away from the corresponding oxygen of water 133 of CrTRXh1. In both structures, these equivalent water molecules hydrogen bond with Asp32/31 (CrTRXf2/CrTRXh1 numbering respectively) and Cys$_N$, and this was previously characterized as a determinant for lowering the pK$_a$ of Cys$_C$ [50].

The third cysteine of CrTRXf2, located at position 65, is perfectly conserved amongst f-type TRXs (Figure 1b) and was shown to be modified by S-glutathionylation [62]. In our structure, Cys65 is likely in the thiol form, since its side chain points inward to a hydrophobic pocket formed by Phe33, Val63 and Ile78. Redox modification of Cys65 thus requires a local rearrangement of the TRX surface that is possible if loop 65–69 adopts an alternate conformation. Consistently, the equivalent loop on CrTRXh1 is flipped by 7 Å towards the domain core relative to the CrTRXf2 position, a conformation that is correlated with an additional flip of the N-terminal loop of CrTRXh1 by 4 Å in the same direction. These alternate conformations argue in favor of a flexibility of this region that may condition the redox regulation of Cys65 of CrTRXf2.

3.4. Comparison of the Secondary Structures of Oxidized and Reduced CrTRXf2

The far-UV circular dichroism (CD) spectra of the reduced and oxidized forms of CrTRXf2 (Figure 3) differ slightly, with the former showing a more intense negative band at 220 nm, and an additional shoulder centered at ~210 nm. Nevertheless, the overall CD profiles of CrTRXf2 in both redox states are similar to those previously reported for other TRXs [63–65]. The secondary structure estimation given by the BeStSel algorithm predicts a lower percentage of α-helices (reduced: 18%; oxidized: 11%) and a slightly higher content in β-strands (reduced: 34%; oxidized: 29%), compared to the secondary structure of the crystal structure of oxidized CrTRXf2 (α-helix: 33%; β-strand; 19%) as calculated using the database of Define Secondary Structure of Proteins (DSSP) web server (https://swift.cmbi.umcn.nl/gv/dssp/ [66]), based on the full sequence of the His-tagged enzyme (125 residues). Even though some divergence can be expected between in-solution and solid-state protein structures [67], the observed variations probably have a different explanation, as detailed below.

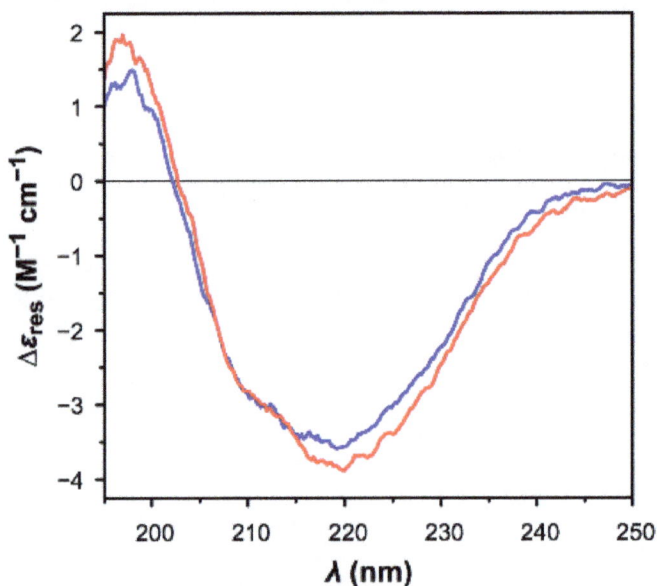

Figure 3. Far-ultraviolet circular dichroism spectra of reduced (blue line) and oxidized (red line) *Chlamydomonas reinhardtii* thioredoxin f2 (9.94 µM for reduced CrTRXf2, 9.28 µM for oxidized CrTRXf2) in Tris-HCl buffer (30 mM; pH 7.9).

On the experimental side, the absorption cut-off of the CD measurements did not allow to collect data below the 195 nm threshold, limiting the accuracy of the estimation. On the theoretical side, mixed α/β proteins still represent a tough challenge for the algorithms available for secondary structure estimations by CD spectroscopy, despite the recent and encouraging improvements provided by new methods. A textbook example of the huge discrepancies of results obtained by these approaches was indeed reported for TRXs [63]. The BeStSel fold recognition analysis, nevertheless, correctly predicts that both samples are structurally related to the class of mixed α/β proteins organized in a 3-layer ($\alpha/\beta/\alpha$) sandwich arrangement, in agreement with the typical tertiary structure of TRX (CATH classification 3.40.30.10; http://www.cathdb.info [68]).

3.5. Surface Specificities of CrTRXf2

CrTRXf2 cleft analysis revealed its 10 largest surface grooves, which include volumes of 785, 375, 295, 393, 262, 251, 206, 144, 127, and 138 Å^3 for a total volume of 2976 Å^3. An equivalent CrTRXh1 cleft analysis revealed its 10 largest surface grooves, which included volumes of 483, 462, 429, 452, 292, 300, 172, 81, 78, and 70 Å^3 for a total volume of 2819 Å^3. Hence, the surface topography of CrTRXf2 appears rougher than that of CrTRXh1, mainly because of its top single cleft of 785 Å^3. Cavity sizes may control the hydrogen bonding of water molecules at the surface of the protein, thus modifying the local flexibility of TRX [69].

CrTRXf2 Cys38-Cys41 site is surrounded by basic side chains of six lysines (Lys14, Lys50, Lys67, Lys70, Lys79, Lys100) and two hydrogen bond donor side chains (Asn66, Asn69) (Figure 2D). These positions align with positively charged (Lys22, Lys58, Lys78, Arg87, Lys108) or hydrogen bond donor side chains (Asn74, Asn77) of the spinach TRX f ortholog (PDB ID: 1FAA, 1F9M) [60] (Figure 2E). CrTRXf2 electropositive patches locate on the α-helix 3 side of the disulfide, and on the β-strand 2 side of the disulfide. These eight positions form an electropositive crown around the active site, conserved in both plant and algal enzymes (Figure 2D,E). In striking opposition, the corresponding residues in CrTRXh1 are negatively charged (Asp9, Asp66), electronegative (Thr49, Thr78, Ser99), or neutral (Ala69, Ala70) (Figure 2F). CrTRXh1 corresponding surface displays an electronegative potential on the α-helix 3 side of the disulfide and of negligible polarity on the β-strand 2 side of the disulfide. Hence, despite the strong conservation bias for a common fold and active site composition, CrTRXf2 and CrTRXh1 display distinct electrostatic potential on their solvent accessible surface (Figure 4B,C).

The electropositive surface of CrTRXf2 compares with the equivalent regions of modelled CrTRXf1, although the latter possess a more neutral character (Figure 4A,B). If experimentally confirmed, this may explain a more stringent specificity of CrTRXf2 than CrTRXf1 for equivalent targets. This f-type electropositive character is maintained in land plant enzymes (Figure 4J–L). CrTRXh1 and the modelled CrTRXh2 both present mixed polarities, confirming the significant difference of these cytoplasmic isoforms compared to the f-type CrTRXs (Figure 4C,D). The surface of chloroplast CrTRXm is closer to the h- than f-type CrTRXs (Figure 4E), while modelled CrTRXx, CrTRXy, and CrTRXz all present neutral or electronegative surfaces around the catalytic cysteines (Figure 4G–I). The modelled mitochondrial CrTRXo appears similar to CrTRXh1 and m (Figure 4F).

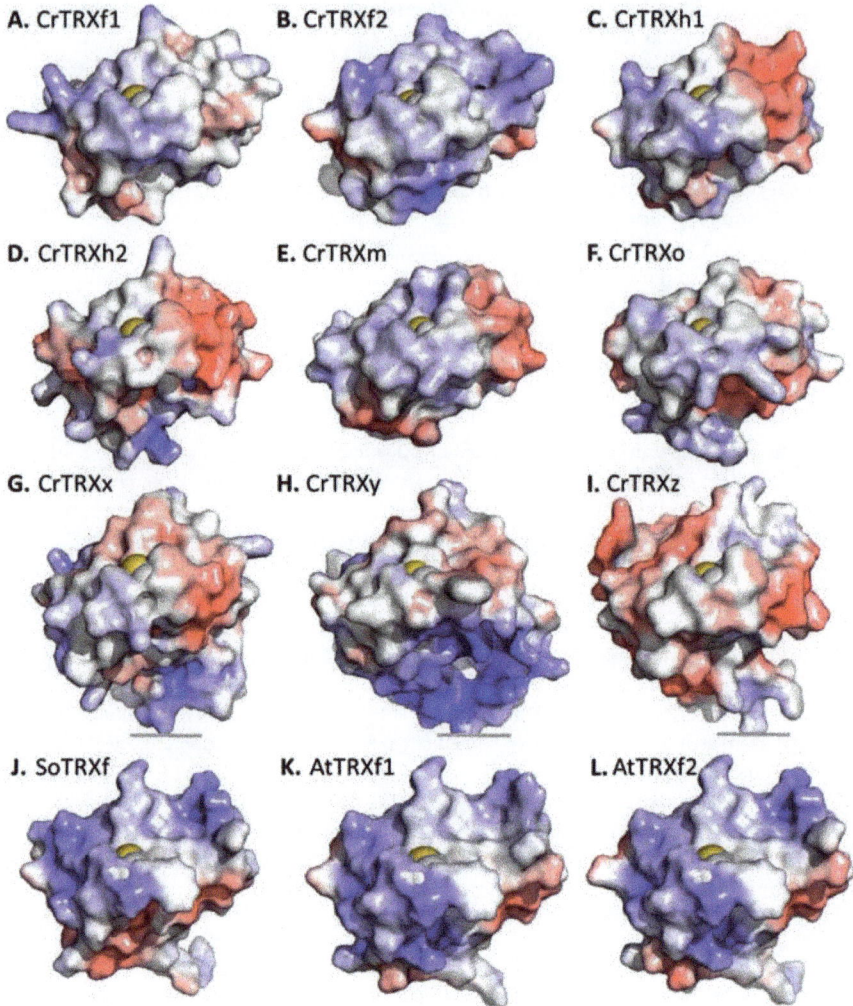

Figure 4. Electrostatic surface variation of plant thioredoxins. (**A**) PHYRE2 [47] homology model of chloroplast CrTRXf1. (**B**) Crystal structure of chloroplast CrTRXf2 (this study). (**C**) Crystal structure of cytosolic CrTRXh1 (this study). (**D**) PHYRE2 homology model of cytosolic CrTRXh2. (**E**) Nuclear magnetic resonance structure of chloroplast CrTRXm (Protein Data Bank identifier: 1DBY) [61]. (**F**) PHYRE2 homology model of mitochondrial CrTRXo. (**G**) PHYRE2 homology model of chloroplast CrTRXx. (**H**) PHYRE2 homology model of chloroplast CrTRXy. (**I**) PHYRE2 homology model of chloroplast CrTRXz. Electrostatic surface potentials were computed with the Protein Data Bank Japan webserver eF-surf [51] (red for electronegative, white for neutral, blue for electropositive). (**J**) Crystal structure of *Spinacia oleracea* TRXf (PDB ID: 1F9M). (**K**) PHYRE2 homology model of chloroplast *Arabidopsis thaliana* TRXf1. (**L**) PHYRE2 homology model of chloroplast *Arabidopsis thaliana* TRXf2. Electrostatic surface potentials were computed with the Adaptive Poisson-Boltzmann Solver (APBS) Electrostatics plugin in PyMOL software (red for electronegative, white for neutral, blue for electropositive). All structures were aligned in PyMOL. N-terminal active site Cys (Cys38 in CrTRXf2) side chain is displayed as spheres at the center of the projection (gold for thiol sulfur, white for carbon beta).

3.6. Surface Specificities of TRXf Targets

Protein–protein interaction with specific CrTRXf2 targets would involve a complementary electronegative surface. Indeed, we observed an extended continuous electronegative surface around the cysteines of the two f-type TRXs targets, FBPase and SBPase (Figure 5). Moreover, the electronegative character of targets was identically observed in orthologues from both vascular (*Pisum sativum* [70], Figure 5A) and non-vascular land plants (*Physcomitrella patens* [32]), accounting for the conservation of this structural feature over speciation and evolution. We applied molecular docking simulations to orient possible interactions of CrTRXf2 with pea FBPase [70]. The FRODOCK algorithm [71] suggested 10 docking models of CrTRXf2 on the pea FBPase surface, four placed Cys_N of CrTRXf2 in the vicinity of target Cys153. These plausible solutions all bring the electropositive surface of α-helix 3 in contact with the target electronegative surface. Contrarily, the alignment of CrTRXh1 at the docked positions unfavorably joins the negative patches of both surfaces.

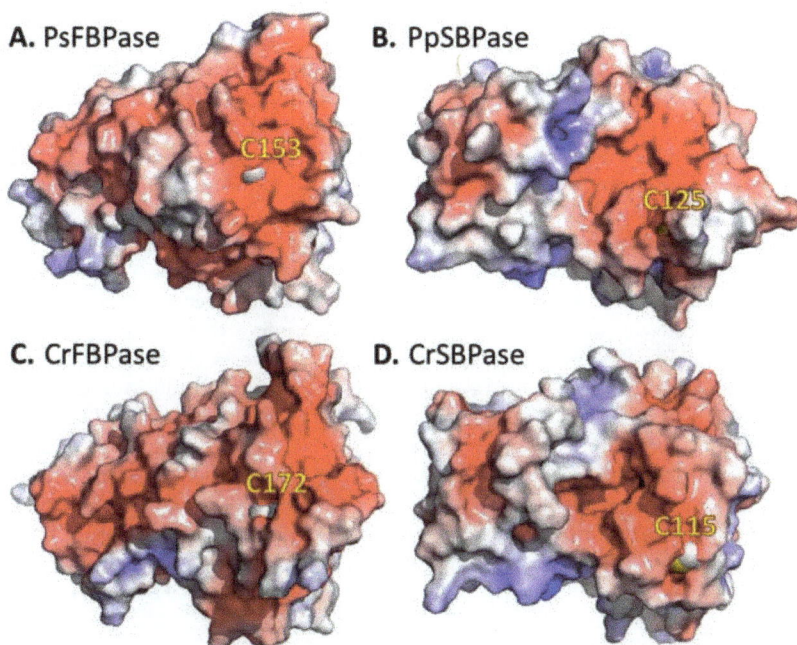

Figure 5. Electrostatic surface potential of thioredoxin f targets. (**A**) Crystal structure of *Pisum sativum* FBPase (PsFBPase, PDB ID: 1D9Q, [70]). (**B**) Crystal structure of *Physcomitrella patens* SBPase (PpSBPase, PDB ID: 5IZ1, [32]). (**C**) PHYRE2 homology model of *Chlamydomonas reinhardtii* FBPase (CrFBPase). (**D**) PHYRE2 homology model of *Chlamydomonas reinhardtii* SBPase (CrSBPase). Target cysteines are displayed as spheres (white for carbon beta, gold for thiol sulfur). Electrostatic surface potentials were computed with the APBS Electrostatics plugin in PyMOL (red for electronegative, white for neutral, blue for electropositive).

Upon the recognition of electro-complementary surfaces, the actual reduction of target disulfide requires a rearrangement of TRX to bring Cys38 in bonding distance to target Cys153. Alignment of CrTRXf2 structure with *Hordeum vulgare* TRXh2 complexed to the model target barley α-amylase/subtilisin inhibitor (BASI) [58,72] suggests that loop 35–38, loop 65–69, and α-helix 1 undergo most of the conformational variation, while the rest of the protein remains unaffected. Molecular structures determined by X-ray crystallography attribute an isotropic displacement B-factor to each atom of the refined model, the value of which quantifies the thermal vibration during data collection, and the variation of the atom position in the unit cell. The β-sheet core of both CrTRXf2 and CrTRXh1 display the lowest B-factor values, in accordance with a stably (thermostable) folded globular domain [73,74]. Both the N-terminal and C-terminal residues of the two TRXs display high B-factors, accounting for the poor resolution at the extremity of the modelled electron density. In the solved CrTRXf2 crystal structure, the complete α-helix 2, the first turn of α-helix 4, and the loop downstream of α-helix 1 are formed by atoms of the highest B-factor in the model. Meanwhile, the solved crystal structure of CrTRXh1 displays the highest B-factors on the complete α-helix 2 and the first turn of α-helix 4, but not at the loop downstream of α-helix 1. The pentapeptide ^{23}QQQDT^{27} of CrTRXf2 appears as a specific site of local flexibility. This hinge at the basis of α-helix 1 would appropriately support its movements upon target recognition.

4. Discussion and Conclusions

The resolution of the crystal structure of CrTRXf2 confirms the highly conserved structure of the TRX fold, despite it having a low sequence identity (Figures 1 and 2). The newly characterized Chlamydomonas TRX f2 has the same secondary structure composition and wiring diagram as CrTRXh1 when used as a reference. In addition, CD spectra revealed minor conformational changes in CrTRXf2, when analyzed under both oxidized and reduced forms. The active site is centered on the pair of Cys38 and 41 near the solvent-exposed surface of the protein. In the CrTRXf2 structure, these cysteines are disulfide bonded. The Cys pair points inward to a peptidic arch composed of the conserved ^{37}WCPGCK^{42} motif. The motif contributes to restrict the accessibility of the disulfide for its reduction by ferredoxin–thioredoxin reductase and its oxidation by TRX targets.

Despite the high functional and structural similarity in the TRX family, chloroplastic f-type TRXs specifically or preferentially activate Calvin–Benson cycle enzymes (i.e., FBPase, GAPDH, SBPase, PRK, and PGK). The structures of these target enzymes have been solved, but not in complex with TRX f, which limits our understanding of the molecular interactions and contact sites between the two molecules. Nevertheless, the solved structure of a complex between barley TRX h2 and the α-amylase/subtilisin inhibitor (BASI) stabilized through a mixed-disulfide bond corresponding to a reaction intermediate serves as a working model [72]. The structure of CrTRXf2 aligns with barley TRXh2 complexed to BASI (RMSD = 1.303 Å). Cys$_N$ is situated slightly away from its internal orientation compared to free TRX, towards an outward exposure that allows for its interaction with target Cys148. The target protein attacks the Tyr45 side of TRX in the deeper pocket of the ^{37}WCGPCK^{42} arch. In the course of complex formation, water molecules 104, 105, 106, 110, and 160 of CrTRXf2 model will leave the surface, to allow for target accommodation. Such an entropic effect of enzyme–substrate complex formation may be tested by in vitro experiments such as isothermal titration calorimetry [75].

Detailed comparative analysis of the crystal structures of CrTRXf2 with other TRXs revealed that the f-type specifically (i) forms a rougher surface with larger cavities, (ii) orients a crown of electropositive patches on the two opposite sides of the active site disulfide, and (iii) adopts a local flexible hinge downstream of α helix 1. These specific structural features of CrTRXf2 should guide the recognition of specific targets by facilitating target space accommodation, flexibility, and electrostatic interactions. These results are consistent with previous studies that suggested a major role for electrostatic interactions in the TRX–target interactions in chloroplasts [29,76]. Modelled CrTRXf1 displays a surface of lower electropositive potential than CrTRXf2, suggesting that CrTRXf1 is less efficient at targeting Calvin–Benson enzymes for reduction. This hypothesis should be tested by comparing CrTRXf1 and CrTRXf2 activities towards Calvin–Benson enzymes, and by the determination of CrTRXf1 experimental structure. To gain further insights into the structural determinants of TRX specificity, future studies should be aimed at solving the structure of TRX-target complexes and engineering the different TRX types, notably by altering the distribution of charges around the active site. Such knowledge may allow predicting the TRX dependence of the numerous putative targets identified by proteomics [23], and possibly rationalize the design of TRXs with predictable specificities.

To test these hypotheses in vivo in *Chlamydomonas reinhardtii*, new tools are available [77,78] that should accelerate prototyping of artificial TRX.

Author Contributions: Conceptualization, S.D.L., M.Z., and J.H.; Methodology, D.T., L.M., M.Z., and J.H.; Investigation, S.D.L., D.T., M.Z., and J.H.; Writing—Original Draft Preparation, S.D.L., M.Z. and J.H.; Writing—Review & Editing, S.D.L., D.T., P.C., S.F., M.Z. and J.H.

Funding: This work was supported in part by the Centre national de la recherche scientifique (CNRS) and Sorbonne Université, by Agence Nationale de la Recherche Grant 17-CE05-0001 CalvinDesign (to J.H., S.D.L. and P.C.), by LABEX DYNAMO ANR-LABX-011 and EQUIPEX CACSICE ANR-11-EQPX-0008, notably through funding of the crystallography platform of the Institut de Biologie Physico-Chimique (IBPC), and by the University of Bologna (grant FARB2012 to S.F. and M.Z.).

Acknowledgments: The authors thank the European Synchrotron Radiation Facility (ESRF, Grenoble, France) beamlines BM30-FIP, ID29, ID30B, and SOLEIL (Saint-Aubin Gif-sur-Yvette, France) beamline Proxima-1 for access to X-ray diffraction facility. The authors thank the crystallography discussion group at the IBPC for fruitful discussions. D.T. and M.Z. gratefully acknowledge supporting of their work by the University of Bologna (Alma Idea 2017 program).

Conflicts of Interest: The authors declare no conflict of interest. The funders had no role in the design of the study, in the collection, analyses, or interpretation of data, in the writing of the manuscript, and in the decision to publish the results.

References

1. Martin, J.L. Thioredoxin—A fold for all reasons. *Structure* **1995**, *3*, 245–250. [CrossRef]
2. Katti, S.K.; LeMaster, D.M.; Eklund, H. Crystal structure of thioredoxin from *Escherichia coli* at 1.68 A resolution. *J. Mol. Biol.* **1990**, *212*, 167–184. [CrossRef]
3. Dyson, H.J.; Gippert, G.P.; Case, D.A.; Holmgren, A.; Wright, P.E. Three-dimensional solution structure of the reduced form of *Escherichia coli* thioredoxin determined by nuclear magnetic resonance spectroscopy. *Biochemistry* **1990**, *29*, 4129–4136. [CrossRef] [PubMed]
4. Forman-Kay, J.D.; Clore, G.M.; Wingfield, P.T.; Gronenborn, A.M. High-resolution three-dimensional structure of reduced recombinant human thioredoxin in solution. *Biochemistry* **1991**, *30*, 2685–2698. [CrossRef] [PubMed]
5. Collet, J.F.; Messens, J. Structure, function, and mechanism of thioredoxin proteins. *Antioxid. Redox Signal.* **2010**, *13*, 1205–1216. [CrossRef] [PubMed]
6. Buchanan, B.B.; Holmgren, A.; Jacquot, J.P.; Scheibe, R. Fifty years in the thioredoxin field and a bountiful harvest. *Biochim. Biophys. Acta* **2012**, *1820*, 1822–1829. [CrossRef] [PubMed]
7. Hanschmann, E.M.; Godoy, J.R.; Berndt, C.; Hudemann, C.; Lillig, C.H. Thioredoxins, glutaredoxins, and peroxiredoxins—Molecular mechanisms and health significance: From cofactors to antioxidants to redox signaling. *Antioxid. Redox Signal.* **2013**, *19*, 1539–1605. [CrossRef] [PubMed]
8. Lee, S.; Kim, S.M.; Lee, R.T. Thioredoxin and thioredoxin target proteins: From molecular mechanisms to functional significance. *Antioxid. Redox Signal.* **2013**, *18*, 1165–1207. [CrossRef] [PubMed]
9. Toledano, M.B.; Delaunay-Moisan, A.; Outten, C.E.; Igbaria, A. Functions and cellular compartmentation of the thioredoxin and glutathione pathways in yeast. *Antioxid. Redox Signal.* **2013**, *18*, 1699–1711. [CrossRef] [PubMed]
10. Lemaire, S.D.; Michelet, L.; Zaffagnini, M.; Massot, V.; Issakidis-Bourguet, E. Thioredoxins in chloroplasts. *Curr. Genet.* **2007**, *51*, 343–365. [CrossRef] [PubMed]
11. Meyer, Y.; Belin, C.; Delorme-Hinoux, V.; Reichheld, J.P.; Riondet, C. Thioredoxin and glutaredoxin systems in plants: Molecular mechanisms, crosstalks, and functional significance. *Antioxid. Redox Signal.* **2012**, *17*, 1124–1160. [CrossRef] [PubMed]
12. Serrato, A.J.; Fernandez-Trijueque, J.; Barajas-Lopez, J.D.; Chueca, A.; Sahrawy, M. Plastid thioredoxins: A "one-for-all" redox-signaling system in plants. *Front. Plant Sci.* **2013**, *4*, 463. [CrossRef] [PubMed]
13. Geigenberger, P.; Thormahlen, I.; Daloso, D.M.; Fernie, A.R. The Unprecedented Versatility of the Plant Thioredoxin System. *Trendsplant Sci.* **2017**, *22*, 249–262.
14. Balsera, M.; Uberegui, E.; Schurmann, P.; Buchanan, B.B. Evolutionary development of redox regulation in chloroplasts. *Antioxid. Redox Signal.* **2014**, *21*, 1327–1355. [CrossRef] [PubMed]

15. Jacquot, J.P.; Eklund, H.; Rouhier, N.; Schurmann, P. Structural and evolutionary aspects of thioredoxin reductases in photosynthetic organisms. *Trends Plant Sci.* **2009**, *14*, 336–343. [CrossRef] [PubMed]

16. Michelet, L.; Zaffagnini, M.; Morisse, S.; Sparla, F.; Perez-Perez, M.E.; Francia, F.; Danon, A.; Marchand, C.H.; Fermani, S.; Trost, P.; et al. Redox regulation of the Calvin-Benson cycle: Something old, something new. *Front. Plant Sci.* **2013**, *4*, 470. [CrossRef] [PubMed]

17. Schurmann, P.; Buchanan, B.B. The ferredoxin/thioredoxin system of oxygenic photosynthesis. *Antioxid. Redox Signal.* **2008**, *10*, 1235–1274. [CrossRef] [PubMed]

18. Wolosiuk, R.A.; Buchanan, B.B. Thioredoxin and glutathione regulate photosynthesis in chloroplasts. *Nature* **1977**, *266*, 565–567. [CrossRef]

19. Jacquot, J.-P.; Vidal, J.; Gadal, P.; Schürmann, P. Evidence for the existence of several enzyme-specific thioredoxins in plants. *FEBS Lett.* **1978**, *96*, 243–246. [CrossRef]

20. Scheibe, R.; Anderson, L.E. Dark modulation of NADP-dependent malate dehydrogenase and glucose-6-phosphate dehydrogenase in the chloroplast. *Biochim. Biophys. Acta* **1981**, *636*, 58–64. [CrossRef]

21. Vaseghi, M.J.; Chibani, K.; Telman, W.; Liebthal, M.F.; Gerken, M.; Schnitzer, H.; Mueller, S.M.; Dietz, K.J. The chloroplast 2-cysteine peroxiredoxin functions as thioredoxin oxidase in redox regulation of chloroplast metabolism. *eLife* **2018**, *7*. [CrossRef] [PubMed]

22. Ojeda, V.; Perez-Ruiz, J.M.; Cejudo, F.J. 2-Cys Peroxiredoxins Participate in the Oxidation of Chloroplast Enzymes in the Dark. *Mol. Plant* **2018**, *11*, 1377–1388. [CrossRef] [PubMed]

23. Perez-Perez, M.E.; Mauries, A.; Maes, A.; Tourasse, N.J.; Hamon, M.; Lemaire, S.D.; Marchand, C.H. The Deep Thioredoxome in *Chlamydomonas reinhardtii*: New Insights into Redox Regulation. *Mol. Plant* **2017**, *10*, 1107–1125. [CrossRef] [PubMed]

24. Ingles-Prieto, A.; Ibarra-Molero, B.; Delgado-Delgado, A.; Perez-Jimenez, R.; Fernandez, J.M.; Gaucher, E.A.; Sanchez-Ruiz, J.M.; Gavira, J.A. Conservation of protein structure over four billion years. *Structure* **2013**, *21*, 1690–1697. [CrossRef] [PubMed]

25. Mouaheb, N.; Thomas, D.; Verdoucq, L.; Monfort, P.; Meyer, Y. In vivo functional discrimination between plant thioredoxins by heterologous expression in the yeast *Saccharomyces cerevisiae*. *Proc. Natl. Acad. Sci. USA* **1998**, *95*, 3312–3317. [CrossRef] [PubMed]

26. Issakidis-Bourguet, E.; Mouaheb, N.; Meyer, Y.; Miginiac-Maslow, M. Heterologous complementation of yeast reveals a new putative function for chloroplast m-type thioredoxin. *Plant J.* **2001**, *25*, 127–135. [CrossRef] [PubMed]

27. Morisse, S.; Michelet, L.; Bedhomme, M.; Marchand, C.H.; Calvaresi, M.; Trost, P.; Fermani, S.; Zaffagnini, M.; Lemaire, S.D. Thioredoxin-dependent redox regulation of chloroplastic phosphoglycerate kinase from *Chlamydomonas reinhardtii*. *J. Biol. Chem.* **2014**, *289*, 30012–30024. [CrossRef] [PubMed]

28. Huppe, H.C.; de Lamotte-Guery, F.; Jacquot, J.P.; Buchanan, B.B. The ferredoxin-thioredoxin system of a green alga, *Chlamydomonas reinhardtii*: Identification and characterization of thioredoxins and ferredoxin-thioredoxin reductase components. *Planta* **1990**, *180*, 341–351. [PubMed]

29. Collin, V.; Issakidis-Bourguet, E.; Marchand, C.; Hirasawa, M.; Lancelin, J.M.; Knaff, D.B.; Miginiac-Maslow, M. The Arabidopsis plastidial thioredoxins: New functions and new insights into specificity. *J. Biol. Chem.* **2003**, *278*, 23747–23752. [CrossRef] [PubMed]

30. Collin, V.; Lamkemeyer, P.; Miginiac-Maslow, M.; Hirasawa, M.; Knaff, D.B.; Dietz, K.J.; Issakidis-Bourguet, E. Characterization of plastidial thioredoxins from Arabidopsis belonging to the new y-type. *Plant Physiol.* **2004**, *136*, 4088–4095. [CrossRef] [PubMed]

31. Schurmann, P.; Buchanan, B.B. Role of ferredoxin in the activation of sedoheptulose diphosphatase in isolated chloroplasts. *Biochim. Biophys. Acta* **1975**, *376*, 189–192. [CrossRef]

32. Gutle, D.D.; Roret, T.; Muller, S.J.; Couturier, J.; Lemaire, S.D.; Hecker, A.; Dhalleine, T.; Buchanan, B.B.; Reski, R.; Einsle, O.; et al. Chloroplast FBPase and SBPase are thioredoxin-linked enzymes with similar architecture but different evolutionary histories. *Proc. Natl. Acad. Sci. USA* **2016**, *113*, 6779–6784. [CrossRef] [PubMed]

33. Marri, L.; Zaffagnini, M.; Collin, V.; Issakidis-Bourguet, E.; Lemaire, S.D.; Pupillo, P.; Sparla, F.; Miginiac-Maslow, M.; Trost, P. Prompt and easy activation by specific thioredoxins of calvin cycle enzymes of Arabidopsis thaliana associated in the GAPDH/CP12/PRK supramolecular complex. *Mol. Plant* **2009**, *2*, 259–269. [CrossRef] [PubMed]

34. Trost, P.; Fermani, S.; Marri, L.; Zaffagnini, M.; Falini, G.; Scagliarini, S.; Pupillo, P.; Sparla, F. Thioredoxin-dependent regulation of photosynthetic glyceraldehyde-3-phosphate dehydrogenase: Autonomous vs. CP12-dependent mechanisms. *Photosynthesis Res.* **2006**, *89*, 263–275. [CrossRef] [PubMed]

35. Fermani, S.; Sparla, F.; Marri, L.; Thumiger, A.; Pupillo, P.; Falini, G.; Trost, P. Structure of photosynthetic glyceraldehyde-3-phosphate dehydrogenase (isoform A4) from *Arabidopsis thaliana* in complex with NAD. *Acta Crystallogr. Sect. F Struct. Biol. Cryst. Commun.* **2010**, *66 Pt 6*, 621–626. [CrossRef]

36. Wolosiuk, R.A.; Buchanan, B.B. Regulation of chloroplast phosphoribulokinase by the ferredoxin/thioredoxin system. *Arch. Biochem. Biophys.* **1978**, *189*, 97–101. [CrossRef]

37. Setterdahl, A.T.; Chivers, P.T.; Hirasawa, M.; Lemaire, S.D.; Keryer, E.; Miginiac-Maslow, M.; Kim, S.K.; Mason, J.; Jacquot, J.P.; Longbine, C.C.; et al. Effect of pH on the oxidation-reduction properties of thioredoxins. *Biochemistry* **2003**, *42*, 14877–14884. [CrossRef] [PubMed]

38. Hirasawa, M.; Schurmann, P.; Jacquot, J.-P.; Manieri, W.; Jacquot, P.; Keryer, E.; Hartman, F.C.; Knaff, D.B. Oxidation-reduction properties of chloroplast thioredoxins, ferredoxin:thioredoxin reductase, and thioredoxin f-regulated enzymes. *Biochemistry* **1999**, *38*, 5200–5205. [CrossRef] [PubMed]

39. Nikkanen, L.; Toivola, J.; Diaz, M.G.; Rintamaki, E. Chloroplast thioredoxin systems: Prospects for improving photosynthesis. *Philos. Trans. R. Soc. Lond. Ser. B Biol. Sci.* **2017**, *372*, 20160474. [CrossRef] [PubMed]

40. Emanuelsson, O.; Nielsen, H.; von Heijne, G. ChloroP, a neural network-based method for predicting chloroplast transit peptides and their cleavage sites. *Protein Sci.* **1999**, *8*, 978–984. [CrossRef] [PubMed]

41. Pasquini, M.; Fermani, S.; Tedesco, D.; Sciabolini, C.; Crozet, P.; Naldi, M.; Henri, J.; Vothknecht, U.; Bertucci, C.; Lemaire, S.D.; et al. Structural basis for the magnesium-dependent activation of transketolase from *Chlamydomonas reinhardtii*. *Biochim. Biophys. Acta* **2017**, *1861*, 2132–2145. [CrossRef] [PubMed]

42. Pace, C.N.; Vajdos, F.; Fee, L.; Grimsley, G.; Gray, T. How to measure and predict the molar absorption coefficient of a protein. *Protein Sci.* **1995**, *4*, 2411–2423. [CrossRef] [PubMed]

43. Zaffagnini, M.; Michelet, L.; Massot, V.; Trost, P.; Lemaire, S.D. Biochemical characterization of glutaredoxins from *Chlamydomonas reinhardtii* reveals the unique properties of a chloroplastic CGFS-type glutaredoxin. *J. Biol. Chem.* **2008**, *283*, 8868–8876. [CrossRef] [PubMed]

44. McWilliam, H.; Li, W.; Uludag, M.; Squizzato, S.; Park, Y.M.; Buso, N.; Cowley, A.P.; Lopez, R. Analysis Tool Web Services from the EMBL-EBI. *Nucleic Acids Res.* **2013**, *41*, W597–600. [CrossRef] [PubMed]

45. Sievers, F.; Wilm, A.; Dineen, D.; Gibson, T.J.; Karplus, K.; Li, W.; Lopez, R.; McWilliam, H.; Remmert, M.; Soding, J.; et al. Fast, scalable generation of high-quality protein multiple sequence alignments using Clustal Omega. *Mol. Syst. Biol.* **2011**, *7*, 539. [CrossRef] [PubMed]

46. Zwart, P.H.; Afonine, P.V.; Grosse-Kunstleve, R.W.; Hung, L.W.; Ioerger, T.R.; McCoy, A.J.; McKee, E.; Moriarty, N.W.; Read, R.J.; Sacchettini, J.C.; et al. Automated structure solution with the PHENIX suite. *Methods Mol. Biol.* **2008**, *426*, 419–435. [PubMed]

47. Kelley, L.A.; Mezulis, S.; Yates, C.M.; Wass, M.N.; Sternberg, M.J. The Phyre2 web portal for protein modeling, prediction and analysis. *Nat. Protoc.* **2015**, *10*, 845–858. [CrossRef] [PubMed]

48. Afonine, P.V.; Grosse-Kunstleve, R.W.; Echols, N.; Headd, J.J.; Moriarty, N.W.; Mustyakimov, M.; Terwilliger, T.C.; Urzhumtsev, A.; Zwart, P.H.; Adams, P.D. Towards automated crystallographic structure refinement with phenix.refine. *Acta Crystallogr. Sect. D Biol. Crystallogr.* **2012**, *68 Pt 4*, 352–367. [CrossRef]

49. Emsley, P.; Lohkamp, B.; Scott, W.G.; Cowtan, K. Features and development of Coot. *Acta Crystallogr. Sect. D Biol. Crystallogr.* **2010**, *66 Pt 4*, 86–501.

50. Menchise, V.; Corbier, C.; Didierjean, C.; Saviano, M.; Benedetti, E.; Jacquot, J.P.; Aubry, A. Crystal structure of the wild-type and D30A mutant thioredoxin h of *Chlamydomonas reinhardtii* and implications for the catalytic mechanism. *Biochem. J.* **2001**, *359 Pt 1*, 65–75. [CrossRef]

51. Kinoshita, K.; Nakamura, H. eF-site and PDBjViewer: Database and viewer for protein functional sites. *Bioinformatics* **2004**, *20*, 1329–1330. [CrossRef] [PubMed]

52. Micsonai, A.; Wien, F.; Kernya, L.; Lee, Y.H.; Goto, Y.; Refregiers, M.; Kardos, J. Accurate secondary structure prediction and fold recognition for circular dichroism spectroscopy. *Proc. Natl. Acad. Sci. USA* **2015**, *112*, E3095–E3103. [CrossRef] [PubMed]

53. Micsonai, A.; Wien, F.; Bulyaki, E.; Kun, J.; Moussong, E.; Lee, Y.H.; Goto, Y.; Refregiers, M.; Kardos, J. BeStSel: A web server for accurate protein secondary structure prediction and fold recognition from the circular dichroism spectra. *Nucl. Acids Res.* **2018**, *46*, W315–W322. [CrossRef] [PubMed]

54. Menchise, V.; Corbier, C.; Didierjean, C.; Jacquot, J.P.; Benedetti, E.; Saviano, M.; Aubry, A. Crystal structure of the W35A mutant thioredoxin h from *Chlamydomonas reinhardtii*: The substitution of the conserved active site Trp leads to modifications in the environment of the two catalytic cysteines. *Biopolymers* **2000**, *56*, 1–7. [CrossRef]

55. De Lamotte-Guery, F.; Pruvost, C.; Minard, P.; Delsuc, M.A.; Miginiac-Maslow, M.; Schmitter, J.M.; Stein, M.; Decottignies, P. Structural and functional roles of a conserved proline residue in the alpha2 helix of *Escherichia coli* thioredoxin. *Protein Eng.* **1997**, *10*, 1425–1432. [CrossRef] [PubMed]

56. Iqbal, A.; Gomes-Neto, F.; Myiamoto, C.A.; Valente, A.P.; Almeida, F.C. Dissection of the water cavity of yeast thioredoxin 1: The effect of a hydrophobic residue in the cavity. *Biochemistry* **2015**, *54*, 2429–2442. [CrossRef] [PubMed]

57. Ladbury, J.E.; Wynn, R.; Hellinga, H.W.; Sturtevant, J.M. Stability of oxidized Escherichia coli thioredoxin and its dependence on protonation of the aspartic acid residue in the 26 position. *Biochemistry* **1993**, *32*, 7526–7530. [CrossRef] [PubMed]

58. Bjornberg, O.; Maeda, K.; Svensson, B.; Hagglund, P. Dissecting molecular interactions involved in recognition of target disulfides by the barley thioredoxin system. *Biochemistry* **2012**, *51*, 9930–9939. [CrossRef] [PubMed]

59. Marchand, C.H.; Fermani, S.; Rossi, J.; Gurrieri, L.; Tedesco, D.; Henri, J.; Sparla, F.; Trost, P.; Lemaire, S.D.; Zaffagnini, M. Structural and biochemical insights into the reactivity of thioredoxin h1 from Chlamydomonas reinhardtii. *Antioxidants* **2018**, *7*, under review.

60. Capitani, G.; Markovic-Housley, Z.; DelVal, G.; Morris, M.; Jansonius, J.N.; Schurmann, P. Crystal structures of two functionally different thioredoxins in spinach chloroplasts. *J. Mol. Biol.* **2000**, *302*, 135–154. [CrossRef] [PubMed]

61. Lancelin, J.M.; Guilhaudis, L.; Krimm, I.; Blackledge, M.J.; Marion, D.; Jacquot, J.P. NMR structures of thioredoxin m from the green alga *Chlamydomonas reinhardtii*. *Proteins* **2000**, *41*, 334–349. [CrossRef]

62. Michelet, L.; Zaffagnini, M.; Marchand, C.; Collin, V.; Decottignies, P.; Tsan, P.; Lancelin, J.M.; Trost, P.; Miginiac-Maslow, M.; Noctor, G.; et al. Glutathionylation of chloroplast thioredoxin f is a redox signaling mechanism in plants. *Proc. Natl. Acad. Sci. USA* **2005**, *102*, 16478–16483. [CrossRef] [PubMed]

63. Reutimann, H.; Straub, B.; Luisi, P.L.; Holmgren, A. A conformational study of thioredoxin and its tryptic fragments. *J. Biol. Chem.* **1981**, *256*, 6796–6803. [PubMed]

64. Collet, J.F.; D'Souza, J.C.; Jakob, U.; Bardwell, J.C. Thioredoxin 2, an oxidative stress-induced protein, contains a high affinity zinc binding site. *J. Biol. Chem.* **2003**, *278*, 45325–45332. [CrossRef] [PubMed]

65. Ado, K.; Taniguchi, Y. Pressure effects on the structure and function of human thioredoxin. *Biochim. Biophys. Acta* **2007**, *1774*, 813–821. [CrossRef] [PubMed]

66. Kabsch, W.; Sander, C. Dictionary of protein secondary structure: Pattern recognition of hydrogen-bonded and geometrical features. *Biopolymers* **1983**, *22*, 2577–2637. [CrossRef] [PubMed]

67. Sikic, K.; Tomic, S.; Carugo, O. Systematic comparison of crystal and NMR protein structures deposited in the protein data bank. *Open Biochem. J.* **2010**, *4*, 83–95. [CrossRef] [PubMed]

68. Dawson, N.L.; Lewis, T.E.; Das, S.; Lees, J.G.; Lee, D.; Ashford, P.; Orengo, C.A.; Sillitoe, I. CATH: An expanded resource to predict protein function through structure and sequence. *Nucl. Acids Res.* **2017**, *45*, D289–D295. [CrossRef] [PubMed]

69. Matthews, B.W.; Liu, L. A review about nothing: Are apolar cavities in proteins really empty? *Protein Sci.* **2009**, *18*, 494–502. [CrossRef] [PubMed]

70. Chiadmi, M.; Navaza, A.; Miginiac-Maslow, M.; Jacquot, J.P.; Cherfils, J. Redox signalling in the chloroplast: Structure of oxidized pea fructose-1,6-bisphosphate phosphatase. *EMBO J.* **1999**, *18*, 6809–6815. [CrossRef] [PubMed]

71. Garzon, J.I.; Lopez-Blanco, J.R.; Pons, C.; Kovacs, J.; Abagyan, R.; Fernandez-Recio, J.; Chacon, P. FRODOCK: A new approach for fast rotational protein-protein docking. *Bioinformatics* **2009**, *25*, 2544–2551. [CrossRef] [PubMed]

72. Maeda, K.; Hagglund, P.; Finnie, C.; Svensson, B.; Henriksen, A. Structural basis for target protein recognition by the protein disulfide reductase thioredoxin. *Structure* **2006**, *14*, 1701–1710. [CrossRef] [PubMed]

73. Richardson, J.M.; 3rd Lemaire, S.D.; Jacquot, J.P.; Makhatadze, G.I. Difference in the mechanisms of the cold and heat induced unfolding of thioredoxin h from *Chlamydomonas reinhardtii*: Spectroscopic and calorimetric studies. *Biochemistry* **2000**, *39*, 11154–11162. [CrossRef] [PubMed]

74. Lemaire, S.D.; Richardson, J.M.; Goyer, A.; Keryer, E.; Lancelin, J.M.; Makhatadze, G.I.; Jacquot, J.P. Primary structure determinants of the pH- and temperature-dependent aggregation of thioredoxin. *Biochim. Biophys. Acta* **2000**, *1476*, 311–323. [CrossRef]

75. Palde, P.B.; Carroll, K.S. A universal entropy-driven mechanism for thioredoxin-target recognition. *Proc. Natl. Acad. Sci. USA* **2015**, *112*, 7960–7965. [CrossRef] [PubMed]

76. Balmer, Y.; Koller, A.; Val, G.D.; Schurmann, P.; Buchanan, B.B. Proteomics uncovers proteins interacting electrostatically with thioredoxin in chloroplasts. *Photosynthesis Res.* **2004**, *79*, 275–280. [CrossRef] [PubMed]

77. Crozet, P.; Navarro, F.J.; Willmund, F.; Mehrshahi, P.; Bakowski, K.; Lauersen, K.J.; Perez-Perez, M.E.; Auroy, P.; Gorchs Rovira, A.; Sauret-Gueto, S.; et al. Birth of a Photosynthetic Chassis: A MoClo Toolkit Enabling Synthetic Biology in the Microalga *Chlamydomonas reinhardtii*. *ACS Synth. Biol.* **2018**, *7*, 2074–2086. [CrossRef] [PubMed]

78. Greiner, A.; Kelterborn, S.; Evers, H.; Kreimer, G.; Sizova, I.; Hegemann, P. Targeting of Photoreceptor Genes in *Chlamydomonas reinhardtii* via Zinc-Finger Nucleases and CRISPR/Cas9. *Plant Cell* **2017**, *29*, 2498–2518. [CrossRef] [PubMed]

antioxidants

MDPI

Article

Characterization of TrxC, an Atypical Thioredoxin Exclusively Present in Cyanobacteria

Luis López-Maury * [iD]**, Luis G. Heredia-Martínez and Francisco J. Florencio**[iD]

Instituto de Bioquímica Vegetal y Fotosíntesis, Universidad de Sevilla-CSIC, Av/Américo Vespucio 49,
E-41092 Sevilla, Spain; heredia@ibvf.csic.es (L.G.H.-M.); floren@us.es (F.J.F.)
* Correspondence: llopez1@us.es; Tel.: +34-954489500 (ext. 909102)

Received: 11 October 2018; Accepted: 11 November 2018; Published: 13 November 2018

Abstract: Cyanobacteria form a diverse group of oxygenic photosynthetic prokaryotes considered to be the antecessor of plant chloroplast. They contain four different thioredoxins isoforms, three of them corresponding to *m*, *x* and *y* type present in plant chloroplast, while the fourth one (named TrxC) is exclusively found in cyanobacteria. TrxC has a modified active site (WCGLC) instead of the canonical (WCGPC) present in most thioredoxins. We have purified it and assayed its activity but surprisingly TrxC lacked all the classical activities, such as insulin precipitation or activation of the fructose-1,6-bisphosphatase. Mutants lacking *trxC* or over-expressing it were generated in the model cyanobacterium *Synechocystis* sp. PCC 6803 and their phenotypes have been analyzed. The Δ*trxC* mutant grew at similar rates to WT in all conditions tested although it showed an increased carotenoid content especially under low carbon conditions. Overexpression strains showed reduced growth under the same conditions and accumulated lower amounts of carotenoids. They also showed lower oxygen evolution rates at high light but higher Fv'/Fm' and Non-photochemical-quenching (NPQ) in dark adapted cells, suggesting a more oxidized plastoquinone pool. All these data suggest that TrxC might have a role in regulating photosynthetic adaptation to low carbon and/or high light conditions.

Keywords: cyanobacteria; thioredoxin; photosynthesis

1. Introduction

Thioredoxin are small (~12 kDa) evolutionary conserved proteins with redox activity that present a conserved three-dimensional structure denominated "Trx fold", composed of five β strands surrounded by four α helices [1]. They serve as redox carriers and catalyse reduction of other proteins activating/deactivating them. Thioredoxins contain a conserved disulphide active site in the form WCGPC that undergoes oxidation-reduction cycles. Most organisms contain at least one thioredoxin gene although their numbers are expanded in photosynthetic organisms [2,3]. *Arabidopsis* contains at least twenty thioredoxin isoforms, ten are present in *Chlamydomonas reinhardtii* in contrast to humans or *Escherichia coli* that only contain two. *Arabidopsis* thioredoxins are classified in seven groups (Trx *h*, Trx *o*, Trx *f*, Trx *z*, Trx *m*, Trx *x* and Trx *y*) and all of them are also present in *Chlamydomonas*. Of these, Trx *m*, Trx *x* and Trx *y* have a cyanobacterial origin, and together with Trx *f* and Trx *z* are located to the chloroplast [4,5]. In cyanobacteria, four thioredoxins classes have been found along the genome of more than 300 species, three of them corresponding to *m*, *x* and *y* types present in plant chloroplast, while the fourth one is exclusively found in cyanobacteria (TrxC) [6].

Thioredoxins are reduced by thioredoxin reductases of which at least two different families exist: NADPH-dependent thioredoxin reductases (NTR) and ferredoxin-dependent thioredoxin reductases (FTR). NTR are present from bacteria to humans while FTR is an Fe-S protein that uses ferredoxin and is present in photosynthetic organisms [2]. Thioredoxin reduction system is also diverse in cyanobacteria, most of them containing the FTR system (except *Gloeobacter* and *Prochlorococcus* groups) [2,6]. Some cyanobacteria also contain NTRC, a protein that contains a NTR module fused to a thioredoxin module in a single polipeptide, which seems to function as a reducing system for the 2-cys peroxiredoxin in *Anabaena* sp. PCC 7120 (hereafter *Anabaena*) [7–9]. They can also contain two other NTR related proteins that were initially annotated as NADPH dependent enzymes. However, these two proteins have recently been shown not to use or bind NADPH, as they lack the aminoacid signature characteristic of the NADPH binding domain, and have been renamed DTR (for Deeply rooted bacterial Thioredoxin Reductase) and DDOR (Diflavin-linked Disulfide OxidoReductase) [10,11]. Dithionite reduced DTR is able to reduce thioredoxins although its physiological reductant is still unknown [10]. In contrast, DDOR shows a new structure containing two to Flavin Adenine Dinucleotide (FAD) per monomer, does not reduce thioredoxins and probably functions as an oxidase [11]. These proteins are scattered distributed in cyanobacteria but strains lacking FTR usually contain a gene coding for DTR. Of these two proteins, only DDOR is present in *Synechocystis* sp. PCC 6803 (hereafter *Synechocystis*) although some cyanobacteria (such as *Gloeobacter violaceus* sp. PCC 7421) contains both [2,10,11].

Putative thioredoxins targets have been studied in cyanobacteria by several proteomics approaches [12–17]. A high degree of overlapping targets for the different isoforms have been found, highlighting that these in vitro proteomics approaches identify reactive cysteine that could be subjected to redox regulation in vivo and that overlapping and redundant roles are expected for the different thioredoxins in vivo. The role of the different thioredoxins has been also analyzed by generating mutants in their corresponding genes. Trx *m* (*trxA*) has been shown to be essential in unicellular cyanobacteria and therefore no specific phenotypes have been associated to it [18,19]. More recently, mutants in *trxM* genes have been described in *Anabaena* which possess two genes coding for Trx *m*: *trxM1* and *trxM2*. Both genes are dispensable although *trxM1* mutant showed a pleiotropic phenotype and was unable to grow in diazotrophic conditions, while *trxM2* lack any appreciable phenotype [20,21]. In plants *trxM* are involved in several functions related to fluctuating light conditions, cyclic electron flow or meristem maintenance [4,5,22]. In *Synechocystis* *trxB⁻* mutant strains (lacking *x* type) showed sensitivity towards HL and to the presence of DTT in the culture media [23], while *trxQ⁻* mutant (lacking *y* type) showed sensitivity to oxidative stress induced by methyl viologen [23]. All three proteins were able to interact and reduce the different peroxiredoxins present in *Synechocystis* although with different efficiencies, with TrxQ being the most efficient in vitro [24]. The fourth group, TrxC, is only found in cyanobacteria. No activity or function has been ascribed to this class of Trx in *Synechocystis* as mutants lacking it did not show any phenotype [25], although *Anabaena trxC⁻* mutant showed more oxidative stress than WT in nitrate grown cultures [20]. Furthermore, *Anabaena* TrxC seems to be inactive in reducing OpcA or G6PDH [21]. Here we describe the characterization of *Synechocystis'* TrxC protein and of mutants either lacking *trxC* or overexpressing it.

2. Materials and Methods

2.1. Strains and Culture Conditions

Synechocystis cells were grown photoautotrophically on BG11 or BG11C [26] at 30 °C under continuous illumination (50 to 180 μMol photons m^{-2} s^{-1}) and bubbled with a stream of 1% (v/v) CO_2 in air as indicated. BG11 media was buffered with 10 mM TES-NaOH pH 7.5. For plate cultures, medium was supplemented with 1% (wt/vol) agar. All media contained the standard copper concentration (0.3 μM) except in Figure 2 in which BG11-Cu was used. Kanamycin and nourseothricin, were added to a final concentration of 50 μg mL^{-1}. *Synechocystis* strains and their relevant genotypes are described in Table 1. *E. coli* DH5α or BL21 cells were grown in Luria broth medium and supplemented with 100 μg mL^{-1} ampicillin, 50 μg mL^{-1} kanamycin and 50 μg mL^{-1} nourseothricin when required.

Table 1. *Synechocystis* strains used in this work.

Strain	Relevant Genotype	Mutated ORFS	Source
WT	*Synechocystis* sp PCC 6803	-	Lab collection
STXC2	*ΔtrxC::C.K1*	*sll1057*	[25]
WTOE	*glnN::PpetE:histrxC:Nat*	*slr0288*	This study
STXCOE	*ΔtrxC::C.K1 glnN::PpetE:histrxC:Nat*	*sll1057, slr0288*	This study

2.2. Western Blotting

Crude extracts were prepared using glass beads and vigorous vortexing using a minibead-beater. Cells (corresponding to 20 OD_{750nm}) were resuspended in 300 μL of buffer A (50 mM Tris-HCl pH 8.0, 50 mM NaCl) and subjected to 2 cycles of 1 min vortexing separated by 5 min of cooling on ice. Cell extracts were recovered from the beads by piercing a hole in bottom of the tube and samples were clarified by two sequential centrifugations: 5′ at $3000 \times g$ to eliminate cells debris and twice 15 min at $18,000 \times g$ to collect membranes. Protein concentration in cell-free extracts by the method of Lowry, using Bovine Serum Albumin as a standard and the specified amounts of proteins were separated on sodium dodecyl sulfate polyacrylamide gel electrophoresis (SDS-PAGE). Gels were transferred to nitrocellulose membranes (Bio-Rad, Düsseldorf, Germany catalog #162-0115), blocked in phosphate-buffered saline (PBS) containing 0.1% Tween 20 and 5% of skimmed milk and incubated with antibodies against TrxA (1:3000), TrxB (1:3000), TrxQ (1:2000), TrxC (1:1000), DDOR (1:5000), GrxA (1:3000), GrxC (1:3000) and 2cysprx (sll0755; 1:5000). The ECL Prime immunoblotting system (GE Healthcare, Little Chalfont) was used to detect the different antigens with goat anti-rabbit conjugated to horseradish peroxidase (Sigma St Louis, USA catalog #A6154) diluted to 1:25,000.

2.3. Cloning and Purification of TrxC

TrxC gene was cloned from *Synechocystis'* DNA after PCR amplification with oligonucleotides Syn_TrxC_BamHI_NdeI and SynTrxC_R_SalI and cloned into NdeI-SalI digested pET28 or *BamHI-SalI* digested pGEX6P to generate pET_trxC or pGST_trxC respectively. To generate site directed TrxCL32P mutant a 366 pb fragment was amplified by two-step PCR using oligonucleotides TrxC_NdeI-trxC_L32P_Rv and trxC_L32P_Fw-TrxC_NotI, that introduced the desired mutation, and cloned into pGEMT generating pG_TrxCL32P. The plasmid was cut with *NdeI-NotI* and a 347 pb fragment was ligated to NdeI-Not digested pET28 to generate pET_TrxCL32P. Sequence of all oligonucleotides are provided in Table S1.

TrxC fusion proteins were expressed in *E. coli* BL21. 200 mL of culture was grown in Luria broth medium to an optical density at 600 nm of 0.6, cooled to 4 °C, induced with 0.5 mM isopropyl-β-D-thiogalactopyranoside for 2.5 h and harvested by centrifugation. For his tagged TrxC purification, cells were resuspended in 50 mM Tris-HCl pH 8.0, 500 mM NaCl, 1 mM PMSF and disrupted (20 kHz, 75 W) on ice for 3 min (in 30-s periods) in a Branson sonicator. Lysates were clarified by centrifugation $20,000 \times g$ for 30 min. Supernatants were supplemented with imidazole to a final concentration of 25 mM and loaded onto a 2 mL Ni-NTA agarose (IBA, Goettingen, Germany catalog #2-3201-025) column for affinity chromatography purification. Column was washed with 50 mM Tris-HCl pH 8.0, 500 mM NaCl, 25 mM imidazole until no protein was detected and bound recombinant proteins were eluted with 250 mM of imidazole in 50 mM Tris-HCl pH 8.0, 500 mM NaCl. For GST and GST-TrxC purification cell pellets were resuspended in 5 mL of PBS buffer (150 mM NaCl, 16 mM Na_2HPO_4, 4 mM NaH_2PO_4, 1 mM phenylmethylsulfonyl fluoride, 7 mM β-mercaptoethanol) supplemented with 0.1% Triton X-100. Cells extract were prepared as above, mixed with 1 mL of glutathione agarose beads (GE Healthcare, Little Chalfont catalog #17-0756-01) and incubated for 2 h at 4 °C with gentle agitation. Then beads were transferred to a column and washed extensively with PBS buffer until no more protein was eluted from the column. GST and GST-TrxC were eluted in 50 mM Tris HCl pH 8.0, 10 mM GSH. TrxC was excised by digestion with PreScission Protease (GE Healthcare, Little Chalfont catalog #27084301) following manufacturer's instructions.

2.4. Mutant Construction

To generate site STXC2 mutant, a 1170 pb fragment was amplified using oligonucleotides trxCHIII and trxCXhoI and cloned into pBS digested with the same enzymes generating pSTXC1. A CK1 cassette was cloned into pSTXC1 digested with *Hinc*II (deleting 345 pb and expanding the whole ORF) generating pSTXC2. This plasmid was used to transform *Synechocystis* generating STXC2 strain. For the overexpression strains a *Xba*I-*Not*I fragment from pET_trxC was cloned into pN:PpetE:Nat (a plasmid containing *glnN* gene in which the *petE* regulatory region, a multiple cloning site and a Nourseothricin resistance cassette has been cloned [27]) digested in the same way generating pNPpetEtrxC. This plasmid was used to transform both WT and STXC2 generating WTOE and STXCOE strains (Table 1).

2.5. FBPase Activation Assay

Recombinant pea FBPase was purified by Ni nickel-affinity column as previously described [28] and the eluted protein was desalted with a PD10 column in 50 mM Tris HCl pH 8.0, 150mM NaCl. 2 μg of recombinant pea FBPase was incubated with DTT (10 mM or 0.1 mM) and 3–30 μg of thioredoxins for 10 min in 175 μL containing 30 mM Tris·HCl pH 8.0, 7 mM $MgCl_2$ at 30 °C. Then, 25 μL of fructose-1,6-bisphosphate 50 mM was added and the reaction was incubated for 30 min at 30 °C. Also, 1 mL LPi mix (2.5% sulfuric acid, 7.5 mM ammonium heptamolybdate, 100 mM $FeSO_4$) was added to stop the reaction and the Pi released was measured at 660 nm [29]. A calibration curve with known concentrations of Na_2PO_4 was used to calculate Pi concentration.

2.6. Insulin Reduction Assay

Thioredoxin insulin reduction assay was carried out using 3 μg of recombinant HisTrxA or HisTrxC in 1 mL of 0.1 M potassium phosphate pH 7.0, 2 mM EDTA, 1 mM DTT and 1mg/ml insulin at 30 °C. The same buffer without DTT was used as reference. Thioredoxin activity was measured by the increase of turbidity at OD_{650nm} due to insulin precipitation [30]. Recombinant HisTrxA was expressed and purified as described in [13,14].

2.7. PAM

A pulse amplitude modulated fluorometer DUAL-PAM-100 (Walz, Effeltrich, Germany) was used to monitor chlorophyll, a fluorescence in intact cells adjusted to $OD_{750nm} = 1$. Measurements were performed in 1 cm × 1 cm cuvettes at 30 °C. Red (620 nm) actinic lights was used as background light and saturating pulses (10,000 μMol photons m^{-2} s^{-1}, 635 nm, 300 ms) were applied to transiently close all PSII centers. The maximal photochemical efficiency of PSII (Fv/Fm) was measured in the presence of 3-(3,4-dichlorophenyl)-1,1-dimethylurea (DCMU). Fm'$_{dark}$ is defined as the Fm of cells adapted to dark conditions for 10 min. Non-photochemical-quenching (NPQ) was calculated as Fm-Fm'$_{dark}$/Fm'$_{dark}$ as described in [31].

2.8. Oxygen Evolution

Oxygen evolution was measured in Clark-type oxygen electrode (Hansatech Chlorolab 2) using mid-logarithmic ($OD_{750nm} = 0.8–1$) cultures adjusted to $OD_{750nm} = 0.5$ in BG11 pH 7.5 media supplemented with 20 mM $NaHCO_3$ using white LED light.

3. Results

3.1. TrxC Is an Atypical Thioredoxin Exclusively Found in Cyanobacteria

Of the four thioredoxins found in cyanobacteria, only TrxC is not homologous to thioredoxins present in plant chloroplasts. Furthermore, analysis of its sequence shows an altered catalytic site WCGL/V/IC that is highly conserved in cyanobacteria (Figure 1A). Despite the N terminal part of the sequence being similar to other Trx, most of the protein shows a clear divergence from other thioredoxin in cyanobacteria and less conservation in the C-terminal sequence of the protein (Figure 1A). There is no other thioredoxin sequence with a similar active site in any other organism. Moreover, all cyanobacterial groups except *Prochlorococcus* and *Gloeobacteria* contain at least one species that presents a *trxC* gene (Table S2) existing in more than 200 strains. This distribution suggests that this protein may have an important role in cyanobacterial environmental adaptation or metabolism.

A

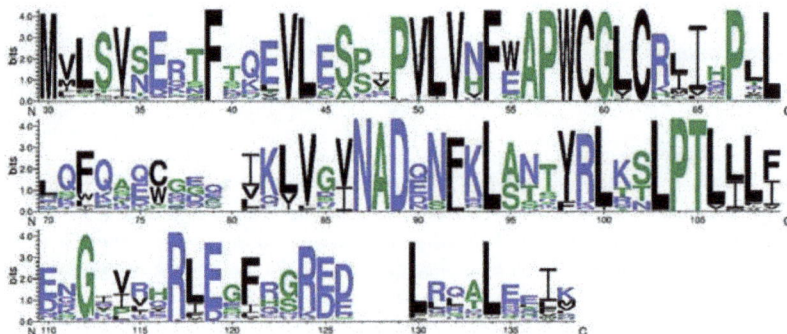

B

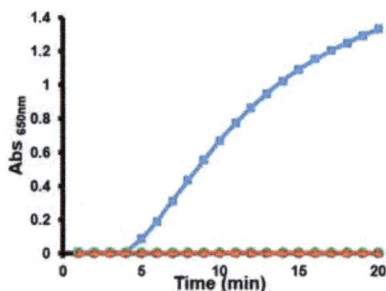

C

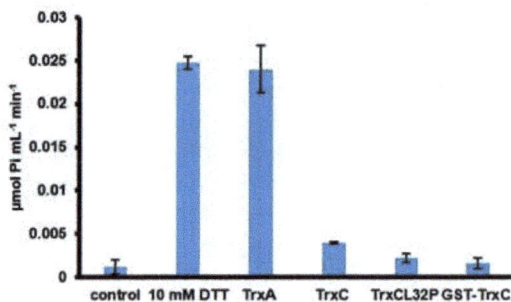

Figure 1. TrxC is an atypical thioredoxin. (**A**) Sequence logo of TrxC proteins form cyanobacteria. TrxC proteins were identified using blast at NCBI and manually curated to retain only those with a WCGL/V/IC sequence (269 sequences). This sequences were aligned using muscle and the alignment was submitted to weblogo3 to generate the consensus sequence shown. (**B**) Insulin reduction assay. 3 μM of recombinant TrxA (■), TrxC (●) and TrxCL32P (▲) were incubated with insulin in the presence of 1 mM of DTT. Insulin precipitation was measured as an increase in absorbance at 650 nm. Three independent purification were assayed for TrxC and two for TrxCL32P with identical results to the one shown. (**C**) FBPase activation assay. Oxidased pea FBPase was preincubated for 30 min with 100 μM DTT (control), 10 mM DTT or 100 μM DTT and 3 μM of TrxA, TrxC, TrxCL32P or 30 μM GST-TrxC. Data are the mean and standard error of 3 independent assays.

In order to characterize TrxC function we have produced it as recombinant His-tagged version in *E. coli*. The protein was expressed at high levels although it was difficult to purify as it was mostly in the insoluble fraction, probably forming inclusion bodies. Only low amounts were recovered in the soluble fraction and this protein was prone to precipitation when concentrated, but we purified enough protein to assay its activity. In contrast to TrxA (*m*-type), TrxC did not reduce insulin in the presence of DTT (Figure 1B). To further characterize the protein, we analyzed chloroplast pea FBPase activation by TrxC. In the same way as in the insulin reduction assay, TrxA was able to activate chloroplast FBPase, while TrxC did not activate it. As TrxC presents an unconventional active site WCGLC, this was changed to a canonical Trx active site (WGCPC) by site directed mutagenesis generating TrxCL32P. The protein was purified and used in both assays with identical results to the WT protein (Figure 1B,C). To further confirm these results we generate a recombinant version of the protein fussed to GST (GST-TrxC), which allowed us to purify higher amounts of soluble recombinant protein. Although TrxC can be excised from GST by protease digestion, the TrxC portion precipitated rapidly after digestion and therefore only GST-TrxC was used. GST and GST-TrxC were used to assay chloroplast FBPase activation at higher concentrations (30 µM) but neither of the proteins activated FBPase. These data suggest that TrxC is not active in the classical thioredoxin assays and that this is not related to its unconventional active site.

3.2. TrxC Mutant and Overexpression Strains Characterization

To investigate TrxC physiological function, a mutant lacking *trxC* (STXC2 mutant strain) and mutants over-expressing a His-tagged version of the protein (HisTrxC) were constructed. For overexpression, the His-tagged *trxC* gene was placed under control of the copper inducible *petE* promoter in a WT and STXC2 backgrounds, generating WTOE and STXCOE strains, respectively (Table 1). All mutants were verified by PCR analysis and shown to be completely segregated (Figure 2A,B). All strains presented similar growth rates to the WT in BG11C plates and liquid media under our standard growth conditions ([25]; data not shown). TrxC protein levels were analyzed in the mutants by western blot using TrxC antibodies in whole cell extracts. TrxC was detected in WT and WTOE strains but not in STXC2 and STXCOE, while a band corresponding to the recombinant HisTrxC was detected in both WTOE and STXCOE strains. Furthermore, HisTrxC expression was regulated by the presence of copper in these strains, the amount of HisTrxC increased after copper addition in both strains. Surprisingly HisTrxC expression levels were higher than endogenous TrxC levels in the WT strain even in the absence of copper in the media in both WTOE and STXCOE strains and was further elevated in the presence of copper (Figure 2C).

To study the TrxC impact on redox regulation in *Synechocystis*, we have analyzed the expression levels of other redox related proteins in WT and *trxC* mutant strains. All strains were grown to exponential phase in copper containing media and levels of different proteins were analyzed by western blot. Levels of the other thioredoxins (TrxA, TrxB and TrxQ), glutaredoxins (GrxA and GrxC) and 2-cys peroxiredoxin did not change in the mutants (Figure 3). In contrast, levels of DDOR (*slr0600*) increased in both the WTOE y STXCOE strains (Figure 3) in which TrxC levels were also increased (Figure 2C). These data showed that TrxC did not affect levels of other thioredoxins and glutaredoxins. The changes in DDOR protein levels suggest that these two proteins could be functionally related.

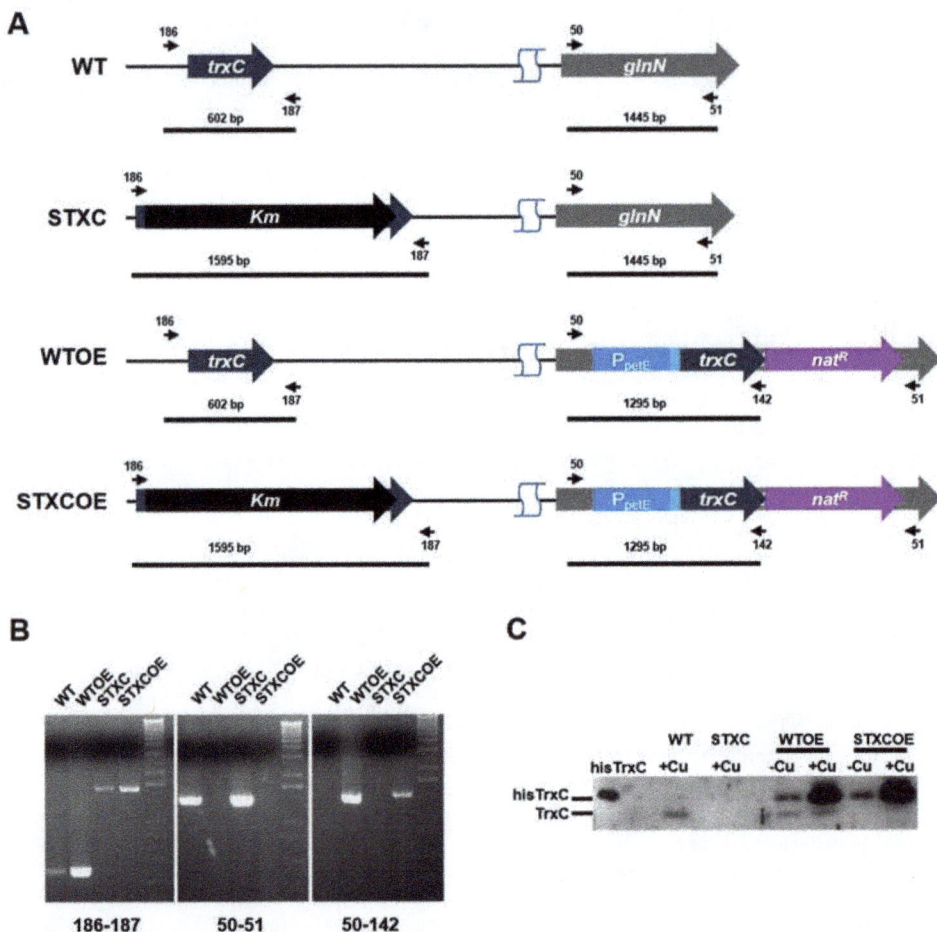

Figure 2. Construction of WT, WT_OE, STXC and STXCOE strains. (**A**) Schematic representation of the *trxC* and *glnN loci* in the WT and mutant strains. (**B**) PCR analysis of the mutant strains using the oligonucleotides indicated in (**A**). (**C**) Western-blot analysis of TrxC protein levels in the WT, WTOE, STXC and STXCOE strains. Cells were grown in BG11 supplemented with 1% CO_2 to mid-log growth phase and 1 OD_{750nm} was collected before and after 2 h of 0.3 µM Cu addition. The pellet was resuspended in 1× Laemmli buffer and boiled for 10 min, then 20 µL of the boiled cell suspension were loaded on and gel and analyzed by western blot.

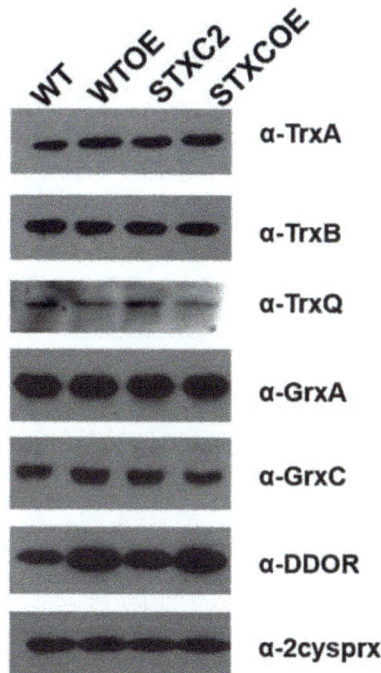

Figure 3. Redox related proteins in WT, WTOE, STXC2 and STXCOE strains. Western-blot analysis of TrxA, TrxB, TrxQ, GrxA, GrxC, DDOR and 2-cysprx in the WT, WTOE, STXC2 and STXCOE strains. Cells were grown in BG11 supplemented with 1% CO_2 to mid-log growth phase and collected. Cells were broken and 20 μg of total protein from soluble extracts were separated by SDS PAGE and analysed by western blot to detect the different proteins using specific antibodies. Experiments were repeated at least two (for TrxQ antibody) or three (all other antibodies) times with biological independent samples. SDS PAGE: sodium dodecyl sulfate polyacrylamide gel electrophoresis.

3.3. Growth of TrxC Mutant Strains under Low CO_2 Conditions

In order to investigate the physiological role of TrxC, *trxC* mutant strains were analyzed under several growth conditions. No difference between strains was detected in BG11C (containing NaHCO₃). We have only detected differences in growth and/or pigmentation under moderate light intensities and low carbon conditions (BG11 pH 7.5 bubble with air + 1% CO_2 or BG11 pH 7.5 bubbled with air). We have selected an intermediate light intensity (180 μMol m^{-2} s^{-1}) and air bubbled cultures (low carbon availability) as these were the most consistent growth conditions in which we were able to detect differences. At higher light intensities (500 μMol m^{-2} s^{-1}), even the WT strain showed impaired growth and in some cases died. All strains were previously adapted to BG11 pH 7.5 and high carbon (bubbled with air + 1% CO_2) in low light (50 μMol m^{-2} s^{-1}) until they reached late exponential phase (OD$_{750nm}$ = 1–2) and then were diluted to OD$_{750nm}$ = 0.2 and shifted 180 μMol m^{-2} s^{-1} in either high carbon (air + 1% CO_2) or low carbon conditions (air); this adaptation step was necessary for the strains to show consistent growth under this condition. In cultures bubbled with air + 1% CO_2 all strains grew at similar rates but differences in color were observed (Figure 4A). These are clearly visible in whole cell absorption spectra in which STXC2 showed a higher absorption at 485 nm, which corresponds to carotenoids absorption maxima, than the WT (Figure 4B). Moreover, both overexpression strains showed lower carotenoid contents (Figure 4B) making them to appear bluish when compared to WT or STXC2 strains. These phenotypes were exacerbated in air bubbled cultures in which carbon availability is further reduced. Under these conditions, all strains showed a reduced growth rate (Figure 4C), although growth of WTOE and STXCOE strains was reduced more than that of WT and STXC2 strains (Figure 4C). Despite the STXC2 strain growth rate being very similar to the WT, it contained even higher carotenoids which gave a yellow appearance to the cultures (Figure 4D,E). Overexpression strains showed similar carotenoids contents that were lower than WT levels (Figure 4D). Furthermore, the STXC2 strain showed higher chlorophyll contents than other strains in both 1% CO_2 (4.1 ± 0.3 µg chl $OD_{750nm}{}^{-1}$ for STXC2 vs 3.5 ± 0.5, 3.5 ± 0.2 and 3.4 ± 0.3 µg chl $OD_{750nm}{}^{-1}$ for WT, WTOE and STXCOE, respectively) and air bubbled conditions (3.83 ± 0.5 µg chl $OD_{750nm}{}^{-1}$ for STXC2 vs. 3.1 ± 0.3, 3.5 ± 0.3 and 3.5 ± 0.2 µg chl $OD_{750nm}{}^{-1}$ for WT, WTOE and STXCOE, respectively). All these results suggest that TrxC is involved in adaptation to low carbon conditions and that it could be mediated by regulating pigment content in *Synechocystis*.

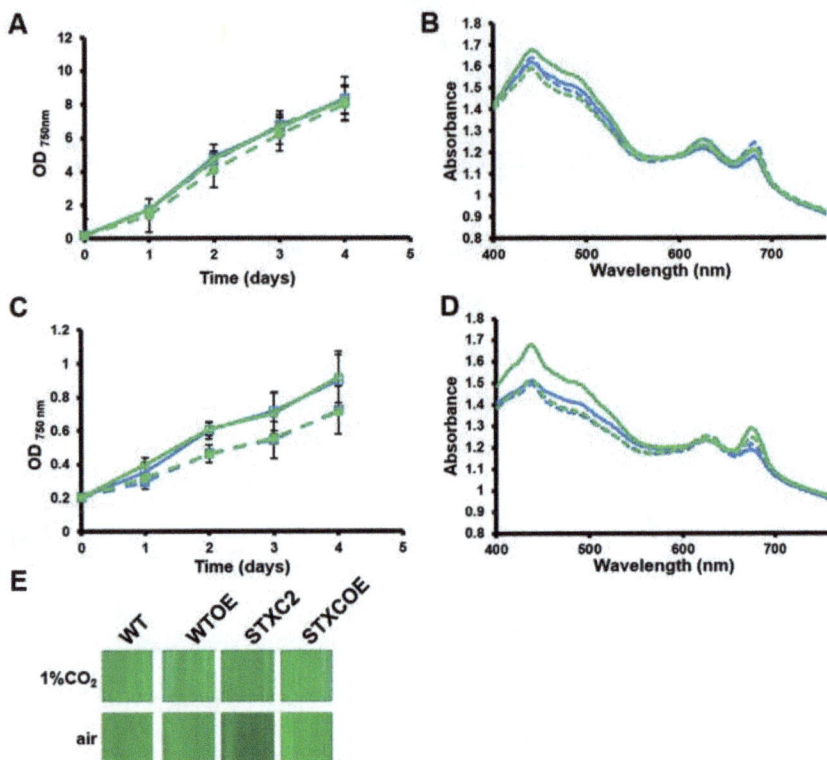

Figure 4. Overexpression of *trxC* slows growth in low carbon conditions. (**A**) Growth of *trxC* mutant strains in high carbon conditions. WT (□), STXC2 (○), WTOE (■) and STXCOE (●) were grown in BG11 pH 7.5 under low light until the exponential phase, diluted to 0.2 OD_{750nm} and shifted to 180 µMol m^{-2} s^{-1} light intensity bubbled with air + 1% CO_2. Growth was monitored by measuring OD_{750nm}. Data represented are the mean and standard error of 3–4 (depending on the time point) biological independent cultures. (**B**) Whole cell spectra of WT (blue solid line), STXC2 (green solid line), WTOE (blue dashed line) and STXCOE (green dashed line) grown as in (A). (**C**) Growth of *trxC* mutant strains in low carbon conditions. WT (□), STXC2 (○), WTOE (■) and STXCOE (●) were grown in BG11 pH 7.5 under low light until the exponential phase, diluted to 0.2 OD_{750nm} and shifted to 180 µMol m^{-2} s^{-1} light intensity and bubbled with air. Growth was monitored by measuring OD_{750nm}. Data represented are the mean and standard error of 3–4 (depending on the time point) biological independent cultures. (**D**) Whole cell spectra of WT (blue solid line), STXC2 (green solid line), WTOE (blue dashed line) and STXCOE (green dashed line) grown as in (C). (**E**) Photograph of WT, STXC2, WTOE and STXCOE cultures grown in BG11 pH 7.5 bubbled with air + 1% CO_2 or air.

3.4. Photosynthetic Characterization of TrxC Mutant Strains

To further characterize these strains, we determined different photosynthetic parameters using a Clark type oxygen electrode and DUAL-PAM 100 fluorimeter in air bubbled cultures. A light saturation curve was performed in exponentially growing cells of WT, WTOE, STXC2 and STXCOE in the oxygen electrode. WT and STXC2 showed similar light saturation curves saturating at around 500 μMol photons m^{-2} s^{-1} and reaching 20 μMol O$_2$ min^{-1} per OD$_{750nm}$ (Figure 5). Both overexpression strains also showed a similar behavior between them, saturating at the same light intensity of WT and STXC2 strains but reaching only 15 μMol O$_2$ min^{-1} per OD$_{750nm}$ (Figure 5). This lower photosynthetic capacity correlates with growth of these strains under this condition (Figure 4C). When photosynthesis was analyzed using a DUAL PAM100 fluorimeter, opposite results were obtained (Table 2). Fv' in the dark-adapted state was higher for the overexpression strains than in WT and STXC2, indicating a higher fraction of open PSII centers in these strains. In contrast, Fv (measured in the presence of DCMU) was much more similar in all strains. These suggest that the amount of open PSII reaction centers is higher under physiological conditions (Fv') but not when the photosynthetic electron flow is blocked (Fv). These data suggest that there are differences in the reduction state of plastoquinone pool. This is reinforced when NPQ (in dark adapted cells) is calculated as both overexpression strains showed lower NPQ.

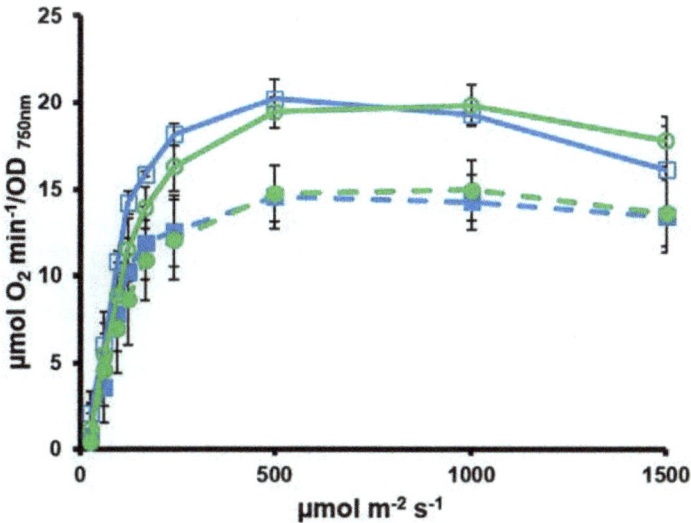

Figure 5. Overexpression of *trxC* causes lower photosynthetic efficiency. Oxygen evolution was measured in a Clark electrode at increasing light intensities in exponential growing cultures (OD$_{750nm}$ = 0.5–1) of WT (○), STXC2 (○), WTOE (■) and STXCOE (●) grown in BG11 pH 7.5 at 180 μMol m^{-2} s^{-1} light intensity and bubbled with air.

Table 2. Photosynthetic parameters calculated using a DUAL-PAM 100 fluorimeter from cultures grown as in Figure 4.

STRAIN	F_0	Fm DCMU	Fv	Fv/Fm	Fm' (Dark)	Fv' (Dark)	Fv'/Fm' (Dark)	NPQ (Dark)
WT	0.363 ± 0.028	0.563 ± 0.033	0.200	0.356	0.427 ± 0.031	0.064	0.150	0.319
WTOE	0.424 ± 0.049	0.617 ± 0.043	0.193	0.313	0.505 ± 0.044	0.080	0.160	0.223
STXC2	0.395 ± 0.066	0.597 ± 0.066	0.202	0.339	0.463 ± 0.076	0.069	0.148	0.288
STXC2OE	0.390 ± 0.028	0.601 ± 0.028	0.211	0.351	0.482 ± 0.027	0.091	0.190	0.248

4. Discussion

Cyanobacteria contain a complex redox proteome which include four thioredoxin classes, at least three different types of thioredoxin reductases, four glutaredoxin classes and the GSH/glutathione reductase system [2,6,32]. Of the four thioredoxins, only TrxC is exclusively present in these groups of organisms and the fact that is present in most of them (with the exception of *Gloeobacteria* and marine *SynPro* clade) suggests that it has an important role in their cell physiology. Interestingly, the two groups in which TrxC is not present also lack FTR, Trx *x* and Trx *y* sequences in their genomes, indicating a reduction of their redox regulatory network. Although TrxC clearly belongs to the thioredoxin family, it shows an altered active site (WCGLCR) that is otherwise invariable for different thioredoxins from cyanobacteria, plants, bacteria or human. This sequence is missing the conserved P in the active site but two other prolines important for Trx structure are conserved in all TrxC sequences. The first proline conserved is five residues from the active site (P67 in Figure 1A) and the second one is from the cis-proline loop which also contains an adjacent threonine that is also conserved (positions 105 and 105 in Figure 1A) [1,33,34]. Other key residues in thioredoxins are also conserved such as phenylalanines in the N-terminal part of the sequence (F39 and F54 in Figure 1A) or an aspartate that is located opposite to the active site (D89 in Figure 1 A), although only one of the two conserved glycines in the C-terminal part of the protein are conserved (G112 in Figure 1A) [1]. The biochemical characterization of the protein has shown that *Synechocystis'* TrxC is inactive in two classical thioredoxin activity assays (Figure 1B,C), in agreement with the data available for *Anabaena's* TrxC that is unable to reduce OpcA [21]. Although there are no known thioredoxin that present a similar active site to the one in TrxC, site directed mutagenesis of the proline in the active site of *E. coli* Trx1 (P34H) or *Staphylococcus aureus* Trx (P31T or P31S) changed the redox potential of these proteins to more oxidizing [1,35–37]. This may be the reason that makes TrxC inactive in the classical thioredoxin assays although it is clearly not the only reason as the L32P mutant was also inactive (Figure 1B). Therefore, it will be worth determining redox potential and structure of this protein to clarify its function. Several other thioredoxins have been shown to be inactive in insulin reduction assays and function and targets of these proteins are not known [30,38]. Recently it has been shown that in plants Trx-fold proteins (TRXL1/2 and ACHT4) are involved in oxidizing, rather than in reducing, proteins during the night or low light conditions allowing to fine tune metabolism in response to changes in light availability [39,40]. Furthermore, DDOR was induced in both WTOE and STXCOE strains, that also showed elevated TrxC protein levels (Figures 2 and 3), suggesting that these proteins might be functionally related. As DDOR is probably an oxidase [11] and changes in proline of the active site in thioredoxins (like the one present in TrxC) make them more oxidizing [1,37], it would be possible that DDOR-TrxC can function as a redox couple during stress in a similar way as the TRXL2/2cys-peroxiredoxin works in *Arabidopsis* [39].

The physiological characterisation of *trxC* mutants has shown that TrxC could be involved in adaptation to light and/or carbon availability because mutants in *trxC* showed a differential phenotype under conditions that change these two parameters. The STXC2 strain, although growing at similar rates to wild type, showed an increased carotenoid content in cultures bubbled with 1% CO_2 or air (Figure 4), suggesting increased photoprotection. In contrast, both overexpression strains showed lower carotenoid contents, suggesting that TrxC somehow regulates pigment accumulation. The lower carotenoid content in the overexpression strains can also explain the lower photosynthetic activity exhibited by these strains at higher light intensities (Figures 4 and 5) when carotenoids are important to maintain photosynthetic activity by preventing oxidative damage to the photosynthetic machinery [41]. It also explains their reduced growth rate under low carbon conditions in which light absorbed by the photosystems cannot be used as efficiently for CO_2 fixation and therefore more Reactive oxygen species (ROS) are produced. In contrast, when photosynthesis was analyzed by DUAL-PAM 100, overexpression strains showed more open PSII reaction centers in the dark (Fv'_{dark}/Fm'_{dark}) than WT and STXC2 strains, and therefore more photosynthetic efficiency. When we measured the same parameter in the presence of DCMU, all strains showed a similar value indicating that the photosynthetic machinery is similar in all strains indicating that there is no difference in total photosynthetic machinery. The higher Fv'$_{dark}$/Fm'$_{dark}$ indicated a more oxidized plastoquinone pool. Several mechanisms can explain this, such as lower cyclic electron transport, higher Mehler-like reactions catalyzed by flavodiiron proteins or increased respiration [42] which will be seen as reduced oxygen evolution due to enhanced photoreduction. These could also explain the lower growth rate of these strains under this condition as they imply a drain of electrons from the photosynthetic electron transfer chain. Another explanation is that these strains contained more phycobilisome (PBS), which can be suggested by the apparent color changes of the overexpression strains that appear bluish. Nevertheless, whole cell spectra showed that the difference in color can be ascribed to changes in carotenoids and not in PBS (Figure 4B,D). In agreement with these data, a *trxC⁻* mutant strain in *Anabaena* also showed altered pigment contents [20]. In this case it contained less chlorophyll and PBS together with less structured thylakoids membranes [20]. Furthermore, it also showed lower catalase activity and higher lipid peroxidation suggesting a redox imbalance [20], although whether the redox stress is caused by the photosynthetic defect or vice versa was not elucidated. Although our data shows that TrxC regulates pigment contents and photosynthetic activity, further characterization of the mutant strains is required to understand the molecular mechanism of these differences.

Finally, analysis of the genomic context has shown that *trxC* is adjacent to *nnrU* gene in many genomes (and is actually annotated as NnrU associated thioredoxin). NnrU is a membrane protein, and although its function is unknown, it is possible that it could functionally interact with TrxC. In *Nostocales trxC* is not only associated to *nnrU* but also to *ndhFM* genes which are part of NDH-1L complex involved in cyclic electron flow and respiration [43,44]. It is possible that some of the phenotypes observed such as slow growth in low carbon condition, lower photosynthetic activity or more oxidized plastoquinone pool in the dark can be ascribed to partially non-functional or non-regulated NDH-1L complex. In fact, *ndhF1* mutants in *Synechocystis* show a similar phenotype to WTOE and STXOE strains with lower oxygen-evolving activity but higher Fv'/Fm' than the WT [45]. This is reinforced as *Anabaena trxC⁻* mutant showed higher levels of NdhF1 protein and other electron transport proteins involved in cyclic electron transport [20]. All these data suggest that TrxC could modulate negatively NdhF1 (and therefore NDH-1L complex) activity and or assembly, although further experiments are needed to confirm this hypothesis.

5. Conclusions

In summary, here we have characterized TrxC, an unusual thioredoxin that is present exclusively in cyanobacteria, and showed that it is inactive in classical thioredoxin assays. Furthermore, we have analyzed both *trxC* knockout and overexpression mutants and showed that these are affected in pigment composition, growth and photosynthetic activity, although the mechanisms remain unknown and will require further characterization of the mutant strains.

Supplementary Materials: The following are available online at http://www.mdpi.com/2076-3921/7/11/164/s1, Table S1: Oligonucleotides used in this work. Table S2: Conservation of *trxC* in different cyanobacteria.

Author Contributions: Conceptualization, L.L.-M. and F.J.F.; investigation, L.L.-M. and L.G.H.-M.; writing—original draft preparation, L.L.-M., L.G.H.-M. and F.J.F.; writing—review and editing, L.L.-M., L.G.H.-M. and F.J.F.; funding acquisition, F.J.F.

Funding: This research was funded by Ministerio de Economía y Competividad (MINECO) grant number BIO2016-75634-P, and by Junta de Andalucía grant number P12-BIO-1119 and Group BIO-284, co-financed by European Regional Funds (FEDER) to Francisco Javier Florencio.

Acknowledgments: We thank Manuel J. Mallén, Raquel M. García, Sandra Díaz-Troya and María J. Huertas for critical reading the manuscript. Pea FBPase expression plasmid was a kind gift from Mariam Sahrawy.

Conflicts of Interest: The authors declare no conflict of interest.

References

1. Collet, J.-F.; Messens, J. Structure, Function, and Mechanism of Thioredoxin Proteins. *Antioxid. Redox Signal.* **2010**, *13*, 1205–1216. [CrossRef] [PubMed]
2. Balsera, M.; Uberegui, E.; Schürmann, P.; Buchanan, B.B. Evolutionary development of redox regulation in chloroplasts. *Antioxid. Redox Signal.* **2014**, *21*, 1327–1355. [CrossRef] [PubMed]
3. Buchanan, B.B. The Path to Thioredoxin and Redox Regulation Beyond Chloroplasts. *Plant Cell Physiol.* **2017**, *58*, 1826–1832. [CrossRef] [PubMed]
4. Meyer, Y.; Belin, C.; Delorme-Hinoux, V.; Reichheld, J.-P.; Riondet, C. Thioredoxin and Glutaredoxin Systems in Plants: Molecular Mechanisms, Crosstalks, and Functional Significance. *Antioxid. Redox Signal.* **2012**, *17*, 1124–1160. [CrossRef] [PubMed]
5. Serrato, A.J.; Fernández-Trijueque, J.; Barajas-López, J.-D.; Chueca, A.; Sahrawy, M. Plastid thioredoxins: A "one-for-all" redox-signaling system in plants. *Front. Plant Sci.* **2013**, *4*, 463. [CrossRef] [PubMed]
6. Florencio, F.J.; Pérez-Pérez, M.E.; López-Maury, L.; Mata-Cabana, A.; Lindahl, M. The diversity and complexity of the cyanobacterial thioredoxin systems. *Photosynth. Res.* **2006**, *89*, 157–171. [CrossRef] [PubMed]
7. Sánchez-Riego, A.M.; Mata-Cabana, A.; Galmozzi, C.V.; Florencio, F.J. NADPH-Thioredoxin Reductase C Mediates the Response to Oxidative Stress and Thermotolerance in the Cyanobacterium *Anabaena* sp. PCC7120. *Front. Microbiol.* **2016**, *7*, 1283. [CrossRef] [PubMed]
8. Mihara, S.; Yoshida, K.; Higo, A.; Hisabori, T. Functional Significance of NADPH-Thioredoxin Reductase C in the Antioxidant Defense System of Cyanobacterium *Anabaena* sp. PCC 7120. *Plant Cell Physiol.* **2017**, *58*, 86–94. [CrossRef] [PubMed]
9. Pascual, M.B.; Mata-Cabana, A.; Florencio, F.J.; Lindahl, M.; Cejudo, F.J. A comparative analysis of the NADPH thioredoxin reductase C-2-Cys peroxiredoxin system from plants and cyanobacteria. *Plant Physiol.* **2011**, *155*, 1806–1816. [CrossRef] [PubMed]
10. Buey, R.M.; Galindo-Trigo, S.; López-Maury, L.; Velázquez-Campoy, A.; Revuelta, J.L.; Florencio, F.J.; de Pereda, J.M.; Schürmann, P.; Buchanan, B.B.; Balsera, M. A New Member of the Thioredoxin Reductase Family from Early Oxygenic Photosynthetic Organisms. *Mol. Plant* **2017**, *10*, 212–215. [CrossRef] [PubMed]
11. Buey, R.M.; Arellano, J.B.; López-Maury, L.; Galindo-Trigo, S.; Velázquez-Campoy, A.; Revuelta, J.L.; de Pereda, J.M.; Florencio, F.J.; Schürmann, P.; Buchanan, B.B.; et al. Unprecedented pathway of reducing equivalents in a diflavin-linked disulfide oxidoreductase. *Proc. Natl. Acad. Sci. USA* **2017**, *114*, 12725–12730. [CrossRef] [PubMed]
12. Lindahl, M.; Florencio, F.J. Systematic screening of reactive cysteine proteomes. *Proteomics* **2004**, *4*, 448–450. [CrossRef] [PubMed]
13. Lindahl, M.; Florencio, F.J. Thioredoxin-linked processes in cyanobacteria are as numerous as in chloroplasts, but targets are different. *Proc. Natl. Acad. Sci. USA* **2003**, *100*, 16107–16112. [CrossRef] [PubMed]
14. Mata-Cabana, A.; Florencio, F.J.; Lindahl, M. Membrane proteins from the cyanobacterium *Synechocystis* sp. PCC 6803 interacting with thioredoxin. *Proteomics* **2007**, *7*, 3953–3963. [CrossRef] [PubMed]
15. Pérez-Pérez, M.E.; Florencio, F.J.; Lindahl, M. Selecting thioredoxins for disulphide proteomics: Target proteomes of three thioredoxins from the cyanobacterium *Synechocystis* sp. PCC 6803. *Proteomics* **2006**, *6*, S186–S195. [CrossRef] [PubMed]
16. Motohashi, K.; Romano, P.G.N.; Hisabori, T. Identification of Thioredoxin Targeted Proteins Using Thioredoxin Single-Cysteine Mutant-Immobilized Resin. In *Plant Signal Transduction*; Methods in Molecular Biology; Humana Press: Totowa, NJ, USA, 2009; Volume 479, pp. 117–131.
17. Nomata, J.; Maeda, M.; Isu, A.; Inoue, K.; Hisabori, T. Involvement of thioredoxin on the scaffold activity of NifU in heterocyst cells of the diazotrophic cyanobacterium *Anabaena* sp. strain PCC 7120. *J. Biochem.* **2015**, *158*, 253–261. [CrossRef] [PubMed]
18. Navarro, F.; Florencio, F.J. The cyanobacterial thioredoxin gene is required for both photoautotrophic and heterotrophic growth. *Plant Physiol.* **1996**, *111*, 1067–1075. [CrossRef] [PubMed]
19. Muller, E.G.; Buchanan, B.B. Thioredoxin is essential for photosynthetic growth. The thioredoxin m gene of Anacystis nidulans. *J. Biol. Chem.* **1989**, *264*, 4008–4014. [PubMed]

20. Deschoenmaeker, F.; Mihara, S.; Niwa, T.; Taguchi, H.; Wakabayashi, K.; Hisabori, T.; Ikeda, K.; Niwa, T.; Taguchi, H.; Hisabori, T. The absence of thioredoxin m1 and thioredoxin C in *Anabaena* sp. PCC 7120 leads to oxidative stress. *Plant Cell Physiol.* **2018**. [CrossRef] [PubMed]

21. Mihara, S.; Wakao, H.; Yoshida, K.; Higo, A.; Sugiura, K.; Tsuchiya, A.; Nomata, J.; Wakabayashi, K.-I.; Hisabori, T. Thioredoxin regulates G6PDH activity by changing redox states of OpcA in the nitrogen-fixing cyanobacterium *Anabaena* sp. PCC 7120. *Biochem. J.* **2018**, *475*, 1091–1105. [CrossRef] [PubMed]

22. Geigenberger, P.; Thormählen, I.; Daloso, D.M.; Fernie, A.R. The Unprecedented Versatility of the Plant Thioredoxin System. *Trends Plant Sci.* **2017**, *22*, 249–262. [CrossRef] [PubMed]

23. Pérez-Pérez, M.E.; Martín-Figueroa, E.; Florencio, F.J. Photosynthetic Regulation of the Cyanobacterium *Synechocystis* sp. PCC 6803 Thioredoxin System and Functional Analysis of TrxB (Trx *x*) and TrxQ (Trx *y*) Thioredoxins. *Mol. Plant* **2009**, *2*, 270–283. [CrossRef] [PubMed]

24. Perez-Perez, M.E.; Mata-Cabana, A.; Sanchez-Riego, A.M.; Lindahl, M.; Florencio, F.J. A Comprehensive Analysis of the Peroxiredoxin Reduction System in the Cyanobacterium *Synechocystis* sp. Strain PCC 6803 Reveals that All Five Peroxiredoxins Are Thioredoxin Dependent. *J. Bacteriol.* **2009**, *191*, 7477–7489. [CrossRef] [PubMed]

25. Díaz-Troya, S.; López-Maury, L.; Sánchez-Riego, A.M.; Roldán, M.; Florencio, F.J. Redox regulation of glycogen biosynthesis in the cyanobacterium *Synechocystis* sp. PCC 6803: Analysis of the AGP and glycogen synthases. *Mol. Plant* **2014**, *7*, 87–100. [CrossRef] [PubMed]

26. Stanier, R.Y.; Deruelles, J.; Rippka, R.; Herdman, M.; Waterbury, J.B. Generic Assignments, Strain Histories and Properties of Pure Cultures of Cyanobacteria. *Microbiology* **1979**, *111*, 1–61. [CrossRef]

27. Giner-Lamia, J.J.; López-Maury, L.; Florencio, F.J.; López-Maury, L.; Florencio, F.J. CopM is a novel copper-binding protein involved in copper resistance in *Synechocystis* sp. PCC 6803. *Microbiologyopen* **2015**, *4*, 167–185. [CrossRef] [PubMed]

28. Serrato, A.J.; Romero-Puertas, M.C.; Lázaro-Payo, A.; Sahrawy, M. Regulation by S-nitrosylation of the Calvin-Benson cycle fructose-1,6-bisphosphatase in Pisum sativum. *Redox Biol.* **2018**, *14*, 409–416. [CrossRef] [PubMed]

29. Gütle, D.D.; Roret, T.; Müller, S.J.; Couturier, J.; Lemaire, S.D.; Hecker, A.; Dhalleine, T.; Buchanan, B.B.; Reski, R.; Einsle, O.; et al. Chloroplast FBPase and SBPase are thioredoxin-linked enzymes with similar architecture but different evolutionary histories. *Proc. Natl. Acad. Sci. USA* **2016**, *113*, 6779–6784. [CrossRef] [PubMed]

30. Susanti, D.; Wong, J.H.; Vensel, W.H.; Loganathan, U.; DeSantis, R.; Schmitz, R.A.; Balsera, M.; Buchanan, B.B.; Mukhopadhyay, B. Thioredoxin targets fundamental processes in a methane-producing archaeon, Methanocaldococcus jannaschii. *Proc. Natl. Acad. Sci. USA* **2014**, *111*, 2608–2613. [CrossRef] [PubMed]

31. Ogawa, T.; Misumi, M.; Sonoike, K. Estimation of photosynthesis in cyanobacteria by pulse-amplitude modulation chlorophyll fluorescence: Problems and solutions. *Photosynth. Res.* **2017**, *133*, 63–73. [CrossRef] [PubMed]

32. Couturier, J.; Jacquot, J.-P.; Rouhier, N. Evolution and diversity of glutaredoxins in photosynthetic organisms. *Cell. Mol. Life Sci.* **2009**, *66*, 2539–2557. [CrossRef] [PubMed]

33. Ren, G.; Stephan, D.; Xu, Z.; Zheng, Y.; Tang, D.; Harrison, R.S.; Kurz, M.; Jarrott, R.; Shouldice, S.R.; Hiniker, A.; et al. Properties of the thioredoxin fold superfamily are modulated by a single amino acid residue. *J. Biol. Chem.* **2009**, *284*, 10150–10159. [CrossRef] [PubMed]

34. Roderer, D.J.A.; Schärer, M.A.; Rubini, M.; Glockshuber, R. Acceleration of protein folding by four orders of magnitude through a single amino acid substitution. *Sci. Rep.* **2015**, *5*, 11840. [CrossRef] [PubMed]

35. Lundström, J.; Krause, G.; Holmgren, A. A Pro to His mutation in active site of thioredoxin increases its disulfide-isomerase activity 10-fold. New refolding systems for reduced or randomly oxidized ribonuclease. *J. Biol. Chem.* **1992**, *267*, 9047–9052. [PubMed]

36. Krause, G.; Lundström, J.; Barea, J.L.; Pueyo de la Cuesta, C.; Holmgren, A. Mimicking the active site of protein disulfide-isomerase by substitution of proline 34 in Escherichia coli thioredoxin. *J. Biol. Chem.* **1991**, *266*, 9494–9500. [PubMed]

37. Roos, G.; Garcia-Pino, A.; Van belle, K.; Brosens, E.; Wahni, K.; Vandenbussche, G.; Wyns, L.; Loris, R.; Messens, J. The Conserved Active Site Proline Determines the Reducing Power of Staphylococcus aureus Thioredoxin. *J. Mol. Biol.* **2007**, *368*, 800–811. [CrossRef] [PubMed]

38. Sharma, A.; Sharma, A.; Dixit, S.; Sharma, A. Structural insights into thioredoxin-2: A component of malaria parasite protein secretion machinery. *Sci. Rep.* **2011**, *1*, 179. [CrossRef] [PubMed]

39. Yoshida, K.; Hara, A.; Sugiura, K.; Fukaya, Y.; Hisabori, T. Thioredoxin-like2/2-Cys peroxiredoxin redox cascade supports oxidative thiol modulation in chloroplasts. *Proc. Natl. Acad. Sci. USA* **2018**. [CrossRef] [PubMed]

40. Eliyahu, E.; Rog, I.; Inbal, D.; Danon, A. ACHT4-driven oxidation of APS1 attenuates starch synthesis under low light intensity in Arabidopsis plants. *Proc. Natl. Acad. Sci. USA* **2015**, *112*, 12876–12881. [CrossRef] [PubMed]

41. Zakar, T.; Laczko-Dobos, H.; Toth, T.N.; Gombos, Z. Carotenoids Assist in Cyanobacterial Photosystem II Assembly and Function. *Front. Plant Sci.* **2016**, *7*, 295. [CrossRef] [PubMed]

42. Allahverdiyeva, Y.; Isojärvi, J.; Zhang, P.; Aro, E.-M.; Allahverdiyeva, Y.; Isojärvi, J.; Zhang, P.; Aro, E.-M. Cyanobacterial Oxygenic Photosynthesis is Protected by Flavodiiron Proteins. *Life* **2015**, *5*, 716–743. [CrossRef] [PubMed]

43. Battchikova, N.; Eisenhut, M.; Aro, E.-M. Cyanobacterial NDH-1 complexes: Novel insights and remaining puzzles. *Biochim. Biophys. Acta Bioenerg.* **2011**, *1807*, 935–944. [CrossRef] [PubMed]

44. Zhao, J.; Gao, F.; Fan, D.-Y.; Chow, W.S.; Ma, W. NDH-1 Is Important for Photosystem I Function of *Synechocystis* sp. Strain PCC 6803 under Environmental Stress Conditions. *Front. Plant Sci.* **2018**, *8*, 2183. [CrossRef] [PubMed]

45. Ogawa, T.; Harada, T.; Ozaki, H.; Sonoike, K. Disruption of the ndhF1 Gene Affects Chl Fluorescence through State Transition in the Cyanobacterium *Synechocystis* sp. PCC 6803, Resulting in Apparent High Efficiency of Photosynthesis. *Plant Cell Physiol.* **2013**, *54*, 1164–1171. [CrossRef] [PubMed]

antioxidants

MDPI

Article

Structural and Biochemical Insights into the Reactivity of Thioredoxin h1 from *Chlamydomonas reinhardtii*

Christophe H. Marchand [1,†], Simona Fermani [2,†,*], Jacopo Rossi [3], Libero Gurrieri [3],
Daniele Tedesco [4], Julien Henri [1], Francesca Sparla [3], Paolo Trost [3], Stéphane D. Lemaire [1]
and Mirko Zaffagnini [3,*]

[1]   Laboratoire de Biologie Moléculaire et Cellulaire des Eucaryotes, Institut de Biologie Physico-Chimique,
      Unité Mixte de Recherche 8226 CNRS Sorbonne Université, 13 rue Pierre et Marie Curie, 75005 Paris, France;
      christophe.marchand@ibpc.fr (C.H.M.); julien.henri@ibpc.fr (J.H.); stephane.lemaire@ibpc.fr (S.D.L.)
[2]   Department of Chemistry "Giacomo Ciamician", University of Bologna, via Selmi 2, 40126 Bologna, Italy
[3]   Laboratory of Molecular Plant Physiology, Department of Pharmacy and Biotechnology, University of
      Bologna, via Irnerio 42, 40126 Bologna, Italy; jacopo.rossi13@studio.unibo.it (J.R.);
      libero.gurrieri2@unibo.it (L.G.); francesca.sparla@unibo.it (F.S.); paolo.trost@unibo.it (P.T.)
[4]   Bio-Pharmaceutical Analysis Section (Bio-PhASe), Department of Pharmacy and Biotechnology, University
      of Bologna, via Belmeloro 6, 40126 Bologna, Italy; daniele.tedesco@unibo.it
*   Correspondence: simona.fermani@unibo.it (S.F.); mirko.zaffagnini3@unibo.it (M.Z.);
      Tel.: +39-051-209-9475 (S.F.); +39-051-209-1314 (M.Z.)
†   These authors contributed equally to the work.

Received: 17 November 2018; Accepted: 18 December 2018; Published: 1 January 2019

Abstract: Thioredoxins (TRXs) are major protein disulfide reductases of the cell. Their redox activity relies on a conserved Trp-Cys-(Gly/Pro)-Pro-Cys active site bearing two cysteine (Cys) residues that can be found either as free thiols (reduced TRXs) or linked together by a disulfide bond (oxidized TRXs) during the catalytic cycle. Their reactivity is crucial for TRX activity, and depends on the active site microenvironment. Here, we solved and compared the 3D structure of reduced and oxidized TRX h1 from *Chlamydomonas reinhardtii* (CrTRXh1). The three-dimensional structure was also determined for mutants of each active site Cys. Structural alignments of CrTRXh1 with other structurally solved plant TRXs showed a common spatial fold, despite the low sequence identity. Structural analyses of CrTRXh1 revealed that the protein adopts an identical conformation independently from its redox state. Treatment with iodoacetamide (IAM), a Cys alkylating agent, resulted in a rapid and pH-dependent inactivation of CrTRXh1. Starting from fully reduced CrTRXh1, we determined the acid dissociation constant (pK_a) of each active site Cys by Matrix-assisted laser desorption/ionization-time of flight (MALDI-TOF) mass spectrometry analyses coupled to differential IAM-based alkylation. Based on the diversity of catalytic Cys deprotonation states, the mechanisms and structural features underlying disulfide redox activity are discussed.

Keywords: *Chlamydomonas reinhardtii*; cysteine alkylation; cysteine reactivity; MALDI-TOF mass spectrometry; thioredoxin; X-ray crystallography

1. Introduction

Thioredoxins (TRXs) are small oxidoreductases that contribute, in most living organisms, to the control of cellular redox homeostasis. Indeed, they reduce disulfide bonds on numerous proteins involved in several processes, including antioxidant defense mechanisms, photosynthetic carbon metabolism, and protein folding [1–4]. In addition, TRXs have been suggested to control two additional Cys-based post-translational modifications, namely S-glutathionylation and S-nitrosylation, by catalyzing the reduction of glutathione-mixed disulfides (–SSGs) and *S*-nitrosothiols (–SNOs), respectively [5–11]. The redox activity of TRXs relies on the presence of two Cys residues in the conserved Trp-Cys-(Gly/Pro)-Pro-Cys (WC(G/P)PC) active site motif. Regardless of the reducing activity in which TRXs are involved, the N-terminal active site Cys (CysN) is responsible for the first nucleophilic attack on oxidized targets. While the subsequent catalytic steps underlying TRX-dependent protein deglutathionylation and denitrosylation are yet undefined, the reduction of protein disulfides implies the formation of a mixed-disulfide bond between the CysN and a Cys on target protein. Following the formation of the TRX-target complex, the C-terminal active site Cys (CysC) of TRX performs a second nucleophilic attack on the TRX-target mixed-disulfide, yielding a reduced target and oxidized TRX [12–14].

Thiol groups are weak acids, and the reactivity of Cys residues depends on their acid dissociation constant (pK_a) that controls thiol deprotonation state under physiological pH conditions [15]. Noteworthy, a deprotonated Cys (i.e., Cys thiolate, $-S^-$) is much more reactive than its protonated counterpart (i.e., Cys thiol, $-SH$). In general, multiple factors related to the Cys microenvironment contribute to its reactivity. The proximity of charged residues (e.g., Arg, Lys, His, Asp), a hydrogen-bond network, and the N-terminal position in α-helix, contribute to increase Cys reactivity by lowering its pK_a [15,16]. In 1997, Holmgren and colleagues established the importance of the active site microenvironment for TRX catalysis [17]. In this study, the mutation of charged residues (Asp26 and Lys57), which are located nearby the catalytic Cys thiols, caused a drastic decline of *Escherichia coli* TRX activity [17]. Whereas the role of the conserved Lys residue is still unclear, the Asp residue was suggested to be involved in the deprotonation of active site CysC in both *E. coli* TRX and *Chlamydomonas reinhardtii* (CrTRXh1) [18,19]. By contrast, in TRX from *S. aureus*, the Asp residue did not appear to be critical for CysC deprotonation [20]. To date, only pK_a values of the two catalytic Cys from non-plant TRXs have been determined ($pK_a = 6.7-7.4$ and >9.5 for CysN and CysC, respectively) [16]. These pK_a values are mainly determined by a combination of interactions with charged residues and the hydrogen-bonding network.

In photosynthetic eukaryotes, the TRX family counts 7 classes (f, h, m, o, x, y, and z) comprising multiple isoforms (9 and 21 in *Chlamydomonas reinhardtii* and *Arabidopsis thaliana*, respectively) localized in plastids, mitochondria, and cytoplasm [1,3,4,21]. The catalytic role of plant TRXs in thiol switching mechanisms (e.g., dithiol/disulfide exchange reactions) has been extensively investigated both in vitro and in vivo (for reviews, see [3,4]. In addition, proteomic-based approaches allowed the identification of hundreds of putative TRX targets in plants [22,23], and more than 1000 targets in *Chlamydomonas* [24], suggesting that TRXs can control multiple cellular pathways by regulating protein redox states in photosynthetic organisms. Although plant TRXs have been thoroughly characterized at the functional level, structural data are still limited to only 11 three-dimensional structures available so far [4]. This list includes TRX h1 and m from *Chlamydomonas reinhardtii* (CrTRXh1, [19]; CrTRXm, [25], TRX h1, o1, and o2 from *Arabidopsis thaliana* (AtTRXh1, [26]; AtTRXo1 and AtTRXo2, [27]), TRX h1 and h2 from *Hordeum vulgare* (HvTRXh1 and HvTRXh2, [28]), TRX h1 and h4 from *Populus trichocarpa* (PtTRXh1, [29]; PtTRX h4, [30]), and TRX f and m from *Spinacia oleracea* (SoTRXf and SoTRXm, [31]). In all cases, TRXs structures were solved in the oxidized form, except for plastidial SoTRXm, and cytoplasmic HvTRXh2 for which the 3D structure of the reduced forms is available.

In the present study, we deepen the structural features of CrTRXh1, providing the 3D structures of both the reduced form and variants in which catalytic cysteines were mutated. Extensive comparison of CrTRXh1 with structurally solved plant TRXs uncovers a conserved 3D folding despite the low sequence identity. Further analysis suggests that plant TRX classes exhibit peculiar structural features for substrate recognition. Structural alignments and circular dichroism (CD) analysis of reduced and oxidized CrTRXh1 reveal an identical folding independent on the redox state. The analysis of the catalytic site microenvironment from the crystal structure of reduced CrTRXh1 indicates that a hydrogen-bonding network is the major factor determining CysN reactivity. Through mass spectrometry (MS) analysis coupled to pH-dependent iodoacetamide (IAM)-based alkylation, we experimentally determined the pK_a

of both active site cysteines. Based on the diversity of catalytic Cys deprotonation states, we discussed the mechanisms and structural features underlying disulfide redox activity of TRX.

2. Materials and Methods

2.1. Material and Enzymes

All reagents were purchased from Sigma-Aldrich (St Louis, Missouri, MO, USA) unless otherwise indicated. Production and purification of recombinant CrTRXh1 (wild-type, C36S and C39S mutants) and nicotinamide adenine dinucleotide phosphate reduced (NADPH)-thioredoxin reductase B from *Arabidopsis thaliana* (AtNTRB) were carried out as previously reported [32,33].

2.2. Enzymatic Assay

Activity of CrTRXh1 was assayed by following the reduction of 5,5′-dithiobis-2-nitrobenzoic acid (DTNB). The reaction mixture contained 50 mM Tris-HCl (pH 7.9), 1 mM ethylenediaminetetraacetic acid (EDTA), 0.2 mM DTNB, 0.22 μM AtNTRB, 0.2 mM NADPH, and recombinant CrTRXh1 in a final volume of 1 mL. The reaction was started by adding CrTRXh1 at indicated concentration, and the reduction of DTNB was monitored by following the Abs_{412} increase associated with the formation of TNB^- (molar extinction coefficient at 412 nm = 14,150 M^{-1} cm^{-1}). The reaction rates were corrected with a reference rate without CrTRXh1.

2.3. Inactivation of CrTRXh1 by IAM-Dependent Alkylation Treatments

Before each treatment, CrTRXh1 (20 or 100 μM) was incubated for 15 min at room temperature in 40 mM ammonium bicarbonate buffer (pH 8.5) in the presence of 5-fold molar excess of dithiothreitol (DTT). After incubation, the reduced protein (20 μM) was treated with 0.05 or 0.1 mM IAM without removal of the excess of DTT. At the indicated time, an aliquot of the sample (20 μL) was withdrawn for the assay of enzyme activity. To determine the pH-dependence of the inactivation of CrTRXh1 by IAM, aliquots of the reduced protein sample (100 μM) were diluted 5-fold in 100 mM sodium citrate (pH 5.0 and pH 6.0) or in 100 mM Bis-Tris (pH 6.5 and 7.0) or in 100 mM Tris-HCl (pH 8.0), and incubated with 0.1 mM IAM. After 2 min incubation, an aliquot of the sample (20 μL) was withdrawn for the assay of enzyme activity.

2.4. pK_a Determination by Matrix-Assisted Laser Desorption/Ionization-Time of Flight (MALDI-TOF) MS

For reduction of active site cysteines, 230 μM CrTRXh1 was incubated beforehand in 50 mM phosphate buffer (pH 7.1) for at least one hour at 25 °C with 2 mM Tris (2-carboxyethyl) phosphine (TCEP), a chemical reductant that does not react with Cys alkylating agents like IAM. For the pK_a determination of CysN, 2.1 μL of pre-reduced CrTRXh1 was directly diluted in 6.9 μL of reaction buffer at the indicated pH, and 1 μL of 1 mM IAM was added to give an IAM/TRX ratio of 2. After incubation (105 seconds), the reaction was quenched by adding 90 μL of a saturated sinapinic acid matrix solution freshly prepared in 30% acetonitrile and 0.3% trifluoroacetic acid. For pK_a determination of the CysC, pre-reduced CrTRXh1 (200 μM) was incubated (2 min) in 40 mM phosphate buffer (pH 7.1) with 2-fold excess of IAM to allow complete monoalkylation. Subsequently, 2.4 μL of pre-alkylated CrTRXh1 was directly diluted in 6.6 μL of reaction buffer at the indicated pH and 1 μL of 3.85 mM IAM was added to give a final IAM/TRX ratio of 10. After 2 min incubation, the reaction was quenched as described above. Reaction buffers used consisted in 200 mM acetate/acetic acid buffer for pH between 3.5 and 5.5; 100 mM phosphate buffer for pH between 6.0 and 8.0; 100 mM HEPES-NaOH for pH between 7.5 and 8.5, and 100 mM glycine buffer for pH values above 8.5.

For MALDI-TOF analyses, 1.8 μL of quenched mixtures were spotted on a sample plate and dried under a gentle air stream. Spectra were acquired in positive linear mode with an Autoflex MALDI-TOF/TOF mass spectrometer using FlexAnalysis 3.3 software (Bruker Daltonics, Bremen, Germany) after calibration on mono- and di-charged ions of equine cytochrome c. Peak area of reduced, mono- and dialkylated forms of CrTRXh1 were determined using DataAnalysis software (Bruker Daltonics, Bremen, Germany).

2.5. Identification of the Reactive Cysteine within the Monoalkylated Form of CrTRXh1

The monoalkylated form of CrTRXh1 was generated at pH 7.1 as described above and, after incubation, 0.5 mM DTT was added to quench the excess of IAM and thus avoid further alkylation during trypsin digestion. Alkylated CrTRXh1 (0.6 µg) was placed in 20 mM ammonium bicarbonate buffer (pH 8.5) and digested in a final volume of 6 µL with 50 ng of trypsin Gold (Promega) for 6 h. One microliter of this mixture was mixed with a half-saturated solution of α-cyano-4-hydroxycinnamic acid prepared in 50% acetonitrile/0.3% trifluoroacetic acid and then spotted on a MALDI sample plate. Peptide mass fingerprints were acquired on a Performa Axima mass spectrometer (Shimadzu, Manchester, UK) in positive reflectron mode, whereas collision-induced dissociation MS/MS spectra were acquired after selection of the alkylated cysteine-containing peptide ion and collision with helium gas.

2.6. Circular Dichroism Analysis

Circular dichroism (CD) analysis was performed on red- and ox-CrTRXh1 at room temperature using a Jasco J-810 spectropolarimeter (Tokyo, Japan). Far- ultraviolet (UV) CD measurements were performed in the 250−195 nm spectral range, using a 0.5 mm path length (high quality quartz cell; Hellma, Milan, Italy), a 20 nm min^{-1} scanning speed, a 4 nm response, a 2 nm spectral bandwidth, and an accumulation cycle of 3; solvent-corrected CD spectra were converted to molar units per residue ($\Delta\varepsilon_{res}$). Samples for CD analysis were prepared at a nominal concentration of 10 µM in 30 mM Tris-HCl buffer (pH 7.9). Reduced CrTRXh1 was obtained following 30 min incubation in the presence of a 10-fold molar excess of TCEP. The exact concentration was determined from the absorbance at 280 nm (1 cm path length) based on the theoretical molar absorption coefficients of 13,980 M^{-1} cm^{-1} and 14,105 M^{-1} cm^{-1} for red- and ox-CrTRXh1, respectively [34]. Reduced CrTRXh1 was obtained by incubation with a 10-fold molar excess of TCEP for 30 min.

2.7. Crystallization and Data Collection

Protein samples, i.e., red- and ox-CrTRXh1, C36S and C39S mutants, were prepared in 50 mM Tris-HCl (pH 7.9), 1 mM EDTA, and concentrated at 18 mg/mL in the presence of 5 mM TCEP except for the oxidized form. They were crystallized by the hanging drop vapor diffusion method at 293 K. The drop composed of 2 µL of protein solution and an equal volume of reservoir, and was equilibrated against 750 µL of reservoir solution. The reservoir solutions were elaborated starting from crystallization conditions reported in Menchise et al. [19]. Crystals of all protein samples grew after one week from various solutions containing 20% (*w/v*) polyethylene glycol (PEG) 8K, or 20% (*w/v*) PEG 10K, or 10% (*w/v*) PEG 8K, or 10% (*w/v*) PEG 10K as precipitant buffered with 0.1 M sodium cacodylate or MES (2-(N-morpholino)ethanesulfonic acid) at pH 6.5, or 0.1 M HEPES-NaOH at pH 7.5, or 0.1 M Tris-HCl at pH 8.5.

Crystals were mounted from the crystallization drop into cryoloops, briefly soaked in a solution containing 22% (*w/v*) PEG 8K or 20% (*w/v*) PEG 10K or 11% (*w/v*) PEG 8K, and 11% (*w/v*) PEG 10K plus 20% (*v/v*) PEG400 as cryoprotectant, then frozen in liquid nitrogen. Diffraction images for red-, ox-CrTRXh1, and C36S mutant were recorded at 100 K at the at the European Synchrotron Radiation Facility (Grenoble, France, beam line ID23-1) using a wavelength of 1.0 Å, a Δφ of 0.15° and a detector distance of 290.04 mm (red-CrTRXh1), 294.66 mm (ox-CrTRXh1), 122.66 mm (monomeric C36S mutant), and 189.10 mm (dimeric C36S mutant). Diffraction images for C39S mutant were collected at Elettra synchrotron radiation source (Trieste, Italy, beam line XRD1) at a wavelength of 1.0 Å, a Δφ of 0.3°, and a detector distance of 200.00 mm.

Diffraction data were analyzed with XDS [35] for data reduction, with POINTLESS [36] for space group determination, and with SCALA [36] for scaling and merging. Statistics of data collection are reported in Table 1.

Table 1. X-ray data collection and refinement statistics.

	Red-CrTRXh1	Ox-CrTRXh1	Monomeric C36S	Dimeric C36S	C39S
Data collection					
Unit cell (Å)	$a = b = 48.77, c = 143.93$	$a = b = 48.68, c = 143.70$	$a = 60.78, b = 34.82, c = 48.16$	$a = b = 48.74, c = 143.19$	$a = b = 48.43, c = 143.66$
Space group	$P3_121$	$P3_121$	$P2_12_12$	$P3_121$	$P3_121$
N° molecules ASU	2	2	1	2	2
Resolution range * (Å)	47.98–1.70	42.16–1.57	48.16–0.94	42.21–1.22	41.94–1.81
	(1.76–1.70)	(1.63–1.57)	(0.96–0.94)	(1.24–1.22)	(1.84–1.81)
Unique reflections	22,494 (2165)	28,348 (2646)	64,505 (1579)	59,565 (2934)	18,676 (1044)
Completeness * (%)	99.3 (98.5)	99.5 (96.7)	96.5 (47.1)	99.7 (98.7)	99.6 (97.0)
R_{merge} *	0.145 (0.858)	0.072 (0.804)	0.043 (0.369)	0.063 (1.108)	0.111 (0.740)
R_{pim} *	0.064 (0.398)	0.034 (0.428)	0.033 (0.367)	0.028 (0.507)	0.057 (0.345)
$CC_{1/2}$ *	0.980 (0.683)	0.998 (0.661)	0.997 (0.625)	0.998 (0.635)	0.990 (0.715)
$I/(I)$ *	7.9 (2.2)	10.0 (1.5)	18.9 (1.7)	14.7 (1.7)	9.7 (1.9)
Multiplicity *	6.5 (6.6)	5.7 (4.9)	4.1 (1.6)	6.8 (6.6)	5.8 (5.7)
Refinement					
Resolution range * (Å)	42.24–1.70	42.16–1.57	48.16–0.94	42.21–1.22	41.94–1.81
	(1.78–1.70)	(1.62–1.57)	(0.95–0.94)	(1.23–1.22)	(1.85–1.81)
Reflection used *	22,475 (2724)	28,277 (2671)	64,420 (1042)	59,506 (3597)	18,626 (2581)
R/R_{free}	0.189/0.224	0.187/0.216	0.157/0.166	0.168/0.174	0.219/0.266
rmsd from ideality (Å, °)	0.006, 0.743	0.006, 0.820	0.006, 1.007	0.007, 1.079	0.002, 0.438
N° atoms					
Non-hydrogen atoms	1867	1832	1063	2015	1799
Protein atoms	1675	1660	856	1675	1637
Solvent molecules	192	172	207	346	162
Hydrogens	/	/	897	1724	/
B value (Å²)					
Mean	26.8	26.1	12.6	16.2	24.7
Wilson	25.8	23.6	8.4	12.6	24.0
Protein atoms	26.1	25.5	9.3	14.6	24.2
Solvent molecules	32.8	32.6	20.7	23.5	30.1
Heteroatoms	/	/	33.2	37.8	/
Ramachandran plot (%) §					
Most favored	99.1	98.6	99.1	97.7	98.6
Allowed	0.9	1.4	0.9	2.3	1.4
Disallowed	0	0	0	0	0

* Values in parentheses refer to the last resolution shell; § As defined by MolProbity [37]. /: not determined.

2.8. Structure Solution and Refinement

The lower resolution crystal structure of oxidized CrTRXh1 reported by [19] was directly used for refinement in the case of crystals isomorphous with the chosen model (Table 1), i.e., red- and ox-CrTRXh1, and dimeric C36S and C39S mutants. The same structure was used as a model to solve the structure of monomeric C36S mutant crystal (Table 1) by molecular replacement, using the software MOLREP [38]. The structures were refined with REFMAC 5.8.0135 [39] by selecting 5% of reflections for R_{free} calculation, and manually rebuild with Coot [40]. Water molecules automatically added with Coot [40], were visually checked and kept into the model if the relative electron density value in the $(2F_o - F_c)$ map was higher than 1.0 σ, and if stabilized by hydrogen bonds with the protein. Structures were further refined with PHENIX.REFINE [41]. The statistics of the refinement are shown in Table 1. Figures showing structures were prepared using PyMOL (The PyMOL Molecular Graphics System, Schrödinger, LLC).

2.9. Data Availability

The coordinates of the structural models and the experimental data (structure factors) are deposited in the Protein Data Bank (PDB). The assigned accession codes are: 6Q46 for red-CrTRXh1, 6Q47 for ox-CrTRXh1, 6Q6U for C39S mutant, and 6Q6T and 6Q6V for monomeric and dimeric C36S mutant, respectively.

3. Results

3.1. Sequence and Structural Comparison of Plant TRXs

Photosynthetic eukaryotes contain a large number of TRXs but, to date, only 11 isoforms were structurally characterized. These proteins belong to chloroplast f- and m-classes, cytoplasmic h-class, and mitochondrial o-class, and are from different photosynthetic organisms, such as the green alga *Chlamydomonas reinhardtii* (CrTRXh1 and m) and several land plants, such as *Arabidopsis* (AtTRXh1 and AtTRXo1/o2), barley (HvTRXh1/h2), poplar (PtTRXh1/h4), and spinach (SoTRXm and f). To achieve a better understanding of structure-function relationships among plant TRXs, we performed a primary sequence analysis coupled to structural alignment. As shown in Figure 1a, TRXs exhibit pairwise sequence identities ranging from ~21% to ~75%. The highest identity was observed between isoforms of mitochondrial o-class (AtTRXo1 and AtTRXo2), whereas the lowest identity was found between AtTRXo2 and SoTRXf. Among selected TRXs, sequence conservation is restricted to 9 residues, including the WC(G/P)PC active site motif, one stretch comprising two residues (Pro80-Thr81, numbered according to CrTRXh1), and two single amino acids (Pro44 and Gly96, numbered according to CrTRXh1) (Figure 1a).

To assess the structural similarity of plant TRXs, a 3D structure alignment was performed using PyMOL 2.1 and CrTRXh1 as template. This protein shares low identity with analyzed TRXs (31–53%) even compared with other h-class TRXs, PtTRXh1 being the closest homolog (Figure 1a). Despite the low sequence identity, the enzymes displayed highly similar 3D structures with a root mean square deviation (rmsd) ranging from 0.84 Å (CrTRXh1 versus HvTRXh1) to 2.06 Å (CrTRXh1 versus AtTRXo2). In-depth analysis of secondary structures revealed that the content of α-helices and β-sheets is also greatly conserved (Figure 1a). In CrTRXh1, around 45% and 26% of the total amino acids are involved in the formation of the α-helices and β-sheets, respectively. These values are substantially maintained in the other TRXs, with only the exception of SoTRXm and CrTRXm, which display a slight decrease of α-helix content (~35% and ~38%, respectively). This difference is mainly ascribed to the length of α-helix 1 (Figure 1a), which is composed by 4 residues in SoTRXm and 7 residues in CrTRXm and SoTRXf, while CrTRXh1 contains 14 residues (Figure 1a,b). All other h- and o-classes TRXs show, like CrTRXh1, a α-helix 1 of 11–14 residues (Figure 1a,b). The key structural element α-helix 2, which contains the active site motif WC(G/P)PC, was found to contain 16 residues in all TRXs except for CrTRXm (17 residues), and AtTRXo1/o2 (14 and 13 residues, respectively) (Figure 1a,b).

The conservation of α-helix 2 is sensibly higher than the rest of TRX sequences, being in the ~39–81% range. Four amino acids are fully conserved, whereas the other residues, though not conserved, share common physicochemical properties (Figure 1a), suggesting their importance for proper folding of α-helix 2.

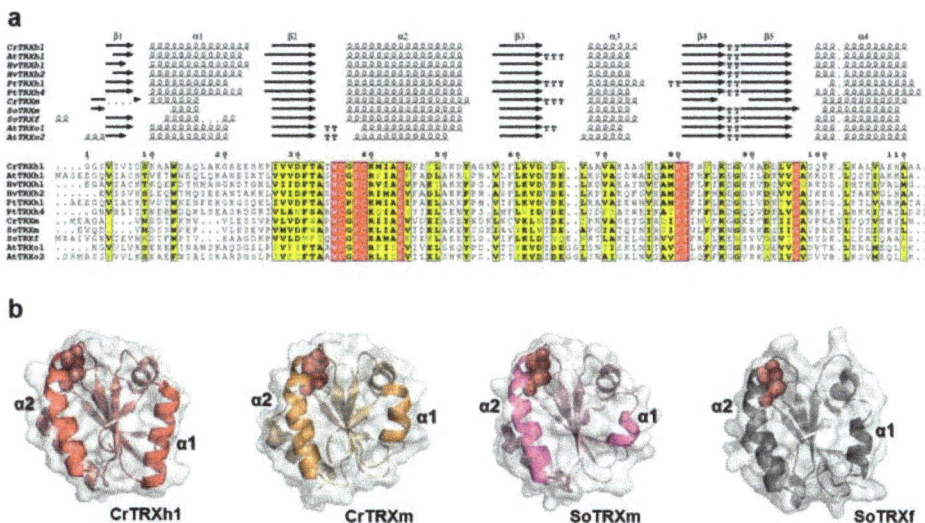

Figure 1. Sequence analysis and structural representation of plant TRXs. (**a**) Multiple sequence alignment of structurally solved plant TRX isoforms was performed with ESPript (http://espript. ibcp.fr, [42]) using *Chlamydomonas reinhardtii* (Cr) TRX h1 (Protein Data Bank (PDB) ID: 1EP7, [19]) and TRX m (PDB ID: 1DBY, [25]), *Arabidopsis thaliana* (At) TRX h1 (PDB ID: 1XFL, [26]), TRX o1 (PDB ID: 6G61, [27]), and TRX o2 (PDB ID: 6G62, [27], *Hordeum vulgare* (Hv) TRX h1 (PDB ID: 2VM1, [28]) and TRX h2 (PDB ID: 2IWT, [28]); *Populus trichocarpa* (Pt) TRX h1 (PDB ID: 1TI3, [29]) and TRX h4 (PDB ID: 3D21, [30]), and *Spinacia oleracea* (So) TRX f (PDB ID: 1F9M, [31]) and TRX m (PDB ID: 1FB0, [31]). Conserved residues are highlighted in white on red boxes, whereas residues with similar physicochemical properties are written in black on yellow boxes. The sequence identities were calculated with Clustal Omega [43]. (**b**) Ribbon representation of the three-dimensional structures of CrTRXh1, CrTRXm, SoTRXm, and SoTRXf. The structural elements α-helix 1 and α-helix 2 are labeled; and active site cysteines are shown as red balls.

3.2. CrTRXh1 Shares the Substrate Recognition Loop with Other h-Class TRXs

The global conservation of aligned plant TRX sequences is lower than 10%, a value that increases to 21% if sequences of h-class TRXs only are compared (Figure 1a). This difference is due to the presence of additional conserved residues randomly distributed along the sequences (Figure 1a), including two stretches composed by Ala–Met–Pro–Thr–Phe (AMPTF) and Val–Gly–Ala (VGA) (Figures 1a and 2a).

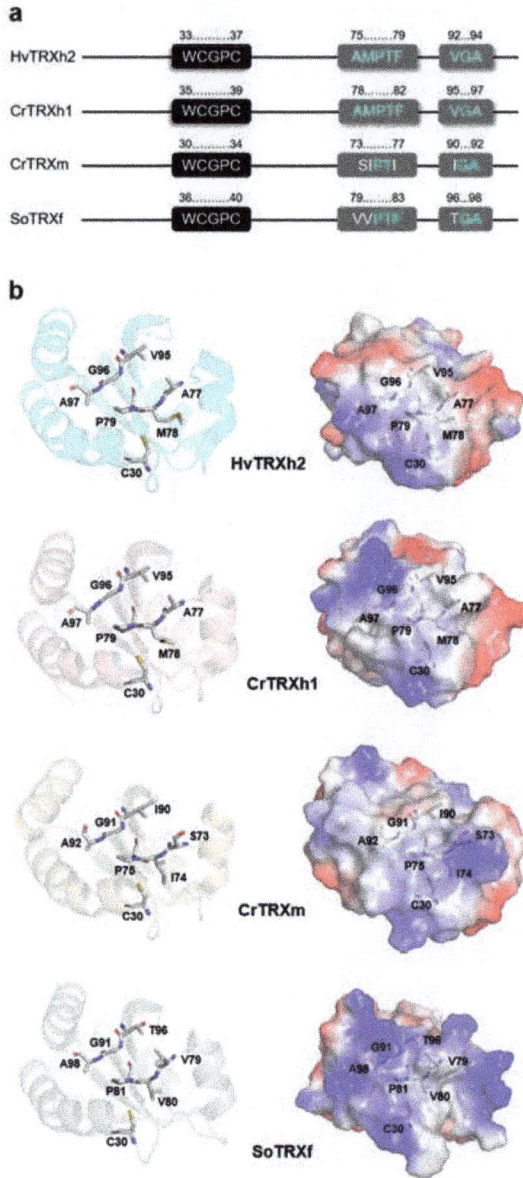

Figure 2. Sequence and structural analysis of the substrate recognition loop (SRL). (**a**) Schematic representation of the substrate recognition loop in HvTRXh2, CrTRXh1, CrTRXm, and SoTRXf. The active site motifs are highlighted in black, whereas the *cis*-Pro and Gly loops are highlighted in gray, with conserved and non-conserved residues indicated in cyan and white, respectively. The size of the strings is not proportional to the length in amino acids. (**b**) Ribbon representation of HvTRXh2, CrTRXh1, CrTRXm, and SoTRXf highlighting the residues forming the *cis*-Pro and Gly loops. (**c**) Electrostatic surface potentials were computed with the Adaptive Poisson-Boltzmann Solver (APBS) Electrostatics plugin in PyMOL (red for electronegative, white for neutral, blue for electropositive).

These stretches include residues conserved in all TRXs (Pro–Thr and Gly, underlined, Figure 1a) but also other residues with little or no variations among h-class TRXs (Figure 2a and Figure S1). These stretches were recognized to be involved in target recognition by Maeda and colleagues [44]. In this study, the crystallographic structure of HvTRXh2 in complex with the target α-amylase/subtilisin inhibitor (BASI) was solved. In the HvTRXh2-BASI complex, the authors observed that the substrate recognition occurs through the WCGPC motif of HvTRXh2 along with residues Ala–Met–Pro (AMP, part of the conserved AMPTF sequence) and Val–Gly–Ala (VGA). The latter regions were named *cis*-Pro and Gly loops, respectively, and together with the active site motif form, the so-called substrate recognition loop (SRL). Interestingly, both *cis*-Pro and Gly loops are hydrophobic regions allowing van der Waals interactions and hydrogen-bonding with the backbone surrounding the target cysteine [45]. In their respective loops, the Pro supports correct positioning of Ala and Met, whereas the Gly confers flexibility to the flanking amino acids [44,45].

As shown in Figure 2a, CrTRXh1 shares with HvTRXh2 the SRL region, which is also partially conserved in other h-class TRXs (Figure S1). When we considered the other structurally solved TRXs (AtTRXo1, AtTRXo2, CrTRXm, SoTRXm, and SoTRXf), the aligned SRL region differs from h-class TRXs (Figure 2a and Figure S1). The major difference was observed in the Pro loop (AMP), in which Ala–Met are substituted by Ser–Ile and Val–Val in m-class TRXs and SoTRXf, respectively (Figure 2a,b, and Figure S1). A single substitution is observed in o-class TRXs, where a Val replaced the Met (Figure S1). By contrast, the Gly loop (VGA) only differs for the first residue Val that is replaced by Ile (m-class TRXs) or Thr (SoTRXf), or the last residue Ala substituted by a Val in AtTRXo2 (Figure 2a,b, and Figure S1). Taking into account the different residue substitutions, we can consider the replacement of hydrophobic residues (Ala and Val) with polar ones (Ser or Thr) as the most effective in modifying the electrostatic surface potentials (Figure 2b), that likely modulate the structural constraints involved in target recognition. Intriguingly, amino acid substitutions are comparable among TRX isoforms from the same class, suggesting that the SRL region could contribute to specifically guiding the recognition of protein targets by each TRX class.

3.3. Structure Analysis of Reduced and Oxidized CrTRXh1

The structure of wild-type CrTRXh1 in its reduced and oxidized forms (red- and ox-CrTRXh1, respectively) were solved at a resolution of 1.70 and 1.57 Å, respectively. Crystals of red- and ox-CrTRXh1 are isomorphous to each other (Table 1), and to that previously reported for ox-CrTRXh1 structure (PDB ID: 1EP7, [19]). The asymmetric unit is composed by a non-covalent dimer, with the two monomers (chains A and B) related by a non-crystallographic two-fold axis. The total accessible surface area (ASA) is similar in the two enzyme forms (10,329 and 10,300 Å^2 in red- and ox-CrTRXh1, respectively), while the buried surface area (BSA) is slightly lower in the reduced form (1117 Å^2) with respect to the oxidized one (1169 Å^2). Similarly, the calculated interface area between chains A and B is 556 Å^2 in red-CrTRXh1 and 556 Å^2 in ox-CrTRXh1. The interface is formed by 16 residues from each chain (Glu72–Met79, Ala33–Gly37, Val64–Asp65, and the single residue Lys40), and stabilized by two salt bridges, several hydrogen bonds, and hydrophobic interactions, which involve mainly Trp35 from both chains (Figure 3a) (Table S1).

The two independent chains A and B are very similar, and their superimposition gave an rmsd of 0.49 Å (on 112 aligned C$_\alpha$ atoms) and 0.48 Å (on 111 aligned C$_\alpha$ atoms) for red- and ox-CrTRXh1, respectively.

The CD spectra of red- and ox-CrTRXh1 are almost identical (Figure S2), suggesting a strong similarity in secondary structure between the two forms of CrTRXh1 in solution. In agreement with this observation, the protein redox state does not alter the overall protein folding in the crystal as well. Indeed, the superimposition of the dimers and the single monomers of red- and ox-CrTRXh1 resulted in rmsd values of 0.167 Å (on 222 C$_\alpha$ atoms) and 0.128 Å (on 111 C$_\alpha$ atoms), respectively. At structural level, CrTRXh1 folds in a 4-stranded β-sheet surrounded by 4 α-helices. The first β-strand, typically observed at the N-terminal end of plant TRXs (Figure 1a), is replaced by a long random coil region that precedes α-helix 1 (Figure 3b).

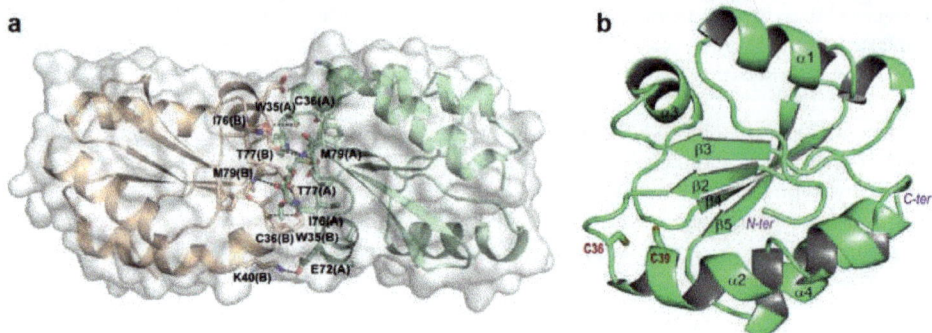

Figure 3. Crystal structure of red-CrTRXh1. (**a**) Cartoon representation of the asymmetric unit dimer. The dimer surface is shown in light gray. The interface interactions are shown, and the corresponding distances are indicated in Table S1. (**b**) Cartoon representation of the monomer structure. The secondary structure elements are shown, and the catalytic Cys36 (C36) and Cys39 (C39) are represented as sticks.

3.4. Active Site of Reduced and Oxidized CrTRXh1

The catalytic site Cys36 and Cys39 (corresponding to CysN and CysC, respectively) are located at the N-terminal end of α-helix 2 protruding from the protein surface of the monomer, but completely embedded in the interface if the crystallographic dimer is considered (Figure 3a). In the red-CrTRXh1 structure, cysteines are fully reduced, and the thiol groups stand at a distance of 3.6 Å (Figure 4a). By contrast, the $2F_o - F_c$ electron density map of ox-CrTRXh1 clearly shows a double conformation of Cys36 (Figure 4b), indicating that a portion of the molecules forming the crystal packing is not oxidized, even if the crystallization was performed under non-reducing conditions.

In red-CrTRXh1, Cys36 is the most exposed, showing an ASA of ~16.0 Å², and its thiol group is hydrogen-bonded to the carbonyl group of Met79 (3.5 Å) and the thiol (3.6 Å), and the amide group (4.0 Å) of Cys39 (Figure 4a). If the crystallographic dimer is considered, the accessibility of Cys36 is drastically reduced (ASA = 1.8 Å²) by residues Ile76 and Thr77 of the adjacent protein chain interacting with the thiol group and contributing with Pro38, Met79, Pro80, and Trp35 to form a hydrophobic cavity (Figure 4a). Cys39 is buried in the active site cavity (ASA = 2.5 Å²) and surrounded by hydrophobic residues, such as Pro38, Ile42, Phe46, Met79, Pro80, and Pro82. However, the thiol group of this residue is involved in hydrogen bonds with Thr32, Ala33, Cys36, and the highly conserved Asp30 mediated by a water molecule (W14) (Figure 4a). This water molecule is structurally conserved in SoTRXm and SoTRXf [31], CrTRXf2 [46], and TRX from *E. coli* [47].

Taken together, these results indicate that both catalytic cysteines, though having different solvent exposure and surrounding microenvironment, form several conserved hydrogen bonds with neighboring residues. However, the deprotonation state of each catalytic Cys cannot be easily derived from structural observations and the real contribution of these interactions in the formation and stabilization of Cys thiolate(s) has to be experimentally determined.

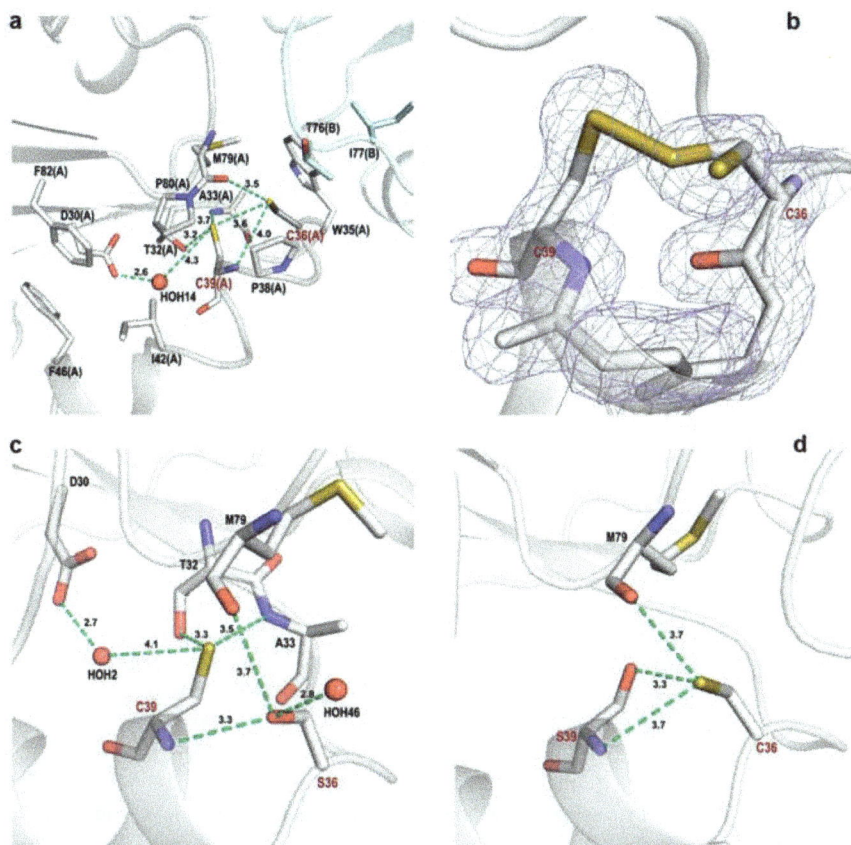

Figure 4. Active site of CrTRXh1. (**a**) Representation of red-CrTRXh1 active site. Residues are shown as sticks. The interactions of Cys36 (C36) and Cys39 (C39) are shown and the corresponding distances are indicated. (**b**) Representation of ox-CrTRXh1 active site and the corresponding $2F_o - F_c$ electron density map (contoured at 1.5 σ). The catalytic cysteines are represented as sticks. (**c**) Representation of C36S mutant active site. Residues are shown as sticks. The interactions of Ser36 (S36) and Cys39 (C39) are shown and the corresponding distances are indicated. (**d**) Representation of C39S mutant active site. Residues are shown as sticks. The interactions of Cys36 (C36) are shown and the corresponding distances are indicated.

3.5. C36S and C39S Mutants vs Wild-Type Reduced CrTRXh1

To gain further insights into the active site microenvironment, we structurally characterized Cys-to-Ser mutants of each catalytic Cys. Crystals of C36S (hereafter defined dimeric C36S) and C39S mutants were isomorphous to the wild-type protein crystal, and their 3D structures were solved at a resolution of 1.22 Å and 1.81 Å, respectively (Table 1). A new crystal form (orthorhombic, Table 1) was obtained, uniquely, for C36S mutant. This polymorph (hereafter defined monomeric C36S) diffracted at a maximum resolution of 0.94 Å, and it contains a single protein chain in the asymmetric unit. This result is quite unusual, since h-type crystal structures from different organisms, such as HvTRXh1 and h2 [28] and PtTRXh4 [30], show a non-covalent dimer in the asymmetric unit, except in the case of C61S mutant of PtTRXh4 [30]. Two chains in the asymmetric unit are also found in the crystal structure of other TRX isoforms from photosynthetic organisms, such as SoTRXf (short form; [31]), ox- and red-CrTRXm [31], and CrTRXf2, described in a companion paper of this journal issue [46], while one monomer is observed in the case of SoTRXf (long form; [31]) and cyanobacterial TRX 2 from *Anabaena* [48]. For the spinach enzyme, the different crystal packing was attributed to an inherent flexibility of its active site and to its interaction with the flexible N-terminal portion. Moreover, it has been proposed that in certain tissues where TRX concentration increases, the enzyme dimerization may have a regulatory role [49].

The different crystal packing does not affect the overall structure of C36S mutant. Indeed, the superimposition between dimeric and monomeric C36S give a rmsd ranging between 0.54 and 0.57 Å (on 110 aligned C_α atoms). A major difference is observed at the C-terminal part of α-helix 1 (residues 18–25) and in the linker region between α2 and β3 (residues 50–57), which are both involved in the interaction with symmetry-related molecules, as well as in the linker regions between β1 and α2 (residues 33–35), β3 and α3 (residues 63–70), and α4 and β4 (residues 77–80), all involved in the interface of the crystallographic dimer (Figure 3b). The comparison between red-CrTRXh1 and the mutants indicates higher differences in the case of C36S (rmsd values ranging from 0.49 and 0.55 Å for monomeric C36S, and from 0.15 and 0.69 Å for dimeric C36S) with respect to C39S (rmsd values ranging from 0.16 and 0.47 Å). These observations indicate that the mutation of Cys36 causes a higher perturbation of protein fold compared to Cys39 mutation. Main deviations are localized at the interface regions in the case of dimeric C36S, while at the C-terminal part of α-helix 1 and in the linker region between α2 and β3 in the case of monomeric C36S.The side chain of Ser36 is involved in several intramolecular hydrogen bonds with the carbonyl group of Met79 and the thiol and the amide group of Cys39, and it is further stabilized by a water molecule in the case of monomeric C36S (Figure 4c). Therefore, it appears that the mutation of Cys36 side chain with a more hydrophilic group perturbs the dimer interface favoring monomeric form.

Consistently with the minor structural alterations observed in Cys mutants compared to red-CrTRXh1, the position and the hydrogen-bond network of the unchanged catalytic Cys thiol groups are retained (Figure 4c,d).

3.6. CrTRXh1 Alkylation by Iodoacetamide is pH-Sensitive

In order to evaluate the presence of Cys thiolate(s) in reduced CrTRXh1, we first examined the sensitivity to iodoacetamide (IAM), a Cys alkylating agent that preferentially alkylates thiolates rather than protonated Cys [8,50]. As shown in Figure 5a, we observed a rapid and complete inhibition of TRX activity (i.e., NADPH/NTR/TRX-dependent DTNB reduction) after exposure to IAM used at 2.5- or 5-fold molar excess. A semi-logarithmic plot reveals that, in our conditions, inactivation kinetics were linear in the 0–3 min range (Figure 5a, inset).

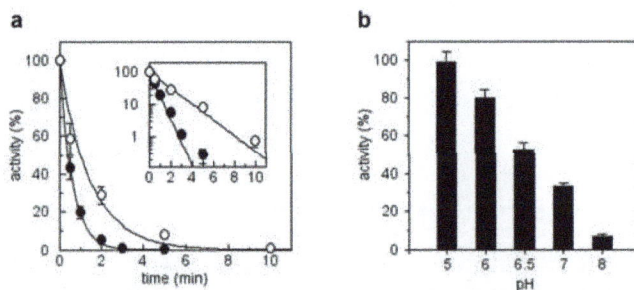

Figure 5. Alkylation treatment of CrTRXh1. (**a**) Reduced enzyme (20 µM) was incubated with 0.05 or 0.1 mM IAM (open and closed circles, respectively). Aliquots of the incubation mixtures were withdrawn at the indicated time points and the remaining TRX-dependent activity was determined. Data are reported as mean ± standard deviation (SD) (*n* = 3). When error bars are not visible, they are within the symbol. (**b**) The reduced enzyme was incubated at the indicated pH values in the presence of 0.1 mM. After 2 min incubation, an aliquot of the sample was withdrawn for the assay of enzyme activity. Data are reported as mean ± SD (*n* = 3).

To determine the pH-dependence of CrTRXh1 inactivation by IAM, we measured residual TRX activity following incubation with 5-fold molar excess IAM at different pH values. As expected, the inhibition of CrTRXh1 activity increased with increasing pH (i.e., increasing deprotonation of catalytic Cys) (Figure 5b). At pH 5, we observed no inhibition, while CrTRXh1 retained ~10% residual activity when alkylation occurred at pH 8 (Figure 5b). In the 6–7 pH range, the residual activity of CrTRXh1 progressively decreased, being around 50% at pH 6.5 (Figure 5b). Taken together, these results indicate that CrTRXh1 active site contains at least one Cys thiolate whose alkylation triggers protein inactivation.

3.7. pK_a Determination of CysN and CysC by Quantitative MALDI-TOF Mass Spectrometry

MALDI-TOF mass spectrometry (MS) was already used to determine pK_a values of both active site cysteines of *E. coli* TRX [51]. Nevertheless, this method is based on MS analysis of tryptic-digested peptides after alkylation using the isotope-coded *N*-phenyl iodoacetamide. Recombinant CrTRXh1 contains only two cysteines and reaction of the fully reduced protein with IAM should increase its parental mass (11,712.6 Da) by 57 and 114 Da for single and double alkylation, respectively. Here, we employed MALDI-TOF MS coupled to IAM treatment to separate precisely at the protein level the parental and alkylated forms of CrTRXh1 with no further steps (Figure 6a).

In a first set of experiments, reduced CrTRXh1 was alkylated with a 10-fold excess of IAM at three different pH values (5.0, 7.0, and 9.0) (Figure S3). In agreement with activity assays (Figure 5), CrTRXh1 was almost insensitive to alkylation at pH 5.0 (Figure S3a). By contrast, at neutral pH, we observed that one active site cysteine was fully alkylated even for short incubation times (30 s), whereas the alkylation of the second residue appeared after 5 min, and progressively increased at a very slow rate (Figure S3b). Consistently, monoalkylation was extremely rapid at pH 9.0, and the peak corresponding to the doubly alkylated form started accumulating after 30 s, being around 80% and 100% after 5 and 20 min, respectively (Figure S3c). These preliminary data showed that CrTRXh1 could be entirely mono- or dialkylated, indicating that quantitative data for non-alkylated forms of CrTRXh1 were not biased by oxidized forms (i.e., non-reactive towards IAM). Moreover, these results indicate that the IAM/TRX ratio should be adapted to quantify, separately, the deprotonation state of the two active site cysteines. Considering that monoalkylation of CrTRXh1 was extremely rapid at both pH 7.0 and 9.0 in the presence of 10-fold excess of IAM, we evaluated the alkylation profile at IAM/TRX ratio of 2. Under these conditions, the occurrence of monoalkylated CrTRXh1 was linear within the 0–2 min range (Figure S4a,c).

The quantitative analyses of non-alkylated and alkylated forms (i.e., reduced and single or double alkylated species, respectively) at different IAM/TRX ratios were then exploited to determine the pH-dependence of CysN/CysC alkylation, and thus derive pK_a values. Before proceeding toward pK_a determination, we verified the Cys site in monoalkylated CrTRXh1 using MS and MS/MS analyses after trypsin digestion. As shown in Figure 6b, we demonstrated that the carbamidomethylated Cys in the monoalkylated form corresponded exclusively to CysN. By extension, we can firmly assume that alkylation of CysC occurs at higher IAM/TRX ratio correlating with the appearance of the dialkylated forms.

In order to measure the reactivity of CysN, CrTRXh1 was incubated for 105 seconds at varying pH values using an IAM/TRX ratio of 2 (Figure 7a). By plotting the percentage of reduced CrTRXh1 (i.e., non-alkylated) against the pH, a sigmoidal titration curve was obtained allowing extrapolation of a pK_a value for CysN of 6.63 ± 0.13 (Figure 7b).

a

b

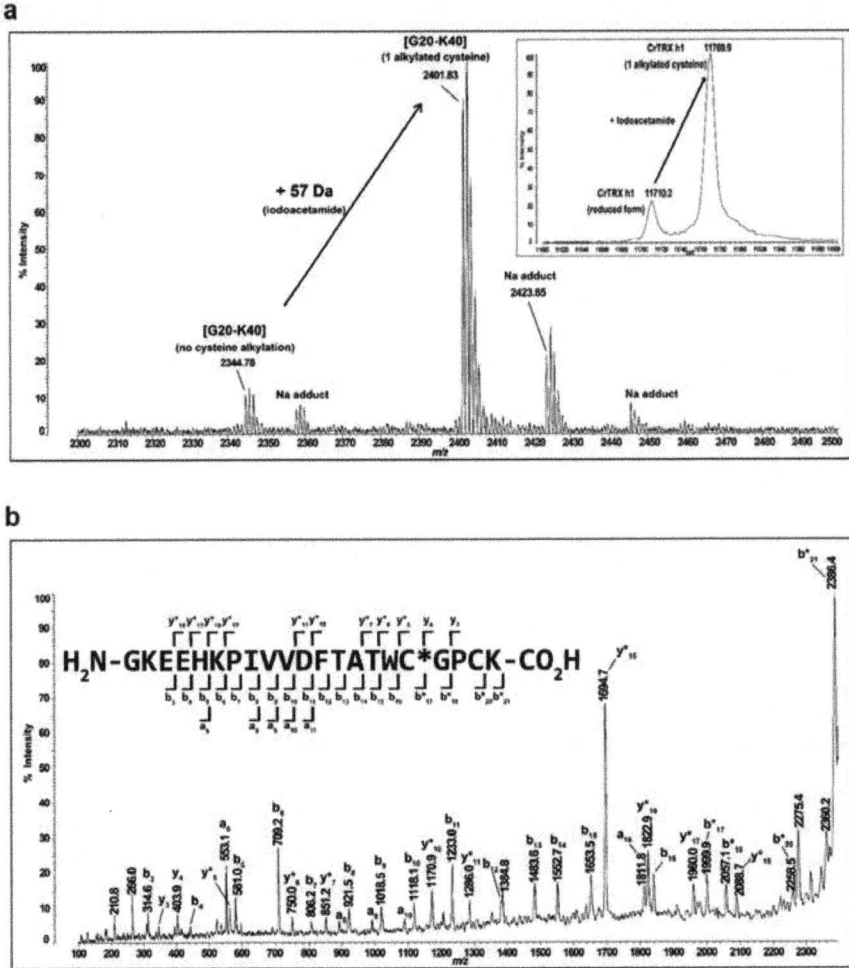

Figure 6. Mass spectrometry analysis of monoalkylated CrTRXh1. (**a**) Matrix-assisted laser desorption/ionization time-of-flight (MALDI-TOF) spectrum after tryptic digestion was determined for reduced CrTRXh1 after treatment with 2-fold molar excess of IAM at pH 7.1 (incubation time: 105 s). The active site containing peptide (Gly20–Lys40) is indicated (experimental mass of 2344.78 Da). The 57 Da shift after treatment with iodoacetamide (IAM) indicates monoalkylation of CrTRXh1 (experimental mass of 2401.83 Da). Inset, MALDI-TOF spectrum of intact protein was determined for reduced CrTRXh1 after treatment with IAM as described above. The experimental masses of 11,710.2 Da and 11,769.9 Da correspond to parental and monoalkylated enzyme, respectively. (**b**) Fragmentation spectrum of the monoalkylated peptide (Gly20–Lys40). The peptide ion (precursor mass at 2401.83 Da) was selected for fragmentation within the MALDI-TOF/TOF mass spectrometer by high-energy collision-induced dissociation using helium gas as collider. Fragment ions are annotated using the classical convention (y-ions when the charge is retained at the C-terminal side, b-ions when the charge is retained at the N-terminal side of the peptide sequence and a-ions correspond to a −28 Da carbonyl loss compared to b-ions). An asterisk (*) is used for fragment ions comprising the carbamidomethylated cysteine. The mass difference between y^*_5 and y_4 ions, on the one hand, and b^*_{17} and b_{16} ions, on the other hand, correspond exactly to 160 Da. This identifies unambiguously the CysN as carbamidomethylated.

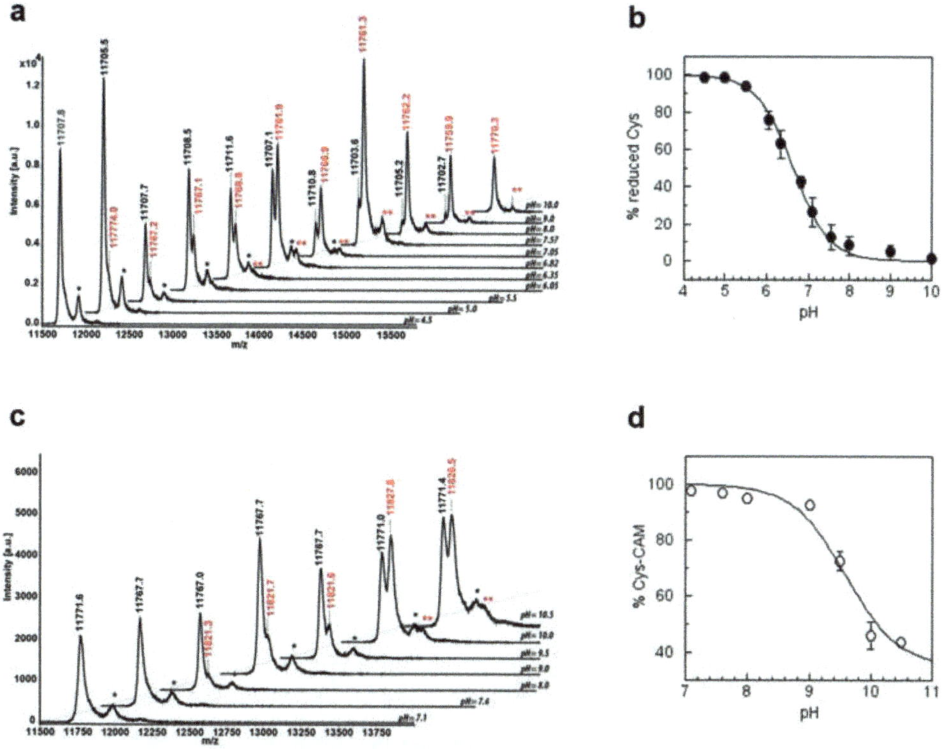

Figure 7. pK_a determination of active site cysteines. (**a**) Reduced CrTRXh1 was incubated with 2-fold excess of IAM at pH values ranging from 4.5 to 10 for 105 s before the reaction was quenched (see Material and Methods). Parental/monoalkylated (CrTRXh1-SH/CrTRXh1-CAM) protein ratios were determined by digital integration of MALDI-TOF mass spectrometry peaks labeled in black (CrTRXh1-SH) and red (CrTRXh1-CAM). The spectra are displayed as a function of pH. Matrix adducts for CrTRXh1-SH (one black asterisk) and CrTRXh1-CAM (two red asterisks) are indicated. (**b**) The fraction of reduced CrTRXh1 was measured by mass spectrometry as described in (**a**). The pK_a values was obtained by nonlinear regression using an adaptation of the Henderson–Hasselbalch equation [50]. Experimental data are displayed as mean ± SD (*n* = 3) and fitted to a full sigmoid. (**c**) Reduced CrTRXh1 was treated with 2-fold excess of IAM at pH 7 prior to incubation at pH values ranging from 4.5 to 10 for 2 min in the presence of 10-fold excess of IAM. After incubation, the reaction was quenched as in (**a**). Monoalkylated/dialkylated (CrTRXh1-CAM/CrTRXh1-CAM-CAM) protein ratios were determined by digital integration of MALDI-TOF mass spectrometry peaks labeled in black (CrTRXh1-CAM) and red (CrTRXh1-CAM-CAM). The spectra are represented as a function of pH. Matrix adducts for CrTRXh1-CAM (one black asterisk) and CrTRXh1-CAM-CAM (two red asterisks) are indicated. (**d**) The fraction of monoalkylated CrTRXh1 was measured by mass spectrometry as described in (**c**). The pK_a values were obtained by nonlinear regression as described above. Experimental data are displayed as mean ± SD (*n* = 2) and fitted to a full sigmoid.

A similar procedure was exploited to evaluate the reactivity of CysC. The protein was first monoalkylated (2 min incubation at pH 7.0 with 2-fold excess of IAM), and then shifted to different pH buffers for additional 2 min in the presence of 10-fold excess of IAM (Figure 7c). The percentage of monoalkylated CrTRXh1 plotted against pH yielded a partial sigmoid titration curve from which a pK_a value of 9.53 ± 0.12 could be estimated for CysC (Figure 7d). Overall, these results indicate that CysN is mainly found in the nucleophilic thiolate form under physiological conditions, while CysC is fully protonated. The lack of intrinsic reactivity of CysC would suggest that the acquisition of a certain nucleophilicity could be related to conformational changes likely occurring during TRX–target complex formation.

4. Discussion

In all free-living organisms, TRXs play a conserved function that mainly consists of reducing protein disulfides. The catalytic activity of these enzymes depends on multiple factors strictly correlated to the structural features that control both the reactivity of active site cysteines and the specific recognition of target proteins. In order to get insight into structural-related catalytic features, we compared the 3D structures of plant TRXs and we observed an identical spatial folding, despite the low sequence conservation. This striking structural similarity is also accompanied by an almost equal content of secondary structures with only a few exceptions, mainly related to the length of α-helix 1. By focusing on the structural characterization of CrTRXh1, we observed that the folding of the protein remains almost unchanged between red- and ox-CrTRXh1, indicating that the protein conformation is not influenced by the redox state. An identical situation is also observed by comparing the 3D structures of red-CrTRXh1 and catalytic cysteine mutants. Nevertheless, the mutation of Cys36 alters the surface hydrophilicity of the catalytic site, hindering the formation of the non-covalent dimer always observed in crystals of wild-type CrTRXh1 and C39S mutant. This observation suggests that the catalytic CysN might be essential for the monomer–dimer equilibrium possibly involved in TRX regulation [49].

Based on the catalytic mechanism underlying TRX-dependent disulfide reduction, the reactivity of the Cys couple is crucial for TRX activity. Indeed, CysN performs a nucleophilic attack on the target disulfide forming a transient mixed disulfide with the target cysteine, whereas CysC acts as resolving Cys by reducing the TRX–target mixed-disulfide, yielding the reduced target and oxidized TRX. The deprotonation state of reactive Cys depends upon multiple factors, including the proximity of basic/acid residues, a hydrogen-bond network, and the dipole of α-helices [15]. In order to get insight into the microenvironment surrounding the catalytic cysteines, we analyzed the active site microenvironment of red-CrTRXh1. In the red-CrTRXh1, Cys36 is involved in three hydrogen bonds that are supposed to increase its reactivity. The pK_a value of Cys36 was experimentally determined to be ~6.6, using both activity assays and mass spectrometry analyses (Figures 5 and 7). Therefore, we can assume that the deprotonation state strongly supports the ability of Cys36 to act as a thiolate nucleophile under physiological conditions, and that the hydrogen-bond network accounts for the low pK_a measured for Cys36 [16]. By contrast, we showed that Cys39 has pK_a value of ~9.5, though its thiol group is involved in four hydrogen bonds with neighboring residues. Based on these observations, we can question the assumption that the hydrogen-bond network is a major determinant of cysteine reactivity. Indeed, Cys39 constitutes an exception since the surrounding hydrophobic residues negatively affect the thiol proton abstraction typically favored by hydrogen bonds. Considering the unreactive nature of Cys39, we can hypothesize that its participation to the catalysis requires structural alterations of its microenvironment. Localized conformational changes, likely occurring after the formation of the TRX–target complex, could increase the reactivity of this residue [20,52,53].

Conservation of the three-dimensional fold and active site residues account for the functional redundancy of plant TRX family members. However, definite structural features are presumably designed to provide specificity towards target proteins. Mounting evidence supports that the specificity of TRXs towards target proteins relies on protein–protein interactions that are dependent on specific substrate recognition regions (e.g., SRLs) and complementary electrostatic surface potentials [46]. Consistently, we observed that each TRX has distinctive structural features that should guide the recognition of target proteins by facilitating protein–protein interactions. Despite the structural constraints involved in substrate recognition, we can assume that sequential catalytic steps ending with the nucleophilic attack of CysC to the TRX–target mixed disulfide control the catalytic activity of

TRX. How the nucleophilicity (i.e., deprotonation) of this residue can be modulated by the formation of the binary complex between TRX and different protein partners is still an open question. A recent work by Messens and coworkers investigated, by molecular dynamics, the conformational changes occurring after the formation of TRX–target complex [20]. Although modest, the backbones of CysN and its preceding Trp slightly change and interact with the thiol of CysC with a final effect to decrease its pK_a by ~0.6 unit (from 8.9 to 8.3). Considering the crucial role of CysC in closing the catalytic cycle, the extent of deprotonation is thus fundamental to determine the overall catalytic efficiency of disulfide reductase activity of TRXs.

5. Conclusions

In conclusion, TRXs share an overall structural homology responsible for their oxidoreductase activity that primarily consists of controlling protein dithiol/disulfide exchange reactions. Despite this, each TRX isoform displays an accurate recognition of partner proteins that determines a regulatory control of specific metabolic pathways. Structural and biochemical features contribute to determine TRX specificity towards target proteins, and further studies are indeed required to shed light on the interconnected TRX features controlling the specific recognition of oxidized targets by one or multiple TRX isoforms and TRX-target interactions.

Supplementary Materials: The following are available online at http://www.mdpi.com/2076-3921/8/1/10/s1, Figure S1: Schematic representation of the Substrate Recognition Loop in AtTRXh1, HvTRXh1, PtTRXh1, PtTRXh4, AtTRXo1, AtTRXo2, and SoTRXm, Figure S2: Far-UV CD spectra of reduced and oxidized CrTRXh1 in Tris-HCl buffer (30 mM; pH 7.9), Figure S3: MALDI-TOF MS analysis of CrTRXh1 after IAM-dependent alkylation, Figure S4: Kinetic of alkylation of CysN and CysC from CrTRXh1, Table S1: Interface interactions including salt bridges and hydrogen bonds in the reduced and oxidized CrTRX h1 (cut-off distance 4.5 Å).

Author Contributions: Conceptualization, C.H.M., S.F., and M.Z.; Investigation, C.H.M., S.F., J.R., L.G., D.T., and M.Z.; Writing-Original Draft Preparation, C.H.M., S.F., J.R., L.G., and M.Z.; Writing—Review & Editing, C.H.M., S.F., J.R., L.G., D.T., J.H., F.S., P.T., S.D.L., and M.Z.

Funding: This work was supported by the University of Bologna Grant FARB2012 (to S.F., F.S., and M.Z.), by CNRS and Sorbonne Université, by Agence Nationale de la Recherche Grant 17-CE05-0001 CalvinDesign (to C.H.M., J.H., S.D.L.), by LABEX DYNAMO ANR-LABX-011 and EQUIPEX CACSICE ANR-11-EQPX-0008, notably through funding of the mass spectrometry platform of Institut de Biologie Physico-Chimique.

Acknowledgments: The authors thank the IBPC proteomic platform as well as the Necker Hospital Proteomic Platform for providing access to MALDI-TOF mass spectrometers, and European Synchrotron Radiation Facility (ESRF, Grenoble) and Elettra (Trieste) for allocation of X-ray diffraction beam time. M.Z. and D.T. gratefully acknowledge supporting of their work by the University of Bologna (Alma Idea 2017 Program). S.F. thanks the Consorzio Interuniversitario di Ricerca in Chimica dei Metalli nei Sistemi Biologici (CIRCMSB). M.Z. is indebted to Stephan Olomun and Carl Cox for helpful suggestions and stimulating discussions.

Conflicts of Interest: The authors declare no conflict of interest.

Abbreviations

ASA	accessible surface area
BSA	buried surface area
CD	circular dichroism
Cys	cysteine(s)
DTNB	5,5′-dithiobis-2-nitrobenzoic acid
IAM	iodoacetamide
MALDI-TOF	matrix-assisted laser desorption/ionization-time of flight
MS	mass spectrometry
MS/MS	tandem mass spectrometry
SRL	substrate recognition loop
TCEP	Tris(2-carboxyethyl)phosphine
TRX	thioredoxin

References

1. Lemaire, S.D.; Michelet, L.; Zaffagnini, M.; Massot, V.; Issakidis-Bourguet, E. Thioredoxins in chloroplasts. *Curr. Genet.* **2007**, *51*, 343–365. [CrossRef] [PubMed]
2. Michelet, L.; Zaffagnini, M.; Morisse, S.; Sparla, F.; Pérez-Pérez, M.E.; Francia, F.; Danon, A.; Marchand, C.H.; Fermani, S.; Trost, P.; et al. Redox regulation of the Calvin-Benson cycle: Something old, something new. *Front. Plant. Sci.* **2013**, *4*, 470. [CrossRef] [PubMed]
3. Geigenberger, P.; Thormählen, I.; Daloso, D.M.; Fernie, A.R. The Unprecedented Versatility of the Plant Thioredoxin System. *Trends Plant. Sci.* **2017**, *22*, 249–262. [CrossRef] [PubMed]
4. Zaffagnini, M.; Fermani, S.; Marchand, C.H.; Costa, A.; Sparla, F.; Rouhier, N.; Geigenberger, P.; Lemaire, S.D.; Trost, P. Redox Homeostasis in Photosynthetic Organisms: Novel and Established Thiol-Based Molecular Mechanisms. *Antioxid. Redox Signal.* **2018**. [CrossRef] [PubMed]
5. Jung, C.H.; Thomas, J.A. S-glutathiolated hepatocyte proteins and insulin disulfides as substrates for reduction by glutaredoxin, thioredoxin, protein disulfide isomerase, and glutathione. *Arch. Biochem. Biophys.* **1996**, *335*, 61–72. [CrossRef] [PubMed]
6. Benhar, M.; Forrester, M.T.; Hess, D.T.; Stamler, J.S. Regulated protein denitrosylation by cytosolic and mitochondrial thioredoxins. *Science* **2008**, *320*, 1050–1054. [CrossRef] [PubMed]
7. Greetham, D.; Vickerstaff, J.; Shenton, D.; Perrone, G.G.; Dawes, I.W.; Grant, C.M. Thioredoxins function as deglutathionylase enzymes in the yeast *Saccharomyces cerevisiae*. *BMC Biochem.* **2010**, *11*, 3. [CrossRef]
8. Bedhomme, M.; Adamo, M.; Marchand, C.H.; Couturier, J.; Rouhier, N.; Lemaire, S.D.; Zaffagnini, M.; Trost, P. Glutathionylation of cytosolic glyceraldehyde-3-phosphate dehydrogenase from the model plant *Arabidopsis thaliana* is reversed by both glutaredoxins and thioredoxins in vitro. *Biochem. J.* **2012**, *445*, 337–347. [CrossRef] [PubMed]
9. Kneeshaw, S.; Gelineau, S.; Tada, Y.; Loake, G.J.; Spoel, S.H. Selective protein denitrosylation activity of Thioredoxin-h5 modulates plant Immunity. *Mol. Cell* **2014**, *56*, 153–162. [CrossRef]
10. Berger, H.; De Mia, M.; Morisse, S.; Marchand, C.H.; Lemaire, S.D.; Wobbe, L.; Kruse, O. A Light Switch Based on Protein S-Nitrosylation Fine-Tunes Photosynthetic Light Harvesting in Chlamydomonas. *Plant Physiol.* **2016**, *171*, 821–832. [CrossRef]
11. Subramani, J.; Kundumani-Sridharan, V.; Hilgers, R.H.; Owens, C.; Das, K.C. Thioredoxin Uses a GSH-independent Route to Deglutathionylate Endothelial Nitric-oxide Synthase and Protect against Myocardial Infarction. *J. Biol. Chem.* **2016**, *291*, 23374–23389. [CrossRef] [PubMed]
12. Kallis, G.B.; Holmgren, A. Differential reactivity of the functional sulfhydryl groups of cysteine-32 and cysteine-35 present in the reduced form of thioredoxin from Escherichia coli. *J. Biol. Chem.* **1980**, *255*, 10261–10265. [PubMed]
13. Brandes, H.K.; Larimer, F.W.; Geck, M.K.; Stringer, C.D.; Schurmann, P.; Hartman, F.C. Direct identification of the primary nucleophile of thioredoxin f. *J. Biol. Chem.* **1993**, *268*, 18411–18414. [PubMed]
14. Holmgren, A. Thioredoxin structure and mechanism: Conformational changes on oxidation of the active-site sulfhydryls to a disulfide. *Structure* **1995**, *3*, 239–243. [CrossRef]
15. Trost, P.; Fermani, S.; Calvaresi, M.; Zaffagnini, M. Biochemical basis of sulphenomics: How protein sulphenic acids may be stabilized by the protein microenvironment. *Plant. Cell Environ.* **2017**, *40*, 483–490. [CrossRef] [PubMed]
16. Roos, G.; Foloppe, N.; Messens, J. Understanding the pKa of redox cysteines: The key role of hydrogen bonding. *Antioxid. Redox Signal.* **2013**, *18*, 94–127. [CrossRef]
17. Dyson, H.J.; Jeng, M.F.; Tennant, L.L.; Slaby, I.; Lindell, M.; Cui, D.S.; Kuprin, S.; Holmgren, A. Effects of buried charged groups on cysteine thiol ionization and reactivity in Escherichia coli thioredoxin: Structural and functional characterization of mutants of Asp 26 and Lys 57. *Biochemistry* **1997**, *36*, 2622–2636. [CrossRef]
18. Chivers, P.T.; Raines, R.T. General acid/base catalysis in the active site of *Escherichia coli* thioredoxin. *Biochemistry* **1997**, *36*, 15810–15816. [CrossRef]
19. Menchise, V.; Corbier, C.; Didierjean, C.; Saviano, M.; Benedetti, E.; Jacquot, J.P.; Aubry, A. Crystal structure of the wild-type and D30A mutant thioredoxin h of Chlamydomonas reinhardtii and implications for the catalytic mechanism. *Biochem. J.* **2001**, *359*, 65–75. [CrossRef]
20. Roos, G.; Foloppe, N.; Van Laer, K.; Wyns, L.; Nilsson, L.; Geerlings, P.; Messens, J. How thioredoxin dissociates its mixed disulfide. *PLoS Comput. Biol.* **2009**, *5*, e1000461. [CrossRef]

21. Meyer, Y.; Belin, C.; Delorme-Hinoux, V.; Reichheld, J.P.; Riondet, C. Thioredoxin and Glutaredoxin Systems in Plants: Molecular Mechanisms, Crosstalks, and Functional Significance. *Antioxid. Redox Signal.* **2012**. [CrossRef]

22. Lindahl, M.; Mata-Cabana, A.; Kieselbach, T. The Disulfide Proteome and Other Reactive Cysteine Proteomes: Analysis and Functional Significance. *Antioxid. Redox Signal.* **2011**, *14*, 2581–2642. [CrossRef] [PubMed]

23. Buchanan, B.B.; Holmgren, A.; Jacquot, J.P.; Scheibe, R. Fifty years in the thioredoxin field and a bountiful harvest. *Biochim. Biophys. Acta* **2012**, *1820*, 1822–1829. [CrossRef]

24. Pérez-Pérez, M.E.; Mauriès, A.; Maes, A.; Tourasse, N.J.; Hamon, M.; Lemaire, S.D.; Marchand, C.H. The Deep Thioredoxome in *Chlamydomonas reinhardtii*: New Insights into Redox Regulation. *Mol. Plant* **2017**, *10*, 1107–1125. [CrossRef] [PubMed]

25. Lancelin, J.M.; Guilhaudis, L.; Krimm, I.; Blackledge, M.J.; Marion, D.; Jacquot, J.P. NMR structures of thioredoxin m from the green alga *Chlamydomonas reinhardtii*. *Proteins* **2000**, *41*, 334–349. [CrossRef]

26. Peterson, F.C.; Lytle, B.L.; Sampath, S.; Vinarov, D.; Tyler, E.; Shahan, M.; Markley, J.L.; Volkman, B.F. Solution structure of thioredoxin h1 from Arabidopsis thaliana. *Protein Sci.* **2005**, *14*, 2195–2200. [CrossRef]

27. Zannini, F.; Roret, T.; Przybyla-Toscano, J.; Dhalleine, T.; Rouhier, N.; Couturier, J. Mitochondrial Arabidopsis thaliana TRXo Isoforms Bind an Iron(-)Sulfur Cluster and Reduce NFU Proteins In Vitro. *Antioxidants* **2018**, *7*. [CrossRef]

28. Maeda, K.; Hägglund, P.; Finnie, C.; Svensson, B.; Henriksen, A. Crystal structures of barley thioredoxin h isoforms HvTrxh1 and HvTrxh2 reveal features involved in protein recognition and possibly in discriminating the isoform specificity. *Protein Sci.* **2008**, *17*, 1015–1024. [CrossRef]

29. Coudevylle, N.; Thureau, A.; Hemmerlin, C.; Gelhaye, E.; Jacquot, J.P.; Cung, M.T. Solution structure of a natural CPPC active site variant, the reduced form of thioredoxin h1 from poplar. *Biochemistry* **2005**, *44*, 2001–2008. [CrossRef]

30. Koh, C.S.; Navrot, N.; Didierjean, C.; Rouhier, N.; Hirasawa, M.; Knaff, D.B.; Wingsle, G.; Samian, R.; Jacquot, J.P.; Corbier, C.; et al. An atypical catalytic mechanism involving three cysteines of thioredoxin. *J. Biol. Chem.* **2008**, *283*, 23062–23072. [CrossRef]

31. Capitani, G.; Markovic-Housley, Z.; DelVal, G.; Morris, M.; Jansonius, J.N.; Schürmann, P. Crystal structures of two functionally different thioredoxins in spinach chloroplasts. *J. Mol. Biol.* **2000**, *302*, 135–154. [CrossRef] [PubMed]

32. Zaffagnini, M.; Michelet, L.; Massot, V.; Trost, P.; Lemaire, S.D. Biochemical characterization of glutaredoxins from *Chlamydomonas reinhardtii* reveals the unique properties of a chloroplastic CGFS-type glutaredoxin. *J. Biol. Chem.* **2008**, *283*, 8868–8876. [CrossRef] [PubMed]

33. Bedhomme, M.; Zaffagnini, M.; Marchand, C.H.; Gao, X.H.; Moslonka-Lefebvre, M.; Michelet, L.; Decottignies, P.; Lemaire, S.D. Regulation by glutathionylation of isocitrate lyase from *Chlamydomonas reinhardtii*. *J. Biol. Chem.* **2009**, *284*, 36282–36291. [CrossRef] [PubMed]

34. Pace, C.N.; Vajdos, F.; Fee, L.; Grimsley, G.; Gray, T. How to measure and predict the molar absorption coefficient of a protein. *Protein Sci.* **1995**, *4*, 2411–2423. [CrossRef] [PubMed]

35. Kabsch, W. Xds. *Acta Crystallogr. D Biol. Crystallogr.* **2010**, *66*, 125–132. [CrossRef] [PubMed]

36. Evans, P. Scaling and assessment of data quality. *Acta Crystallogr. D Biol. Crystallogr.* **2006**, *62*, 72–82. [CrossRef] [PubMed]

37. Chen, V.B.; Arendall, W.B., 3rd; Headd, J.J.; Keedy, D.A.; Immormino, R.M.; Kapral, G.J.; Murray, L.W.; Richardson, J.S.; Richardson, D.C. MolProbity: All-atom structure validation for macromolecular crystallography. *Acta Crystallogr. D Biol. Crystallogr.* **2010**, *66*, 12–21. [CrossRef]

38. Vagin, A.; Teplyakov, A. Molecular replacement with MOLREP. *Acta Crystallogr. D Biol. Crystallogr.* **2010**, *66*, 22–25. [CrossRef]

39. Murshudov, G.N.; Vagin, A.A.; Dodson, E.J. Refinement of macromolecular structures by the maximum-likelihood method. *Acta Crystallogr. D Biol. Crystallogr.* **1997**, *53*, 240–255. [CrossRef]

40. Emsley, P.; Cowtan, K. Coot: Model-building tools for molecular graphics. *Acta Crystallogr. D Biol. Crystallogr.* **2004**, *60*, 2126–2132. [CrossRef]

41. Adams, P.D.; Afonine, P.V.; Bunkoczi, G.; Chen, V.B.; Davis, I.W.; Echols, N.; Headd, J.J.; Hung, L.W.; Kapral, G.J.; Grosse-Kunstleve, R.W.; et al. PHENIX: A comprehensive Python-based system for macromolecular structure solution. *Acta Crystallogr. D Biol. Crystallogr.* **2010**, *66*, 213–221. [CrossRef] [PubMed]

42. Robert, X.; Gouet, P. Deciphering key features in protein structures with the new ENDscript server. *Nucleic Acids Res.* **2014**, *42*, W320–W324. [CrossRef]

43. Li, W.; Cowley, A.; Uludag, M.; Gur, T.; McWilliam, H.; Squizzato, S.; Park, Y.M.; Buso, N.; Lopez, R. The EMBL-EBI bioinformatics web and programmatic tools framework. *Nucleic Acids Res.* **2015**, *43*, W580–W584. [CrossRef] [PubMed]

44. Maeda, K.; Hägglund, P.; Finnie, C.; Svensson, B.; Henriksen, A. Structural basis for target protein recognition by the protein disulfide reductase thioredoxin. *Structure* **2006**, *14*, 1701–1710. [CrossRef] [PubMed]

45. Bjornberg, O.; Maeda, K.; Svensson, B.; Hagglund, P. Dissecting molecular interactions involved in recognition of target disulfides by the barley thioredoxin system. *Biochemistry* **2012**, *51*, 9930–9939. [CrossRef] [PubMed]

46. Lemaire, S.D.; Tedesco, D.; Crozet, P.; Michelet, L.; Fermani, S.; Zaffagnini, M.; Henri, J. Crystal Structure of Chloroplastic Thioredoxin f2 from *Chlamydomonas reinhardtii* Reveals Distinct Surface Properties. *Antioxidants* **2018**, *7*. [CrossRef] [PubMed]

47. Katti, S.K.; LeMaster, D.M.; Eklund, H. Crystal structure of thioredoxin from Escherichia coli at 1.68 A resolution. *J. Mol. Biol.* **1990**, *212*, 167–184. [CrossRef]

48. Saarinen, M.; Gleason, F.K.; Eklund, H. Crystal structure of thioredoxin-2 from Anabaena. *Structure* **1995**, *3*, 1097–1108. [CrossRef]

49. Weichsel, A.; Gasdaska, J.R.; Powis, G.; Montfort, W.R. Crystal structures of reduced, oxidized, and mutated human thioredoxins: Evidence for a regulatory homodimer. *Structure* **1996**, *4*, 735–751. [CrossRef]

50. Zaffagnini, M.; Bedhomme, M.; Marchand, C.H.; Couturier, J.R.; Gao, X.H.; Rouhier, N.; Trost, P.; Lemaire, S.D. Glutaredoxin S12: Unique properties for redox signaling. *Antioxid. Redox Signal.* **2012**, *16*, 17–32. [CrossRef]

51. Nelson, K.J.; Day, A.E.; Zeng, B.B.; King, S.B.; Poole, L.B. Isotope-coded, iodoacetamide-based reagent to determine individual cysteine pK(a) values by matrix-assisted laser desorption/ionization time-of-flight mass spectrometry. *Anal. Biochem.* **2008**, *375*, 187–195. [CrossRef] [PubMed]

52. Jeng, M.F.; Campbell, A.P.; Begley, T.; Holmgren, A.; Case, D.A.; Wright, P.E.; Dyson, H.J. High-resolution solution structures of oxidized and reduced *Escherichia coli* thioredoxin. *Structure* **1994**, *2*, 853–868. [CrossRef]

53. Qin, J.; Clore, G.M.; Gronenborn, A.M. The high-resolution three-dimensional solution structures of the oxidized and reduced states of human thioredoxin. *Structure* **1994**, *2*, 503–522. [CrossRef]

antioxidants

MDPI

Article

Mitochondrial *Arabidopsis thaliana* TRXo Isoforms Bind an Iron–Sulfur Cluster and Reduce NFU Proteins In Vitro

Flavien Zannini [1], Thomas Roret [1,2], Jonathan Przybyla-Toscano [1,3], Tiphaine Dhalleine [1], Nicolas Rouhier [1] and Jérémy Couturier [1,*]

[1]   Université de Lorraine, Inra, IAM, F-54000 Nancy, France; flavien.zannini@univ-lorraine.fr (F.Z.); thomas.roret@sb-roscoff.fr (T.R.); przybylajonathan@orange.fr (J.P.-T.); tiphaine.dhalleine@univ-lorraine.fr (T.D.); nicolas.rouhier@univ-lorraine.fr (N.R.)
[2]   CNRS, LBI2M, Sorbonne Universités, F-29680 Roscoff, France
[3]   Department of Plant Physiology, Umeå Plant Science Centre, Umeå University, S-90187 Umea, Sweden
*   Correspondence: jeremy.couturier@univ-lorraine.fr; Tel.: +33-372-745-159

Received: 11 August 2018; Accepted: 9 October 2018; Published: 13 October 2018

Abstract: In plants, the mitochondrial thioredoxin (TRX) system generally comprises only one or two isoforms belonging to the TRX h or o classes, being less well developed compared to the numerous isoforms found in chloroplasts. Unlike most other plant species, *Arabidopsis thaliana* possesses two TRXo isoforms whose physiological functions remain unclear. Here, we performed a structure–function analysis to unravel the respective properties of the duplicated TRXo1 and TRXo2 isoforms. Surprisingly, when expressed in *Escherichia coli*, both recombinant proteins existed in an apo-monomeric form and in a homodimeric iron–sulfur (Fe-S) cluster-bridged form. In TRXo2, the [4Fe-4S] cluster is likely ligated in by the usual catalytic cysteines present in the conserved Trp-Cys-Gly-Pro-Cys signature. Solving the three-dimensional structure of both TRXo apo-forms pointed to marked differences in the surface charge distribution, notably in some area usually participating to protein–protein interactions with partners. However, we could not detect a difference in their capacity to reduce nitrogen-fixation-subunit-U (NFU)-like proteins, NFU4 or NFU5, two proteins participating in the maturation of certain mitochondrial Fe-S proteins and previously isolated as putative TRXo1 partners. Altogether, these results suggest that a novel regulation mechanism may prevail for mitochondrial TRXs o, possibly existing as a redox-inactive Fe-S cluster-bound form that could be rapidly converted in a redox-active form upon cluster degradation in specific physiological conditions.

Keywords: mitochondria; thioredoxin; iron–sulfur cluster; redox regulation

---

1. Introduction

Mitochondria are important organelles being notably the site of production of cellular energy in the form of adenosine triphosphate (ATP) through the process of oxidative phosphorylation and being also an important site for the amino acid and lipid metabolisms and for the biosynthesis of many crucial vitamins and cofactors. For instance, mitochondria are crucial for the synthesis of the iron–sulfur (Fe-S) clusters found in numerous essential proteins present in the matrix but also in the cytosol and the nucleus [1–3]. Despite the existence of several energy-dissipating systems, the over-reduction of the mitochondrial electron transport chain (ETC) releases reactive oxygen species (ROS) notably at the level of the complexes I and III [4]. Scavenging enzymes for the superoxide ion and H_2O_2 are present in the mitochondrial matrix. This includes superoxide dismutases and ascorbate- and thiol-dependent peroxidases belonging to the peroxiredoxin (PRX) or glutathione-peroxidase-like

(GPXL) families [4]. These thiol peroxidases use reactive cysteine residues for catalysis and rely either on a glutathione/glutaredoxin (GSH/GRX) system or on a thioredoxin (TRX) system for their regeneration [5–8]. Despite the existence of these scavenging systems, ROS-mediated protein oxidation occurs in this compartment. At the level of some sensitive protein cysteine residues, this leads to reversible modifications such as the formation of sulfenic acid, of disulfide bonds, or of glutathionylated or nitrosylated cysteines. In this respect, it is surprising that the systems devoted to the reduction of these oxidized cysteine forms are not very developed in mitochondria. Among the 30 genes coding for GRXs [9] and the 30 to 45 genes encoding TRX or TRX-like proteins in angiosperms [10,11], only one GRX named GRXS15 [12] and one or two TRXs are found in mitochondria, the majority being present in plastids and in the cytosol/nucleus. The mitochondrial TRX isoforms belong to the TRXh or TRXo classes [13,14]. Unlike most other plant species, *A. thaliana* possesses two TRXo isoforms which add to TRXh2 [15]. Their regeneration should be in principle dependent on the nicotinamide adenine dinucleotide phosphate reduced (NADPH)-thioredoxin reductase (NTR) A and/or B [13,16]. Concerning GRXS15, no in vitro GSH-dependent reductase activity was observed for GRXS15 so far whereas it possesses an oxidase activity being able to oxidize the reduction-oxidation sensitive green fluorescent protein 2 (roGFP2) at the expense of oxidized glutathione (GSSG) [17]. The physiological relevance of this activity is unclear because it is rather thought that GRXS15 participates to the maturation of Fe-S proteins serving as a transfer protein for the exchange of Fe-S clusters from scaffold proteins, on which de novo synthesis occurs, either to other maturation factors or directly to acceptor proteins [12,18]. Hence, TRXs o and h likely represent the major disulfide reductases in plant mitochondria.

Both *A. thaliana* TRXs o have the typical characteristics of TRXs i.e., a molecular weight around 13 kDa and a conserved WCGPC motif that comprises the catalytic cysteines. In their reduced forms, TRXs are rather competent for reducing disulfide bonds in target proteins. Nevertheless, as demonstrated for some peculiar TRXs, they might also reduce S-nitrosylated or S-glutathionylated proteins [19,20]. If TRXs o also possessed such a property, this would help understanding the absence of GRXs with deglutathionylation activity in mitochondria. The molecular function of these TRXs o is unclear and only *A. thaliana* TRXo1 and pea TRXo were studied so far. TRXs o may function in the regeneration of peroxiredoxin IIF [7,8], in the activation by reduction of citrate synthase [21], alternative oxidase (AOX) [22,23], and isocitrate dehydrogenase [24], but also in the deactivation of both mitochondrial succinate dehydrogenase and fumarase [25]. However, at the macromolecular level, the phenotypes of *A. thaliana* mutants for *trxo1* and *trxo2* single mutants or for the double mutant are extremely mild. In one study, no growth defect was observed for these mutants grown on soil under long-day conditions for four weeks [26]. In other studies, the *A. thaliana trxo1* mutant showed an accelerated germination in the presence of salt [27], and a significant reduction in the fresh weight of shoots was visible during the first four weeks of growth whereas the root growth was not affected [25]. At the cellular level, the activity of enzymes of the tricarboxylic acid (TCA) cycle, or associated with it, is deregulated is the *A. thaliana trxo1* mutant and this is accompanied by changes in the amounts of some metabolites, notably citrate, malate, and pyruvate [25]. In fact, several experiments aiming at identifying mitochondrial partners of TRXs identified more than 100 putative targets including all enzymes of the TCA cycle and enzymes involved in many other processes [22,23,28]. In addition, one should also consider as putative partners, proteins forming intra- or intermolecular disulfide bridges [29], and proteins subject to other redox post-translational modifications (sulfenylation, nitrosylation, glutathionylation) as recently repertoried for photorespiratory and associated enzymes [30].

In order to understand whether the two mitochondrial TRXs o from *A. thaliana* could be distinguished by their biochemical properties, we have performed a structure–function analysis of both isoforms. We observed that both proteins bind an Fe-S cluster when expressed as recombinant proteins in *Escherichia coli*. The Fe-S cluster ligation in TRXo2 depends on the cysteines found in the conserved WCGPC motif. The physiological relevance of this observation remains unclear. Another observation that may give hints towards a possible connection with the mitochondrial Fe-S cluster assembly machinery is that apo-TRXs o have the capacity to reduce oxidized NFU4 or NFU5, proteins previously isolated as putative TRX partners and known to participate in the maturation of certain mitochondrial Fe-S proteins [1–3]. Although solving their respective 3D structures indicated that marked differences in the surface charge distribution exist between both proteins, no difference between TRXo1 and TRXo2 was observed in the capacity to reduce NFU proteins.

2. Materials and Methods

2.1. Cloning and Site-Directed Mutagenesis

The sequences coding for the presumed mature form (i.e., devoid of N-terminal targeting sequences) of *A. thaliana* TRXo1 (At2g35010) and TRXo2 (At1g31020) were cloned into the *Nde*I and *Bam*HI restriction sites of both pET12a and pET15b. The cysteines of TRXo2 were individually substituted into serines by site-directed mutagenesis by primer extension using two complementary mutagenic primers [31,32]. In a first round of PCR, two fragments with overlapping ends are generated using TRXo2 forward and mutagenic reverse primers and TRXo2 reverse and mutagenic forward primers, respectively. For the second round of PCR, these fragments were mixed with TRXo2 forward and reverse primers added after 10 PCR cycles to obtain the final product with the desired mutation. The corresponding variants were named TRXo2 C37S and C40S. The sequences coding for the presumed mature forms of AtNFU4 (At3g20970) and AtNFU5 (At1g51390) were cloned into the *Nco*I and *Bam*HI restriction sites of pET3d and the sequence coding for *E. coli* IscS was cloned into the *Nde*I and *Bam*HI restriction sites of pET12a. All primers used in this study are listed in Table S1.

2.2. Heterologous Expression in Escherichia coli and Purification of Recombinant Proteins

For protein production, the *E. coli* BL21 (DE3) strain containing the pSBET plasmid, which allows expression of the transfer ribonucleic acid (tRNA) needed to recognize the AGG and AGA rare codons [33], was co-transformed with recombinant pET3d and pET12a plasmids to produce untagged proteins and with recombinant pET15b plasmids to produce N-terminal His-tagged proteins. The volumes of cultures of *E. coli* transformed cells were progressively increased up to 2.4 L in LB medium at 37 °C supplemented with 50 µg/mL of ampicillin and kanamycin. Protein expression was induced at exponential phase by adding 100 µM isopropyl β-D-thiogalactopyranoside for 4 h at 37 °C. Cultures were then centrifuged for 20 min at 6318 g and the cell pellets were resuspended in about 20 mL of Tris NaCl (30 mM Tris-HCl pH 8.0, 200 mM NaCl) for untagged proteins or TI NaCl (50 mM Tris-HCl pH 8.0, 10 mM imidazole, 300 mM NaCl) for His-tagged proteins and conserved at −20 °C. Cell lysis was performed by sonication (3 × 1 min with intervals of 1 min) and the soluble and insoluble fractions were separated at 4 °C by centrifugation for 30 min at 27,216× *g*.

For His-tagged TRXs o, the soluble fraction was loaded on Ni^{2+} affinity columns (Sigma-Aldrich, St Louis MO, USA). After extensive washing, proteins were eluted by adding 50 mM Tris-HCl pH 8.0, 300 mM NaCl, 250 mM imidazole. The recombinant proteins were concentrated by ultrafiltration under nitrogen pressure and dialyzed (Amicon, YM10 membrane, Merck, Berlington MA, USA) and stored in a 30 mM Tris-HCl pH 8.0 buffer supplemented with 50% glycerol at −20 °C. The purification of His-tagged proteins was performed in aerobic or anaerobic conditions. Anaerobic manipulations were performed in a Jacomex glovebox under a nitrogen atmosphere with oxygen levels below 1 ppm.

For untagged proteins, the soluble fraction was first precipitated by ammonium sulfate from 0 to 40% and then to 80% of the saturation. Both TRXo and NFU proteins precipitated mostly between 40 and 80% of ammonium sulfate saturation and *E. coli* IscS between 0 and 40%. The fractions of interest were subjected to size exclusion chromatography (ACA44 for TRXo and NFU proteins, ACA34 for IscS) equilibrated with a Tris NaCl buffer. After dialysis against a 30 mM Tris-HCl pH 8.0 buffer and concentration, the interesting fractions were loaded to a DEAE (diethylaminoethyl) sepharose column equilibrated with the same buffer. The recombinant TRXo proteins that passed through the DEAE column, were concentrated by ultrafiltration under nitrogen pressure (Amicon cells, YM10 membrane), and were stored in the same buffer in the presence of 50% glycerol at −20 °C. On the contrary, NFU and IscS proteins were retained and eluted using a linear 0–0.4 M NaCl gradient. The purest fractions as judged by sodium dodecyl sulfate polyacrylamide gel electrophoresis (SDS-PAGE) analysis were pooled and dialyzed against 30 mM Tris-HCl pH 8.0 buffer as described above.

Protein concentrations were determined spectrophotometrically using a molecular extinction coefficient at 280 nm of 11,585 M^{-1} cm^{-1} for TRXo1, TRXo2, and NFU5; of 11,460 M^{-1} cm^{-1} for TRXo2 cysteine variants; of 13,075 M^{-1} cm^{-1} for NFU4; and of 41,495 M^{-1} cm^{-1} for IscS, as determined from the molar extinction coefficients of individual tyrosines, tryptophans, and cystines (1490, 5500, and 125 M^{-1} cm^{-1}, respectively) [34].

The other recombinant proteins used in this work e.g., NTRB and GRXS15 from *A. thaliana* have been purified as described previously [35,36].

2.3. Preparation of Apo-TRXs o and IscS-Mediated In Vitro Fe-S Cluster Reconstitution of TRXs o

Under strictly anaerobic conditions in a glove box, 200 μM untagged apo-TRXo isoforms (proteins obtained after aerobic purification) were incubated with 100-fold excess of dithiothreitol (DTT) in Tris NaCl buffer for 1 h. The Fe-S cluster reconstitution was then initiated by the addition of catalytic amount of EcIscS (0.05 fold, 10 μM), five-fold excess of ferrous ammonium sulfate citrate (1 mM) and finally five-fold excess of L-cysteine (1 mM). After 30 min of incubation, the Fe-S cluster-loaded TRXo proteins were desalted on a G25 column equilibrated with Tris NaCl buffer.

2.4. Determination of the Oligomerization State

The oligomerization state of TRXo isoforms was analyzed by size-exclusion chromatography. Samples containing 100 to 300 μg of protein were loaded onto Superdex S200 10/300 columns equilibrated with 30 mM Tris-HCl pH 8.0, 200 mM NaCl and connected to an Äkta purifier system (GE Healthcare). The flow rate was fixed at 0.5 mL min^{-1}, and detection was recorded at 280 and 420 nm. The column was calibrated using a molecular weight standard from Sigma. Protein names, molecular weights and elution volumes are as follows: thyroglobulin 669 kDa, 8.47 mL; apoferritin 443 kDa, 10.51 mL; β-amylase 200 kDa, 11.92 mL; bovine serum albumin 66 kDa, 13.56 mL; carbonic anhydrase 29 kDa, 15.54 mL; cytochrome c 12.4 kDa, 16.82 mL and aprotinin 6.5 kDa, 18.16 mL.

2.5. GRX- and TRX-Mediated Reduction of NFUs

Around 3 mg of NFU proteins were reduced using 10 mM DTT in 200 μL of 30 mM Tris-HCl pH 8.0 buffer for 1 h at 25 °C. The reduced proteins were then desalted on a G25 column pre-equilibrated with 30 mM Tris-HCl pH 8.0 buffer. The redox state of untreated and reduced NFUs was determined by electrospray ionization mass spectrometry analysis as described previously [37], and by a cysteine alkylation assay, which also served for assessing reduction by the TRX and GRX reducing systems. The TRX system reconstituted in vitro comprised 200 μM NADPH, 100 nM NTRB, and 1 or 10 μM untagged TRXo1 or o2. The GSH/GRX system was composed of 200 μM NADPH, 0.1 units GR from bakers'yeast (Sigma-Aldrich, St Louis MO, USA), 1 mM GSH and 1 or 10 μM GRXS15. Each reducing system was incubated at 25 °C for 15 min prior addition of 10 μM of oxidized NFU4/5 in 50 μL of 30 mM Tris-HCl pH 8.0 buffer. The reaction was stopped after 15 min by the addition of one volume of 20% TCA. Protein free thiol groups have been alkylated with methoxyl-PEG (mPEG)-maleimide of 2 kDa as described previously [38], before separating the protein mixtures on non-reducing 15% SDS-PAGE.

2.6. Crystallization, Diffraction Data Collection, Processing, Structure Solution, and Refinement

Crystallization at 4 °C was performed by a microbatch-under-oil method, in which 1 μL of the protein solutions were mixed with an equal volume of the precipitant solution containing 100 mM HEPES pH 7.5 and 20% PEG 8000 for TRXo1 and 100 mM citrate buffer pH 5.6, 30% PEG 4000 and 100 mM ammonium sulfate for TRXo2. Solutions of TRXo1 and TRXo2 had concentrations of

20 mg mL^{-1} and 13 mg mL^{-1}, respectively. Suitable crystals were soaked briefly in crystallization solution supplemented with 20% glycerol and then flash-frozen immediately in a nitrogen stream before data collection. X-ray diffraction experiments were carried out at 100 K on the French Beamline for Investigation of Proteins–Beam Magnet 30A (FIP-BM30A) at the European Synchrotron Radiation Facility, Grenoble, France. The data sets at 1.80 and 1.50 Å for TRXo1 and TRXo2, respectively, were indexed and processed using XDS [39], and scaled and merged with Aimless from the CCP4 program package [40]. The initial protein model of TRXo2 was built in ARP/wARP [41]. The crystal structure of TRXo1 was solved by molecular replacement with the CCP4 suite program (MOLREP) [42], using the structure of TRXo2 as a search model. Structures were refined (Table 1) with phenix.refine implemented in the PHENIX package [43], with visual inspection and manual correction in Coot [44]. Structures of TRXo1 and TRXo2 were refined to a crystallographic R_{work} and R_{free} of 0.22/0.24 and 0.17/0.18, respectively, by using standard protocols, in which 5% of the reflections were used to continuously monitor the R_{free}. The refined structures maintained excellent geometry and showed no outlier in Ramachandran plots. The validation of the crystal structures was performed with MolProbity [45].

Table 1. Data collection and refinement statistics.

Data Collection	AtTRXo1	AtTRXo2
Beam line	FIP-BM30A	
Space group	$P2_12_12_1$	$P6_5$
Cell dimensions		
a, b, c (Å)	37.15; 39.24; 79.34	70.81; 70.81; 35.75
α, β, γ (°)	$\alpha = \beta = \gamma = 90°$	$\alpha = \beta = 90°$ $\gamma = 120°$
Resolution (Å)	39.67−1.80 (1.84−1.80)	35.40−1.50 (1.53−1.50)
R_{merge}	0.122 (0.242)	0.070 (0.432)
R_{meas}	0.131 (0.283)	0.075 (0.500)
R_{pim}	0.049 (0.145)	0.028 (0.251)
No. unique reflections	11,316 (653)	16,473 (814)
Mean I/σI	13.1 (3.7)	20.0 (3.1)
$CC_{1/2}$	0.996 (0.961)	0.997 (0.942)
Completeness (%)	100.0 (100.0)	99.4 (100.0)
Average redundancy	11.4 (7.0)	12.0 (7.3)
Refinement		
Resolution (Å)	39.67−1.80	35.40−1.50
R_{free}/R_{work}	23.96/21.96	17.93/16.74
Total number of atoms	1848	1882
Water	130	92
Crystallographic B-factor		
Overall B-factor (Å^2)	29.16	35.36
B-factor: molecule (Å^2)	28.88	35.35
B-factor: water (Å^2)	32.89	35.47
R.m.s deviations		
Bonds	0.002	0.003
Angles	0.474	0.528
MolProbity analysis		
Clashscore, all atoms	2.91 (99%)	0.00 (100%)
MolProbity score	1.40 (96%)	0.50 (100%)
Protein Data Bank entry	6G61	6G62

Values in parentheses refer to data in the highest-resolution shell. AtTRXo: *Arabidopsis thaliana* thioredoxin-o; FIP-BM30A: French beamline for Investigation of Proteins-Bending Magnet section.

3. Results

3.1. Arabidopsis thaliana TRXo Isoforms Exist in Two Forms upon Expression in E. coli

To characterize the structure–function relationship of TRXo1 and TRXo2, the mature forms of these proteins were expressed in *E. coli* as untagged and His-tagged recombinant proteins by removing respectively 82 and 47 residues at the N-terminus constituting their putative mitochondrial targeting sequence. As observed for several related proteins of the GRX family [37,46], lysed bacterial cells exhibited a slight but visible brownish color typical of the presence of an Fe-S cluster. After purification in aerobic conditions, His-tagged recombinant proteins purified in one step displayed a residual brownish color unlike untagged versions (data not shown). The reason was obviously associated to the oxygen lability of the cluster and to the length of the purification procedure. These observations prompted us to purify both His-tagged TRXs in anaerobic conditions to avoid chromophore degradation. TRXo2 exhibited an UV–vis absorption spectrum comprising a broad shoulder centered at around 400 nm and a band at around 300 nm that prolonged the polypeptide absorption band at 280 nm (Figure 1A). This spectrum is usually typical of the presence of a [4Fe-4S] cluster [47,48]. Analytical gel filtration analyses demonstrated that TRXo2 (theoretical molecular mass of ca 14.5 kDa) separated into two peaks, a major one with an estimated molecular mass of 12.9 kDa and a minor one of 27.7 kDa suggesting that the purified protein solution contained both monomeric and dimeric forms (Figure 1B). Furthermore, the Fe-S cluster, detected by the absorbance at 420 nm, was only associated with the dimeric form (Figure 1B). Using a similar protocol, TRXo1 displayed a nearly identical UV–vis absorption spectrum, albeit absorption bands in the visible region were much less intense (Figure 1C). Moreover, TRXo1 (theoretical mass of ca 14.6 kDa) also separated into two peaks when analyzed by analytical gel filtration, a major one with an estimated molecular mass of 12.9 kDa and a minor one of 24.8 kDa which corresponded to a apomonomer and a holodimer form, respectively (Figure 1D). Altogether, these results indicate that both TRXo isoforms incorporate an Fe-S cluster after heterologous expression in *E. coli*. Despite these similarities, it seems that the Fe-S cluster in TRXo1 is either less-well maturated by the *E. coli* Fe-S cluster assembly machineries or more oxygen-labile as cell lysis is performed in the presence of oxygen.

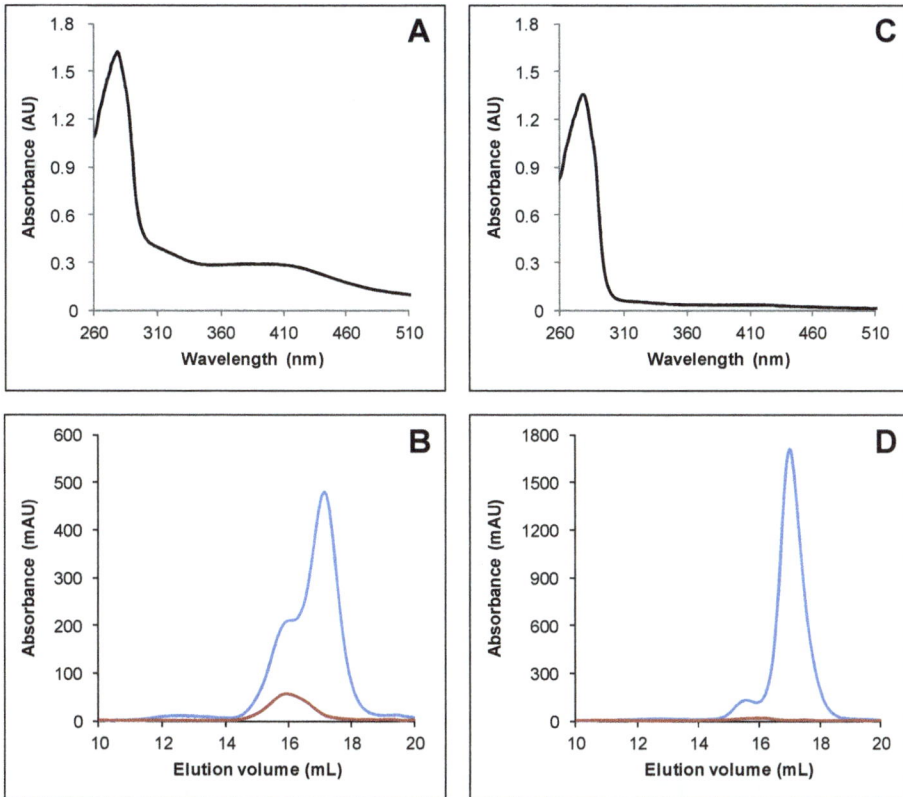

Figure 1. Arabidopsis TRXo (thioredoxin-*o*) isoforms incorporate an Fe-S cluster after heterologous expression in *E. coli*. UV–visible absorption spectra of His-tagged recombinant TRXo2 (**A**) and TRXo1 (**C**). Spectra were recorded after an anaerobic purification in 30 mM Tris-HCl pH 8.0. Analytical gel filtration of His-tagged recombinant TRXo2 (**B**) and TRXo1 (**D**), purified in anaerobic conditions, was performed by loading 100 to 300 µg of protein onto a Superdex S200 10/300 column. The presence of the polypeptide and of the Fe-S cluster have been detected by the absorbance at 280 nm (blue line) and 420 nm (red line), respectively.

3.2. IscS-Mediated In Vitro Fe-S Cluster Reconstitution in Arabidopsis TRXo Isoforms

As purified recombinant Arabidopsis TRXo isoforms existed predominantly under apoforms, we sought to prepare samples by performing in vitro enzymatic Fe-S cluster reconstitution experiments in the presence of the *E. coli* cysteine desulfurase IscS. Untagged apo-TRXo1 and TRXo2 (Figure S1) were incubated anaerobically with an excess of L-cysteine and ferrous ammonium sulfate, and a catalytic amount of IscS for 1 h. Both reconstituted TRXo1 and TRXo2 exhibited an UV–visible absorption spectrum (Figure 2A,B) similar to the one observed for as-purified proteins but with more intense absorption bands around 300 and 400 nm relative to the one at 280 nm. Analytical gel filtration analyses demonstrated that both TRXo1 and TRXo2 still separated into two peaks, one with an estimated molecular mass of 25 kDa and another one of 12.9 kDa (Figure 2C,D). The smaller peak around 14 mL corresponds to IscS protein as deduced from the comparison with the analytical gel filtration performed with an IscS concentration similar to the one used for reconstitution (Figure S2). Although the peak containing the dimeric TRX holoforms was more prominent, monomeric forms were still present in the reconstituted samples. The presence of an absorbance at 420 nm associated with the monomeric TRXo1 may be due to the fact that thiol groups of DTT present in the reconstitution mixture served as ligands. These results indicated that IscS-mediated in vitro reconstitution increased the content of Fe-S clusters in both TRXo isoforms but we could not find conditions to get fully-repleted Fe-S cluster-loaded forms that would allow an accurate analytical titration of Fe and labile sulfide contents. A rough evaluation of the Fe-S cluster content in TRXs o was performed from the absorption band at 410 nm by comparison with data obtained with a fully replete ferredoxin-thioredoxin reductase (FTR) an enzyme, which also contains a [4Fe-4S] center. Indeed, the UV–visible absorption spectra of FTRs purified from different organisms consistently exhibited an A_{410}/A_{280} ratio of 0.42 and molar extinction coefficient values at 410 nm of 17,400 and 15,000 M^{-1} cm^{-1} were determined for spinach and corn FTRs [49]. From the comparable A_{410}/A_{280} ratio in the UV–visible absorption spectra of TRXs o and FTR and an averaged molar extinction coefficient value at 410 nm of 16,000 M^{-1} cm^{-1} as determined for FTRs, we could deduce that there is ca 30 μM of Fe-S cluster in each TRX o sample formed by a mixture of monomer and dimer. On the other hand, using the theoretical molar extinction coefficients at 280 nm, we estimated that TRX o samples contained approximately 85–90 μM TRXo monomers (Figure 2A,B). Hence, considering that the Fe-S cluster is bridged into TRX o dimers, we concluded that about two-thirds of the proteins are replete with an Fe-S cluster.

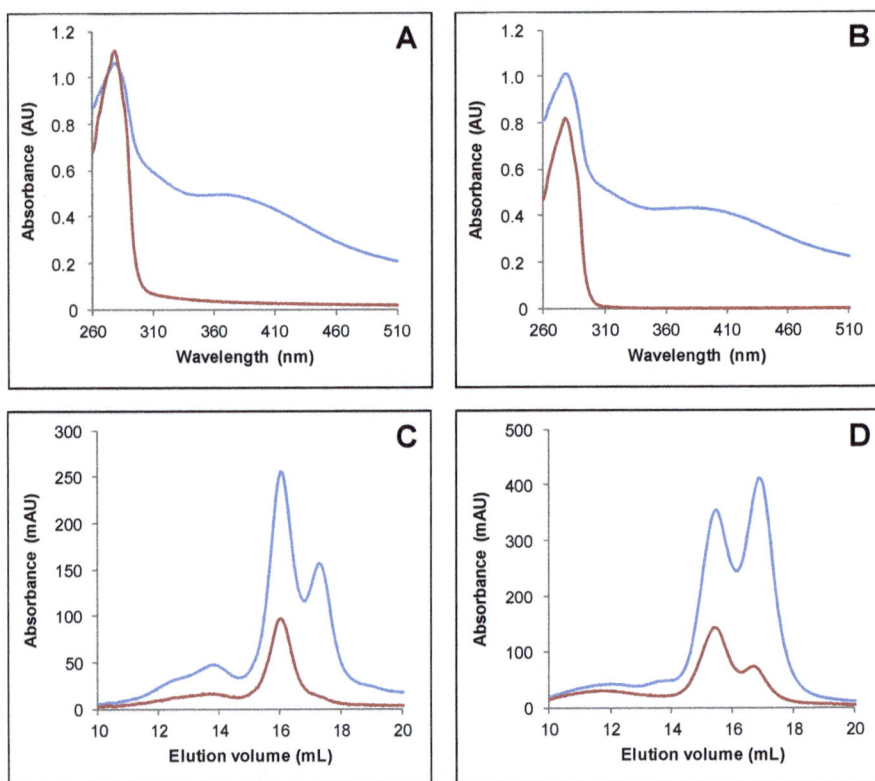

Figure 2. IscS-mediated in vitro Fe-S cluster reconstitution of Arabidopsis TRXo isoforms. UV–visible absorption spectra of TRXo2 (**A**) and TRXo1 (**C**) before (red line) and after (blue line) an anaerobic reconstitution performed in the presence of IscS in Tris NaCl buffer. Analytical gel filtration of reconstituted TRXo2 (**B**) and TRXo1 (**D**) was performed by loading 100 to 300 μg of protein (including 10 μM of EcIscS) onto a Superdex S200 10/300 column. The presence of the polypeptide and of the Fe-S cluster have been detected by the absorbance at 280 nm (blue line) and 420 nm (red line), respectively.

3.3. Both Active Site Cysteines of TRXo2 Are Required for Fe-S Cluster Incorporation

The existence of Fe-S cluster-bridged TRX isoforms were previously observed for the atypical TRX isoform (IsTRP) from the human pathogen *Echinococcus granulosus* [50], and *E. coli* TrxA variants with CACC and CACA active site motifs [51]. The cysteines of the active site signature of IsTRP proved to be essential for cluster binding. Among potential Fe-S cluster ligands, both TRXs o have in common the two active site cysteines but no extra cysteine. To investigate the role of these cysteines for Fe-S cluster ligation, His-tagged C37S and C40S variants of TRXo2 were purified under anaerobiosis. Unlike TRXo2, no coloration was visible on as-purified TRXo2 C37S and C40S variants and accordingly no other absorption band than the one centered at 280 nm was detected on their UV–vis absorption spectra (Figure 3A). Analytical gel filtration experiments revealed that the TRXo2 C37S variant separated as a single peak with an apparent molecular mass of 12.2 kDa likely corresponding to an apomonomer, whereas a small additional peak (15.8 mL, 24.9 kDa) was observed for the TRXo2 C40S variant, which likely contained a covalent dimeric form with Cys^{37} of two monomers forming an intermolecular disulfide (Figure 3B,C). Altogether these results suggest that the substitution of Cys^{37} and Cys^{40} hampered Fe-S cluster incorporation in TRXo2 underlying that both cysteines are required for Fe-S cluster binding.

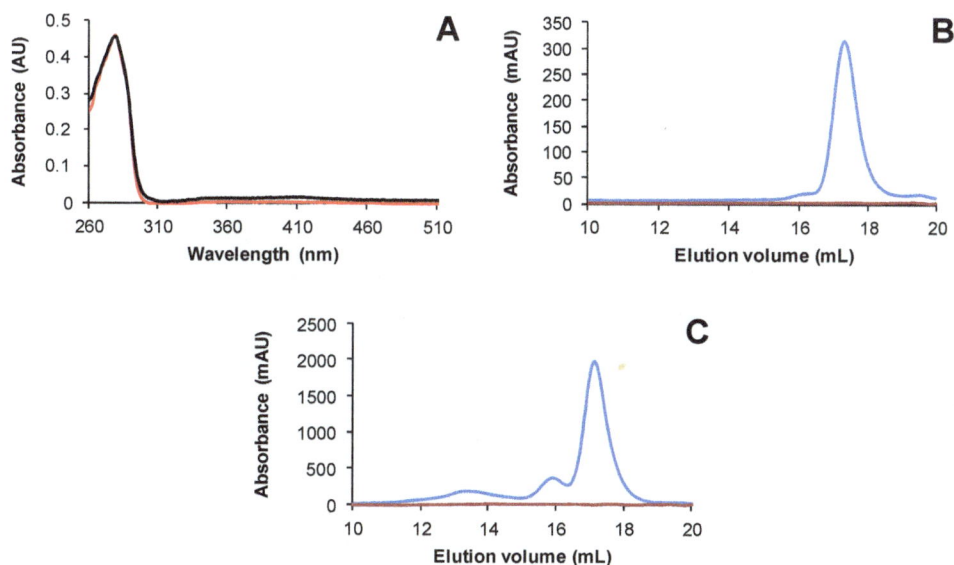

Figure 3. Monocysteinic variants of Arabidopsis TRXo2 isoform are mostly apo-monomers. UV–visible absorption spectra of His-tagged recombinant TRXo2 C37S (red line) and C40S (black line) (**A**) recorded after an anaerobic purification in 30 mM Tris-HCl pH 8.0. Analytical gel filtration of His-tagged recombinant TRXo2 C37S (**B**) and C40S (**C**) purified in anaerobic conditions, was performed by loading 100 to 300 μg of protein onto a Superdex S200 10/300 column. The presence of the polypeptide and of the Fe-S cluster have been detected by the absorbance at 280 nm (blue line) and 420 nm (red line), respectively.

3.4. TRXo1 and TRXo2 Possess the Structural Properties of TRX Family Members

To definitely confirm the identity of the Fe-S cluster ligands and decipher the other structural features of TRXo isoforms for which there is no known 3D structure solved so far, we tried to crystallize under nitrogen atmosphere both TRXo1 and TRXo2 holoforms. Crystals have been obtained for both proteins but it turned out that they were not colored and contained only apoforms. TRXo1 crystallized in the orthorhombic space group $P2_12_12_1$ whereas TRXo2 crystallized in the hexagonal system and belonged to space group $P6_5$. Well-diffracting crystals of TRXo1 and TRXo2 were obtained with one molecule per asymmetric unit (a.s.u) with Matthews's coefficient (Vm) values of 2.3 $Å^3$ Da^{-1} and 2.0 $Å^3$ Da^{-1} corresponding to solvent contents of 46% and 40%, respectively. Their structures were determined by molecular replacement at high resolution of 1.80 and 1.50 Å, respectively. In both protein structures, almost all residues were well defined in the electron density map, except the first two residues (Met^1 and Glu^2) of TRXo1, which were disordered (amino acid numbering refers to the mature protein without its mitochondrial presequence).

Both proteins adopt the overall TRX fold ($\beta\alpha\beta\alpha\beta\alpha\beta\beta\alpha$) with all regular secondary structure elements preserved and consisting of a central core with a twisted five-stranded β-sheet in which β4 is antiparallel to the others (Figure 4A,B). This β-sheet is capped on one side by α1 and α3 helices and on the other side by α2 and α4 helices (Figure 4B). All secondary structures of these proteins (TRXo1 numbering) are β1 (Val^5-Val^8), α1 (Ser^{10}-Gln^{22}), β2 (Ser^{28}-Thr^{33}), α2 (Gly^{38}-Tyr^{54}), β3 (Thr^{58}-Asp^{63}), α3 (Gly^{68}-Leu^{76}), β4 (Thr^{83}-Lys^{88}), β5 (Ser^{91}-Val^{97}), and α4 (Asp^{100}-Lys^{113}). TRXs generally contain a buried Asp^{31} near Cys^{40} residue, which is replaced by Tyr^{31} in *Arabidopsis* TRXo isoforms as in *Anabaena* Trx2 [52]. The position of the sidechain hydroxyl of Tyr^{31} corresponds very closely to a water-mediated hydrogen bonding between the classical Asp and the Cys^{40} Sγ. In TRXo1 and TRXo2 X-ray structures, the distance between the sidechain hydroxyl of the Tyr^{31} and the Cys^{40} Sγ atom are 3.76 and 5.51 Å, respectively, too far away for an interaction.

The Cys^{37} and Cys^{40} residues are found at the end of the α2-helix, the Cys^{40} residue being buried compared to Cys^{37}. Close examination of the resultant Fo-Fc electron density maps revealed a negative density between Cys^{37} and Cys^{40} in TRXo2, and additional positive density in the opposite sides of Sγ atoms. Refinements, using PHENIX and Coot in an iterative manner, indicated that this effect is an artifact of data collection (radiation damage), due to partial reduction of the disulfide bond by the X-ray beam during data collection. The distances between Cys^{37} and Cys^{40} Sγ atoms are 2.03 and 2.05 Å in TRXo1 and TRXo2, respectively. They represent the typical length of disulfide bonds, suggesting that both TRXo isoforms were crystallized in an oxidized form. Hence, we modeled all the cysteine pairs as forming intramolecular disulfide bonds (Figure 4C).

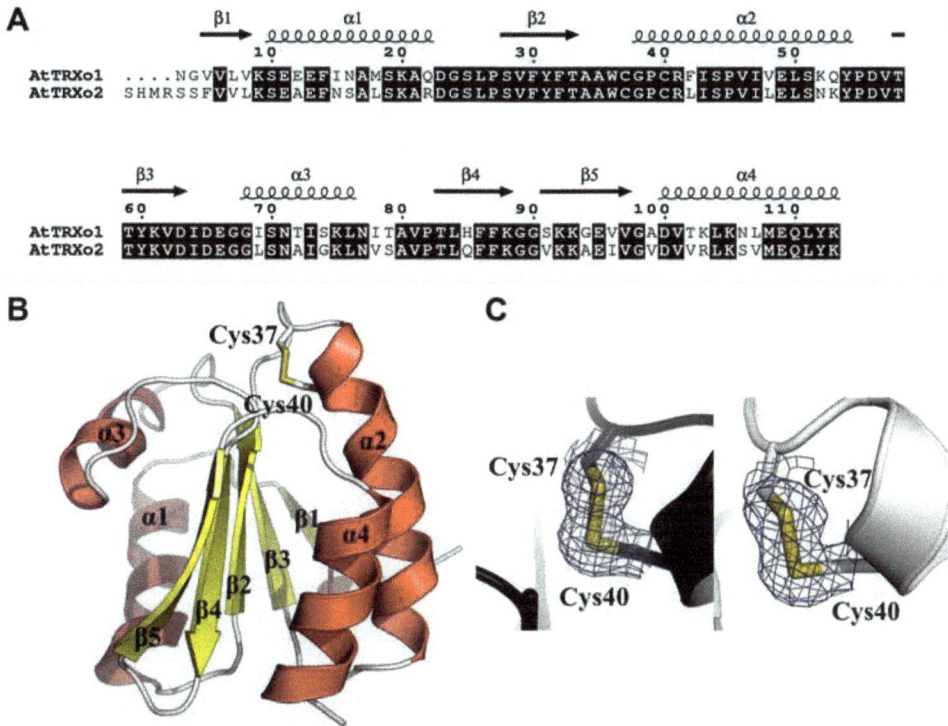

Figure 4. Three-dimensional structure and sequence conservation of oxidized Arabidopsis TRXo isoforms. (**A**) Structure-based sequence alignment of Arabidopsis TRXo isoforms. Conserved residues are highlighted in black. (**B**) Three-dimensional structure of AtTRXo2 at 1.50 Å resolution. The X-ray structure of AtTRXo2 is shown as a ribbon representation with helices in red and strands in yellow. In addition, the side chains of Cys^{37} and Cys^{40} residues are shown as sticks. (**C**) Electron density around the Cys^{37}-Cys^{40} disulfide bond. The maps shown are σA-weighted $2mF_o$-DF_c maps contoured at 1.2σ (0.07 and 0.38 e/Å^3 for AtTRXo1 and AtTRXo2, respectively). AtTRXo1 and AtTRXo2 are colored in white and black, respectively.

3.5. TRXo1 and TRXo2 Structures Are Not Strictly Superimposable

TRXo1 and TRXo2 share high levels of sequence identity and similarity (74% and 97%, respectively) (Figure 4A) and their crystal structures are quite similar and overlay well (RMSD of 0.9 Å) (Figure 5A). Nevertheless, a fine comparison of both structures revealed two regions presenting significant structural differences. The first area corresponds to the active site loop, which protrudes on one side of the molecule to the N-terminus of α2-helix (Figure 5A). The region containing Trp^{36}-Phe^{42} residues including the WCGPC signature in TRXo1 has a different conformation to that in TRXo2 and is shifted by approximately 1.9 Å. Trp^{36} and Arg^{41} residues are involved in contacts with symmetric molecules through hydrogen bond stabilization. In both proteins, the carbonyl oxygen of Trp^{36} forms a hydrogen bond with the Nζ atom of Lys^{105} (2.81 Å in TRXo1 and 3.15 Å in TRXo2). The NH1 guanidinium nitrogen of Arg^{41} is hydrogen-bonded to the carbonyl oxygen of Ser^{74} (2.85 Å) in TRXo1 and to the carbonyl oxygen of Ala^{12} (3.50 Å) in TRXo2. This regional difference around the active site is near crystal contacts and may be attributable to different crystal packing interactions. On another note, the fully refined TRXo1 structure contains 130 water molecules whereas the TRXo2 structure includes 92 water molecules, of which 51 have equivalent positions in TRXo1. Despite the high degree of conservation of both primary and tertiary structures between TRXo1 and TRXo2, only 39% and

55% of the water molecule positions are conserved indicating probable differences in the charge state distribution. Even though TRXo1 was crystallized at pH 7.5 whereas TRXo2 was crystallized at pH 5.6, which may explain differences in the solvation shell, TRXo2 would be more positively charged than TRXo1 in a pH range between 5 and 10 (Figure 5B). With a pH value of 8.1 ± 0.2 for the mitochondrial matrix of *A. thaliana* [53], the global charge of TRXo1 and TRXo2 in vivo should be quite different with values of -1 and $+2$, respectively. The differences in the distribution of the electrostatic potential on the TRXo1 and TRXo2 surfaces at pH 8.1 are however not located in the area surrounding the active site (Figure 5C). The second divergent area involves Gly^{68}-Thr^{79} residues (Figure 5A) with an RMSD of 2.6 Å. This area surrounds the α3-helix upstream of the conserved *cis*-Pro^{82} residue, which is located at the N-terminus of β4-strand. The increase of the crystallographic B-factor along the amino acid chain for TRXo1 (65.33 Å^2) and TRXo2 (68.91 Å^2) (Figure 5D), indicates that this part is flexible. Altogether, this may hint to a capacity of TRXo isoforms to interact with different partner proteins.

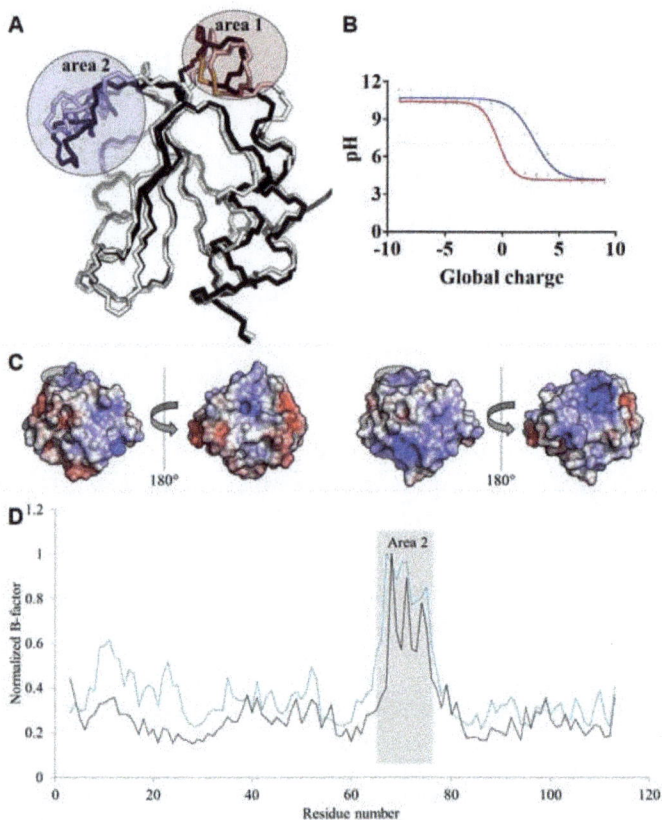

Figure 5. Structural features *vs.* charge and crystallographic B-factor distribution. (**A**) Structure superposition of the backbone trace of Arabidopsis TRXo isoforms. AtTRXo1 and AtTRXo2 are colored in white and black, respectively. The two divergent areas are circled in red and blue, respectively. (**B**) Global protein charge of AtTRXo1 (red) and AtTRXo2 (blue) as a function of pH as indicated by the PDB2PQR Server [54]. The dot line corresponds to pH 7.0. (**C**) Electrostatic potential mapped onto AtTRXo1 (left) and AtTRXo2 (right) structures at pH 8.1. The WCGPC signature residues are circled in black. (**D**) Flexibility of AtTRXo1 (black) and AtTRXo2 (blue) related to the crystallographic B-factor.

3.6. Oxidized NFU4 and NFU5 Are Reduced by TRXo Isoforms but Not by GRXS15

With the aim of confirming novel TRXo partners, we examined the 101 putative TRXo1 partners identified in a previous proteomic analysis performed by affinity chromatography using a TRXo1 C40S variant and mitochondrial protein extracts [22]. The DTT-dependent elution of these proteins suggests that these TRX partners were trapped because they existed under oxidized forms, which led to the formation of mixed-disulfide intermediates with the TRXo1 variant. Among these proteins, we focused our attention on two proteins, NFU4 and NFU5, involved in the late steps of the maturation of Fe-S proteins and acting as Fe-S cluster transfer proteins. These proteins possess a CxxC motif, the cysteines of which being responsible of the transient ligation of the Fe-S cluster. Thus, they have to be reduced to receive the Fe-S cluster from other maturation factors.

The mature forms of both NFU proteins have been expressed in *E. coli* and purified aerobically. As expected, signs for the presence of an Fe-S cluster were initially visible, but both NFU4 and NFU5 were found as apoproteins at the end of the purification. They have three cysteines including those of the CxxC motif. To determine their oxidation state, mass spectrometry analyses were performed with NFU4 and NFU5 either untreated or reduced by a DTT excess and dialyzed (Figures S3 and S4). A single species was obtained for both untreated proteins but it presented a mass decrease of ca 131 Da compared to their theoretical molecular masses (Table 2). This difference corresponds undoubtedly to the cleavage of the first methionine as expected from the presence of an alanine at the second position. An increase of around 2 Da in the molecular masses of reduced proteins likely corresponded to the gain of two protons indicating the existence of an intramolecular disulfide bond in as-purified proteins (Table 2).

Table 2. Electrospray ionization mass spectrometry analysis of untreated and reduced NFU proteins

Protein	Theoretical Size (Da)	Theoretical Size without Met (Da)	Untreated	Treated with DTT	Mass Difference upon Reduction (Da)
NFU4	22,167.9	22,036.7	22,035.1	22,037.4	+2.3
NFU5	21,823.5	21,692.3	21,690.9	21,693.0	+2.1

The mass accuracy is generally ± 0.5 Da.

To examine the ability of the various reducing systems found in mitochondria to reduce oxidized NFU proteins, untreated NFU4 and NFU5 were incubated with TRX or GSH/GRX reducing systems. Subsequent alkylation of thiol groups with 2 kDa mPEG maleimide and separation on non-reducing SDS-PAGE allowed visualizing the redox state of the proteins (Figure 6). After 15 min reaction, the NADPH/GR/GSH system was clearly unable to reduce oxidized NFU proteins and adding the sole mitochondrial GRX, GRXS15, did not improve the reduction. On the contrary, in the presence of NADPH and NTR, both TRXo isoforms reduced completely oxidized NFU proteins when added at equimolar concentrations (Figure 6). Adding more catalytic amounts of TRXs o by decreasing the relative concentrations of TRXs o vs. NFUs to 1:10 still allowed an efficient reduction of both NFUs, although this was not complete (Figure S5). In the presence of a NADPH/NTR regeneration system, GRXS15 was still not able to reduce oxidized NFUs (Figure S5). These in vitro data indicated that the reduction of oxidized mitochondrial NFU isoforms, would this oxidation occur under specific physiological conditions, would depend on the TRX system but not on GSH/GRX or NTR/GRX systems. No difference between both TRXs o was visible using this assay.

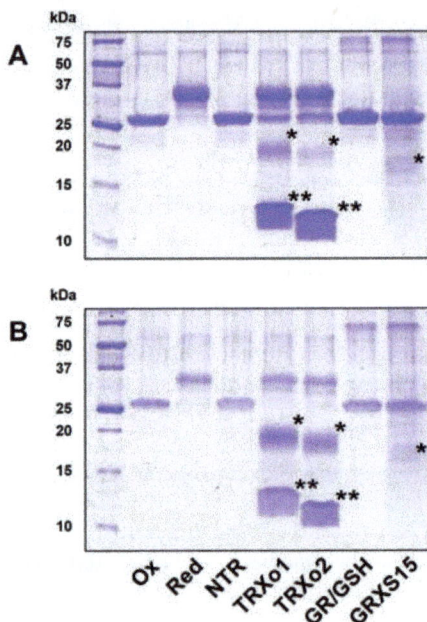

Figure 6. The disulfide bridge of mitochondrial NFUs (nitrogen-fixation-subunit-Us) is reduced by TRXs o but not GRXS15. The reduction of as-purified, oxidized forms of NFU4 (**A**) or NFU5 (**B**) was assessed after a 15 min incubation in the presence of the following reducing systems: NTR: NADPH + NTR; TRXo1: NADPH + NTR + TRXo1; TRXo2: NADPH + NTR + TRXo2; GR/GSH: NADPH + GR + GSH ; GRXS15: NADPH + GR + GSH + GRXS15. After alkylation with 2 kDa mPEG maleimide, proteins were separated on non-reducing SDS-PAGE (sodium dodecyl sulfate polyacrylamide gel electrophoresis). Reduced (Red) and oxidized (Ox) proteins served as controls. The stars indicate the alkylated (*) and non-alkylated (**) forms of the oxidoreductases in the respective regeneration systems when visible.

4. Discussion and Conclusions

Compared to plastidial TRX isoforms, the molecular and physiological roles of mitochondrial TRXo isoforms are uncertain although numerous mitochondrial proteins are known or assumed to undergo reversible redox post-translational modifications and a hundred of putative TRXo1 partners have been identified [22]. In *A. thaliana*, TRXo1 is the major TRXo isoform but *A. thaliana trxo1* and *trxo2* single and double mutants have no pronounced phenotype [25,26]. Because previous studies did not provide information about the redundancy or specificity of these isoforms, we have decided to investigate the biochemical and structural properties of both proteins. Given that TRXo1 and TRXo2 have been previously purified as recombinant proteins [22], and possess the regular WCGPC signature found in many TRXs, it was completely unexpected that both recombinant proteins formed homodimers bridging a [4Fe-4S] cluster using cysteines of the active site signature either upon expression in *E. coli* or upon in vitro Fe-S cluster reconstitution experiments. To our knowledge, this is the first report that a non-modified TRX incorporates a [4Fe-4S] cluster. Up to now, only a few reports pointed to the capacity of proteins of the TRX superfamily to bind an Fe-S cluster but never a [4Fe-4S] cluster in regular TRXs. A thioredoxin-related protein from the human pathogen *Echinococcus granulosus*, IsTRP, was shown to bind a [2Fe-2S] cluster using cysteines present in an atypical active site signature (NCFAC) [50]. Two atypical PDI-A isoforms from poplar and Arabidopsis, formed by a single domain and possessing a WCKHC signature, are able to bind a [2Fe-2S] cluster into a homodimer using the cysteines present in this active site signature [55,56]. Several glutaredoxins can bind a [2Fe-2S] cluster into homodimers using the first cysteine of the CxxC/S signature and the thiol group of two glutathione molecules [37,46,57–59] Interestingly, *S. cerevisiae* GRX5 also binds a [4Fe-4S] cluster in the absence of glutathione, using a cysteine present in the C-terminal part [48].

Besides these naturally-existing isoforms, several variants of *E. coli* or human TRX1 were shown to bind various types of Fe-S cluster. By introducing two mutations, W28C and I75C, a mononuclear [FeS4] cluster can be incorporated in *E. coli* Trx1 both in vitro and in vivo using also the two active site cysteines, Cys^{32} and Cys^{35} [60]. By searching proteins that could restore disulfide bond formation when exported to the periplasm of strains lacking the entire periplasmic oxidative pathway, two *E. coli* Trx1 variants with CACC and CACA active site motifs were shown to form Fe-S cluster-bridged homodimers with the cysteines forming the CxC motif being essential for cluster ligation [51]. The structure of an apo Trx1 CACA variant, where the cysteines thought to serve as ligands of the Fe-S cluster are engaged in two intermolecular disulfide bonds, has been solved [61]. It seems that the exposure of the second active site cysteine, generally buried in TRXs, caused by the unraveling of $\alpha2$ helix may be sufficient to enable Fe-S cluster binding. Whether a similar change occurs in the structures of TRXo is unknown. Last but not least, a [4Fe-4S] cluster can be assembled into the hydrophobic core of a monomeric *E. coli* Trx1 [62]. The polypeptide contained several substitutions for introducing the Fe-S cluster binding residues (L24C, L42C, V55C, and L99C) and removing the two catalytic cysteines. The incorporation was achieved into a fully reduced, partially unfolded protein (2 M urea treatment) from a synthetic, preformed tetranuclear Fe-S cluster. In other words, it required drastic changes and conditions. Very interestingly, in human TRX1, which possesses a regular WCGPC signature, the simple mutation of the *cis*-Pro (P75S/T/A) found in the active site of most TRX superfamily members is sufficient to allow incorporation of a [2Fe-2S] cluster into homodimers [63]. Among the five cysteines present in human TRX1, it seems that Cys^{32} and the specific Cys^{73} are important for iron atom binding. Introducing a glutaredoxin active site (CSYC) into human TRX1, without substituting the *cis*-Pro, allowed Fe-S cluster incorporation. Hence, all these mutagenesis studies performed using *E. coli* Trx1 and human TRX1 suggested that the combination of a WCGPC active site motif and a *cis*-Pro prevented the active site cysteines from binding an Fe-S cluster. However, both Arabidopsis TRXs o have these two features and the capacity to bind an Fe-S cluster indicating that other factors come into play. In order to understand the structural factors that favor Fe-S cluster incorporation into TRXs o, we sought to solve their structures but could only get information on the apoforms so far.

Although both TRXs o display globally similar structures, interesting observations have been made when the 28 residues varying between Arabidopsis TRXo isoforms were mapped onto AtTRXo1 X-ray structure. These non-conserved residues are solvent exposed and outside the highly conserved area (Figure 7A) comprising the WCGPC active site motif and residues that surround it. However, half of these non-conserved residues is potentially involved in protein–protein interactions (Figure 7B) as determined by a comparison with already characterized complexes involving thioredoxin homologs. These 14 non-conserved residues are mainly located on the same face of the protein (comprising $\alpha3$-helix, the loop between $\alpha3$ and $\beta4$, $\beta5$-strand, and $\alpha4$-helix) with marked differences in the surface charge distribution (Figure 5C, left panels). The fact that (i) this region is involved in the interaction between TRXs and other proteins [64], (ii) that at mitochondrial pH, TRXo1 (calculated pI of 6.22) would be negatively charged whereas TRXo2 (calculated pI of 10.12) would be positively charged and (iii) that interactions between TRX and their partners are mostly electrostatic could point to a distinct ability of both TRXs o to interact with their partners. However, whether TRXo2 could have specific partners is unknown as only TRXo1 was used to isolate putative partners so far and both isoforms have not been systematically tested with the same proteins. Using the two identified mitochondrial Fe-S cluster maturation factors, NFU4 and NFU5, we could not demonstrate a difference between TRXo1 and TRXo2, both being able to efficiently reduce the intramolecular disulfide formed on NFUs, unlike GRXS15. However, regardless whether a specificity exists, the fact that several ISC factors (NFS1, ISCA4, ISU1) involved in the maturation of Fe-S proteins in mitochondria were identified as putative partners of TRXo1 [22], raises the question of whether TRXo isoforms are involved in the reduction of these proteins in specific conditions where they can become oxidized. Indeed, most proteins involved in the maturation of Fe-S proteins, including NFU4 and NFU5, possess critical cysteine residues that have to be under a reduced state to bind their Fe-S cluster. Interestingly, the recent work that aimed at isolating TRX partners in *Chlamydomonas reinhardtii* identified several proteins belonging to the Fe-S cluster assembly machinery in chloroplasts (SUF machinery) such as NFU1/2/3, SUFB/C/D, and SUFS but also the mitochondrial NFS1 and several Fe-S proteins [65]. This may be fortuitous because these proteins have exposed and reactive cysteines when they do not bind Fe-S clusters but a control of their redox state by TRX isoforms might add a light-dependent regulation to the maturation of Fe-S proteins in chloroplasts. For instance, the reduction of an oxidized form of Arabidopsis GRXS16, a presumed SUF component, is achieved by the plastidial FTR/TRX system [66]. Hence, a TRX-dependent control of the redox state of proteins involved in Fe-S cluster biogenesis may be a unified picture among organelles and organisms.

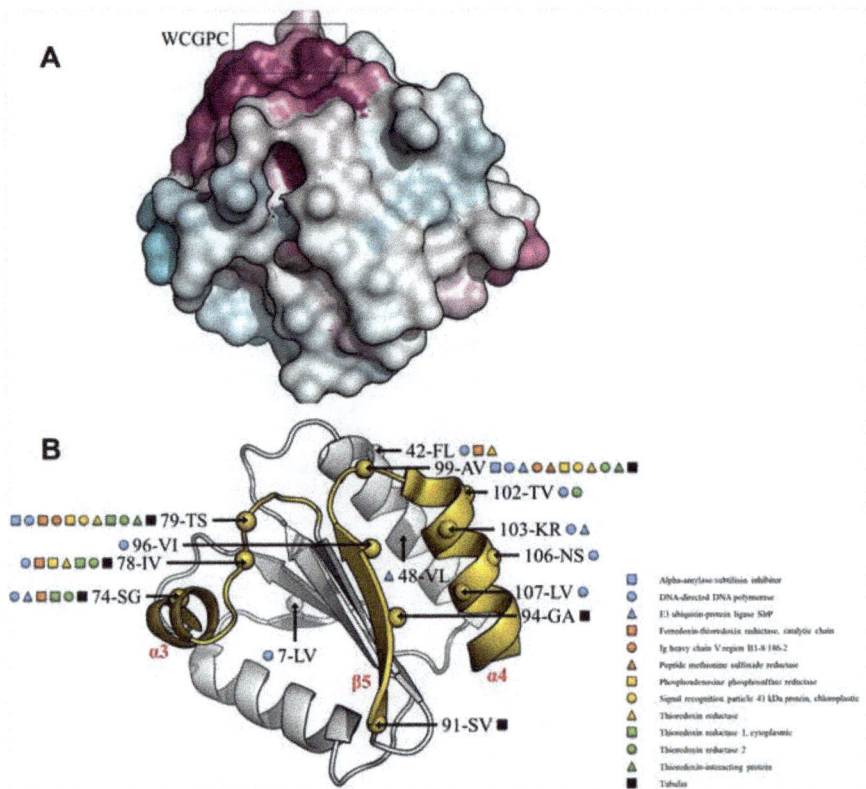

Figure 7. Thioredoxin interfacing residues. (**A**) Residue conservation among 500 thioredoxin orthologs of AtTRXo1 and AtTRXo2 using the ConSurf server [67] with the UniRef90 database (www.uniprot.org/uniref/). Residues are colored in white to purple, for least to most conserved residues. The conserved WCGPC motif is circled in black. (**B**) AtTRXo1 and AtTRXo2 sequences were blasted against the PDB to find protein–protein interactions involving thioredoxin homologs. 44 complexes were found using an E-Value Cutoff of 0.001. AtTRXo1 and AtTRXo2 non-conserved residues potentially involved in protein–protein interactions were mapped onto the X-ray structure of AtTRXo1. In each case the residue position, and the amino acids in one letter code for AtTRXo1 and AtTRXo2 are shown (ex: 78-IV; position 78, isoleucine and valine found in AtTRXo1 and AtTRXo2, respectively). The area comprising α3-helix, the loop between α3 and β4, β5-strand, and α4-helix is colored in yellow.

In summary, this study might point to a connection between TRXo isoforms and the ISC assembly machinery. Although the *E. coli* ISC machinery seems able to perform the maturation of a [4Fe-4S] cluster on both Arabidopsis TRXs o, the maturation is not complete, even though these TRXs are expressed at moderate levels, at least comparable to some Fe-S proteins that are completely maturated. Evidence that the plant ISC system is also competent for this maturation and that an Fe-S cluster is present on these TRXs in a cellular context are now necessary to give more credence to a physiological role of holoforms of TRXs o. The capacity to interact with NFU4/5 and ISCA2 [22], which are the maturation factors responsible for the delivery and insertion of preformed [4Fe-4S] clusters into client proteins, is already a good indication. As the catalytic cysteines constitute the ligands of the Fe-S cluster, TRXo holoforms should not exhibit reductase activity, which was confirmed for IsTRP that is unable to reduce efficiently insulin [52]. Thus, as previously proposed for human Grx2 [68], the formation of an Fe-S cluster on TRXs o may be a convenient regulation mechanism of their reductase activity, at least for a certain pool of proteins. In the absence of a light control as for chloroplastic TRXs, having an inactive pool of labile Fe-S cluster-bridged TRXs may be a rapid and convenient way to adjust the redox metabolism during adverse conditions. In addition to other confirmed stress-responsive proteins—such as PRX IIF [7] and AOX [22]—targets that require TRXo reduction may include ISC maturation factors as NFU4/5.

Supplementary Materials: The following are available online at http://www.mdpi.com/2076-3921/7/10/142/s1, FigureS1: Purity of recombinant AtTRXo1 and AtTRXo2, Figure S2: Analytical gel filtration of *E. coli* IscS, Figure S3: Electrospray ionization mass spectrometry analysis of NFU4, Figure S4: Electrospray ionization mass spectrometry analysis of NFU5, Figure S5: The disulfide bridge of mitochondrial NFUs is reduced by NTR/TRXo system but not by NTR/GRXS15 system, Table S1: Primers used for cloning and site-directed mutagenesis experiments.

Author Contributions: Conceptualization, N.R. and J.C.; Investigation, F.Z., T.R., J.P.-T., and T.D.; Supervision, N.R. and J.C.; Writing—Original draft, F.Z. and T.R.; Writing—Review & Editing, N.R. and J.C.

Funding: This work was supported by a grant overseen by the French National Research Agency (ANR) as part of the "Investissements d'Avenir" program (ANR-11-LABX-0002-01, Lab of Excellence ARBRE).

Acknowledgments: Technical support from Fabien Lachaud of the "Service Commun de Spectrométrie de Masse et Chromatographie" of the Université de Lorraine is gratefully acknowledged.

Conflicts of Interest: The authors declare no conflict of interest.

References

1. Couturier, J.; Touraine, B.; Briat, J.F.; Gaymard, F.; Rouhier, N. The iron–sulfur cluster assembly machineries in plants: Current knowledge and open questions. *Front. Plant Sci.* **2013**, *4*, 259. [CrossRef] [PubMed]
2. Braymer, J.J.; Lill, R. Iron–sulfur cluster biogenesis and trafficking in mitochondria. *J. Biol. Chem.* **2017**, *292*, 12754–12763. [CrossRef] [PubMed]
3. Ciofi-Baffoni, S.; Nasta, V.; Banci, L. Protein networks in the maturation of human iron–sulfur proteins. *Metallomics* **2018**, *10*, 4972. [CrossRef] [PubMed]
4. Navrot, N.; Rouhier, N.; Gelhaye, E.; Jacquot, J.P. Reactive oxygen species generation and antioxidant systems in plant mitochondria. *Physiol. Plant.* **2007**, *129*, 185–195. [CrossRef]
5. Navrot, N.; Collin, V.; Gualberto, J.; Gelhaye, E.; Hirasawa, M.; Rey, P.; Knaff, D.B.; Issakidis, E.; Jacquot, J.P.; Rouhier, N. Plant glutathione peroxidases are functional peroxiredoxins distributed in several subcellular compartments and regulated during biotic and abiotic stresses. *Plant Physiol.* **2006**, *142*, 1364–1379. [CrossRef] [PubMed]
6. Gama, F.; Keech, O.; Eymery, F.; Finkemeier, I.; Gelhaye, E.; Gardeström, P.; Dietz, K.J.; Rey, P.; Jacquot, J.P.; Rouhier, N. The mitochondrial type II peroxiredoxin from poplar. *Physiol. Plant* **2007**, *129*, 196–206. [CrossRef]
7. Finkemeier, I.; Goodman, M.; Lamkemeyer, P.; Kandlbinder, A.; Sweetlove, L.J.; Dietz, K.J. The mitochondrial type II peroxiredoxin F is essential for redox homeostasis and root growth of *Arabidopsis thaliana* under stress. *J. Biol. Chem.* **2005**, *280*, 12168–12180. [CrossRef] [PubMed]
8. Barranco-Medina, S.; Krell, T.; Bernier-Villamor, L.; Sevilla, F.; Lázaro, J.J.; Dietz, K.J. Hexameric oligomerization of mitochondrial peroxiredoxin PrxIIF and formation of an ultrahigh affinity complex with its electron donor thioredoxin Trx-o. *J. Exp. Bot.* **2008**, *59*, 3259–3269. [CrossRef] [PubMed]

9.   Couturier, J.; Jacquot, J.P.; Rouhier, N. Evolution and diversity of glutaredoxins in photosynthetic organisms. *Cell. Mol. Life Sci.* **2009**, *66*, 2539–2557. [CrossRef] [PubMed]

10.   Chibani, K.; Wingsle, G.; Jacquot, J.P.; Gelhaye, E.; Rouhier, N. Comparative genomic study of the thioredoxin family in photosynthetic organisms with emphasis on *Populus trichocarpa*. *Mol. Plant* **2009**, *2*, 308–322. [CrossRef] [PubMed]

11.   Plomion, C.; Aury, J.M.; Amselem, J.; Leroy, T.; Murat, F.; Duplessis, S.; Faye, S.; Francillonne, N.; Labadie, K.; Le Provost, G.; et al. Oak genome reveals facets of long lifespan. *Nat. Plants* **2018**, *1*. [CrossRef] [PubMed]

12.   Moseler, A.; Aller, I.; Wagner, S.; Nietzel, T.; Przybyla-Toscano, J.; Mühlenhoff, U.; Lill, R.; Berndt, C.; Rouhier, N.; Schwarzländer, M.; et al. The mitochondrial monothiol glutaredoxin S15 is essential for iron–sulfur protein maturation in *Arabidopsis thaliana*. *Proc. Natl. Acad. Sci. USA* **2015**, *112*, 13735–13740. [CrossRef] [PubMed]

13.   Laloi, C.; Rayapuram, N.; Chartier, Y.; Grienenberger, J.M.; Bonnard, G.; Meyer, Y. Identification and characterization of a mitochondrial thioredoxin system in plants. *Proc. Natl. Acad. Sci. USA* **2001**, *98*, 14144–14149. [CrossRef] [PubMed]

14.   Gelhaye, E.; Rouhier, N.; Jacquot, J.P. The thioredoxin h system of higher plants. *Plant Physiol. Biochem.* **2004**, *42*, 265–271. [CrossRef] [PubMed]

15.   Meng, L.; Wong, J.H.; Feldman, L.J.; Lemaux, P.G.; Buchanan, B.B. A membrane-associated thioredoxin required for plant growth moves from cell to cell, suggestive of a role in intercellular communication. *Proc. Natl. Acad. Sci. USA* **2010**, *107*, 3900–3905. [CrossRef] [PubMed]

16.   Reichheld, J.P.; Meyer, E.; Khafif, M.; Bonnard, G.; Meyer, Y. AtNTRB is the major mitochondrial thioredoxin reductase in *Arabidopsis thaliana*. *FEBS Lett.* **2005**, *579*, 337–342. [CrossRef] [PubMed]

17.   Begas, P.; Liedgens, L.; Moseler, A.; Meyer, A.J.; Deponte, M. Glutaredoxin catalysis requires two distinct glutathione interaction sites. *Nat. Commun.* **2017**, *8*, 14835. [CrossRef] [PubMed]

18.   Ströher, E.; Grassl, J.; Carrie, C.; Fenske, R.; Whelan, J.; Millar, A.H. Glutaredoxin S15 Is Involved in Fe-S Cluster Transfer in Mitochondria Influencing Lipoic Acid-Dependent Enzymes, Plant Growth, and Arsenic Tolerance in Arabidopsis. *Plant Physiol.* **2016**, *170*, 1284–1299. [CrossRef] [PubMed]

19.   Bedhomme, M.; Adamo, M.; Marchand, C.H.; Couturier, J.; Rouhier, N.; Lemaire, S.D.; Zaffagnini, M.; Trost, P. Glutathionylation of cytosolic glyceraldehyde-3-phosphate dehydrogenase from the model plant *Arabidopsis thaliana* is reversed by both glutaredoxins and thioredoxins in vitro. *Biochem. J.* **2012**, *445*, 337–347. [CrossRef] [PubMed]

20.   Kneeshaw, S.; Gelineau, S.; Tada, Y.; Loake, G.J.; Spoel, S.H. Selective protein denitrosylation activity of thioredoxin-h5 modulates plant immunity. *Mol. Cell* **2014**, *56*, 153–162. [CrossRef] [PubMed]

21.   Schmidtmann, E.; König, A.C.; Orwat, A.; Leister, D.; Hartl, M.; Finkemeier, I. Redox regulation of Arabidopsis mitochondrial citrate synthase. *Mol. Plant* **2014**, *7*, 156–169. [CrossRef] [PubMed]

22.   Yoshida, K.; Noguchi, K.; Motohashi, K.; Hisabori, T. Systematic exploration of thioredoxin target proteins in plant mitochondria. *Plant Cell Physiol.* **2013**, *54*, 875–892. [CrossRef] [PubMed]

23.   Martí, M.C.; Olmos, E.; Calvete, J.J.; Díaz, I.; Barranco-Medina, S.; Whelan, J.; Lázaro, J.J.; Sevilla, F.; Jiménez, A. Mitochondrial and nuclear localization of a novel pea thioredoxin: Identification of its mitochondrial target proteins. *Plant Physiol.* **2009**, *150*, 646–657. [CrossRef] [PubMed]

24.   Yoshida, K.; Hisabori, T. Mitochondrial isocitrate dehydrogenase is inactivated upon oxidation and reactivated by thioredoxin-dependent reduction in Arabidopsis. *Front. Environ. Sci.* **2014**, *2*, 38. [CrossRef]

25.   Daloso, D.M.; Müller, K.; Obata, T.; Florian, A.; Tohge, T.; Bottcher, A.; Riondet, C.; Bariat, L.; Carrari, F.; Nunes-Nesi, A.; et al. Thioredoxin, a master regulator of the tricarboxylic acid cycle in plant mitochondria. *Proc. Natl. Acad. Sci. USA* **2015**, *112*, E1392–E1400. [CrossRef] [PubMed]

26.   Yoshida, K.; Hisabori, T. Adenine nucleotide-dependent and redox-independent control of mitochondrial malate dehydrogenase activity in *Arabidopsis thaliana*. *Biochim. Biophys. Acta* **2016**, *1857*, 810–818. [CrossRef] [PubMed]

27.   Ortiz-Espín, A.; Iglesias-Fernández, R.; Calderón, A.; Carbonero, P.; Sevilla, F.; Jiménez, A. Mitochondrial AtTrxo1 is transcriptionally regulated by AtbZIP9 and AtAZF2 and affects seed germination under saline conditions. *J. Exp. Bot.* **2017**, *68*, 1025–1038. [CrossRef] [PubMed]

28.   Balmer, Y.; Vensel, W.H.; Tanaka, C.K.; Hurkman, W.J.; Gelhaye, E.; Rouhier, N.; Jacquot, J.P.; Manieri, W.; Schürmann, P.; Droux, M.; et al. Thioredoxin links redox to the regulation of fundamental processes of plant mitochondria. *Proc. Natl. Acad. Sci. USA* **2004**, *101*, 2642–2647. [CrossRef] [PubMed]

29. Winger, A.M.; Taylor, N.L.; Heazlewood, J.L.; Day, D.A.; Millar, A.H. Identification of intra- and intermolecular disulphide bonding in the plant mitochondrial proteome by diagonal gel electrophoresis. *Proteomics* **2007**, *7*, 4158–4170. [CrossRef] [PubMed]

30. Keech, O.; Gardeström, P.; Kleczkowski, L.A.; Rouhier, N. The redox control of photorespiration: From biochemical and physiological aspects to biotechnological considerations. *Plant Cell Environ.* **2017**, *40*, 553–569. [CrossRef] [PubMed]

31. Higuchi, R.; Krummel, B.; Saiki, R.K. A general method of in vitro preparation and specific mutagenesis of DNA fragments: Study of protein and DNA interactions. *Nucleic Acids Res.* **1988**, *16*, 7351–7367. [CrossRef] [PubMed]

32. Ho, S.N.; Hunt, H.D.; Horton, R.M.; Pullen, J.K.; Pease, L.R. Site-directed mutagenesis by overlap extension using the polymerase chain reaction. *Gene* **1989**, *77*, 51–59. [CrossRef]

33. Schenk, P.M.; Baumann, S.; Mattes, R.; Steinbiss, H.H. Improved high-level expression system for eukaryotic genes in *Escherichia coli* using T7 RNA polymerase and rare ArgtRNAs. *Biotechniques* **1995**, *19*, 196–200. [PubMed]

34. Gill, S.C.; von Hippel, P.H. Calculation of protein extinction coefficients from amino acid sequence data. *Anal. Biochem.* **1989**, *182*, 319–326. [CrossRef]

35. Jacquot, J.P.; Rivera-Madrid, R.; Marinho, P.; Kollarova, M.; Le Marechal, P.; Miginiac-Maslow, M.; Meyer, Y. *Arabidopsis thaliana* NAPHP thioredoxin reductase. cDNA characterization and expression of the recombinant protein in *Escherichia coli*. *J. Mol. Biol.* **1994**, *235*, 1357–1363. [CrossRef] [PubMed]

36. Bandyopadhyay, S.; Gama, F.; Molina-Navarro, M.M.; Gualberto, J.M.; Claxton, R.; Naik, S.G.; Huynh, B.H.; Herrero, E.; Jacquot, J.P.; Johnson, M.K.; et al. Chloroplast monothiol glutaredoxins as scaffold proteins for the assembly and delivery of [2Fe-2S] clusters. *EMBO J.* **2008**, *27*, 1122–1133. [CrossRef] [PubMed]

37. Couturier, J.; Stroher, E.; Albetel, A.N.; Roret, T.; Muthuramalingam, M.; Tarrago, L.; Seidel, T.; Tsan, P.; Jacquot, J.P.; Johnson, M.K.; et al. Arabidopsis chloroplastic glutaredoxin C5 as a model to explore molecular determinants for iron–sulfur cluster binding into glutaredoxins. *J. Biol. Chem.* **2011**, *286*, 27515–27527. [CrossRef] [PubMed]

38. Zannini, F.; Couturier, J.; Keech, O.; Rouhier, N. In Vitro Alkylation Methods for Assessing the Protein Redox State. *Methods Mol. Biol.* **2017**, *1653*, 51–64. [CrossRef] [PubMed]

39. Kabsch, W. XDS. *Acta Crystallogr. D Biol. Crystallogr.* **2010**, *66*, 125–132. [CrossRef] [PubMed]

40. Winn, M.D.; Ballard, C.C.; Cowtan, K.D.; Dodson, E.J.; Emsley, P.; Evans, P.R.; Keegan, R.M.; Krissinel, E.B.; Leslie, A.G.; McCoy, A.; et al. Overview of the CCP4 suite and current developments. *Acta Crystallogr. D Biol. Crystallogr.* **2011**, *67*, 235–242. [CrossRef] [PubMed]

41. Langer, G.; Cohen, S.X.; Lamzin, V.S.; Perrakis, A. Automated macromolecular model building for X-ray crystallography using ARP/wARP version 7. *Nat. Protoc.* **2008**, *3*, 1171–1179. [CrossRef] [PubMed]

42. Vagin, A.; Teplyakov, A. Molecular replacement with MOLREP. *Acta Crystallogr. D Biol. Crystallogr.* **2010**, *66*, 22–25. [CrossRef] [PubMed]

43. Adams, P.D.; Afonine, P.V.; Bunkoczi, G.; Chen, V.B.; Davis, I.W.; Echols, N.; Headd, J.J.; Hung, L.W.; Kapral, G.J.; Grosse-Kunstleve, R.W.; et al. PHENIX: A comprehensive Python-based system for macromolecular structure solution. *Acta Crystallogr. D Biol. Crystallogr.* **2010**, *66*, 213–221. [CrossRef] [PubMed]

44. Emsley, P.; Lohkamp, B.; Scott, W.G.; Cowtan, K. Features and development of Coot. *Acta Crystallogr. D Biol. Crystallogr.* **2010**, *66*, 486–501. [CrossRef] [PubMed]

45. Chen, V.B.; Arendall, W.B.; Headd, J.J.; Keedy, D.A.; Immormino, R.M.; Kapral, G.J.; Murray, L.W.; Richardson, J.S.; Richardson, D.C. MolProbity: All-atom structure validation for macromolecular crystallography. *Acta Crystallogr. D Biol. Crystallogr.* **2010**, *66*, 12–21. [CrossRef] [PubMed]

46. Rouhier, N.; Unno, H.; Bandyopadhyay, S.; Masip, L.; Kim, S.K.; Hirasawa, M.; Gualberto, J.M.; Lattard, V.; Kusunoki, M.; Knaff, D.B.; et al. Functional, structural, and spectroscopic characterization of a glutathione-ligated [2Fe-2S] cluster in poplar glutaredoxin C1. *Proc. Natl. Acad. Sci. USA* **2007**, *104*, 7379–7384. [CrossRef] [PubMed]

47. Gao, H.; Subramanian, S.; Couturier, J.; Naik, S.G.; Kim, S.K.; Leustek, T.; Knaff, D.B.; Wu, H.C.; Vignols, F.; Huynh, B.H.; et al. *Arabidopsis thaliana* Nfu2 accommodates [2Fe-2S] or [4Fe-4S] clusters and is competent for in vitro maturation of chloroplast [2Fe-2S] and [4Fe-4S] cluster-containing proteins. *Biochemistry* **2013**, *52*, 6633–6645. [CrossRef] [PubMed]

48. Zhang, B.; Bandyopadhyay, S.; Shakamuri, P.; Naik, S.G.; Huynh, B.H.; Couturier, J.; Rouhier, N.; Johnson, M.K. Monothiol glutaredoxins can bind linear [Fe$_3$S$_4$]$^+$ and [Fe$_4$S$_4$]$^{2+}$ clusters in addition to [Fe$_2$S$_2$]$^{2+}$ clusters: spectroscopic characterization and functional implications. *J. Am. Chem. Soc.* **2013**, *135*, 15153–15164. [CrossRef] [PubMed]

49. Droux, M.; Jacquot, J.P.; Miginac-Maslow, M.; Gadal, P.; Huet, J.C.; Crawford, N.A.; Yee, B.C.; Buchanan, B.B. Ferredoxin-thioredoxin reductase, an iron–sulfur enzyme linking light to enzyme regulation in oxygenic photosynthesis: Purification and properties of the enzyme from C3, C4, and cyanobacterial species. *Arch. Biochem. Biophys.* **1987**, *252*, 426–439. [CrossRef]

50. Bisio, H.; Bonilla, M.; Manta, B.; Graña, M.; Salzman, V.; Aguilar, P.S.; Gladyshev, V.N.; Comini, M.A.; Salinas, G. A new class of thioredoxin-related protein able to bind iron–sulfur clusters. *Antioxid. Redox Signal.* **2015**, *24*, 205–216. [CrossRef] [PubMed]

51. Masip, L.; Pan, J.L.; Haldar, S.; Penner-Hahn, J.E.; DeLisa, M.P.; Georgiou, G.; Bardwell, J.C.; Collet, J.F. An engineered pathway for the formation of protein disulfide bonds. *Science* **2004**, *303*, 1185–1189. [CrossRef] [PubMed]

52. Saarinen, M.; Gleason, F.K.; Eklund, H. Crystal structure of thioredoxin-2 from Anabaena. *Structure* **1995**, *3*, 1097–1108. [CrossRef]

53. Shen, J.; Zeng, Y.; Zhuang, X.; Sun, L.; Yao, X.; Pimpl, P.; Jiang, L. Organelle pH in the Arabidopsis endomembrane system. *Mol. Plant* **2013**, *6*, 1419–1437. [CrossRef] [PubMed]

54. Dolinsky, T.J.; Czodrowski, P.; Li, H.; Nielsen, J.E.; Jensen, J.H.; Klebe, G.; Baker, N.A. PDB2PQR: Expanding and upgrading automated preparation of biomolecular structures for molecular simulations. *Nucleic Acids Res.* **2007**, *35*, 522–525. [CrossRef] [PubMed]

55. Selles, B.; Zannini, F.; Couturier, J.; Jacquot, J.P.; Rouhier, N. Atypical protein disulfide isomerases (PDI): Comparison of the molecular and catalytic properties of poplar PDI-A and PDI-M with PDI-L1A. *PLoS ONE* **2017**, *12*, e0174753. [CrossRef] [PubMed]

56. Remelli, W.; Santabarbara, S.; Carbonera, D.; Bonomi, F.; Ceriotti, A.; Casazza, A.P. Iron Binding Properties of Recombinant Class A Protein Disulfide Isomerase from *Arabidopsis thaliana*. *Biochemistry* **2017**, *56*, 2116–2125. [CrossRef] [PubMed]

57. Feng, Y.; Zhong, N.; Rouhier, N.; Hase, T.; Kusunoki, M.; Jacquot, J.P.; Jin, C.; Xia, B. Structural insight into poplar glutaredoxin C1 with a bridging iron–sulfur cluster at the active site. *Biochemistry* **2006**, *45*, 7998–8008. [CrossRef] [PubMed]

58. Johansson, C.; Roos, A.K.; Montano, S.J.; Sengupta, R.; Filippakopoulos, P.; Guo, K.; von Delft, F.; Holmgren, A.; Oppermann, U.; Kavanagh, K.L. The crystal structure of human GLRX5: Iron–sulfur cluster co-ordination, tetrameric assembly and monomer activity. *Biochem. J.* **2011**, *433*, 303–311. [CrossRef] [PubMed]

59. Banci, L.; Brancaccio, D.; Ciofi-Baffoni, S.; Del Conte, R.; Gadepalli, R.; Mikolajczyk, M.; Neri, S.; Piccioli, M.; Winkelmann, J. [2Fe-2S] cluster transfer in iron–sulfur protein biogenesis. *Proc. Natl. Acad. Sci. USA* **2014**, *111*, 6203–6208. [CrossRef] [PubMed]

60. Benson, D.E.; Wisz, M.S.; Liu, W.; Hellinga, H.W. Construction of a novel redox protein by rational design: conversion of a disulfide bridge into a mononuclear iron–sulfur center. *Biochemistry* **1998**, *37*, 7070–7076. [CrossRef] [PubMed]

61. Collet, J.F.; Peisach, D.; Bardwell, J.C.; Xu, Z. The crystal structure of TrxA(CACA): Insights into the formation of a [2Fe-2S] iron–sulfur cluster in an *Escherichia coli* thioredoxin mutant. *Protein Sci.* **2005**, *14*, 1863–1869. [CrossRef] [PubMed]

62. Coldren, C.D.; Hellinga, H.W.; Caradonna, J.P. The rational design and construction of a cuboidal iron–sulfur protein. *Proc. Natl. Acad. Sci. USA* **1997**, *94*, 6635–6640. [CrossRef] [PubMed]

63. Su, D.; Berndt, C.; Fomenko, D.E.; Holmgren, A.; Gladyshev, V.N. A conserved cis-proline precludes metal binding by the active site thiolates in members of the thioredoxin family of proteins. *Biochemistry* **2007**, *46*, 69036910. [CrossRef] [PubMed]

64. Berndt, C.; Schwenn, J.D.; Lillig, C.H. The specificity of thioredoxins and glutaredoxins is determined by electrostatic and geometric complementarity. *Chem. Sci.* **2015**, *6*, 7049–7058. [CrossRef] [PubMed]

65. Pérez-Pérez, M.E.; Mauriès, A.; Maes, A.; Tourasse, N.J.; Hamon, M.; Lemaire, S.D.; Marchand, C.H. The Deep Thioredoxome in Chlamydomonas reinhardtii: New Insights into Redox Regulation. *Mol. Plant* **2017**, *10*, 1107–1125. [CrossRef] [PubMed]

66. Zannini, F.; Moseler, A.; Bchini, R.; Dhalleine, T.; Meyer, A.J.; Rouhier, N.; Couturier, J. The thioredoxin-mediated recycling of *Arabidopsis thaliana* GRXS16 relies on a conserved C-terminal cysteine. *BBA General Subj.* under revision.

67. Landau, M.; Mayrose, I.; Rosenberg, Y.; Glaser, F.; Martz, E.; Pupko, T.; Ben-Tal, N. ConSurf 2005: The projection of evolutionary conservation scores of residues on protein structures. *Nucleic Acids Res.* **2005**, *33*, W299–W302. [CrossRef] [PubMed]

68. Lillig, C.H.; Berndt, C.; Vergnolle, O.; Lönn, M.E.; Hudemann, C.; Bill, E.; Holmgren, A. Characterization of human glutaredoxin 2 as iron–sulfur protein: a possible role as redox sensor. *Proc. Natl. Acad. Sci. USA* **2005**, *102*, 8168–8173. [CrossRef] [PubMed]

antioxidants

MDPI

Article

Determining the Rate-Limiting Step for Light-Responsive Redox Regulation in Chloroplasts

Keisuke Yoshida *[iD] and **Toru Hisabori** *[iD]

Laboratory for Chemistry and Life Science, Institute of Innovative Research, Tokyo Institute of Technology, Nagatsuta 4259-R1-8, Midori-ku, Yokohama 226-8503, Japan
* Correspondence: yoshida.k.ao@m.titech.ac.jp (K.Y.); thisabor@res.titech.ac.jp (T.H.);
  Tel.: +81-45-924-5234 (T.H.)

Received: 17 October 2018; Accepted: 30 October 2018; Published: 31 October 2018

Abstract: Thiol-based redox regulation ensures light-responsive control of chloroplast functions. Light-derived signal is transferred in the form of reducing power from the photosynthetic electron transport chain to several redox-sensitive target proteins. Two types of protein, ferredoxin-thioredoxin reductase (FTR) and thioredoxin (Trx), are well recognized as the mediators of reducing power. However, it remains unclear which step in a series of redox-relay reactions is the critical bottleneck for determining the rate of target protein reduction. To address this, the redox behaviors of FTR, Trx, and target proteins were extensively characterized in vitro and in vivo. The FTR/Trx redox cascade was reconstituted in vitro using recombinant proteins from *Arabidopsis*. On the basis of this assay, we found that the FTR catalytic subunit and *f*-type Trx are rapidly reduced after the drive of reducing power transfer, irrespective of the presence or absence of their downstream target proteins. By contrast, three target proteins, fructose 1,6-bisphosphatase (FBPase), sedoheptulose 1,7-bisphosphatase (SBPase), and Rubisco activase (RCA) showed different reduction patterns; in particular, SBPase was reduced at a low rate. The in vivo study using *Arabidopsis* plants showed that the Trx family is commonly and rapidly reduced upon high light irradiation, whereas FBPase, SBPase, and RCA are differentially and slowly reduced. Both of these biochemical and physiological findings suggest that reducing power transfer from Trx to its target proteins is a rate-limiting step for chloroplast redox regulation, conferring distinct light-responsive redox behaviors on each of the targets.

Keywords: redox regulation; thioredoxin; ferredoxin-thioredoxin reductase; chloroplast

---

1. Introduction

Thiol-based redox regulation is a post-translational mechanism for the control of enzymatic activity through modification of redox-active Cys residues on the target protein (e.g., formation/cleavage of disulfide bonds). A small protein, thioredoxin (Trx), serves as a key factor in redox regulation. Trx has a highly conserved amino acid sequence of WCGPC at its active site, allowing it to reduce disulfide bonds of its target proteins through the dithiol-disulfide exchange reaction. Trx was first identified in *Escherichia coli* in 1964 as a ribonucleotide reductase cofactor [1]. It is now known that the Trx-dependent redox regulation system is ubiquitously preserved in all kingdoms of life and adjusts biological functions in response to changes in local redox environments.

In plant chloroplasts, the redox regulation system is unique in terms of linking to the excitation of photosynthetic electron transport and, thereby, light. Upon illumination, a part of the reducing power generated from photochemical reactions is transmitted from the photosynthetic electron transport chain to the ferredoxin-thioredoxin reductase (FTR)/Trx redox cascade [2–4]. FTR is a soluble [4Fe-4S] protein and acts as a signaling hub linking photosynthetically reduced ferredoxin (Fd) to Trx [5,6].

A reduced form of Trx, in turn, transfers reducing power to a specific set of redox-sensitive target proteins in chloroplasts. Some Calvin-Benson cycle enzymes, including fructose 1,6-bisphosphatase (FBPase) and sedoheptulose 1,7-bisphosphatase (SBPase), are well-known targets; they are activated upon reduction [7]. Consequently, the FTR/Trx redox cascade plays a key role in switching on photosynthetic carbon metabolism in a light-coordinated manner. This is a canonical pathway for chloroplast redox regulation, pioneered by Buchanan and colleagues [2,3].

Since the turn of the century, our understanding of chloroplast redox regulation has been increasingly extended. Progress in plant genomic and proteomic studies has revealed a large number of factors that constitute the redox regulation system in chloroplasts. For example, five Trx subtypes (*f*-, *m*-, *x*-, *y*-, and *z*-type) have been identified in chloroplasts [8,9]. Despite a common active site motif and structure, they show different biochemical characteristics, such as their protein surface charges and midpoint redox potentials [10–12]. Furthermore, several hundreds of chloroplast proteins, which are involved in a broad spectrum of biological processes, have been suggested as potential targets of Trx [13,14]. These emerging data indicate that the redox regulation system is highly organized in chloroplasts, flexibly controlling an array of functions. A number of studies have been directed toward revealing the molecular basis and physiological significance of the redox-based regulatory network in chloroplasts; however, many aspects remain to be elucidated (for recent reviews, see [15,16]).

As described above, the FTR/Trx redox cascade acts as a pivotal pathway in supporting light-responsive redox regulation in chloroplasts. Although the regulatory system itself has been firmly established, the rate-limiting step in the redox-relay process remains elusive. Answering this question is important both to the field of basic biology and biotechnology of plants, because the rate of target protein reduction is linked directly to the activating kinetics of key photosynthetic reactions (including ATP synthesis and the Calvin–Benson cycle) and may ultimately impact on overall photosynthetic performance. To this end, the redox behaviors of FTR, Trx, and target proteins need to be characterized individually. In this study, we addressed this issue from both in vitro and in vivo standpoints, allowing us to identify the bottleneck in light-responsive redox regulation in chloroplasts. The data from this study provide an insight into the working dynamics of chloroplast redox regulation under varying light environments.

2. Materials and Methods

2.1. Preparation of Arabidopsis Recombinant Proteins

All expression plasmids used in this work were constructed during previous studies [12,17,18]. Each expression plasmid was transformed into *E. coli* strain BL21 (DE3) (for ferredoxin-NADP$^+$ reductase (FNR), FTR heterodimer, Trx-*f*1, Trx-*m*1, Trx-*x*, Trx-*y*2, Trx-*z*, FBPase, and SBPase) or Rosetta (DE3) pLysS (for Rubisco activase (RCA; redox-sensitive isoform)). Transformed cells were cultured at 37 °C. The expression was induced by the addition of 0.5 mM isopropyl-1-thio-β-D-galactopyranoside followed by overnight culture at 21 °C. Cells were disrupted by sonication. After centrifugation (125,000× *g* for 40 min), the resulting supernatant was used to purify the protein. All recombinant proteins contained no affinity tag, and they were purified by a combination of anion-exchange chromatography, hydrophobic-interaction chromatography, and size-exclusion chromatography, as described previously [12,17]. The protein concentration was determined with a BCA protein assay (Pierce, Rockford, USA).

2.2. Reconstitution of FTR/Trx Redox Cascade and Determination of Protein Redox State In Vitro

The FTR/Trx redox cascade was reconstituted in vitro as previously described [17] with modifications. All proteins (FNR, Fd (from spinach; Sigma-Aldrich, St. Louis, MO, USA), FTR heterodimer, the indicated Trx isoform, and the indicated target protein) were incubated at 2 μM in medium containing 50 mM Tris-HCl (pH 7.5 or 8.2) and 50 mM NaCl. Reducing power-transferring reactions were initiated by the addition of 1 mM nicotinamide adenine dinucleotide phosphate

(NADPH). This assay was performed at 25 °C. After the reaction, the redox state of the protein was determined by identifying the thiol status with the use of thiol-modifying reagent 4-acetamido-4′-maleimidylstilbene-2,2′-disulfonate (AMS) as previously described [17].

2.3. Plant Materials, Growth Conditions, and High Light Treatments

Arabidopsis thaliana wild-type plants (Col-0) were grown in soil in a controlled growth chamber (light intensity, 70–80 μmol photons m^{-2} s^{-1}; temperature, 22 °C; relative humidity, 60%; 16 h day/8 h night) for four weeks. For high light (HL) treatments, plants were placed in the dark for 8 h and then irradiated at 650–700 μmol photons m^{-2} s^{-1} for the time period indicated.

2.4. Measurement of Photosynthetic Electron Transport Rate

Chlorophyll fluorescence and absorbance change at 830 nm were measured simultaneously using a Dual-PAM-100 (Walz, Effeltrich, Germany) with the intact leaves. A saturating pulse of red light (800 ms, >5000 μmol photons m^{-2} s^{-1}, 635 nm) was applied to calculate the quantum yields of photosystem (PS)II (Y (II)) and PSI (Y (I)). Relative electron transport rates of PSII (ETR II) and PSI (ETR I) were calculated as Y (II) × light intensity and Y (I) × light intensity, respectively.

2.5. Determination of Protein Redox State In Vivo

Plants were frozen using liquid nitrogen, and the redox state of the protein was determined as previously described [19].

3. Results

3.1. The In Vitro Biochemical Study

Firstly, comparative characterization of the reduction kinetics of FTR, Trx, and target proteins was attempted using an in vitro biochemical procedure. The *Arabidopsis* FTR/Trx redox cascade was reconstituted as previously described [17] with modifications; in this study, all proteins were incubated at an equimolar concentration (2 μM each). These changes to the procedure were introduced in order to determine the bottleneck in reducing power transfer at the protein molecule level. The redox state of the protein was identified using the thiol-modifying reagent AMS. AMS binds to free thiols and then lowers the protein mobility on Sodium dodecyl sulfate polyacrylamide gel electrophoresis. (SDS-PAGE), allowing the determination of redox state as a band shift [20].

Three stromal proteins were prepared for use as targets; these included FBPase, SBPase, and RCA. Previous reports have noted that Trx-*f* is the most effective Trx subtype to reduce these targets in the presence of the chemical reductant dithiothreitol (DTT) [10,12,21]. As a preliminary test, whether the preference for Trx-*f* is also true in the reconstituted FTR/Trx system was investigated (Figure S1). FBPase, SBPase, and RCA were efficiently shifted from oxidized to reduced forms by Trx-*f*. These results were obtained both at pH 7.5 and 8.2. On the basis of these data, Trx-*f* was chosen for the following biochemical assays.

Figure 1 shows the time course for redox change in the FTR catalytic subunit (FTR-C), Trx-*f*, and target proteins. These assays were performed at pH 7.5. Under target-free conditions, FTR-C and Trx-*f* were rapidly shifted from oxidized to reduced forms after the onset of the reaction; the redox state of these proteins reached a plateau within 1 min (Figure 1A). Notably, almost identical redox responses of FTR-C and Trx-*f* were observed even in the presence of target proteins (Figure 1B–D). By contrast, largely different reduction rates were seen in the three target proteins. The redox state of FBPase and RCA reached a plateau at 2–5 and 1–2 min, respectively (Figure 1B,D). SBPase was reduced at a fairly low rate; its reduction level was continuously elevated for 15 min (Figure 1C). Taken together, it was suggested that the rate of target protein reduction is primarily determined by the efficiency of reducing power transfer downstream of Trx-*f*. Similar results were observed under different pH conditions (pH 8.2; Figure S2).

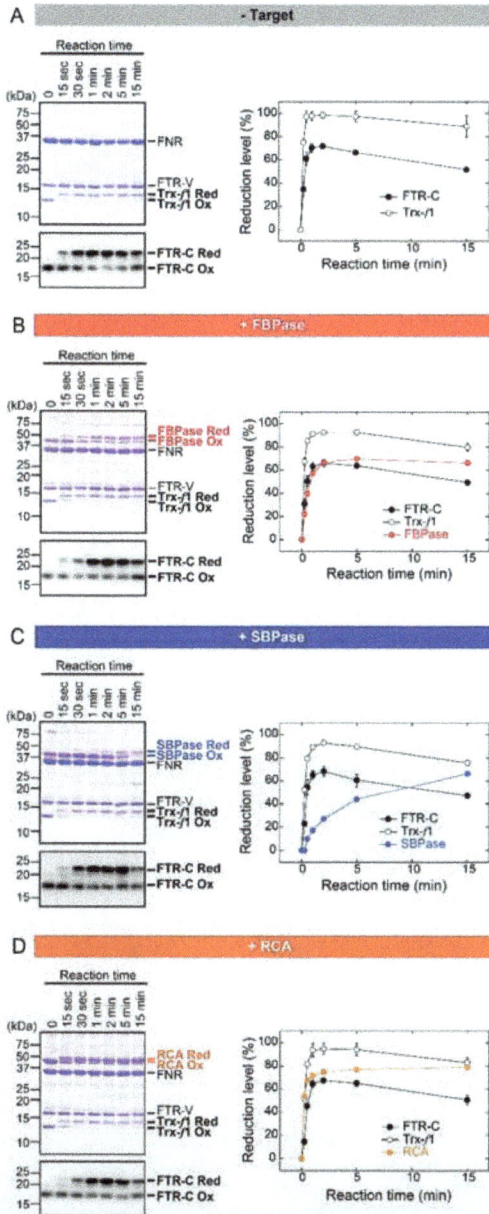

Figure 1. The in vitro redox responses of FTR-C, Trx-*f*1, and target proteins from *Arabidopsis* (pH 7.5). (**A**) Experiments were performed under target-free conditions. (**B–D**) FBPase (**B**), SBPase (**C**), or RCA (**D**) was used as the target. Each of the oxidized proteins (2 μM) was incubated with 1 mM NADPH, 2 μM FNR and 2 μM Fd for the indicated time period. Assays were performed at pH 7.5. Following the reaction, proteins were labeled with AMS and subjected to non-reducing SDS-PAGE. Proteins were then stained with Coomassie Brilliant Blue R-250 (CBB). FTR-C was detected by immunoblotting, as its CBB-derived signal was low. The reduction level of each protein was calculated as the ratio of

the reduced form to the total. Values represent the mean ± SD (n = 3). FTR-V, FTR variable subunit; Red, reduced form; Ox, oxidized form. FTR-C: FTR catalytic subunit; FBPase: Fructose 1,6-bisphosphatase; SBPase: Sedoheptulose 1,7-bisphosphatase; RCA: Rubisco activase; NADPH: Nicotinamide adenine dinucleotide phosphate; FNR: Ferredoxin-NADP$^+$ reductase; AMS: 4-acetamido-4′-maleimidylstilbene-2,2′-disulfonate; SDS-PAGE: Sodium dodecyl sulfate polyacrylamide gel electrophoresis.

3.2. The In Vivo Physiological Study

In order to address another major concern, the in vivo redox behaviors of Trx and target proteins were observed. After dark adaptation, *Arabidopsis* plants were transferred to HL conditions (650–700 µmol photons m^{-2} s^{-1}). The redox state of several chloroplast proteins (some Trx isoforms and target proteins) was sequentially determined as previously described [19].

Four Trx isoforms (Trx-*f*1, Trx-*m*2, Trx-*x*, and Trx-*y*2) were rapidly converted from oxidized to reduced forms after exposure to HL (Figure 2A). Their redox states apparently reached stable levels within 1 min of HL exposure, which was common with all Trx isoforms examined here. For comparison, we analyzed the induction pattern of photosynthetic electron transport (Figure 2B). ETR II and ETR I were gradually elevated upon exposure to HL, and they attained maximal levels at approximately 6 min. These results indicate that Trx can receive reducing power even when the photosynthetic electron transport is not fully activated during dark-light transitions.

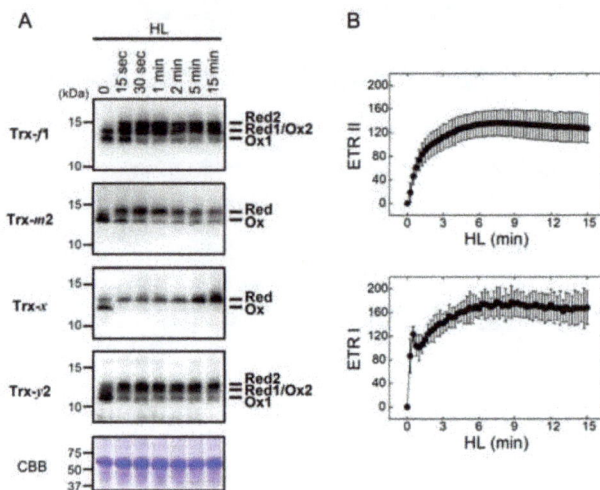

Figure 2. The in vivo redox responses of Trx isoforms and their relationship to photosynthetic electron transport rates in *Arabidopsis*. (**A**) Dark-adapted plants were placed under HL conditions (650–700 µmol photons m^{-2} s^{-1}) for the indicated time period. The redox states of Trx-*f*1, Trx-*m*2, Trx-*x*, and Trx-*y*2 were then determined as previously described [19]. As a loading control, Rubisco large subunit was stained with Coomassie Brilliant Blue R-250 (CBB). Red, reduced form; Ox, oxidized form. (**B**) Dark-adapted plants were irradiated by HL, and relative ETR II and ETR I were sequentially measured. Values represent the mean ± SD (n = 5). HL: High light; ETR II: Electron transport rates of PSII; ETR I: Electron transport rates of PSI.

Figure 3 shows the HL-responsive redox changes in FBPase, SBPase, RCA, and ATP synthase CF$_1$-γ subunit. The shift from oxidized to reduced forms of FBPase, SBPase, and RCA occurred with different kinetics. RCA showed the highest reduction rate of these three proteins, but it took approximately 2 min for RCA to reach a highly reduced state. The reduction rates of FBPase, SBPase, and RCA were clearly slower than those of Trx isoforms (for comparison, see Figure 2A). By contrast, CF$_1$-γ showed an exceptionally high rate of reduction; CF$_1$-γ was almost fully reduced within 15 s of

HL exposure. It was thus evident that, even if reducing power is supplied immediately to Trx upon HL exposure, Trx distributes it to each of the targets at largely different rates.

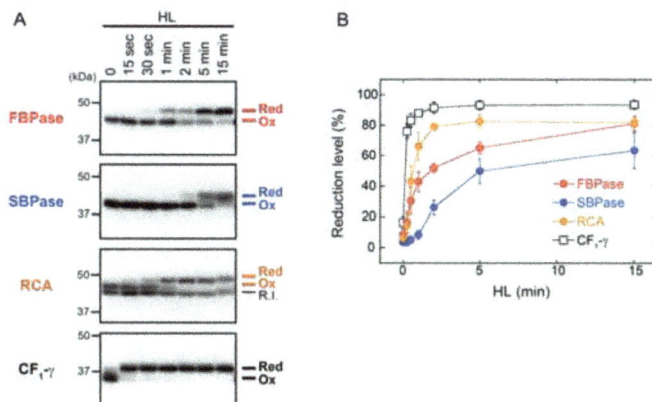

Figure 3. The in vivo redox responses of Trx-targeted proteins in *Arabidopsis*. Dark-adapted plants were placed under HL conditions (650–700 μmol photons $m^{-2} s^{-1}$) for the indicated time period. (**A**) The redox state of FBPase, SBPase, RCA, and ATP synthase CF_1-γ subunit was determined as previously described [19]. Red, reduced form; R.I., redox-insensitive form of RCA; Ox, oxidized form. (**B**) The reduction level of each protein was calculated as the ratio of the reduced form to the total. Values represent the mean ± SD (*n* = 3).

4. Discussion and Conclusions

The aim of this study was to determine the rate-limiting step for light-responsive redox regulation in chloroplasts. The in vitro and in vivo studies carried out here led to a consistent conclusion; reducing power transfer from Trx to its target proteins is a critical bottleneck for determining the rate of target protein reduction. The data also highlight that this step confers distinct light-responsive redox behaviors onto each of the targets.

In the reconstituted FTR/Trx system, FTR-C and Trx-*f* exhibited rapid reduction following the drive of reducing power transfer (Figure 1, Figure S2). In agreement with this, the in vivo study revealed that the Trx family was rapidly reduced upon HL irradiation (Figure 2A). These results suggest the close coupling of FTR and Trx to the activation of photosynthetic electron transport. By referring to the data on PSII and PSI quantum yields (Figure 2B), it was shown that Trx actively receives reducing power even during the induction phase of photosynthetic electron transport. It thus seems likely that, together with other systems [22,23], the FTR/Trx system acts as an electron sink during this period. This idea is supported by a previous study showing that, in Trx-*f*-deficient *Arabidopsis*, the induction of photosynthetic electron transport was hampered because of prolonged electron accumulation in the PSI acceptor side [24]. Therefore, an additional role for the FTR/Trx system as an initiator of photosynthesis can be considered. To evaluate its contribution quantitatively, future studies should be directed toward estimating the extent of electron transfer from Fd to the FTR/Trx redox cascade.

In the in vitro experiments, FBPase, SBPase, and RCA showed different rates of reduction; RCA was most rapidly reduced, followed by FBPase and finally SBPase (Figure 1, Figure S2). Because these experiments were performed with a fixed stoichiometry, differential responses of these three proteins must be attributed to the biochemical properties of each molecule, including the midpoint redox potential, the electrostatic charge on the protein surface, and the affinity to Trx-*f*. Different reduction rates for FBPase, SBPase, and RCA were also observed in vivo (Figure 3), but they appeared to be uniformly slower compared with the in vitro responses (Figure 1, Figure S2). This is possibly related to the fact that various target proteins co-exist in chloroplasts in different amounts [12,25]. It should also be noted that the recently identified protein-oxidizing redox cascade works in vivo [18], which may lower the net rate of target protein reduction. Furthermore, the localization of target proteins within chloroplasts seems to be a key intrinsic determinant for their reduction rates, given that the thylakoid membrane-bound ATP synthase CF_1-γ subunit showed marked rapid reduction upon exposure to HL (Figure 3). In summary, the in vivo reduction kinetics of target proteins is thought to be determined by numerous biochemical and physiological factors in a complex manner. Further studies are needed to clarify the overall mechanisms underlying dynamic and divergent redox behaviors of chloroplast proteins.

Supplementary Materials: The following are available online at http://www.mdpi.com/2076-3921/7/11/153/s1, Figure S1: Trx selectivity for reducing FBPase, SBPase, and RCA in the reconstituted FTR/Trx system in *Arabidopsis*, Figure S2: The in vitro redox responses of FTR-C, Trx-*f*1, and target proteins from *Arabidopsis* (pH 8.2).

Author Contributions: Conceptualization, K.Y. and T.H.; Investigation, K.Y.; Formal Analysis, K.Y. and T.H.; Writing—Original Draft Preparation, K.Y.; Writing—Review and Editing, K.Y. and T.H.

Funding: This study was supported by the Japan Society for the Promotion of Science (JSPP) KAKENHI Grant 16H06556 and by the Dynamic Alliance for Open Innovation Bridging Human, Environment and Materials.

Conflicts of Interest: The authors declare no conflict of interest.

References

1. Laurent, T.C.; Moore, E.C.; Reichard, P. Enzymatic synthesis of deoxyribonucleotides. IV. Isolation and characterization of thioredoxin, the hydrogen donor from *Escherichia coli* B. *J. Biol. Chem.* **1964**, *239*, 3436–3444. [PubMed]
2. Buchanan, B.B. Role of light in the regulation of chloroplast enzymes. *Annu. Rev. Plant Physiol.* **1980**, *31*, 341–374. [CrossRef]
3. Buchanan, B.B.; Schürmann, P.; Wolosiuk, R.A.; Jacquot, J.P. The ferredoxin/thioredoxin system: From discovery to molecular structures and beyond. *Photosynth. Res.* **2002**, *73*, 215–222. [CrossRef] [PubMed]
4. Buchanan, B.B.; Balmer, Y. Redox regulation: A broadening horizon. *Annu. Rev. Plant Biol.* **2005**, *56*, 187–220. [CrossRef] [PubMed]
5. Dai, S.; Schwendtmayer, C.; Schürmann, P.; Ramaswamy, S.; Eklund, H. Redox signaling in chloroplasts: Cleavage of disulfides by an iron-sulfur cluster. *Science* **2000**, *287*, 655–658. [CrossRef] [PubMed]
6. Dai, S.; Friemann, R.; Glauser, D.A.; Bourquin, F.; Manieri, W.; Schürmann, P.; Eklund, H. Structural snapshots along the reaction pathway of ferredoxin-thioredoxin reductase. *Nature* **2007**, *448*, 92–96. [CrossRef] [PubMed]
7. Michelet, L.; Zaffagnini, M.; Morisse, S.; Sparla, F.; Pérez-Pérez, M.E.; Francia, F.; Danon, A.; Marchand, C.H.; Fermani, S.; Trost, P.; Lemaire, S.D. Redox regulation of the Calvin-Benson cycle: Something old, something new. *Front. Plant Sci.* **2013**, *4*, 470. [CrossRef] [PubMed]
8. Lemaire, S.D.; Michelet, L.; Zaffagnini, M.; Massot, V.; Issakidis-Bourguet, E. Thioredoxins in chloroplasts. *Curr. Genet.* **2007**, *51*, 343–365. [CrossRef] [PubMed]
9. Serrato, A.J.; Fernández-Trijueque, J.; Barajas-López, J.D.; Chueca, A.; Sahrawy, M. Plastid thioredoxins: A "one-for-all" redox-signaling system in plants. *Front. Plant Sci.* **2013**, *4*, 463. [CrossRef] [PubMed]
10. Collin, V.; Issakidis-Bourguet, E.; Marchand, C.; Hirasawa, M.; Lancelin, J.M.; Knaff, D.B.; Miginiac-Maslow, M. The *Arabidopsis* plastidial thioredoxins: New functions and new insights into specificity. *J. Biol. Chem.* **2003**, *278*, 23747–23752. [CrossRef] [PubMed]
11. Toivola, J.; Nikkanen, L.; Dahlström, K.M.; Salminen, T.A.; Lepistö, A.; Vignols, H.F.; Rintamäki, E. Overexpression of chloroplast NADPH-dependent thioredoxin reductase in *Arabidopsis* enhances leaf growth and elucidates in vivo function of reductase and thioredoxin domains. *Front. Plant Sci.* **2013**, *4*, 389. [CrossRef] [PubMed]
12. Yoshida, K.; Hara, S.; Hisabori, T. Thioredoxin selectivity for thiol-based redox regulation of target proteins in chloroplasts. *J. Biol. Chem.* **2015**, *290*, 14278–14288. [CrossRef] [PubMed]

13. Hisabori, T.; Motohashi, K.; Hosoya-Matsuda, N.; Ueoka-Nakanishi, H.; Romano, P.G.N. Towards a functional dissection of thioredoxin networks in plant cells. *Photochem. Photobiol.* **2007**, *83*, 145–151. [CrossRef] [PubMed]

14. Montrichard, F.; Alkhalfioui, F.; Yano, H.; Vensel, W.H.; Hurkman, W.J.; Buchanan, B.B. Thioredoxin targets in plants: The first 30 years. *J. Proteom.* **2009**, *72*, 452–474. [CrossRef] [PubMed]

15. Geigenberger, P.; Thormählen, I.; Daloso, D.M.; Fernie, A.R. The unprecedented versatility of the plant thioredoxin system. *Trends Plant Sci.* **2017**, *22*, 249–262. [CrossRef] [PubMed]

16. Nikkanen, L.; Toivola, J.; Diaz, M.G.; Rintamäki, E. Chloroplast thioredoxin systems: Prospects for improving photosynthesis. *Philos. Trans. R. Soc. Lond. B Biol. Sci.* **2017**, *372*, 20160474. [CrossRef] [PubMed]

17. Yoshida, K.; Hisabori, T. Distinct electron transfer from ferredoxin-thioredoxin reductase to multiple thioredoxin isoforms in chloroplasts. *Biochem. J.* **2017**, *474*, 1347–1360. [CrossRef] [PubMed]

18. Yoshida, K.; Hara, A.; Sugiura, K.; Fukaya, Y.; Hisabori, T. Thioredoxin-like2/2-Cys peroxiredoxin redox cascade supports oxidative thiol modulation in chloroplasts. *Proc. Natl. Acad. Sci. USA* **2018**, *115*, E8296–E8304. [CrossRef] [PubMed]

19. Yoshida, K.; Matsuoka, Y.; Hara, S.; Konno, H.; Hisabori, T. Distinct redox behaviors of chloroplast thiol enzymes and their relationships with photosynthetic electron transport in *Arabidopsis thaliana*. *Plant Cell Physiol.* **2014**, *55*, 1415–1425. [CrossRef] [PubMed]

20. Kobayashi, T.; Kishigami, S.; Sone, M.; Inokuchi, H.; Mogi, T.; Ito, K. Respiratory chain is required to maintain oxidized states of the DsbA-DsbB disulfide bond formation system in aerobically growing *Escherichia coli* cells. *Proc. Natl. Acad. Sci. USA* **1997**, *94*, 11857–11862. [CrossRef] [PubMed]

21. Zhang, N.; Portis, A.R., Jr. Mechanism of light regulation of Rubisco: A specific role for the larger Rubisco activase isoform involving reductive activation by thioredoxin-*f*. *Proc. Natl. Acad. Sci. USA* **1999**, *96*, 9438–9443. [CrossRef] [PubMed]

22. Makino, A.; Miyake, C.; Yokota, A. Physiological functions of the water-water cycle (Mehler reaction) and the cyclic electron flow around PSI in rice leaves. *Plant Cell Physiol.* **2002**, *43*, 1017–1026. [CrossRef] [PubMed]

23. Yoshida, K.; Watanabe, C.K.; Hachiya, T.; Tholen, D.; Shibata, M.; Terashima, I.; Noguchi, K. Distinct responses of the mitochondrial respiratory chain to long- and short-term high-light environments in *Arabidopsis thaliana*. *Plant Cell Environ.* **2011**, *34*, 618–628. [CrossRef] [PubMed]

24. Naranjo, B.; Diaz-Espejo, A.; Lindahl, M.; Cejudo, F.J. Type-*f* thioredoxins have a role in the short-term activation of carbon metabolism and their loss affects growth under short-day conditions in *Arabidopsis thaliana*. *J. Exp. Bot.* **2016**, *67*, 1951–1964. [CrossRef] [PubMed]

25. Peltier, J.B.; Cai, Y.; Sun, Q.; Zabrouskov, V.; Giacomelli, L.; Rudella, A.; Ytterberg, A.J.; Rutschow, H.; van Wijk, K.J. The oligomeric stromal proteome of *Arabidopsis thaliana* chloroplasts. *Mol. Cell. Proteom.* **2006**, *5*, 114–133. [CrossRef] [PubMed]

antioxidants

MDPI

Article

Cytosolic Isocitrate Dehydrogenase from *Arabidopsis thaliana* Is Regulated by Glutathionylation

Adnan Khan Niazi [1,2,3], Laetitia Bariat [1,2], Christophe Riondet [1,2], Christine Carapito [4], Amna Mhamdi [5,6,7], Graham Noctor [5] and Jean-Philippe Reichheld [1,2,*]

[1] Laboratoire Génome et Développement des Plantes, Université Perpignan Via Domitia, F-66860 Perpignan, France; niazi@uaf.edu.pk (A.K.N.); laetitia.bariat@univ-perp.fr (L.B.); christophe.riondet@univ-perp.fr (C.R.)
[2] Laboratoire Génome et Développement des Plantes, CNRS, F-66860 Perpignan, France
[3] Centre of Agricultural Biochemistry and Biotechnology, University of Agriculture Faisalabad, 38000 Faisalabad, Pakistan
[4] Laboratoire de Spectrométrie de Masse BioOrganique (LSMBO), IPHC, Université de Strasbourg, CNRS UMR 7178, 67037 Strasbourg, France; ccarapito@unistra.fr
[5] Institute of Plant Sciences Paris Saclay IPS2, Université Paris-Sud, CNRS, INRA, Université Evry, Paris Diderot, Sorbonne Paris-Cité, Université Paris-Saclay, Bâtiment 630, 91405 Orsay, France; amna.mhamdi@psb.vib-ugent.be (A.M.); graham.noctor@u-psud.fr (G.N.)
[6] Department of Plant Biotechnology and Bioinformatics, Ghent University, 9052 Gent, Belgium
[7] Center for Plant Systems Biology, VIB, 9052 Gent, Belgium
[*] Correspondence: jpr@univ-perp.fr; Tel: +33-468-662-225; Fax: +33-468-668-499

Received: 15 November 2018; Accepted: 22 December 2018; Published: 8 January 2019

Abstract: NADP-dependent (Nicotinamide Adénine Dinucléotide Phosphate-dependent) isocitrate dehydrogenases (NADP-ICDH) are metabolic enzymes involved in 2-oxoglutarate biosynthesis, but they also supply cells with NADPH. Different NADP-ICDH genes are found in *Arabidopsis* among which a single gene encodes for a cytosolic ICDH (cICDH) isoform. Here, we show that cICDH is susceptible to oxidation and that several cysteine (Cys) residues are prone to S-nitrosylation upon nitrosoglutathione (GSNO) treatment. Moreover, we identified a single S-glutathionylated cysteine Cys363 by mass-spectrometry analyses. Modeling analyses suggest that Cys363 is not located in the close proximity of the cICDH active site. In addition, mutation of Cys363 consistently does not modify the activity of cICDH. However, it does affect the sensitivity of the enzyme to GSNO, indicating that S-glutathionylation of Cys363 is involved in the inhibition of cICDH activity upon GSNO treatments. We also show that glutaredoxin are able to rescue the GSNO-dependent inhibition of cICDH activity, suggesting that they act as a deglutathionylation system *in vitro*. The glutaredoxin system, conversely to the thioredoxin system, is able to remove S-nitrosothiol adducts from cICDH. Finally, NADP-ICDH activities were decreased both in a *catalase2* mutant and in mutants affected in thiol reduction systems, suggesting a role of the thiol reduction systems to protect NADP-ICDH activities *in planta*. In line with our observations in Arabidopsis, we found that the human recombinant NADP-ICDH activity is also sensitive to oxidation *in vitro*, suggesting that this redox mechanism might be shared by other ICDH isoforms.

Keywords: Isocitrate dehydrogenase; glutathionylation; nitrosylation; glutaredoxin; *Arabidopsis thaliana*

1. Introduction

Isocitrate dehydrogenases (ICDHs) reversibly catalyze the oxidative decarboxylation of isocitrate to 2-oxoglutarate (2-OG), a key compound in ammonia assimilation by the glutamine synthetase/glutamate synthase pathway. Through their catalytic activity, ICDHs reduce NAD$^+$ or NADP$^+$, producing NADH or NADPH respectively [1,2]. In plants, both NAD and NADP-dependent ICDH isoforms are found. The NAD-dependent isoform is restricted to mitochondria, where it takes part in the tricarboxylic acid (TCA) cycle [3]. NADP-dependent ICDH isoforms are found in the cytosol, chloroplasts, mitochondria and peroxisomes [4]. In the dicot plant *Arabidopsis thaliana*, a single isoform is found in each cell compartment. The cytosolic isoform (cICDH) is the most abundant form in leaves, as it is responsible for more than 80% of the extractible ICDH activity [5,6]. Biochemical analyses have shown that both mitochondrial ICDH and cytosolic cICDH activities are dependent on the nitrogen status. As both isoforms are involved in nitrogen assimilation, they have been proposed to have overlapping functions [1,2].

NADH and NADPH produced by ICDH are important in reducing equivalents for the regeneration of thiol reduction enzymes like glutathione, thioredoxin or GSNO-reductases [7,8]. Therefore, ICDH have been proposed to play an antioxidant role against oxidative stress, and damage to ICDH may result in the perturbation of the balance between oxidants and antioxidants, and lead to pro-oxidant conditions. This was shown in mammals, where the cytosolic ICDH isoform is highly reactive to peroxynitrite, affecting its activity by formation of nitrotyrosine and S-nitrosothiol adducts [9,10]. Another study in mammalian cells shows that the cytosolic NADP-ICDH activity is regulated by S-glutathionylation [11].

Moreover in mammals, cysteine residues of ICDH play an essential role in the catalytic function of ICDH. Such regulation was not explored in plant isoforms, but evidence suggested that ICDH isoforms can also be redox regulated. Several ICDH isoforms were found to exhibit sulfenylated [12,13], glutathionylated [14] or nitrosylated [15] Cys, suggesting that ICDH cysteines are redox sensitive. Moreover, ICDHs were identified as interactors of thioredoxin (TRX) or glutaredoxin (GRX) in different proteomic approaches [16–20]. Another piece of evidence came from the analyses of a cICDH knock-out mutant in *Arabidopsis*. cICDH protein is dispensable for plant growth and for the leaf basic metabolism. However, *cicdh* mutants exhibit accumulation of defense gene transcripts in the absence of pathogen attack and exacerbate the phenotype and the redox perturbation of the oxidative stress mutant *catalase2* (*cat2*). *cat2* is inactivated in the major leaf catalase isoform, which increases the availability of H_2O_2 produced in the peroxisomes [6].

Here, we examine the redox sensitivity of the *Arabidopsis* cytosolic cICDH isoform. We show that enzyme activity is affected by oxidative agents like oxidized glutathione (GSSG) and nitrosoglutathione (GSNO), through modifications of conserved Cys residues in the protein. In particular, Cys363 is S-glutathionylated and can be reversed by the glutaredoxin system. Our data indicate that cytosolic ICDH is redox regulated and that this regulation might be shared in other organisms.

2. Materials and Methods

2.1. Plant Materials and Growth Conditions

All *Arabidopsis thaliana* lines used in this study were of Columbia-0 (Col-0) ecotype. The plants were grown in soil in a controlled growth chamber (180 μE m^{-2} s^{-1}, 16 h day/8 h night, 22 °C 55% RH day, 20 °C 60% RH night) up to 3 weeks. Plant mutant lines used in this study *ntra ntrb, cat2, gr1, gr1 cat2, icdh, icdh cat2, gsnor, nox1, noa1, nia1 nia2* and *nia1 nia2 noa 1* were previously described [6,21–28].

2.2. In Vitro Protein-Based Complementation and TRX/GRX Activity Assays

For *in vitro* protein-based TRX complementation assays, 4.38 μM (200 ng/μL) recombinant cICDH protein was incubated in 25 μL for 2 h on ice with 1 mM NADPH, 4.59 μM TRXh3 or TRXh5 and 3.12 μM NADPH-Dependent Thioredoxin Reductase A (NTRA). This reaction mixture was diluted 40 times in 100 mM phosphate buffer (KOH (Potassium hydroxyde), pH 7.5) and the ICDH activity assay was performed as described above.

For *in vitro* protein-based GRX complementation assays, 4.38 µM (200 ng/µL) recombinant cICDH was incubated in 25 µL for 2 h on ice with 1 mM NADPH, 5 µM GRXC1 or GRXC2, 0.8 mM GSH and 5 µM Glutaredoxin Reductase (GR). This reaction mixture was diluted 40 times in 100 mM phosphate buffer (KOH, pH 7.5) and the ICDH activity assay was performed as described above.

2.3. Cloning, Expression and Purification of Recombinant cICDH

The cICDH-coding sequence (At1g65930), with NdeI and BamHI restriction sites at the N- and C-terminal ends was inserted into pET16b vector (Novagen). Point mutation of cysteine 363 to serine was generated using QuikChange II Directed Mutagenesis Kit (Agilent) using primers detailed in Supplemental Table S1. The constructs were transferred into *E. coli* BL21 stain and transformed cells were cultured at 37 °C until A_{600} = 0.7. cICDH expression was induced by the addition of 1 mM isopropyl-1-thio-d-galactopyranoside (IPTG), followed by further culture at 21 °C for 16 h. All the steps for the purification of recombinant His-tagged cICDH were performed at 4 °C as previously described [29]. Briefly, the cells were resuspended in 50 mM Tris-HCl pH 7.5, disrupted by sonication, and centrifuged at 125,000× g for 30 min at 4 °C. The supernatant was loaded onto the Ni^{2+}-Sepharose column equilibrated with 50 mM Tris-HCl, pH 7.5, 200 mM NaCl. The obtained fractions containing the purified protein were collected and evaluated on a SDS-PAGE (Sodium dodecyl sulfate-polyacrylamide gel electrophoresis) gel. Purified recombinant protein samples were pooled and stored at 4 °C.

2.4. Protein Extracts from Arabidopsis Plants and cICDH Enzymatic Assay

To perform enzymatic assays, Arabidopsis leaf protein extracts were prepared as described before [30,31]. Briefly, 100 mg leaves were ground in a mortar in liquid nitrogen and resuspended in 0.5 mL of a buffer containing 50 mM Hepes/KOH, pH 7.4, 1 mM EDTA, 1 mM EGTA, 2 mM Benzamidine, 2 mM ε-aminocaproic acid, 0.5 mM PMSF, 10% glycerol, 0.1% Triton X-100 and 1 tablet of EDTA-free protease inhibitor cocktail (Roche). After 15 min of incubation on ice, samples were centrifuged at 15,000× g and at 4 °C for 15 min to remove tissue debris. NADP-ICDH activity was measured at room temperature from 40 µg of protein extracts in total of 1 mL reaction medium, having 100 mM phosphate buffer (KOH, pH 7.5), 5 mM MgCl$_2$, 250 µM of NADP$^+$ and 2.5 mM DL-isocitric acid. The activity was monitored as a change in absorbance at 340 nm due to the isocitrate-dependent rate of NADP$^+$ reduction. Catalase activity was quantified as described before [6]. Briefly, 20 µg of protein extracts were taken in 50 mM potassium phosphate buffer (pH 7.0) and 10 mM H$_2$O$_2$ at 25 °C in 1 mL reaction mixture. The activity was monitored as a change in absorbance at 240 nm due to breakdown of H$_2$O$_2$. To get the reduced cICDH, it was incubated with 20 mM DTT at 25 °C for 1 h. To remove excess dithiothreitol (DTT), protein samples were passed through a micro bio-spin column (Bio-Rad, Marnes-la-Coquette, France) after each treatment. The concentration of modified cICDH measured by absorbance at 340 nm.

For recombinant cICDH enzymatic assay, the reaction was performed with 800 ng of cICDH at 25 °C in a 1000 µL final volume containing 100 mM phosphate buffer (KOH, pH 7.5), 5 mM MgCl$_2$, 250 µM of NADP$^+$ and 2.5 mM DL-isocitric acid. The decrease in NADP$^+$ absorbance at 340 nm was monitored using a UV-1800 spectrophotometer (Shimadzu, Marne la Vallée, France). The molar extinction coefficient for NADP$^+$ of 6220 M^{-1} cm^{-1} was used for the calculation. To obtain the K$_M$ for Isocitrate or NADP$^+$, progress curves were recorded using varying concentrations of Isocitrate (0–5 mM) or NADP$^+$ (0–1 mM). The initial velocity (*vi*) for each substrate concentration was measured, and the *vi*/E_0 values were plotted and fitted with the Michaelis-Menten equation to obtain the kinetic parameters. Three independent replicates of *vi* were measured for each substrate concentration.

For the reactivation of cICDH by TRXh, cICDH was incubated 30 min with recombinant TRXh (4.59 µM) in the presence of NADPH (0.125 mM) and NTRA (0.5 µM). ICDH activity assay was performed as described earlier after diluting the mix 40-fold. For cICDH reactivation by GRX, it was incubated for 30 min with recombinant GRX (5 µM) in presence of NADPH (0.125 mM), GSH (0.8 mM) and GR (5 µM). Then the mix was diluted 40-fold and used for ICDH activity assay, as described above.

2.5. Gel Filtration Chromatography

His-ICDH recombinant proteins (~10 µg) were fractionated using Superose 12 (GE Healthcare, Buc, France) column equilibrated in Protein Buffer (50 mM Tris pH 7.5, 5 mM MgCl$_2$) containing 150 mM NaCl. The protein standards were aldolase (158 kDa), conalbumin (75 kDa) and ovalbumin (43 kDa) (GE Healthcare).

2.6. Liquid Chromatography-Tandem Mass Spectrometry (LC-MS/MS)

Purified His-ICDH recombinant protein was dissolved in 8 M urea / 0.1 M ammonium bicarbonate buffer and thoroughly vortexed. The urea concentration was then lowered to 1 M by dilution with fresh 0.1 M ammonium bicarbonate, and proteins were digested in solution by addition of trypsin overnight at 37 °C. After acidification with formic acid, desalting and concentration of the peptides were carried out using a C18 Sep-Pak cartridge (Sep-pak Vac 1cc (50mg) tC18 cartridges, Waters, Guyancourt, France).

LC-MS/MS analyses were conducted on a nanoHPLC-Q Exactive Plus system (Thermo Fisher Scientific, Bremen, Germany). Peptide separation was performed on an ACQUITY UPLC BEH130 C18 column (250 mm × 75 µm with 1.7 µm diameter particles, Waters). The solvent system consisted of 0.1% FA in water (solvent A) and 0.1% FA in ACN (solvent B). Peptides were eluted at 450 nL/min with the following gradient of solvent B: from 1 to 8% over 2 min, from 8 to 35% over 28 min and then 90% for 5 min. The system was operated in positive mode using the following settings: MS1 survey scans (m/z 300 to 1800) were performed at a resolution of 70 000 with an AGC target of 3×10^6 and the maximum injection time was set to 50 ms. MS/MS spectra were acquired at a resolution of 17500. The MS2 AGC target was set to 1×10^5 and the maximum injection time was set to 100 ms.

NanoLC-MS/MS data were searched using a local Mascot server (version 2.5.1, MatrixScience, London, UK) in an *Arabidopsis thaliana* protein sequences database downloaded from the TAIR site (TAIR10 version, Phoenix Bioinformatics, Fremont, CA, USA), to which decoy sequences were added using the in-house developed software tool MSDA [32]. Spectra were searched with a mass tolerance of 5 ppm in MS and 0.07 Da in MS/MS mode. One missed cleavage was tolerated. Oxidation of cysteine residues were specified as variable modifications. Spectra and fragmentation tables that have allowed identifying glutathionylated peptides are provided.

2.7. Biotin Labeling of S-Nitrosylated Proteins

The biotinylation of S-nitrosylated proteins resulting from the *in vitro* S-nitrosylation was detected using the Biotin Switch Test (BST) as described by reference [33] and with minor modifications. To achieve this, 2.19 µM of recombinant wild-type cICDH or Cys-mutated version of cICDH (cICDH-C363S) was incubated with 1 mM GSNO at room temperature in a reaction mixture called HEN buffer. HEN buffer contains protease inhibitor cocktail along with 25 mM 4-(2-hydroxyethyl)-1-piperazineethanesulfonic acid (HEPES) [pH 7.7], 1 mM Ethylenediaminetetraacetic acid (EDTA), and 0.1 mM neocuproine. After 30 min of incubation, the reaction mixture was desalted on Pierce™ Zeba spin columns (Thermo Fisher Scientific, Dardilly, France). To perform the denitrosylation, the resulting protein-SNO were incubated for 45 min with either TRXh, NTRA, and NADPH or with GRXC, GSH, GR and NADPH. Protein S-nitrosylation was assessed using BST as described in reference [33], except for 20 mM NEM being used to alkylate free thiols. Residual NEM was removed by centrifugation (minimum 2000 g, 20 min, 4 °C) with 2 volumes of −20 °C acetone (pre-chilled). The supernatant was removed and pellets were resuspended in 0.1 mL of HENS buffer (HEN buffer containing 1% SDS)/mg protein. To achieve biotinylation, the resuspended proteins were incubated at room temperature for 1 h after adding 2 mM biotin-HPDP and 1 mM ascorbate. After removing biotin-HPDP, the precipitated proteins were resuspended in

0.1 mL of HENS buffer/mg of protein and 2 volumes of neutralization buffer (20 mm HEPES, pH 7.7, 100 mm NaCl, 1 mm EDTA, and 0.5% Triton X-100). A total of 15 μL of Streptravidin-agarose/mg of protein were added and incubated for 1 h at RT. The matrix was washed five times with 10 volumes of washing buffer (600 mM NaCl in neutralization buffer). The sample were centrifuged at 200 g for 5 s at room temperature between each wash. Finally, the bound proteins were eluted with 100 mM β-mercaptoethanol in neutralization buffer.

To perform western blot analysis, SDS-PAGE sample buffer was added to agarose beads. SDS-PAGE gel was run after heating the samples to 70 °C for 10 min. The samples separated on SDS-PAGE were transferred to nitrocellulose membranes. Western blots were probed with anti-His antibodies (Merck KGaA, Darmstadt, Germany).

2.8. Bioinformatics Analyses

The gene and protein sequences were obtained from the NCBI website (http://www.ncbi.nlm. nih.gov, Rockville Pike, Bethesda MD, USA). ClustalW2 software (http://www.ebi.ac.uk/Tools/msa/ clustalw2/, EMBL-EBI, Wellcome Genome Campus, Hinxton, Cambridgeshire, UK) was used to perform the multiple sequence alignments. Swiss Model software (http://swissmodel.expasy.org/, Protein Structure Bioinformatics Group, Swiss Institute of Bioinformatics Biozentrum, University of Basel, Switzerland) was used for cICDH modeling by using human cytosolic ICDH sequence structure (PDB ID: 1T0L, [34]), and the Geneious 9.0 software (Biomatters ApS, Aarhus, Denmark) was used for the 3D representation.

3. Results

3.1. Enzymatic Characterization of Cytosolic NADP-ICDH

In order to characterize redox modifications on the cytosolic NADP-ICDH (cICDH, At1g65930), we produced the recombinant cICDH in *E. coli*. The predicted protein shows 77.4% and 85.2% identity with the mito/chloro and peroxisomal NADP-ICDH from *A. thaliana*, respectively. It also shares conserved cysteine residues with most plant and mammalian NADP-ICDH proteins (Supplemental Figure S1). To verify that cICDH is a genuine NADP-ICDH, the cDNA was cloned in an expression vector and purified using an introduced N-terminal His-tag. SDS/PAGE analysis of purified cICDH indicated a monomer corresponding to the predicated molecular mass of 48.3 kDa calculated for cICDH, including the N-terminal His-tag (Supplemental Figure S2). The migration of the protein is not modified when the SDS/PAGE gel is run under reducing or non-reducing conditions. However, size exclusion chromatography indicated that purified cICDH eluted at a molecular weight of ~90 kDa, suggesting that the native cICDH is present as a dimer coordinated by non-covalent interactions (Supplemental Figure S3). Analysis of the catalyzed reaction revealed Michaelis-Menten kinetics with a Km for the substrate isocitrate of 99 μM and a k_{cat} of 4.93 s^{-1} (Figure 1A and Table 1). Km and k_{cat} for NADP$^+$ were estimated at 28 μM and and a k_{cat} of 5.81 s^{-1} (Figure 1B). The optimal pH for cICDH activity was found to be in the range of 7.5 to 8.5 (Figure 1C).

Table 1. Steady-state enzymatic parameters for cICDH and cICDH-C297S.

Recombinant Protein	K_M Isocitrate (μM)	k_{cat} Isocitrate (s^{-1})	k_{cat}/K_M (M^{-1} s^{-1})
cICDH	99	4.93	4.97×10^4
cICDH-C297S	95	4.31	4.54×10^4

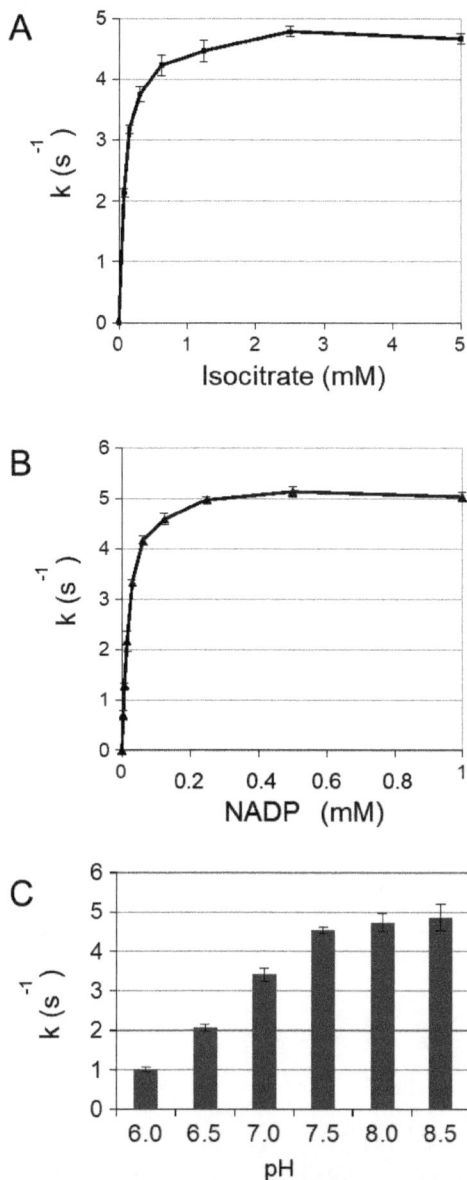

Figure 1. Cytosolic ICDH1 activity. ICDH (Isocitrate dehydrogenase) activity was measured for 800 ng of ICDH by monitoring the change in absorbance at 340 nm due to the isocitrate-dependent rate of $NADP^+$ reduction at 25 °C in 1 mL reaction medium containing 5 mM MgCl2, 250 µM of $NADP^+$ and (**A**) different concentrations (0–5 mM) of DL-isocitric acid or (**B**) different concentrations (0–1 mM) of NADP in 100 mM phosphate buffer (KOH, pH 7.5). (**C**) 2.5 mM DL-isocitric acid in 100 mM phosphate buffer at different pH (6–8.5). Error bars represent SE (Standard Error) (*n* = 3).

Next, we studied the effect of different oxidants on the activity of cICDH. Increasing concentrations of H_2O_2 hardly affected the enzyme activity (Figure 2A). However, both increasing concentrations of GSNO and GSSG progressively inhibited the cICDH activity, as did the combination of H_2O_2 and

GSH (Figure 2). These latter treatments act as GSH donors which might modify cysteine residues by S-glutathionylation. Moreover, GSNO is also an excellent NO donor which can affect cysteine residues by S-nitrosylation.

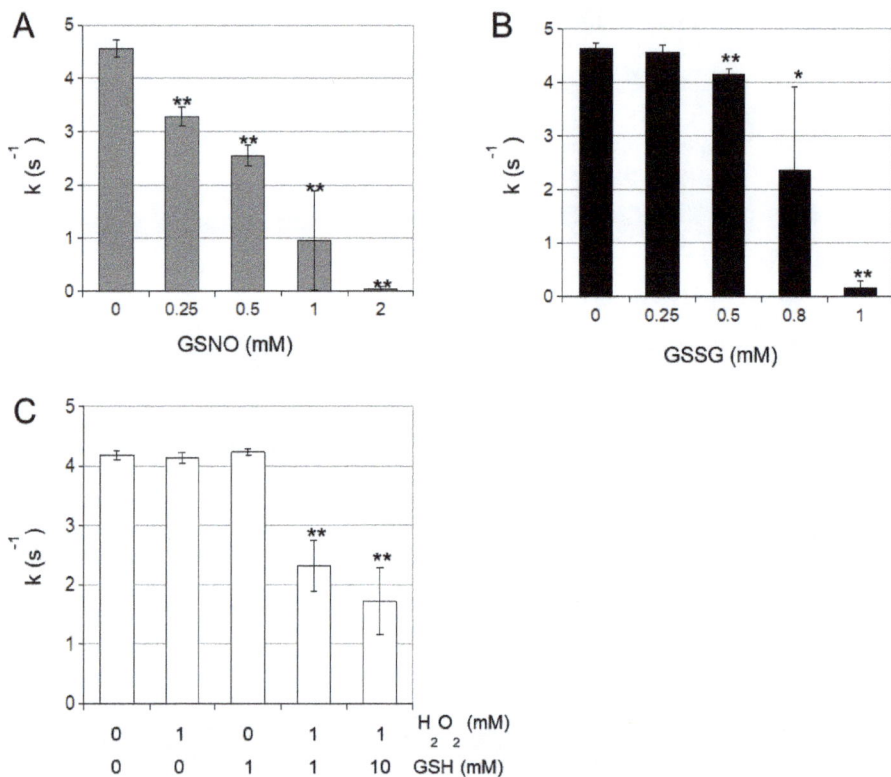

Figure 2. Cytosolic ICDH activity is sensitive to oxidation. 4.38 μM (200 ng/μL) ICDH were incubated with different concentrations of (**A**) H_2O_2/GSH, (**B**) nitrosoglutathione (GSNO) or (**C**) oxidized glutathione (GSSG) for 15 min at 25 °C. This reaction mixture was diluted 250 times in reaction buffer during the assay. ICDH activity was measured for 800 ng of ICDH at 25 °C in 1 mL reaction medium containing 100 mM phosphate buffer (KOH, pH 7.5), 5 mM MgCl2, 250 μM of $NADP^+$ and 2.5 mM DL-isocitric acid by monitoring the change in absorbance at 340 nm due to the isocitrate-dependent rate of $NADP^+$ reduction. Error bars represent SE ($n = 3$). * $p < 0.05$, ** $p < 0.01$ for statistical differences compared to non-treated samples (Student's t test).

We investigated if cICDH can be glutathionylated/nitrosylated by treating His-cICDH protein with GSNO and analyzing samples by mass spectrometry. After trypsin digestion, peptides were analyzed by LC-MS/MS (Figure 3A). This reveals a glutathione adduct on two different Cys363-containing peptides (with and without a missed trypsin cleavage) detected thanks to a 305-Da mass increase of the modified peptides when compared to the non-modified peptide. These experiments attest that Cys363 can be glutathionylated *in vitro*. No other glutathionylated Cys residue was detected. The Cys363 glutathionylated site was confirmed and validated on independent triplicate experiments (Supplemental Figure S4). Then, we produced a cICDH-C363S mutant protein and tested it for glutathionylation in three same experimental conditions. No glutathionylated residue was detected in three different replicate experiments. Therefore, Cys363 is prone to S-glutathionylation upon GSNO treatment. By modeling cICDH, we established that Cys363 is probably not located in the close proximity of the cICDH active site (Figure 3B).

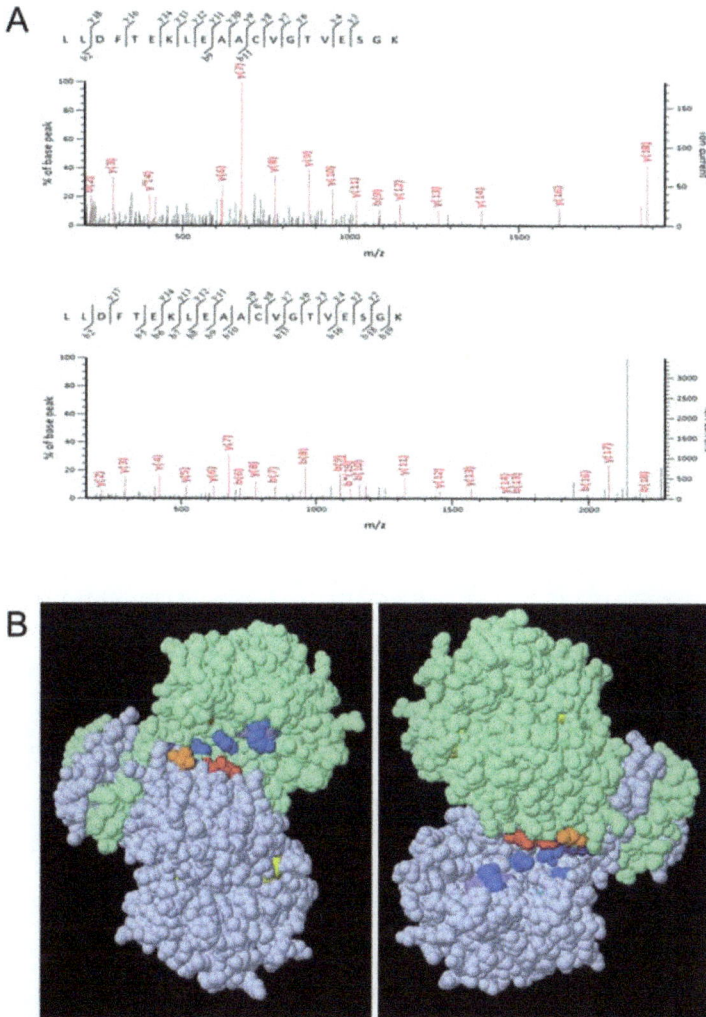

Figure 3. Glutathionylation of cICDH. (**A**) cICDH was treated or not with 1 mM GSNO for 30 min at 25 °C. Samples were trypsin digested and analyzed by nanoLC-MSMS. The panels show fragmentation spectra matching peptides with either unmodified (top) or with glutathionylated C363 (bottom). The same glutathionylated residue was identified in three biological repetitions. (**B**) Modeled cICDH (residues 4–408) homodimer (grey and green chains) based on human cytosolic ICDH. Conserved amino acids (R111, R134, Y141, T214, D252, D279, R314, H315) with the most plant and mammalian NADP-ICDH proteins, in the active site pocket are shown in different colors depending on the nature of the residue: R, blue; Y, purple; T, orange; D, red; H, grey. The conserved cysteine C363 in each chain is represented in yellow (arrowheads). In the right panel, the cICDH homodimer was rotated 180°.

We also performed a biotin-switch experiment to test if cICDH is S-nitrosylated. Treatment with GSNO leads to a specific biotin-switch signal, indicating that cICDH is S-nitrosylated *in vitro* (Figure 4A). To identify the nitrosylated residues, we subjected the cICDH-C363S mutant protein to the same experiment. No decrease of the biotin-switch signal was observed, which did not allow us to identify the S-nitrosylated Cys residues (Figure 4A).

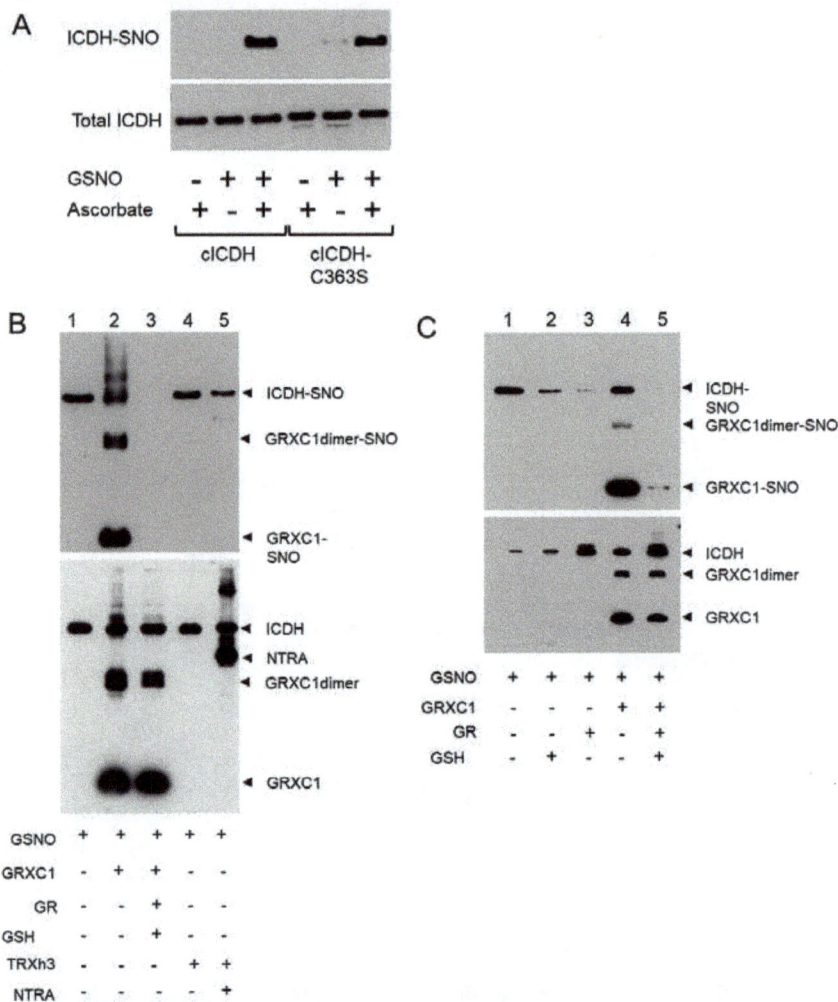

Figure 4. cICDH is S-nitrosylated and is denitrosylated by the GRX system. (**A**) 2.19 μM of recombinant wild-type cICDH or Cys-mutated versions of cICDH (cICDH-C363S) were treated with or without 1 mM GSNO for 30 min at 25 °C, and subjected to the biotin-switch assay in presence or absence of sodium ascorbate. (**B**) After treatment with GSNO (1 mM), the protein was treated with GRXC1 (5 μM) alone (lane 2), GRXC1, GSH (0.8 mM) and GR (0.45 μM) (Lane 3), NTRA (3 μM) alone (lane 4) and NTRA+TRXh3 (4.59 μM) (lane 5) for 30 min at 25 °C and subjected to the biotin-switch assay in the presence of sodium ascorbate. (**C**) The same experimental design in the presence of GSH (0.8 mM) (lane 2), GR (5 μM) (lane 3), GrxC1 (5 μM) (lane 4) and GRXC1+GR+GSH (lane 5). Afterwards, the proteins were separated by reducing SDS-PAGE and transferred onto nitrocellulose membrane. Total ICDH (bottom panel) or S-nitrosylated ICDH (top panel) was detected using an anti-His antibody.

Then, we examined the denitrosylation activity of thiol reductases by biotin switch (Figure 4B). When GSNO treated cICDH was further incubated by the recombinant cytosolic TRXh3 in the presence or absence of its physiological reducer NTRA, no denitrosylation activity was observed, suggesting that the TRX system was not able to remove SNO adducts from cICDH (Figure 4B). However, adding the GRX system to the reaction triggered the disappearing of the biotin switch signal on GSNO-treated cICDH, suggesting that the GRX-dependent thiol reduction system has a denitrosylation activity on cICDH (Figure 4B). By looking closer to this activity, we noticed that the denitrosylation activity needs the full GRX system (NADPH/GR/GSH/GRX) to be optimal (Figure 4C). Interestingly, while performing the biotin switch in presence of GRXC1, we noticed a strong signal on the GRXC1, suggesting that GRXC1 is prone to trans-nitrosylation (Figure 4B,C).

To test the function of Cys363 in cICDH activity, we examined the impact of the C363S point mutation on enzymatic activities. The cICDH C363S mutant had a Km of 95 μM and a k_{cat} of 4.31 s^{-1} for the substrate isocitrate, which is similar to the wild-type protein, suggesting that the C363S mutation does not perturb the enzymatic characteristics of cICDH (Figure 5A and Table 1). However, cICDH C363S was found to be less affected by GSNO treatment than the wild-type enzyme, suggesting that S-nitrosylation/S-glutathionylation of Cys363 play a regulatory function for cICDH activity (Figure 5B).

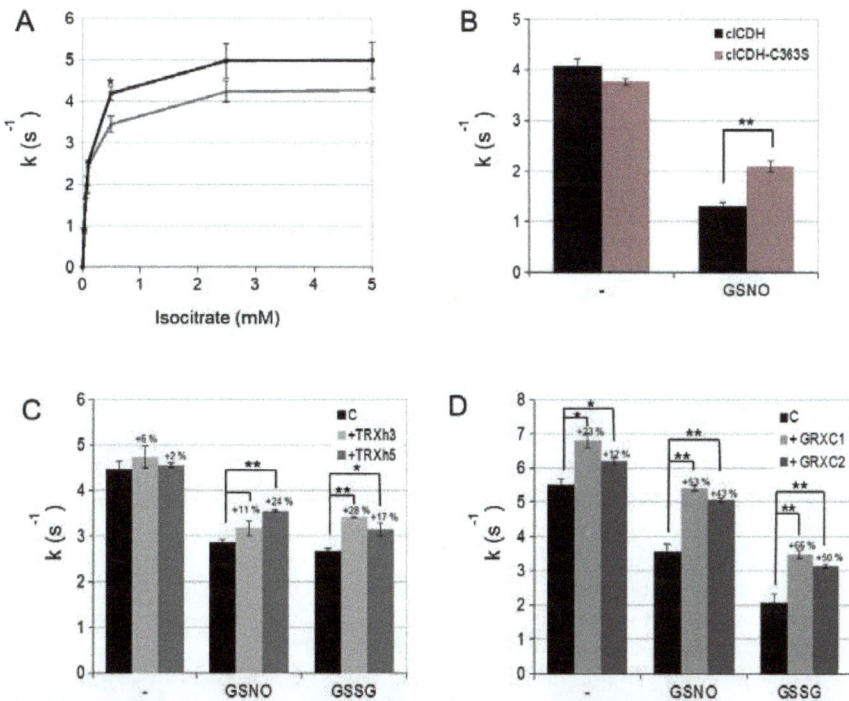

Figure 5. Cysteine-dependent regulation of cICDH activity. (**A**) ICDH activity was measured for 800 ng of ICDH and ICDH-C363S by monitoring the change in absorbance of NADPH at 340 nm due to the isocitrate-dependent rate of NADP$^+$ reduction at 25 °C in 1 mL reaction medium containing 5 mM MgCl$_2$, 250 μM of NADP$^+$ and different (0–5 mM) concentrations of DL-isocitric acid. (**B**) ICDH were incubated with 1 mM GSNO and measured in the same conditions than described previously. (**C**) 800 ng of cICDH were incubated with 0.5 mM GSNO, 0.75 mM GSSG for 15 min at 25 °C. The samples were diluted 2-fold and then incubated with 1 mM NADPH in the presence of NTRA (3 μM) and TRXh3 or TRXh5 (4.59 μM) (**D**) Same experimental design as in (**C**), but in presence of GR (0.45 μM), GSH (0.8 mM) and GRXC1 or GRXC2 (5 μM). Error bars represent SE (*n* = 3). * *p* < 0.01, ** *p* < 0.001 for statistical differences compared to non-treated samples (Student's *t* test).

3.2. cICDH Activity Is Restored by Glutaredoxins

After having established that cICDH activity is affected by glutathionylation on Cyc363, we wished to determine if cICDH activity could be restored by thiol reduction pathways. For this purpose, we subjected GSNO-treated cICDH to different thiol reductases: thioredoxins are major disulfide reductases, but have also been shown to have denitrosylation capacities [35]. However, they are generally not able to reduce S-glutathionylated cysteine adducts. On the contrary, glutaredoxins exhibit a major deglutathionylation activity [8]. We found that cytosolic GRXC1 and GRXC2 are more efficient than TRXh3 and TRXh5 to restore cICDH activity, which is consistent with a deglutathionylation activity of GRXC1 and GRXC2 (Figure 5C,D). Collectively, our data suggest that cICDH activity is inhibited by S-glutathionylation on Cys363 and that GRXC1 and GRXC2 are able to reverse this effect through their deglutathionylation activity.

3.3. Enzyme Activities Are Affected in Mutants

In order to study whether cICDH activity is redox sensitive *in planta*, we measured ICDH activities in leaves from two-weeks old plants of different Arabidopsis mutants. While exhibiting no phenotypic perturbations on the rosette growth, the cytosolic *icdh* KO mutant shows a marked decrease (-60%) of NADP-ICDH activities (Figure 6A). This confirmed previous data showing that the cytosolic ICDH contributed to the major pool of the shoot extractible NADP-ICDH activity. Nevertheless, this contribution seems somehow less than reported previously (-90%), possibly due differences in growth conditions or developmental stage of the plants or to the contribution of organellar NADP-ICDH activities [6]. NADP-dependent ICDH activities were also consistently decreased in the *cat2* mutant (Figure 6A), which is impaired in the major isoform of the peroxisomal H_2O_2 detoxification enzyme Catalase 2 (Supplemental Figure S5), suggesting that NADP-ICDH activities are affected by oxidizing conditions. Interestingly, the steady-state ICDH activity was even more affected (-90%) in the *icdh cat2* double mutant, suggesting that other cellular NADP-ICDH isoforms might be affected by the *cat2* mutation. Moreover, the NADP-ICDH was also slightly decreased in the *gr1* involved in the reduction of the cytosolic glutathione pool [23,36]. The latter mutant accumulates higher levels of oxidized glutathione than wild-type plants, suggesting that NADP-ICDH activities might be sensitive to perturbed glutathione conditions [23]. The NADP-dependent ICDH activities are more strikingly affected in the *cat2 gr1* mutant, which accumulates much higher GSSG levels as *cat2* and *gr1* single mutants [24].

We also examined NADP-ICDH activity in different mutants affected in the NO metabolism [22,26–28]. NADP-ICDH activities were decreased in the *gsnor* mutant which accumulates a high level of GSNO. As expected from *in vitro* experiments, treatment of wild-type protein extracts with GSNO strikingly inhibits the activity. Surprisingly, NADP-ICDH steady-state activity is not affected in the *nox1* mutant which overaccumulates NO, suggesting a specific impact of GSNO on ICDH (Figure 6B). This was further confirmed by treating wild-type protein extracts with the NO donor sodium nitroprusside (SNP), which did not affect the NADP-ICDH activities. Finally, the activity is not perturbed in *noa1, nia1 nia2* and *noa1, nia1 nia2* mutants in which the biosynthesis of NO is alleviated (Figure 6B).

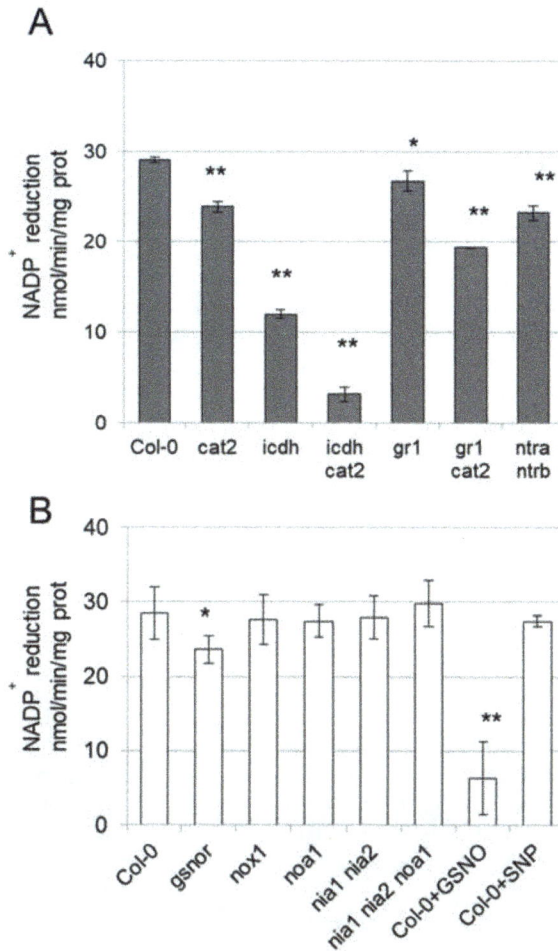

Figure 6. ICDH activity in planta. (**A**) ICDH activity was measured for 40 μg of protein extracts at 25 °C by monitoring the change in absorbance at 340 nm due to the isocitrate-dependent rate of NADP+ reduction. Wild-type (Col-0) or mutant A. thaliana plants were grown in soil in a controlled growth chamber (180 μE m^{-2} s^{-1}, 16 h day/8 h night, 22 °C 55% RH day, 20 °C 60% RH night) for 2 weeks. (**B**) Same design as in (**A**). In Col-O+GSNO and Col-0+SNP, protein extracts of wild-type plants were treated with 1 mM GSNO or SNP for 30 min before performing activity tests. Error bars represent SE (n = 3). * $p < 0.01$, ** $p < 0.001$ for statistical differences compared to non-treated samples (Student's t test).

4. Discussion

Plant ICDH are part of a multigenic family (Supplemental Figure S1). Mitochondrial NAD-dependent ICDH involved in TCA cycle are composed of 6 genes, two of them are encoding catalytic subunits, and the four others for regulatory subunits [37]. NADP-ICDH are found in the cytosol, peroxisome, mitochondria and chloroplast. Each of these compartments only contains a single isoform, one of them being dually targeted to chloroplasts and mitochondria [4,6,38]. Different arguments are in favor of redox regulation of ICDH: (i) ICDHs were found to exhibit sulfenylated [12,13], glutathionylated [14] or nitrosylated [15] Cys upon oxidizing conditions (ii) ICDH isoforms including the cICDH were previously identified as interactors of thiols reductases TRX and GRX in different proteomic analyses [16–20,39]. (iii) conserved Cys residues are found in ICDH isoforms of most organisms, including mammals (iiii) mammalian cytosolic ICDH have been shown to be regulated by S-nitrosylation [9,10], which we also confirmed in our study (Supplemental Figure S6). Supporting these assumptions, we found that cytosolic NADP-ICDH activities are affected in GSNO, GSSG and H_2O_2/GSH-supplied purified cICDH as well as in mutants exhibiting perturbed GSNO, GSSG or H_2O_2 reduction. While these oxidizing conditions seem not to induce intermolecular disulfide bonds in the recombinant enzyme as previously shown for other metabolic enzymes [29], the single Cys363 residue was consistently found to be glutathionylated after GSNO treatments. Interestingly, the Cys363 is also conserved in other ICDH isoforms (mitochondrial, peroxisome and chloroplastic/mitochondrial) as well as in cytosolic ICDH isoforms from other organisms (Supplemental Figure S1), suggesting that Cys363 redox regulation might occur in other ICDH isoforms.

Intriguingly, our ICDH model suggests that Cys363 residue is not located close to the ICDH active site and thus seems not to interfere directly with the catalytic activity. A likely hypothesis would be that upon oxidation, a major conformational change occurs, which might change the activity of the enzyme. The fact that the C363S point mutation does not affect the activity of the recombinant cICDH (Figure 5A) does not really support this hypothesis, assuming that a Cys to Ser mutation mimics Cys oxidation. Another hypothesis is that Cys363 glutathionylation (or nitrosylation) occurs that protects the residue from an irreversible overoxidation triggered by oxidative stress [29].

Interestingly, while having no significant impact on cICDH activity, the C363S mutation does affect the sensitivity of the enzyme to GSNO inhibition, which indicates that Cys363 is targeted by GSNO. We also showed that this single mutation does not fully alleviates the sensitivity to GSNO, suggesting that other residues might be involved. Thus, the mechanism by which oxidation affects cICDH activity needs to be further explored.

It has to be underlined that, while other Cys are found in the cICDH sequence, no other oxidized residues have been identified in our MS/MS experiments under GSNO treatment. Nevertheless, biotin-switch experiments have clearly identified S-nitrosylated residues upon GSNO treatments, suggesting that cICDH is also S-nitrosylated. Whether Cys363 is nitrosylated and if other residues are nitrosylated cannot be concluded from the non-quantitative biotin-switch technique performed on the cICDH-C363S protein and would need additional point mutation experiments. Consistently, an isotope-coded affinity tags (ICAT) approach identified three other S-nitrosylated residues (Cys 75 and Cys269) in cICDH in Arabidopsis [40]. Residue Cys75 of Arabidopsis NADP-ICDH activity has been found to be differentially S-nitrosylated in response to salt stress [41]. And Tyr392 has also been reported to be the only nitrated residue in pea plants and is possibly responsible for the inhibition of catalytic activity following treatment with SIN-1 [42]. These residues are conserved in NADP-ICDH of plants and other organisms (Supplemental Figure S1). While preparing this manuscript, Munos-Vargas et al. (2018) showed that NO donors peroxynitrite ($ONOO^-$), S-nitrosocyteine (CysNO) and DETA-NONOate also inhibited NADP-ICDH activity in sweet pepper. Munos-Vargas et al. (2018) established by *in silico* analysis of the tertiary structure of sweet pepper NADP-ICDH activity (UniProtKB ID A0A2G2Y555) that residues Cys133 and Tyr450 are the most likely potential targets for S-nitrosation and nitration, respectively [43].

Therefore, future mutagenesis experiments should establish the contribution of other residues than Cys363 in the redox regulation of the protein. As S-nitrosylation is a rather unstable modification, the fact that we do not identified S-nitrosylated residues in our LC-MS/MS is not surprising. Interestingly, biotin switch experiments indicate that the TRX system is unable to act as a denitrosylation system for cICDH, as shown for other substrates, including in plants [44–46]. In contrast, the GRX system is efficient and needs the complete NADPH/GR/GSH/GRXC1 system to be optimal at least *in vitro*. This is consistent with the observation that the GRXC1 or GRXC2 thiol reduction system is more efficient than the TRX system in rescuing the inhibitory effect of GSNO or GSSG on cICDH activity, suggesting that deglutathionylation activity of GRX is involved in the redox mechanism of cICDH regulation.

Our NADP-ICDH activity analyses in the *icdh* mutant confirm that cICDH contributes to the major part (~60%) of the NADP-ICDH activity. The other ~40% of the overall extractible activity likely provide from organellar NADP-ICDH activities. These proportions are somehow different than those reported previously, in which cICDH contributed to over 80% of the extractible activity in potato and *Arabidopsis* [5,6]. These differences might be due to different growth conditions (*in vitro* vs soil growth), developmental stages or tissues (10 day-old plantlets vs 3 week-old adult leaves) analyzed. Nevertheless, the observation that the NADP-ICDH activity is more affected in the *cat2 icdh* double mutant compared to the *icdh* single mutant suggests that organellar ICDH isoforms might be affected in by the *cat2* mutation. Nevertheless, we cannot rule out that the decrease in the NADP-ICDH activity in the cat2 mutant also relies on other modifications induced by oxidative distress.

ICDH is an important enzyme in 2-OG synthesis, as the carbon backbone of ammonia to glutamate by the GS/GOGAT pathway [1]. But it also provides reduced NAD(P)H equivalents to redox enzymes like GR or NTR. Interestingly, both *cat2, gr1* and *cat2 gr1* double mutants exhibit decreased NADPH production [23,25], which might rely on a decreased NADP-ICDH activity. Moreover, the lower ICDH activity we found in the cytosolic *gr1* mutant and to a higher extent in the *cat2 gr1* double mutant is possibly due to the high accumulation of oxidized glutathione found in these mutants. Consistently, the *gsnor* mutant impaired in the reduction of GSNO also exhibits lower ICDH activity, although this effect is less pronounced than an exogenous treatment by GSNO. Intriguingly, both in exogenous treatment with SNP and in the *nox1* mutant which accumulate high levels of NO, the ICDH activity is not affected, suggesting a distinct role of GSNO and NO in ICDH regulation. Such observations were reported previously and suggested that GSNO and NO have distinct substrates [45–47]. In the present work, the difference between impact on ICDH activity caused by *gsnor* and *nox1* could result from the fact that GSNO is also an efficient glutathione donor for cICDH S-glutathionylation.

5. Conclusions

Collectively, our data add another evidence of the potential redox regulation of primary carbon and nitrogen metabolism through the regulation of the cytosolic isocitrate dehydrogenase. We also highlight the importance the GRX-dependent thiol reduction systems as a potential actor in fine-tuning the redox regulation of NADP-ICDH activity. While we clearly demonstrate that the enzyme can be modified by glutathionylation *in vitro*, future experiments are needed to demonstrate that this redox regulation actually occurs *in vivo* in response to environmental clues.

Supplementary Materials: The following are available online at http://www.mdpi.com/2076-3921/8/1/16/s1, Figure S1: Alignement of NADP-ICDH sequence from different eukaryotics, Figure S2: Recombinant cICDH and cICDH-C363S, Figure S3: Recombinant cICDH is dimeric, Figure S4: Fragmentation tables of cICDH peptides identified with glutathionylated cysteines on Cys363., Figure S5: Catalase activity in planta, Figure S6: Human cytosolic ICDH activity is sensitive to NO donors, Table S1: Primers used in this study.

Author Contributions: Conceptualization, A.K.N., G.N. and J.-P.R.; Data curation, C.R.; Funding acquisition, G.N. and J.-P.R.; Investigation, A.K.N., L.B. and C.C.; Project administration, J.-P.R.; Resources, A.M. and G.N.; Supervision, J.-P.R.; Writing—Original draft, J.-P.R.; Writing—Review & Editing, A.K.N., C.R., C.C. and G.N.

Funding: This research was funded by [Agence Nationale de la Recherche] grant number [12-BSV6-0011].

Acknowledgments: This work was supported by grants from the Centre National de la Recherche Scientifique and the Agence Nationale de la Recherche (ANR-Blanc Cynthiol 12-BSV6-0011). This work has been supported by LabEx Agro and LabEx TULIP.

Conflicts of Interest: The authors declare no conflicts of interest.

References

1. Hodges, M. Enzyme redundancy and the importance of 2-oxoglutarate in plant ammonium assimilation. *J. Exp. Bot.* **2002**, *53*, 905–916. [CrossRef] [PubMed]

2. Lemaitre, T.; Urbanczyk-Wochniak, E.; Flesch, V.; Bismuth, E.; Fernie, A.R.; Hodges, M. NAD-dependent isocitrate dehydrogenase mutants of Arabidopsis suggest the enzyme is not limiting for nitrogen assimilation. *Plant Physiol.* **2007**, *144*, 1546–1558. [CrossRef] [PubMed]
3. Møller, I.M.; Rasmusson, A.G. The role of NADP in the mitochondrial matrix. *Trends Plant Sci.* **1998**, *3*, 21–27.
4. Galvez, S.; Bismuth, E.; Sarda, C.; Gadal, P. Purification and Characterization of *Chloroplastic* NADP-Isocitrate Dehydrogenase from Mixotrophic Tobacco Cells (Comparison with the *Cytosolic Isoenzyme*). *Plant Physiol.* **1994**, *105*, 593–600. [CrossRef]
5. Kruse, A.; Fieuw, S.; Heineke, D.; Müller-Röber, B. Antisens inhibition of cytosolic NADP-dependent isocitrate dehydrogenase in transgenic potato plants. *Planta* **1998**, *205*, 82–91. [CrossRef]
6. Mhamdi, A.; Mauve, C.; Gouia, H.; Saindrenan, P.; Hodges, M.; Noctor, G. Cytosolic NADP-dependent isocitrate dehydrogenase contributes to redox homeostasis and the regulation of pathogen responses in Arabidopsis leaves. *Plant Cell Environ.* **2010**, *33*, 1112–1123. [PubMed]
7. Benhar, M.; Forrester, M.T.; Stamler, J.S. Protein denitrosylation: Enzymatic mechanisms and cellular functions. *Nat. Rev. Mol. Cell Biol.* **2009**, *10*, 721–732. [CrossRef]
8. Meyer, Y.; Belin, C.; Delorme-Hinoux, V.; Reichheld, J.-P.; Riondet, C. Thioredoxin and glutaredoxin systems in plants: Molecular mechanisms, crosstalks, and functional significance. *Antioxid. Redox Signal.* **2012**, *17*, 1124–1160. [CrossRef]
9. Yang, E.S.; Richter, C.; Chun, J.-S.; Huh, T.-L.; Kang, S.-S.; Park, J.-W. Inactivation of NADP(+)-dependent isocitrate dehydrogenase by nitric oxide. *Free Radic. Biol. Med.* **2002**, *33*, 927–937. [CrossRef]
10. Lee, J.H.; Yang, E.S.; Park, J.-W. Inactivation of NADP+-dependent isocitrate dehydrogenase by peroxynitrite. Implications for cytotoxicity and alcohol-induced liver injury. *J. Biol. Chem.* **2003**, *278*, 51360–51371. [CrossRef]
11. Shin, S.W.; Oh, C.J.; Kil, I.S.; Park, J.-W. Glutathionylation regulates cytosolic NADP+-dependent isocitrate dehydrogenase activity. *Free Radic. Res.* **2009**, *43*, 409–416. [CrossRef] [PubMed]
12. Akter, S.; Huang, J.; Bodra, N.; De Smet, B.; Wahni, K.; Rombaut, D.; Pauwels, J.; Gevaert, K.; Carroll, K.; Van Breusegem, F.; et al. DYn-2 Based Identification of Arabidopsis Sulfenomes. *Mol. Cell. Proteom.* **2015**, *14*, 1183–1200. [CrossRef] [PubMed]
13. Waszczak, C.; Akter, S.; Eeckhout, D.; Persiau, G.; Wahni, K.; Bodra, N.; Van Molle, I.; De Smet, B.; Vertommen, D.; Gevaert, K.; et al. Sulfenome mining in *Arabidopsis thaliana*. *Proc. Natl. Acad. Sci. USA* **2014**, *111*, 11545–11550. [CrossRef] [PubMed]
14. Zaffagnini, M.; Bedhomme, M.; Groni, H.; Marchand, C.H.; Puppo, C.; Gontero, B.; Cassier-Chauvat, C.; Decottignies, P.; Lemaire, S.D. Glutathionylation in the photosynthetic model organism *Chlamydomonas reinhardtii*: A proteomic survey. *Mol. Cell. Proteom.* **2012**, *11*. [CrossRef]
15. Morisse, S.; Zaffagnini, M.; Gao, X.-H.; Lemaire, S.D.; Marchand, C.H. Insight into protein S-nitrosylation in Chlamydomonas reinhardtii. *Antioxid. Redox Signal.* **2014**, *21*, 1271–1284. [CrossRef] [PubMed]
16. Balmer, Y.; Vensel, W.H.; Tanaka, C.K.; Hurkman, W.J.; Gelhaye, E.; Rouhier, N.; Jacquot, J.-P.; Manieri, W.; Schürmann, P.; Droux, M.; et al. Thioredoxin links redox to the regulation of fundamental processes of plant mitochondria. *Proc. Natl. Acad. Sci. USA* **2004**, *101*, 2642–2647. [CrossRef]
17. Marchand, C.; Le Maréchal, P.; Meyer, Y.; Miginiac-Maslow, M.; Issakidis-Bourguet, E.; Decottignies, P. New targets of Arabidopsis thioredoxins revealed by proteomic analysis. *Proteomics* **2004**, *4*, 2696–2706. [CrossRef]
18. Marchand, C.; Le Maréchal, P.; Meyer, Y.; Decottignies, P. Comparative proteomic approaches for the isolation of proteins interacting with thioredoxin. *Proteomics* **2006**, *6*, 6528–6537. [CrossRef]
19. Rouhier, N.; Villarejo, A.; Srivastava, M.; Gelhaye, E.; Keech, O.; Droux, M.; Finkemeier, I.; Samuelsson, G.; Dietz, K.J.; Jacquot, J.-P.; et al. Identification of plant glutaredoxin targets. *Antioxid. Redox Signal.* **2005**, *7*, 919–929. [CrossRef]
20. Pérez-Pérez, M.E.; Mauriès, A.; Maes, A.; Tourasse, N.J.; Hamon, M.; Lemaire, S.D.; Marchand, C.H. The Deep Thioredoxome in Chlamydomonas reinhardtii: New Insights into Redox Regulation. *Mol. Plant* **2017**, *10*, 1107–1125. [CrossRef]
21. He, Y.; Tang, R.-H.; Hao, Y.; Stevens, R.D.; Cook, C.W.; Ahn, S.M.; Jing, L.; Yang, Z.; Chen, L.; Guo, F.; et al. Nitric oxide represses the Arabidopsis floral transition. *Science* **2004**, *305*, 1968–1971. [CrossRef] [PubMed]
22. Lee, U.; Wie, C.; Fernandez, B.O.; Feelisch, M.; Vierling, E. Modulation of nitrosative stress by S-nitrosoglutathione reductase is critical for thermotolerance and plant growth in Arabidopsis. *Plant Cell* **2008**, *20*, 786–802. [CrossRef] [PubMed]

23. Mhamdi, A.; Hager, J.; Chaouch, S.; Queval, G.; Han, Y.; Taconnat, L.; Saindrenan, P.; Gouia, H.; Issakidis-Bourguet, E.; Renou, J.-P.; et al. Arabidopsis GLUTATHIONE REDUCTASE1 plays a crucial role in leaf responses to intracellular hydrogen peroxide and in ensuring appropriate gene expression through both salicylic acid and jasmonic acid signaling pathways. *Plant Physiol.* **2010**, *153*, 1144–1160. [CrossRef]

24. Reichheld, J.-P.; Khafif, M.; Riondet, C.; Droux, M.; Bonnard, G.; Meyer, Y. Inactivation of thioredoxin reductases reveals a complex interplay between thioredoxin and glutathione pathways in Arabidopsis development. *Plant Cell* **2007**, *19*, 1851–1865. [CrossRef]

25. Queval, G.; Issakidis-Bourguet, E.; Hoeberichts, F.A.; Vandorpe, M.; Gakière, B.; Vanacker, H.; Miginiac-Maslow, M.; Van Breusegem, F.; Noctor, G. Conditional oxidative stress responses in the Arabidopsis photorespiratory mutant cat2 demonstrate that redox state is a key modulator of daylength-dependent gene expression, and define photoperiod as a crucial factor in the regulation of H_2O_2-induced cell death. *Plant J.* **2007**, *52*, 640–657. [PubMed]

26. Lozano-Juste, J.; León, J. Enhanced abscisic acid-mediated responses in *nia1nia2noa1-2* triple mutant impaired in NIA/NR- and AtNOA1-dependent nitric oxide biosynthesis in Arabidopsis. *Plant Physiol.* **2010**, *152*, 891–903. [CrossRef] [PubMed]

27. Hu, W.-J.; Chen, J.; Liu, T.-W.; Liu, X.; Chen, J.; Wu, F.-H.; Wang, W.-H.; He, J.-X.; Xiao, Q.; Zheng, H.-L. Comparative proteomic analysis on wild type and nitric oxide-overproducing mutant (*nox1*) of *Arabidopsis thaliana*. *Nitric Oxide* **2014**, *36*, 19–30. [CrossRef] [PubMed]

28. Chen, R.; Sun, S.; Wang, C.; Li, Y.; Liang, Y.; An, F.; Li, C.; Dong, H.; Yang, X.; Zhang, J.; et al. The *Arabidopsis PARAQUAT RESISTANT2* gene encodes an S-nitrosoglutathione reductase that is a key regulator of cell death. *Cell Res.* **2009**, *19*, 1377–1387. [CrossRef]

29. Huang, J.; Niazi, A.K.; Young, D.; Rosado, L.A.; Vertommen, D.; Bodra, N.; Abdelgawwad, M.R.; Vignols, F.; Wei, B.; Wahni, K.; et al. Self-protection of cytosolic malate dehydrogenase against oxidative stress in Arabidopsis. *J. Exp. Bot.* **2018**, *69*, 3491–3505. [CrossRef] [PubMed]

30. Gibon, Y.; Blaesing, O.E.; Hannemann, J.; Carillo, P.; Höhne, M.; Hendriks, J.H.M.; Palacios, N.; Cross, J.; Selbig, J.; Stitt, M. A Robot-based platform to measure multiple enzyme activities in *Arabidopsis* using a set of cycling assays: Comparison of changes of enzyme activities and transcript levels during diurnal cycles and in prolonged darkness. *Plant Cell* **2004**, *16*, 3304–3325. [CrossRef] [PubMed]

31. Daloso, D.M.; Müller, K.; Obata, T.; Florian, A.; Tohge, T.; Bottcher, A.; Riondet, C.; Bariat, L.; Carrari, F.; Nunes-Nesi, A.; et al. Thioredoxin, a master regulator of the tricarboxylic acid cycle in plant mitochondria. *Proc. Natl. Acad. Sci. U.A* **2015**, *112*, E1392–E1400. [CrossRef] [PubMed]

32. Carapito, C.; Burel, A.; Guterl, P.; Walter, A.; Varrier, F.; Bertile, F.; Van Dorsselaer, A. MSDA, a proteomics software suite for in-depth Mass Spectrometry Data Analysis using grid computing. *Proteomics* **2014**, *14*, 1014–1019. [CrossRef] [PubMed]

33. Forrester, M.T.; Foster, M.W.; Benhar, M.; Stamler, J.S. Detection of protein S-nitrosylation with the biotin-switch technique. *Free Radic. Biol. Med.* **2009**, *46*, 119–126. [CrossRef] [PubMed]

34. Xu, X.; Zhao, J.; Xu, Z.; Peng, B.; Huang, Q.; Arnold, E.; Ding, J. Structures of human cytosolic NADP-dependent isocitrate dehydrogenase reveal a novel self-regulatory mechanism of activity. *J. Biol. Chem.* **2004**, *279*, 33946–33957. [CrossRef] [PubMed]

35. Kneeshaw, S.; Keyani, R.; Delorme-Hinoux, V.; Imrie, L.; Loake, G.J.; Le Bihan, T.; Reichheld, J.-P.; Spoel, S.H. Nucleoredoxin guards against oxidative stress by protecting antioxidant enzymes. *Proc. Natl. Acad. Sci. USA* **2017**, *114*, 8414–8419. [CrossRef] [PubMed]

36. Marty, L.; Siala, W.; Schwarzländer, M.; Fricker, M.D.; Wirtz, M.; Sweetlove, L.J.; Meyer, Y.; Meyer, A.J.; Reichheld, J.-P.; Hell, R. The NADPH-dependent thioredoxin system constitutes a functional backup for cytosolic glutathione reductase in Arabidopsis. *Proc. Natl. Acad. Sci. USA* **2009**, *106*, 9109–9114. [CrossRef]

37. Lemaitre, T.; Hodges, M. Expression analysis of *Arabidopsis thaliana* NAD-dependent isocitrate dehydrogenase genes shows the presence of a functional subunit that is mainly expressed in the pollen and absent from vegetative organs. *Plant Cell Physiol.* **2006**, *47*, 634–643. [CrossRef]

38. Corpas, F.J.; Barroso, J.B.; Sandalio, L.M.; Palma, J.M.; Lupiáñez, J.A.; del Río, L.A. Peroxisomal NADP-Dependent Isocitrate Dehydrogenase. Characterization and Activity Regulation during Natural Senescence. *Plant Physiol.* **1999**, *121*, 921–928. [CrossRef]

39.   Montrichard, F.; Alkhalfioui, F.; Yano, H.; Vensel, W.H.; Hurkman, W.J.; Buchanan, B.B. Thioredoxin targets in plants: The first 30 years. *J Proteomics* **2009**, *72*, 452–474. [CrossRef]

40.   Fares, A.; Rossignol, M.; Peltier, J.-B. Proteomics investigation of endogenous S-nitrosylation in *Arabidopsis*. *Biochem. Biophys. Res. Commun.* **2011**, *416*, 331–336. [CrossRef]

41.   Källberg, M.; Wang, H.; Wang, S.; Peng, J.; Wang, Z.; Lu, H.; Xu, J. Template-based protein structure modeling using the RaptorX web server. *Nature Protocols* **2012**, *7*, 1511–1522. [CrossRef] [PubMed]

42.   Begara-Morales, J.C.; Chaki, M.; Sánchez-Calvo, B.; Mata-Pérez, C.; Leterrier, M.; Palma, J.M.; Barroso, J.B.; Corpas, F.J. Protein tyrosine nitration in pea roots during development and senescence. *J. Exp. Bot.* **2013**, *64*, 1121–1134. [CrossRef] [PubMed]

43.   Muñoz-Vargas, M.A.; González-Gordo, S.; Cañas, A.; López-Jaramillo, J.; Palma, J.M.; Corpas, F.J. Endogenous hydrogen sulfide (*H2S*) is up-regulated during sweet pepper (*Capsicum annuum* L.) fruit ripening. In vitro analysis shows that NADP-dependent isocitrate dehydrogenase (ICDH) activity is inhibited by H2S and NO. *Nitric Oxide* **2018**, *81*, 36–45. [CrossRef] [PubMed]

44.   Benhar, M.; Forrester, M.T.; Hess, D.T.; Stamler, J.S. Regulated protein denitrosylation by cytosolic and mitochondrial thioredoxins. *Science* **2008**, *320*, 1050–1054. [CrossRef]

45.   Benhar, M. Application of a Thioredoxin-Trapping Mutant for Analysis of the Cellular Nitrosoproteome. *Meth. Enzymol.* **2017**, *585*, 285–294. [PubMed]

46.   Kneeshaw, S.; Gelineau, S.; Tada, Y.; Loake, G.J.; Spoel, S.H. Selective protein denitrosylation activity of Thioredoxin-*h*5 modulates plant Immunity. *Mol. Cell* **2014**, *56*, 153–162. [CrossRef]

47.   Tada, Y.; Spoel, S.H.; Pajerowska-Mukhtar, K.; Mou, Z.; Song, J.; Wang, C.; Zuo, J.; Dong, X. Plant immunity requires conformational changes [corrected] of NPR1 via S-nitrosylation and thioredoxins. *Science* **2008**, *321*, 952–956. [CrossRef] [PubMed]

antioxidants

MDPI

Article

Redox Regulation of Monodehydroascorbate Reductase by Thioredoxin y in Plastids Revealed in the Context of Water Stress

Hélène Vanacker, Marjorie Guichard [†], Anne-Sophie Bohrer [‡] and Emmanuelle Issakidis-Bourguet *

Institute of Plant Sciences Paris-Saclay (IPS2), UMR Université Paris Sud—CNRS 9213—INRA 1403, Bât. 630, 91405 Orsay CEDEX, France; helene.vanacker@u-psud.fr (H.V.); marjorie.guichard@cos.uni-heidelberg.de (M.G.); bohreras@msu.edu (A.-S.B.)
* Correspondence: emmanuelle.issakidis-bourguet@u-psud.fr; Tel.: +33-1-69-15-33-37
† Present address: Centre for Organismal Studies Heidelberg, Universität Heidelberg, 69120 Heidelberg, Germany.
‡ Present address: Department of Biochemistry and Molecular Biology, Michigan State University, East Lansing, MI 48824, USA.

Received: 31 October 2018; Accepted: 5 December 2018; Published: 6 December 2018

Abstract: Thioredoxins (TRXs) are key players within the complex response network of plants to environmental constraints. Here, the physiological implication of the plastidial y-type TRXs in Arabidopsis drought tolerance was examined. We previously showed that TRXs y1 and y2 have antioxidant functions, and here, the corresponding single and double mutant plants were studied in the context of water deprivation. TRX y mutant plants showed reduced stress tolerance in comparison with wild-type (WT) plants that correlated with an increase in their global protein oxidation levels. Furthermore, at the level of the main antioxidant metabolites, while glutathione pool size and redox state were similarly affected by drought stress in WT and *trxy1y2* plants, ascorbate (AsA) became more quickly and strongly oxidized in mutant leaves. Monodehydroascorbate (MDA) is the primary product of AsA oxidation and NAD(P)H-MDA reductase (MDHAR) ensures its reduction. We found that the extractable leaf NADPH-dependent MDHAR activity was strongly activated by TRX y2. Moreover, activity of recombinant plastid Arabidopsis MDHAR isoform (MDHAR6) was specifically increased by reduced TRX y, and not by other plastidial TRXs. Overall, these results reveal a new function for y-type TRXs and highlight their role as major antioxidants in plastids and their importance in plant stress tolerance.

Keywords: thioredoxin; monodehydroascorbate reductase; water stress; protein oxidation; antioxidants; ascorbate; glutathione

1. Introduction

As sessile organisms plants are continuously exposed to environmental fluctuations. In order to maintain photosynthetic carbon fixation efficiency, especially in varying light conditions, they have evolved diverse adaptive strategies including redox regulation. Indeed, thiol-based redox systems, i.e., glutathione and thioredoxins (TRXs), play major roles in the complex redox regulatory network underlying plant responses to fluctuating environmental cues.

TRXs are small ubiquitous redox proteins catalyzing dithiol–disulfide exchange reactions with their target enzymes thanks to the presence of 2 reactive Cys residues in the conserved WC(G/P)PC motif in their active site. TRXs can fulfill two types of functions, either as redox regulators that usually allow the reductive activation of their target enzymes, or as reducing substrates that provide reducing power for antioxidant systems that detoxify H_2O_2 [1].

Plant genome sequencing data revealed that photosynthetic organisms possess a high number of trx genes including numerous isoforms localized in chloroplasts that were classified into five subtypes. In Arabidopsis (*Arabidopsis thaliana*) 10 plastidial TRXs were found: two TRXf, four TRX m, one TRX x, two TRX y, and one TRX z [2,3]. Biochemical studies enabled functional specificity to be assigned to different plastidial TRXs, leading to a global picture in which the f and m-type TRXs are regulators of enzymes directly or indirectly linked to photosynthetic carbon metabolism, while the x and the y types appear to have antioxidant functions [4–7]. TRX z which displays unique properties among plastidial TRXs [8,9] has been recently validated as a regulator of plastidial gene expression [10]. Recent studies, using Trx mutant plants and over-expressors, allowed some of these functions to be confirmed *in planta*, enabling evaluation of the degree of redundancy between the various TRX types. Most of these studies have investigated the roles of TRXs f and TRXs m in the regulation of stromal enzymes to adjust their activity to varying photosynthetic electron flow under fluctuating light intensities [11–14]. Moreover, Arabidopsis TRXs y1 and y2 were shown to play antioxidant roles whatever their redox state, by serving as reducing substrates [5–7,15], performing oxidative activation of G6PDH [16], and maintaining leaf MSR (Methionine Sulfoxide Reductase) capacity in high light conditions [17].

As a first approach to identify TRX y protein partners, we previously performed a proteomic study of putative targets of y-type TRX in Arabidopsis roots, since Trx y1 is mostly expressed in non-photosynthetic organs [5,9,18]. A monocysteinic mutant of TRX y was used as a bait to trap protein partners by affinity chromatography in a root crude extract. Seventy-two proteins have been identified, functioning mainly in metabolism, detoxification and response to stress, as well as in protein processing and signal transduction. In particular, we identified the plastidial monodehydroascorbate reductase (MDHAR) as a TRX-linked protein [19]. This enzyme is considered to play a role in ROS detoxification, being part of the ascorbate-glutathione pathway using ascorbate (AsA) as reducing substrate. AsA oxidation leads to monodehydroascorbate which can be recycled back to AsA thanks to MDHAR using NAD(P)H as the reductant [20]. Therefore, MDHARs play an important role in the response of plants to oxidative stress by maintaining the intracellular ascorbate redox state mainly in the reduced state.

Drought is one of the most serious environmental stresses affecting plant performance and crop yield and is expected to become more widespread and severe due to climate change [21]. Moreover, it is well established that cell redox homeostasis is disturbed under dehydration stress [22]. In this context, we studied the impact of the TRXs y mutations on the antioxidant response of Arabidopsis plants challenged with drought stress.

2. Materials and Methods

2.1. Reagents

All biochemical reagents were purchased from Sigma-Aldrich (Sigma-Aldrich Chimie, Saint-Quentin Fallavier, France), unless otherwise mentioned.

2.2. Plant Material

All the Arabidopsis (*Arabidopsis thaliana*) mutants used in this study were in the Columbia (Col-0) genetic background. Knock-out plants (T-DNA insertion mutants) in the trx y1 (At1g76760) or/and trx y2 (At1g43560) gene(s), either single (*trxy1-1* and *trxy1-2*, two allelic mutants lines, and *trxy2*) or double (*trxy1y2* obtained from *trxy1-2 and trxy2*) mutant lines, used in this study were previously obtained and described [17].

2.3. Plant Growth Conditions and Water Stress Treatment

15-days old seedlings (obtained under in vitro short day conditions i.e., 8-h photoperiod at 100 μmol photons m^{-2} s^{-1} for 8 h, 20 °C/18 °C (day/night) temperature regime and a relative humidity of 65%, on 1/2 MS agar medium) were individually transferred into a ready-to-use plant multiplication plug system (Fertiss 455.40, FERTIL, Boulogne-Billancourt, France) and further grown in a controlled-environment growth chamber under an 8-h photoperiod at an irradiance of 150 μmol photons $m^{-2} s^{-1}$. The temperature regime was 20 °C/18 °C (day/night) and the relative humidity was 65%. Plants were irrigated twice a week with fertilizing solution (NPK 14.12.32, PLANT-PROD, FERTIL). Plants were grown for 3 weeks and either sampled or further cultivated in control or drought stress conditions and sampled for experiments as indicated. Samples were rapidly frozen in liquid nitrogen and stored at −80 °C until analysis. All data are means ± SD of at least three leaf samples obtained from different plants, and experiments were repeated at least twice. Drought stress was imposed by stopping irrigation of 3-week-old plants.

2.4. Relative Water Content Measurement

Relative water content was calculated according to the equation of RWC (%) = [(FW − DW)/ (TW − DW)] × 100, by measuring fresh weight of excised leaves (FW); turgid weight after dipping in water for 4 h (TW), and dry weight after overnight drying at 80 °C (DW).

2.5. mBBR Labelling and Quantification

The monobromobimane (mBBr) probe was used for detection of reduced proteins since it fluoresces following its covalent interaction with thiols (reduced form of Cys). Leaf samples (100 mg) were ground in Tris-HCl 100 mM pH 7.6 supplemented with the broadly used cocktail of protease inhibitors (special plant from Sigma-Aldrich P9599). Free thiols were directly labeled with 2 mM mBBr included in the extraction buffer (30 min incubation at room temperature). After centrifugation (14,000× g, 10 min, at 4 °C), soluble proteins were quantified (Qubit protein assay kit, Life technologies, Thermo Fisher Scientific, Illkirch, France) and resolved by SDS-PAGE in 4–20% acrylamide gels (25 μg protein sample loaded per well). Reduced proteins were visualized under UV before staining of total proteins with Coomassie blue. The mBBr fluorescence and Coomassie colorimetric signals were quantified using the VisionCapt software (Quantum ST5 from Vilber-Lourmat, Vilber, Marne-La-Vallée, France).

2.6. Protein Carbonylation

The spectrophotometric dinitrophenyl hydrazine (DNPH) method was used for the determination of carbonyl groups in proteins. Centrifugation-clarified leaf samples were prepared as described above (without mBBr) before removal of nucleic acids by precipitation with streptomycin sulphate 1% (w/v) (20 min incubation and centrifugation at 12,000× g at room temperature). Supernatants were mixed with 7.5 mM DNPH final concentration and incubated for 15 min prior to precipitation in presence of TCA 10% (v/v). The pellets were washed five times with ethanol:ethylacetate (1:1), dried and finally dissolved in 6 M guanidine hydrochloride and the absorption at 370 nm was measured. Carbonyl content was calculated using a molar absorption coefficient for aliphatic hydrazones of 22,000 M^{-1} cm^{-1} and protein recovery was estimated by measuring the A276 and corrected using the formula [Protein] = (A276 − 0.43 × A370) established previously [23].

2.7. MDHAR Activity Measurements

For measurement of MDHAR (EC 1.6.5.4) activity in leaves, freshly harvested leaf tissue (in the middle of the light period, 250 mg) was extracted in 1 mL of 50 mM MES-KOH buffer (pH 6.0), containing 40 mM KCl, 2 mM $CaCl_2$, and 1 mM L-ascorbic acid (AsA, freshly prepared). The homogenate was centrifuged at 14,000× g for 10 min at 4 °C, and the supernatant analyzed immediately for MDHAR activity.

MDHAR activity in leaf extracts was assayed spectrophotometrically at 25 °C by the slightly modified method described previously [24]. The MDHAR reaction was started by adding 0.4 unit of ascorbate oxidase (1 unit defined as the amount of enzyme catalyzing the oxidation of 1 μmol ascorbate per min) to generate the monodehydroascorbate radical in the reaction mixture (1 mL) containing 50 mM HEPES-KOH buffer (pH 7.6), 2.5 mM AsA, 0.25 mM NADPH and 15 µM FAD. The activity was determined by following for 2 min the decrease in absorbance at 340 nm due to the oxidation of NADPH using an extinction coefficient of 6.22 mM^{-1} cm^{-1}. The same protocol was used to measure the MDHAR activity of recombinant protein (obtained as described below) following 15 min incubation in 100 mM Tris-HCl pH 7.9 at room temperature (1 µM MDHAR6 in the final 1 mL cuvette assay) in the presence or absence of 10 mM dithiothreitol (DTT), alone or with 10 µM TRX.

2.8. Production and Purification of Recombinant Proteins

The cDNA sequence corresponding to the MDHAR6 (At1g63940) was obtained from the "Arabidopsis Biological Resource Center" (ABRC, DKLAT1G63940; clone U09541) and amplified by PCR (primers used detailed in Supplemental Table S1) and cloned at *NcoI* and *XhoI* restriction sites into the pGENI vector [9] allowing the production of AtMDHAR6 (without transit peptide, starting at Phe39) with a Strep-tag at its C-terminus in BL21 (DE3) *E. coli* cells. Bacteria were cultured, at 37 °C, in Luria-Bertani broth (LB) medium supplemented with 100 µg/mL ampicillin. AtMDHAR6 protein production was induced with 500 µM isopropyl-*β*-D-thiogalactopyranoside (IPTG) for 3 h at 37 °C. Cells were harvested by centrifugation at 8000× *g* for 20 min at 4 °C, resuspended in 30 mM Tris-HCl, pH 7.9 with a cocktail of protease inhibitors (Complete EDTA-free protease inhibitor cocktail, Roche Diagnostics), disrupted by three passages through a French press (10,000 p.s.i.) and soluble extract was cleared by centrifugation at 19,000× *g*, 4 °C for 45 min. The supernatant was loaded onto a *Strep*-Tactin affinity column (*Strep*-Tactin® Sepharose®, IBA GmbH Göttingen, Germany), pre-equilibrated with buffer 100 mM Tris-HCl, pH 8.0, 150 mM NaCl, 1 mM EDTA. After washing with the same buffer, the recombinant protein was eluted with 2.5 mM desthiobiotin and dialysed against 30 mM Tris-HCl, pH 7.9, 1 mM EDTA. Purity and molecular mass of the protein were checked by SDS-PAGE with Coomassie blue staining. Protein concentrations were determined spectrophotometrically at 450 nm, corresponding to the flavin adenine nucleotide (FAD) absorption peak, using a molar extinction coefficient of 11.3 mM^{-1} cm^{-1} [25]. The identity and purity of recombinant AtMDHAR6 protein preparation was confirmed by mass spectrometry. Recombinant Arabidopsis plastidial TRXs (TRX f1, TRX m1, TRX x and TRX y2) were obtained and purified as previously described [4,5].

2.9. Determination of Ascorbate and Glutathione

Antioxidant metabolites were extracted from whole leaves as described previously [26]. The content of AsA and DHA were measured as described previously via the decrease in *A*265 after the addition of AsA oxidase [27]. DHA content was calculated as the difference between total and reduced AsA. Total AsA was measured after incubation of the sample with DTT (2.4 mM) for 15 min. Total glutathione (GSH and GSSG) and GSSG were measured as described previously [28]. Total glutathione was estimated via the increase in *A*412 after the addition of Glutathione Reductase (GR) and NADPH.

2.10. Statistical Analysis

All analyses were performed according to a completely randomized design. Each experiment was repeated 2–4 times. The results were expressed as means and error bars were used to show standard deviation ($\pm$SD). Significant differences between genotypes (WT vs. mutant) or growth conditions (control vs. stress) were compared using Student's *t*-test, with $p < 0.05$ considered as significantly different.

3. Results

3.1. TRX y1 and TRX y2 are Important for Arabidopsis Drought Stress Tolerance

Previous work had suggested that both TRXs of the y-type could play an important role in determining tolerance of Arabidopsis plants to environmental stress [17]. Here, we studied the impact of the TRXs y mutations on the antioxidant response after drought stress. We first studied the tolerance of TRXs y1 and y2 single mutants to dehydration (two allelic mutant lines *trxy1-1* and *trxy1-2* and one *trxy2* mutant line). While in control growth conditions, *trxy1* and *trxy2* mutant plants developed similarly to wild-type (WT) plants (Figure 1A), they showed an increased sensitivity to dehydration as evidenced by their wilted phenotype (Figure 1B). This behavior was confirmed by the lower capacity of mutant plants to recover from a 9-day period of water deprivation (Figure 1C,D). Indeed, 24 h after re-watering, while almost all WT plants were still alive, only 50–60% of the *trxy1* and *trxy2* mutant plants were able to recover from the stress.

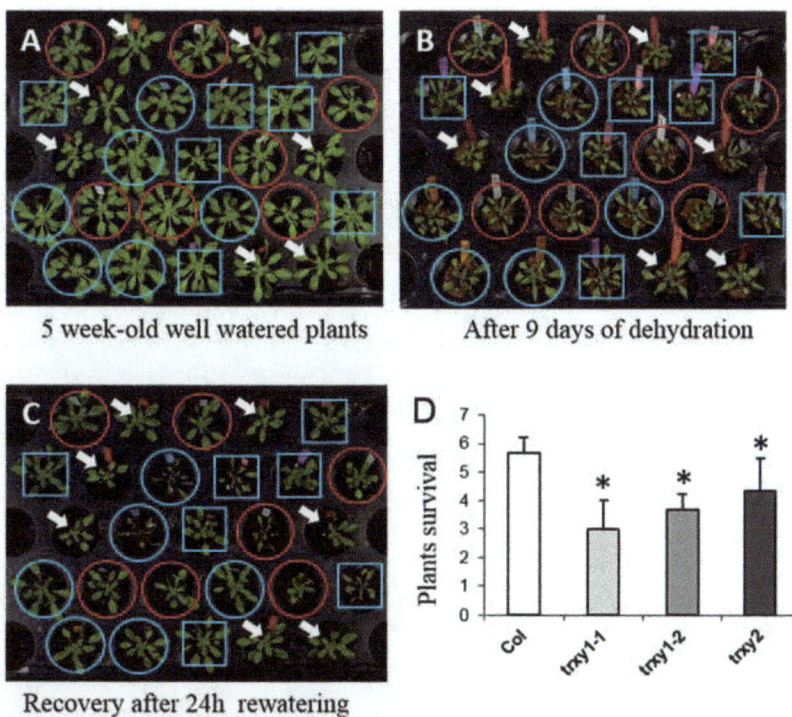

Figure 1. Water stress tolerance of WT and *trxy* mutant plants. After 3 weeks of growth under (**A**) standard conditions; (**B**) plants watering was stopped for 9 days and then (**C**) rehydrated for 24 h before (**D**) plant survival was monitored. Data correspond to means ± SD ($n = 6$). The asterisk (*) indicates a mutant sample significantly different from wild-type ($p < 0.05$). White arrows, blue circles, blue squares and red circles indicate wild-type (Col), *trxy1-1*, *trxy1-2* and *trxy2* plants, respectively.

These preliminary results provided a first indication of a functional role for TRX y1 and TRX y2 in determining the tolerance of Arabidopsis to water deficiency as well as possible redundancy between the two y-type TRX isoforms in this stress context. To further investigate this question, we obtained a *trxy1y2* double mutant line by crossing the *trxy1-2* and *trxy2* single mutants. The *trxy1y2* double mutant did not show any obvious phenotype in optimal growth conditions ([17], and this work). In the double mutant, leaf relative water content (RWC), remained high (ca. 80%) during 9 days of dehydration, but subsequently decreased more markedly than in the WT (Figure 2). Thus, y-type TRXs seemed to be important for the tolerance of Arabidopsis to water deficiency, suggesting an impaired capacity of the corresponding double mutant *trxy1y2* to cope with stress-triggered oxidative effects at the molecular level.

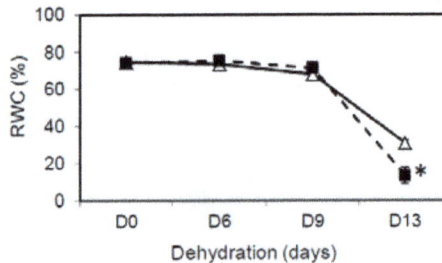

Figure 2. Effect of water stress on the relative water content (RWC, in %) in Col-0 (open triangles) and the double mutant *trxy1y2* (dark squares). Means ± SD are the average of 6 biological repeats from 2 independent experiments (*n* = 6). The asterisk (*) indicates a mutant sample significantly different from WT (*p* < 0.05).

3.2. Global Protein Oxidation is Enhanced in the trxy1y2 Mutant Under Drought Stress

The extent of protein carbonylation, a stress-related PTM considered as a hallmark of protein oxidation was studied in the *trxy1y2* mutant [29]. After 7 days of dehydration, we found that protein carbonylation was higher in mutant plants compared to WT (Figure 3). The correlation between stress tolerance and protein oxidation level was even clearer when the global protein thiol content was quantified (Figure 4). In control conditions, proteins from mutant plant leaves had a lower thiol content of ca. 15% than WT, and this difference was even more pronounced in drought conditions (ca. 33%). These results suggested that loss of y-type TRX functions is accompanied by oxidative stress at the molecular level and that this effect might be a primary reason for the increased sensitivity of the corresponding mutants to water deficit.

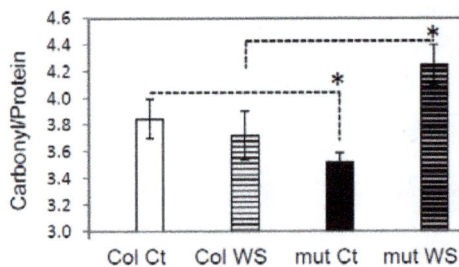

Figure 3. Protein carbonylation levels in wild-type (WT) and *trxy1y2* mutant leaves after 7 days of drought. Protein carbonyls were detected after dinitrophenyl hydrazine (DNPH) labeling and quantification. Ct: control (well watered); WS: water stress (7 days). For each genotype/condition, 4 samples from two independent experiments were analyzed. Data correspond to molar carbonyl/protein ratios, means ± SD (*n* = 4). Samples significantly different (*p* < 0.05) are indicated by an asterisk (*).

(A)

(B)

(C)

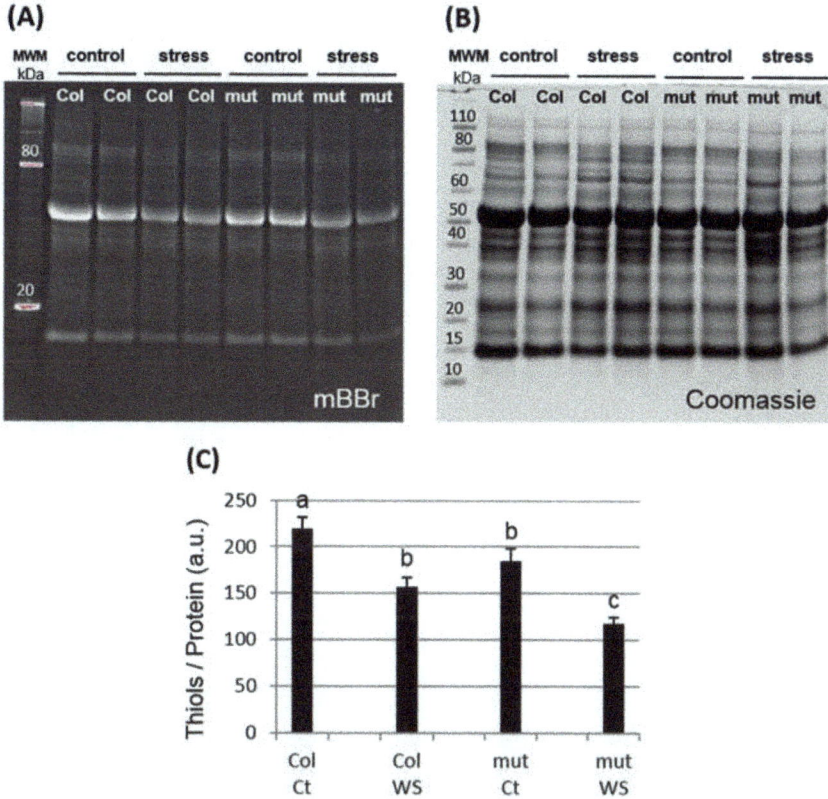

Figure 4. Protein thiols quantification in WT and *trxy1y2* mutant leaves after 7 days of drought. (**A**) mBBr fluorescent labelling of protein thiols after SDS-PAGE; (**B**) Coomassie staining of the same gel; (**C**) Thiol content relative to protein content. Ct: control (well watered); WS: water stress (7 days). MWM: molecular weight marker. a. u.: arbitrary unit. For each genotype/condition, 4 samples from two independent experiments were analyzed. Representative gels are shown in (**A,B**); quantification data were obtained using ImageLab software and means ± SD of fluorescent signal reported to Coomassie signal are shown in (**C**) (*n* = 4). Samples significantly different (*p* < 0.05) are indicated by different letters.

3.3. The Ascorbate Pool is More Oxidized in the trxy1y2 Mutant During Drought Stress

To further characterize the effect of the *trxy1* and *trxy2* mutations on leaf cellular redox homeostasis, we analyzed the redox state and the pool size of glutathione, a metabolite considered as a cellular redox buffer [30]. In both WT and mutant leaves 6 days without watering caused the total glutathione pool to decrease to about 50% of the control value. It was further affected after 9 days of drought but then recovered to initial levels (ca. 0.5 μmol/mg Chl) after 13 days of stress (Figure 5A). This effect was also observed for GSSG, the oxidized form of glutathione, which was strongly increased after 13 days of dehydration (Figure 5B), causing a drastic drop in the glutathione redox state from ca. 90% to ca. 30% (Figure 5C). Hence, the effects of drought on glutathione were comparable in the leaves of WT and mutant plants and could not explain their contrasting capacities to cope with water deprivation.

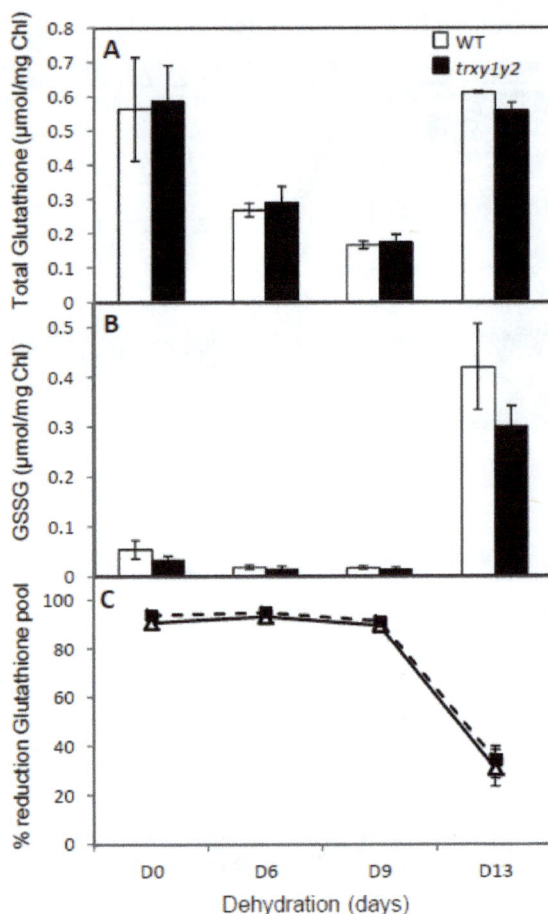

Figure 5. Effect of drought on leaf glutathione pool size and redox state. Glutathione pool after dehydration in WT (open column or triangle) and in *trxy1y2* mutant (dark column or square) leaves. (**A**) Total glutathione content; (**B**) GSSG content; (**C**) redox state of the glutathione pool. Data correspond to means ± SD (*n* = 4, from 2 independent experiments).

Ascorbate (AsA) is another low molecular weight antioxidant metabolite abundant in plant cells that plays an important role in tolerance to environmental constraints. It allows the alleviation of H_2O_2 accumulation through non-enzymatic and enzymatic pathways for scavenging ROS produced during stress [31–33]. Throughout the drought stress experiment, while the total pool size of AsA initially decreased (by ca. 30% at d6) and then remained mainly unchanged for both genotypes (Figure 6A), its oxidized form, DHA (dehydroascorbate) increased earlier and more markedly in mutant leaves compared to WT (Figure 6B). As a consequence, in WT leaves the reduction state of AsA remained high (ca. 95%) at d9 and decreased to ca. 85% at d13 whereas in mutant leaves it reached ca. 85% at d9 and dropped to ca. 65% at d13 (Figure 6C). Clearly, the leaf capacity to avoid AsA oxidation in response to drought seemed to be affected in the *trxy1y2* mutant in comparison to WT.

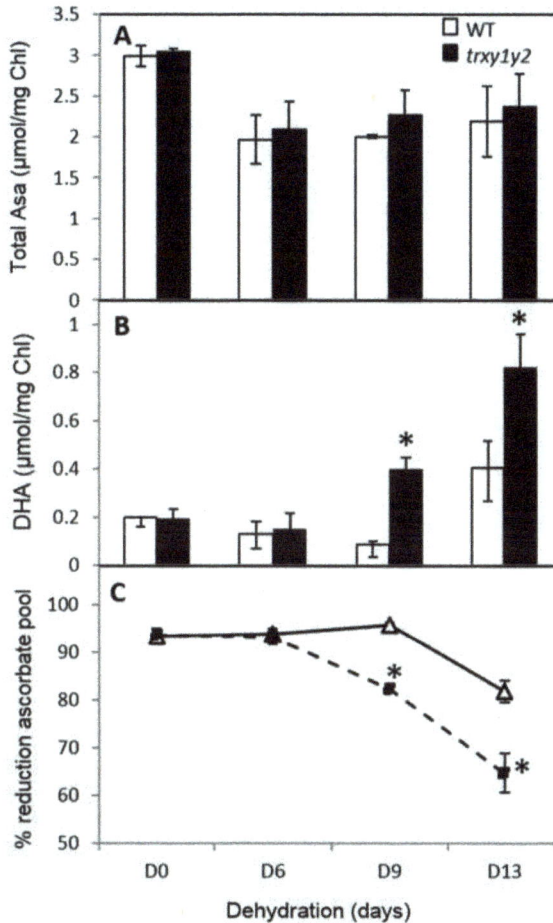

Figure 6. Effect of drought on the ascorbate pool size and redox state. Ascorbate (AsA) pool after dehydration in leaves from Col-0 (open bars or triangles) and in double *trxy1y2* mutant (dark bars or squares). (**A**) Total AsA content; (**B**) Oxidized AsA (DHA) content; (**C**) Redox state of the ascorbate pool. Data correspond to means ± SD ($n = 4$, from 2 independent experiments). Mutant samples significantly different from WT ($p < 0.05$) are indicated by an asterisk (*).

3.4. TRX y Can Increase MDHAR Activity in Leaf Crude Extracts

Monodehydroascorbate (MDA) is the primary product of AsA oxidation occurring during H_2O_2 detoxification by ascorbate peroxidase which uses AsA as a reduced substrate. MDA, if not rapidly reduced, can also be further oxidized to DHA by a non-enzymatic reaction. In leaves, MDA reductase (MDHAR) (EC 1.6.5.4) plays a key role in AsA recycling by catalyzing MDA reduction using NAD(P)H as an electron donor [34]. Therefore, MDHAR plays an important role in the response of plants to oxidative stress by maintaining the intracellular AsA redox state mainly in its reduced state. Past work in the laboratory unraveled that MDHAR was a potential target of TRX y and we found that adding reduced TRX y1 (a TRX isoform expressed in non-photosynthetic tissues) to crude root extracts strongly increased MDHAR activity [19]. Recently, we also showed that leaf extractable MDHAR activity was markedly increased in the photorespiratory *cat2* mutant, deficient in the major catalase isoform of Arabidopsis leaves [35], suggesting redox sensitivity of this enzyme in leaves too [36]. To test this hypothesis and a possible regulatory role of TRXs y towards MDHAR in leaves, we incubated WT leaf extracts in the presence of purified plastidial TRXs. While pre-incubation with the chemical reductant DTT alone or in presence of TRX f1 did not change NADPH-dependent MDHAR activity (which is mainly attributable to the organellar isoform), this activity was increased 4-fold after a 15 min treatment in the presence of reduced TRX y2, the y-type TRX isoform expressed in Arabidopsis leaves (Figure 7). Interestingly, the same reducing treatments showed no effect on NADH-dependent activity attributable to cytosolic isoforms (Figure S1). The incubation in oxidizing conditions, using the strong oxidant trans-4,5-Dihydroxy-1,2-dithiane (DTTox), did not change the NADPH-dependent activity in leaf extracts suggesting a spontaneous complete oxidation of the enzyme upon sample preparation (data not shown). Thus, taken together, these results strongly suggested a role for TRX y in the regulation of NADPH-dependent MDHAR activity in leaves.

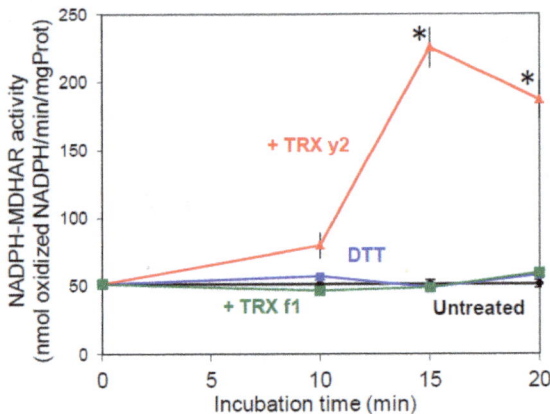

Figure 7. Effect of TRX on NADPH-dependent MDA reductase (MDHAR) activity in leaf extracts. Col-0 leaf protein extracts were incubated at room temperature in absence (black), or in presence of DTT, alone (blue), or with 20 μM TRX y2 (red), or TRX f1 (green) prior measuring NADPH-MDHAR activity. Data correspond to means $\pm$ SD ($n = 4$, from 2 independent experiments). Mutant samples significantly different from untreated ($p < 0.05$) are indicated by an asterisk (*).

3.5. In Vitro Validation of a TRX y-Specific Regulation of AtMDHAR6 Activity

In Arabidopsis, 6 MDHAR isoforms were found and MDHAR6 was shown to be plastid-targeted and to use specifically NADPH as reducing cofactor for catalysis [37]. Therefore, we cloned the cDNA of MDHAR6 from Arabidopsis (AtMDHAR6) into an *E. coli* expression plasmid allowing its production in its mature form with a strep-tag at its C-terminal end. The corresponding recombinant protein was purified to homogeneity (Figure S2) and its activity was tested for sensitivity to redox treatments. In vitro, recombinant AtMDHAR6 activity was increased more than two-fold by reduced TRX y2, while the other TRXs tested (of the f, m, or x types) had no effect on the activity of this enzyme (Figure 8).

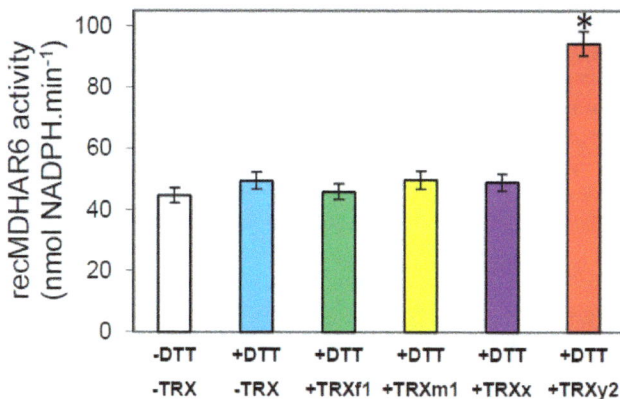

Figure 8. Effect of reducing treatments on AtMDHAR6 activity. A sample of purified recombinant AtMDHAR6 enzyme was incubated for 15 min at room temperature prior to measuring its NADPH-dependent activity. Incubation was in absence of DTT (white bar), or in presence of DTT, alone (blue bar), or with TRX: f1 (green bar), or m1 (yellow bar), or x (purple bar), or y2 (red bar). Data correspond to means $\pm$ SD ($n = 4$, from 2 independent experiments). The treatment significantly increasing NADPH-MDHAR activity (comparison with untreated) ($p < 0.05$) is indicated by an asterisk (*).

4. Discussion and Conclusions

4.1. TRX y Depletion Leads to a Higher Sensitivity of Arabidopsis to Drought Stress

The importance of oxidative stress and the role of ROS in local and systemic signaling in plants in response to drought has been largely documented (see for reviews [38–40]) and the induction of oxidative stress occurring as a drought effect is now widely accepted. This implies that a limited availability of water favors an imbalance between the production of ROS and their elimination. This leads to an increase in ROS levels such as H_2O_2 and singlet oxygen ($^1O_2$) during drought [40], even though the probability of $^1O_2$ production may be low [22]. Besides, previous biochemical studies on plastidial TRXs have revealed that y-type TRXs are preferential substrates for antioxidant enzymes such as Peroxiredoxin (PRX) Q [5,15], Glutathione Peroxidase (GPX1) [6], and Methionine Sulfoxide Reductase (MSR B2) [7]. Furthermore, the involvement of TRX y2 in the maintenance of leaf MSR capacity was demonstrated in vivo and the corresponding mutants both showed altered capacities to grow in high light/long day conditions [17]. Thus, we wondered whether y-type TRXs could be functionally important in the tolerance of Arabidopsis to other pro-oxidative stress conditions such as drought. We first challenged Arabidopsis plants lacking TRX y1 or y2 with water deficiency and found that they had an altered tolerance as compared with wild-type (WT) plants.

Then, we obtained a *trxy1y2* double mutant in which we performed a comparative study of water stress physiological effects and antioxidant responses relative to WT plants. We found that in control growth conditions, while showing no phenotype, the mutant had a lower leaf osmotic potential (−11.25 bar) compared to WT (−9.8 bar, data not shown). The *trxy1y2* mutant exhibited an enhanced sensitivity to drought correlating with a reduced capacity to keep high leaf water content (Figure 2).

4.2. TRX y is Necessary for Antioxidant Responses During Drought Stress

At the molecular level, in stress conditions, this importance of TRX y was underscored by increased levels of protein carbonylation and thiol oxidation in the *trxy1y2* mutant in comparison with WT (Figures 3 and 4). Thus, in mutant leaves the protein global redox status was modified towards oxidation. Moreover, the antioxidant metabolites ascorbate and glutathione, often used as biochemical markers of the cellular redox state, were quite strongly affected in our mild drought conditions. However, in both lines (WT and *trxy1y2* mutant), the glutathione content and redox state, as well as the total ascorbate content followed the same progressive decreasing trend in response to water deficiency (Figures 5 and 6). It is worth mentioning that past studies carried out on how plant antioxidant systems respond to drought revealed a high degree of complexity and a large diversity between plant species, making generalizations difficult [22]. A transcriptomic study on Arabidopsis plants exposed to drought stress drew similar conclusions [41]. The major difference observed in our comparative study was in the redox state of the ascorbate (AsA) pool. Indeed, under water deprivation the *trxy1y2* mutant showed an AsA pool that oxidized more rapidly and markedly than in the WT (Figure 6). Thus, one reason why the TRX mutant is more sensitive to drought could be its limited capacity to regenerate the AsA pool and maintain its antioxidant capacity. Moreover, chloroplast AsA Peroxidases (APXs) have been shown to be highly sensitive to oxidative inactivation in the absence of reduced AsA [42]. Thus, the drought stress sensitivity of the *trxy1y2* mutant could be linked to its impaired capacity to regenerate AsA from the radical MDA, with possible consequences for the antioxidant function of APXs.

4.3. TRX y Controls the Reduction of the Ascorbate Pool in Redox Regulating the Plastidial MDHAR

MDA reductase (MDHAR) is known to play an important role in the response of plants to oxidative stress by maintaining intracellular AsA redox state mainly in its reduced state. Past studies have evidenced a direct relationship between stress tolerance and MDHAR leaf activity [43]. Furthermore, the activity of MDHAR was found to be increased by diverse stresses including drought and salt stress [44–48]. Other studies suggest that MDHAR plays a key role in the regeneration of AsA and in tolerance to oxidative stress in plants [20,49]. Consistent with this notion, overexpression of Arabidopsis MDHAR1 (MDAR1) in tobacco plants enhanced their tolerance to ozone and increased their photosynthetic activity under salt stress [50]. In agreement with the possibility of a functional link between y-type TRXs and MDHAR in Arabidopsis, we previously identified the plastid MDHAR isoform (MDHAR6) as a putative target of TRX y2 [19]. We showed that NADPH-dependent MDHAR activity was enhanced in Arabidopsis root extracts by incubation with reduced TRX y1 whose gene is mostly expressed in non-photosynthetic tissues. In the present study, we have reported that NADPH-MDHAR activity can be also strongly increased in leaf extracts by reduced TRX y2 (Figure 7), while NADH-dependent MDHAR activity cannot and TRX f has no effect on leaf extractible MDHAR activity. Furthermore, we were able to validate in vitro a direct regulation of recombinant plastidial NADPH-MDHAR6 activity by TRX y2 (Figure 8). Since the other plastidial TRXs we tested had no effect, it seems that regulation of NAPH-MDHAR is specific to the y type. It is worth mentioning that the inefficiency of the other TRXs to activate MDHAR cannot be linked to their lower reducing capacity since TRXs y have a less negative redox potential [5,6,15]. Instead, steric and electrostatic complementarities between TRX y and its target enzyme might be determinants of specificity. Interestingly, multiple alignments of higher plant MDHAR primary sequences indicate that 3 out of 4 Cys are conserved in plastid isoforms [19]. Future work will allow identifying the cysteines involved in the activation process of NADPH-MDHAR by TRXs y.

4.4. Physiological Relevance of TRX-Dependent MDHAR Regulation in Chloroplasts

The present work reveals a new function for y-type TRXs and underlines their role as major thiol-based antioxidant proteins in plastids. This newly validated regulation might be physiologically relevant not only under water deprivation but also in other stress conditions where MDHAR capacity has been correlated with plant tolerance, as well as under normal growth conditions when a strong capacity for the reduction of the AsA pool is required, for example at sunrise. Indeed, it has been shown that during the night, the AsA pool size can decrease [51], and that the biosynthesis of AsA requires the activity of the photosynthetic electron transport and is therefore light-dependent [52]. In addition, AsA synthesis and regeneration can be influenced by the quality and the amount of light [53]. Because TRX y2 reduction is directly linked to the photosynthetic electron transport chain by ferredoxin/TRX reductase [9,15], the TRX-dependent regulation of MDHAR may allow its activity to be coordinated with the production of its reductant, NADPH, in chloroplasts exposed to changing conditions such as irradiance.

Supplementary Materials: The following are available online at http://www.mdpi.com/2076-3921/7/12/183/s1, Figure S1. Effect of TRX treatments on NADH-dependent MDHAR activity in leaf extracts. Figure S2. Purification of recMDHAR6. Table S1. List of primers used for cloning AtMDHAR6 cDNA in pGENI expression plasmid.

Author Contributions: E.I.-B., H.V. and A.-S.B. participated in designing the study. H.V., A.-S.B. and M.G. performed the experiments. H.V. and E.I-B. wrote the article.

Funding: This work was supported by a MENRT grant to A.-S.B. and the Saclay Plant Sciences program (SPS, ANR-10-LABX-40).

Acknowledgments: The authors are grateful to Myroslawa Miginiac-Maslow and Graham Noctor for critical reading and editing of the manuscript, and to Paulette Decottignies for validation of recMDHAR6 identity by Mass Spectrometry.

Conflicts of Interest: The authors declare no conflict of interest.

References

1. Lemaire, S.D.; Michelet, L.; Zaffagnini, M.; Massot, V.; Issakidis-Bourguet, E. Thioredoxins in chloroplasts. *Curr. Genet.* **2007**, *51*, 343–365. [CrossRef] [PubMed]
2. Meyer, Y.; Belin, C.; Delorme-Hinoux, V.; Reichheld, J.P.; Riondet, C. Thioredoxin and glutaredoxin systems in plants: Molecular mechanisms, crosstalks, and functional significance. *Antioxid. Redox Signal.* **2012**, *17*, 1124–1160. [CrossRef]
3. Geigenberger, P.; Thormählen, I.; Daloso, D.M.; Fernie, A.R. The unprecedented versatility of the plant thioredoxin system. *Trends Plant Sci.* **2017**, *22*, 249–262. [CrossRef] [PubMed]
4. Collin, V.; Issakidis-Bourguet, E.; Marchand, C.; Hirasawa, M.; Lancelin, J.M.; Knaff, D.B.; Miginiac-Maslow, M. The *Arabidopsis* plastidial thioredoxins: New functions and new insights into specificity. *J. Biol. Chem.* **2003**, *278*, 23747–23752. [CrossRef] [PubMed]
5. Collin, V.; Lamkemeyer, P.; Miginiac-Maslow, M.; Hirasawa, M.; Knaff, D.B.; Dietz, K.J.; Issakidis-Bourguet, E. Characterization of plastidial thioredoxins from Arabidopsis belonging to the new y-type. *Plant Physiol.* **2004**, *136*, 4088–4095. [CrossRef] [PubMed]
6. Navrot, N.; Collin, V.; Gualberto, J.; Gelhaye, E.; Hirasawa, M.; Rey, P.; Knaff, D.B.; Issakidis, E.; Jacquot, J.P.; Rouhier, N. Plant glutathione peroxidases are functional peroxiredoxins distributed in several subcellular compartments and regulated during biotic and abiotic stresses. *Plant Physiol.* **2006**, *142*, 1364–1379. [CrossRef] [PubMed]
7. Vieira Dos Santos, C.; Laugier, E.; Tarrago, L.; Massot, V.; Issakidis-Bourguet, E.; Rouhier, N.; Rey, P. Specificity of thioredoxins and glutaredoxins as electron donors to two distinct classes of *Arabidopsis* plastidial methionine sulfoxide reductases B. *FEBS Lett.* **2007**, *581*, 4371–4376. [CrossRef]
8. Chibani, K.; Tarrago, L.; Schurmann, P.; Jacquot, J.P.; Rouhier, N. Biochemical properties of poplar thioredoxin z. *FEBS Lett.* **2011**, *585*, 1077–1081. [CrossRef]
9. Bohrer, A.S.; Massot, V.; Innocenti, G.; Reichheld, J.P.; Issakidis-Bourguet, E.; Vanacker, H. New insights into the reduction systems of plastidial thioredoxins point out the unique properties of thioredoxin z from Arabidopsis. *J. Exp. Bot.* **2012**, *63*, 6315–6323. [CrossRef]
10. Díaz, M.G.; Hernández-Verdeja, T.; Kremnev, D.; Crawford, T.; Dubreuil, C.; Strand, Å. Redox regulation of PEP activity during seedling establishment in *Arabidopsis thaliana*. *Nat Commun.* **2018**, *9*, 50. [CrossRef]
11. Okegawa, Y.; Motohashi, K. Chloroplastic thioredoxin m functions as a major regulator of Calvin cycle enzymes during photosynthesis in vivo. *Plant J.* **2015**, *84*, 900–913. [CrossRef] [PubMed]
12. Naranjo, B.; Diaz-Espejo, A.; Lindahl, M.; Cejudo, F.J. Type-f thioredoxins have a role in the short-term activation of carbon metabolism and their loss affects growth under short-day conditions in *Arabidopsis thaliana*. *J. Exp. Bot.* **2016**, *67*, 1951–1964. [CrossRef] [PubMed]

13. Yoshida, K.; Hisabori, T. Two distinct redox cascades cooperatively regulate chloroplast functions and sustain plant viability. *Proc. Natl. Acad. Sci. USA* **2016**, *113*, 3967–3976. [CrossRef] [PubMed]

14. Thormählen, I.; Zupok, A.; Rescher, J.; Leger, J.; Weissenberger, S.; Groysman, J.; Orwat, A.; Chatel-Innocenti, G.; Issakidis-Bourguet, E.; Armbruster, U.; et al. Thioredoxins play a crucial role in dynamic acclimation of photosynthesis in fluctuating light. *Mol. Plant.* **2017**, *10*, 168–182. [CrossRef] [PubMed]

15. Yoshida, K.; Hisabori, T. Distinct electrontransfer from ferredoxin-thioredoxin reductase to multiple thioredoxin isoforms in chloroplasts. *Biochem. J.* **2017**, *474*, 1347–1360. [CrossRef] [PubMed]

16. Née, G.; Zaffagnini, M.; Trost, P.; Issakidis-Bourguet, E. Redox regulation of chloroplastic glucose-6-phosphate dehydrogenase: A new role for f-type thioredoxin. *FEBS Lett.* **2009**, *583*, 2827–2832. [CrossRef]

17. Laugier, E.; Tarrago, L.; Courteille, A.; Innocenti, G.; Eymery, F.; Rumeau, D.; Issakidis-Bourguet, E.; Rey, P. Involvement of thioredoxin y2 in the preservation of leaf methionine sulfoxide reductase capacity and growth under high light. *Plant Cell Environ.* **2013**, *36*, 670–682. [CrossRef]

18. Belin, C.; Bashandy, T.; Cela, J.; Delorme-Hinoux, V.; Riondet, C.; Reichheld, J.P. A comprehensive study of thiol reduction gene expression under stress conditions in *Arabidopsis thaliana. Plant Cell Environ.* **2015**, *38*, 299–314. [CrossRef]

19. Marchand, C.; Vanacker, H.; Collin, V.; Issakidis-Bourguet, E.; Le Maréchal, P.; Decottignies, P. Thioredoxins targets in Arabidopsis roots. *Proteomics* **2010**, *10*, 2418–2428. [CrossRef]

20. Tuzet, A.; Rahantaniaina, M.S.; Noctor, G. Analyzing the Function of Catalase and the Ascorbate-Glutathione Pathway in H_2O_2 Processing: Insights from an Experimentally Constrained Kinetic Model. *Antioxid. Redox Signal.* **2018**. [CrossRef]

21. Anjum, N.A.; Khan, N.A.; Sofo, A.; Baier, M.; Kizek, R. Redox homeostasis managers in plants under environmental stresses. *Front. Environ. Sci.* **2016**, *4*, 35. [CrossRef]

22. Noctor, G.; Mhamdi, A.; Foyer, C.H. The roles of reactive oxygen metabolism in drought stress: Not so cut and dried. *Plant Physiol.* **2014**, *164*, 1636–1648. [CrossRef] [PubMed]

23. Wehr, N.B.; Levine, R.L. Quantification of protein carbonylation. *Methods Mol. Biol.* **2013**, *965*, 265–281. [PubMed]

24. Hossain, M.A.; Asada, K. Monodehydroascorbate reductase from cucumber is a flavin adenine dinucleotide enzyme. *J. Biol. Chem.* **1985**, *260*, 12920–12926. [PubMed]

25. Aliverti, A.; Curti, B.; Vanoni, M.A. Identifying and quantitating FAD and FMN in simple and in iron-sulfur-containing flavoproteins. *Methods Mol. Biol.* **1999**, *131*, 9–23. [PubMed]

26. Vanacker, H.; Carver, T.L.W.; Foyer, C.H. Pathogen-induced changes in the antioxidant status of the apoplast in barley leaves. *Plant Physiol.* **1998**, *117*, 1103–1114. [CrossRef] [PubMed]

27. Foyer, C.; Rowell, J.; Walker, D. Measurement of the ascorbate content of spinach leaf protoplasts and chloroplasts during illumination. *Planta* **1983**, *157*, 239–244. [CrossRef]

28. Griffith, O.W. Determination of glutathione and glutathione disulfide using glutathione reductase and 2-vinylpyridine. *Anal. Biochem.* **1980**, *106*, 207–212. [CrossRef]

29. Friso, G.; van Wijk, K.J. Plant posttranslational modifications in plant metabolism. *Plant Physiol.* **2015**, *169*, 1469–1487.

30. Noctor, G.; Mhamdi, A.; Chaouch, S.; Han, Y.; Neukermans, J.; Marquez-Garcia, B.; Queval, G.; Foyer, C.H. Glutathione in plants: An integrated overview. *Plant Cell Environ.* **2012**, *35*, 454–484. [CrossRef]

31. Asada, K. The water-water cycle in chloroplasts: Scavenging of active oxygens and dissipation of excess photons. *Annu. Rev. Plant Physiol. Plant Mol. Biol.* **1999**, *50*, 601–639. [CrossRef] [PubMed]

32. Mittler, R. Oxidative stress, antioxidants and stress tolerance. *Trends Plant Sci.* **2002**, *7*, 405–410. [CrossRef]

33. Sirikhachornkit, A.; Niyogi, K.K. Antioxidants and photo-oxidative stress responses in plants and algae. In *Advances in Photosynthesis and Respiration: Chloroplast: Basics and Applications*; Rebeiz, C.A., Benning, C., Bohnert, H.J., Daniell, H., Hoober, J.K., Lichtenthaler, H.K., Portis, A.R., Tripathy, B.C., Eds.; Springer: Berlin, Germany, 2010; pp. 379–396.

34. Gill, S.S.; Tuteja, N. Reactive oxygen species and antioxidant machinery in abiotic stress tolerance in crop plants. *Plant Physiol. Biochem.* **2010**, *48*, 909–930. [CrossRef] [PubMed]

35. Queval, G.; Issakidis-Bourguet, E.; Hoeberichts, F.A.; Vandorpe, M.; Gakière, B.; Vanacker, H.; Miginiac-Maslow, M.; Van Breusegem, F.; Noctor, G. Conditional oxidative stress responses in the Arabidopsis photorespiratory mutant *cat2* demonstrate that redox state is a key modulator of daylength-dependent gene expression, and define photoperiod as a crucial factor in the regulation of H_2O_2-induced cell death. *Plant J.* **2007**, *52*, 640–657.

36. Rahantaniaina, M.S.; Li, S.; Chatel-Innocenti, G.; Tuzet, A.; Mhamdi, A.; Vanacker, H.; Noctor, G. Glutathione oxidation in response to intracellular H_2O_2: Key but overlapping roles for dehydroascorbate reductases. *Plant Signal. Behav.* **2017**, *12*, e1356531. [CrossRef]

37. Obara, K.; Sumi, K.; Fukuda, H. The use of multiple transcription starts causes the dual targeting of Arabidopsis putative monodehydroascorbate reductase to both mitochondria and chloroplasts. *Plant Cell Physiol.* **2002**, *43*, 697–705. [CrossRef] [PubMed]

38. Smirnoff, N. The role of active oxygen in the response of plants to water deficit and desiccation. *New Phytol.* **1993**, *125*, 27–58. [CrossRef]

39. Miller, G.; Suzuki, N.; Ciftci-Yilmaz, S.; Mittler, R. Reactive oxygen species homeostasis and signalling during drought and salinity stresses. *Plant Cell Environ.* **2010**, *33*, 453–467. [CrossRef]

40. De Carvalho, M.H.C. Drought stress and reactive oxygen species. Production, scavenging and signalling. *Plant Signal. Behav.* **2013**, *3*, 156–165. [CrossRef]

41. Harb, A.; Krishnan, A.; Ambavaram, M.M.R.; Pereira, A. Molecular and physiological analysis of drought stress in Arabidopsis reveals early responses leading to acclimation in plant growth. *Plant Physiol.* **2010**, *154*, 1254–1271. [CrossRef]

42. Ishikawa, T.; Shigeoka, S. Recent advances in ascorbate biosynthesis and the physiological significance of ascorbate peroxidase in photosynthesizing organisms. *Biosci. Biotechnol. Biochem.* **2008**, *72*, 1143–1154. [CrossRef] [PubMed]

43. Gallie, D.R. The role of L-ascorbic acid recycling in responding to environmental stress and in promoting plant growth. *J. Exp. Bot.* **2013**, *64*, 433–443. [CrossRef]

44. Reddy, A.R.; Chaitanya, K.V.; Jutur, P.P.; Sumithra, K. Differential antioxidative responses to water stress among five mulberry (*Morus alba* L.) cultivars. *Environ. Exp. Bot.* **2004**, *52*, 33–42. [CrossRef]

45. Hernandez, J.A.; Ferrer, M.A.; Jimenez, A.; Barcelo, A.R.; Sevilla, F. Antioxidant systems and O_2^-/H_2O_2 production in the apoplast of pea leaves. Its relation with salt-induced necrotic lesions in minor veins. *Plant Physiol.* **2001**, *127*, 817–831. [CrossRef] [PubMed]

46. Kavitha, K.; George, S.; Venkataraman, G.; Parida, A. A salt-inducible chloroplastic monodehydroascorbate reductase from halophyte *Avicennia marina* confers salt stress tolerance on transgenic plants. *Biochimie* **2010**, *92*, 1321–1329. [CrossRef]

47. Shu, S.; Yuan, L.Y.; Guo, S.R.; Sun, J.; Yuan, Y.H. Effects of exogenous spermine on chlorophyll fluorescence, antioxidant system and ultrastructure of chloroplasts in *Cucumis sativus* L. under salt stress. *Plant Physiol. Biochem.* **2013**, *63*, 209–216. [CrossRef] [PubMed]

48. Sudan, J.; Negi, B.; Arora, S. Oxidative stress induced expression of monodehydroascorbate reductase gene in *Eleusine coracana*. *Physiol. Mol. Biol. Plants* **2015**, *21*, 551–558. [CrossRef]

49. Li, F.; Wu, Q.Y.; Sun, Y.L.; Wang, L.Y.; Yang, X.H.; Meng, Q.W. Overexpression of chloroplastic monodehydroascorbate reductase enhanced tolerance to temperature and methyl viologen-mediated oxidative stresses. *Physiol. Plant* **2010**, *139*, 421–434. [CrossRef] [PubMed]

50. Eltayeb, A.E.; Kawano, N.; Badawi, G.H.; Kaminaka, H.; Sanekata, T.; Shibahara, T.; Inanaga, S.; Tanaka, K. Overexpression of monodehydroascorbate reductase in transgenic tobacco confers enhanced tolerance to ozone, salt and polyethylene glycol stresses. *Planta* **2007**, *225*, 1255–1264. [CrossRef]

51. Bartoli, C.G.; Yu, J.; Gomez, F.; Fernandez, L.; McIntosh, L.; Foyer, C.H. Inter-relationships between light and respiration in the control of ascorbic acid synthesis and accumulation in *Arabidopsis thaliana* leaves. *J. Exp. Bot.* **2006**, *57*, 1621–1631. [CrossRef]

52. Yabuta, Y.; Mieda, T.; Rapolu, M.; Nakamura, A.; Motoki, T.; Maruta, T.; Yoshimura, K.; Ishikawa, T.; Shigeoka, S. Light regulation of ascorbate biosynthesis is dependent on the photosynthetic electron transport chain but independent of sugars in Arabidopsis. *J. Exp. Bot.* **2007**, *58*, 2661–2671. [CrossRef] [PubMed]

53. Bartoli, C.G.; Tambussi, E.A.; Diego, F.; Foyer, C.H. Control of ascorbic acid synthesis and accumulation and glutathione by the incident light red/far red ratio in *Phaseolus vulgaris* leaves. *FEBS Lett.* **2009**, *583*, 118–122. [CrossRef] [PubMed]

antioxidants

MDPI

Article

Towards Initial Indications for a Thiol-Based Redox Control of Arabidopsis 5-Aminolevulinic Acid Dehydratase

Daniel Wittmann, Sigri Kløve, Peng Wang and Bernhard Grimm *

Institute of Biology/Plan Physiology, Humboldt-Universität zu Berlin, Philippstr. 13, Building 12, 10155 Berlin, Germany; wittmada@hu-berlin.de (D.W.); sigrik@gmail.com (S.K.); wangp2014@gmail.com (P.W.)
* Correspondence: bernhard.grimm@rz.hu-berlin.de; Tel.: +49-030-2093-6119

Received: 27 September 2018; Accepted: 28 October 2018; Published: 31 October 2018

Abstract: Thiol-based redox control is one of the important posttranslational mechanisms of the tetrapyrrole biosynthesis pathway. Many enzymes of the pathway have been shown to interact with thioredoxin (TRX) and Nicotinamide adenine dinucleotide phosphate (NADPH)-dependent thioredoxin reductase C (NTRC). We examined the redox-dependency of 5-aminolevulinic acid dehydratase (ALAD), which catalyzed the conjugation of two 5-aminolevulinic acid (ALA) molecules to porphobilinogen. ALAD interacted with TRX f, TRX m and NTRC in chloroplasts. Consequently, less ALAD protein accumulated in the *trx f1*, *ntrc* and *trx f1/ntrc* mutants compared to wild-type control resulting in decreased ALAD activity. In a polyacrylamide gel under non-reducing conditions, ALAD monomers turned out to be present in reduced and two oxidized forms. The reduced and oxidized forms of ALAD differed in their catalytic activity. The addition of TRX stimulated ALAD activity. From our results it was concluded that (i) deficiency of the reducing power mainly affected the in planta stability of ALAD; and (ii) the reduced form of ALAD displayed increased enzymatic activity.

Keywords: ALAD; tetrapyrrole biosynthesis; redox control; thioredoxins; posttranslational modification; chlorophyll

1. Introduction

Tetrapyrrole biosynthesis (TBS) in higher plants is a tightly controlled metabolic pathway that requires multiple regulatory mechanisms. In particular, posttranslational modifications ensure rapid modifications of the activity and stability of many committed enzymes in the TBS pathway and their interactions with other enzymes and effectors in response to changing environmental conditions, such as light intensity and temperature variations [1,2]. The constant adjustment of the metabolic flow in the TBS pathway prevents the accumulation of photoreactive tetrapyrrole intermediates and end-products, which could cause severe photo-oxidative damage upon light exposure due to the generation of singlet oxygen [1,3].

Thiol-based redox-regulation is one of the widespread post-translational mechanisms to modulate the protein activity and stability of stromal proteins involved in various metabolic pathways, like the Calvin–Benson cycle [4,5] or TBS [1]. Thiol-based redox regulation relies on different methods of modification by thiol-disulfide cycling of cysteine residues. These redox-dependent modifications result in the formation of intra- and intermolecular disulfide bonds within a protein and between two proteins, respectively. The reduction of the oxidized thiol groups can be accomplished by redox regulators, such as thioredoxins (TRXs) and reduced Nicotinamide adenine dinucleotide phosphate (NADPH)-dependent thioredoxin reductase C (NTRC). TRXs are small 12–14 kDa proteins catalyzing thiol-disulfide exchanges. In chloroplasts, ten typical TRXs act in the redox regulation of chloroplast proteins and are subdivided into five groups: two f-type TRXs (f1-f2), four m-type TRXs (m1-m4), one x-type TRX, two y-type TRXs (y1–y2) and one z-type TRX [6]. The NTRC with a C-terminal TRX domain functions in chloroplasts as a reductant of 2-cysteine peroxiredoxins and other target proteins [7,8]. Its NADPH dependency enables NTRC to provide reducing potential to its target proteins also in darkness [9].

TRX and NTRC play essential roles in maintaining efficient TBS in higher plants. It has been shown that a deficiency of f- and m-type TRX variants and NTRC leads to an obvious pale-green leaf phenotype and multiple defects in the TBS pathway [10,11]. Thus far, four TBS enzymes have been proved to be targets of TRX and NTRC for the reduction of thiol bonds: the rate-limiting enzyme glutamyl-tRNA reductase (GluTR) [10,12], magnesium protoporphyrin IX methyltransferase (CHLM) [10,12–14], the I subunit of magnesium chelatase (CHLI) [11,15] and the Mg–protoporphyrin monomethylester cyclase (CHL27) [16]. Among other TBS enzymes, glutamate 1-semialdehyde aminotransferase (GSAAT), 5-aminolevulinic acid dehydratase (ALAD), protoporphyrinogen oxidase (PPOX) and protochlorophyllide oxidoreductase (POR) were suggested to be the potential interacting partner of TRXs and NTRC [17,18]. Thus, TRXs and NTRC are important regulators for the TBS of higher plants with the potential to activate or stabilize several proteins of the pathway under reducing conditions.

ALAD, also known as porphobilinogen synthase, catalyzes the asymmetric condensation of two linear molecules of 5-aminolevulinic acid (ALA) to the first cyclic intermediate of the pathway, the monopyrrole porphobilinogen (PBG) [19]. *A. thaliana* has two ALAD isoforms encoded by *HEMB1* and *HEMB2*. *HEMB1* is induced during dark-to-light transitions and represents the predominant gene in green seedlings. A T-DNA insertion in *HEMB1* is embryo lethal [20]. Additionally, *HEMB1* was shown to be upregulated by FAR-RED ELONGATED HYPOCOTYL3 (FHY3) and FAR-RED IMPAIRED RESPONSE1 (FAR1), two transcription factors involved in phytochrome a signaling [20].

While ALAD from human, yeast and some bacteria requires the binding of a catalytic Zn^{2+} ion at the active site, which is coordinated by three conserved cysteine residues, the plant ALAD binds Mg^{2+} ions [21]. The catalytic Mg^{2+}-binding amino acid residues are not defined. It was suggested that Mg^{2+} is required for catalysis, acting as an allosteric activator of the multimerization of the enzyme and as a potential inhibitor [22–24]. The pea ALAD was reported to maintain an equilibrium of hexameric and octameric ALAD complexes and to shift reversibly in the presence of Mg^{2+} to the catalytically more active octameric form [23].

In 1983, the activity of extracted radish ALAD was shown to be upregulated with the reducing agent dithiothreitol (DTT), TRXs f and m, with TRX f being the more effective stimulator than TRX m [25]. ALAD has been found to interact with TRXA in *Synechocystis* spec. by using thioredoxin affinity chromatography [18]. In *Arabidopsis*, the steady-state level of ALAD was decreased in the triple TRX m1/m2/m4 gene silencing plants, indicating that TRX is required for the stabilization of ALAD [14]. Based on these observations, it is hypothesized that thiol-based redox regulation and TRXs are important for the stability and enzymatic activation of ALAD. Nevertheless, the redox regulatory mechanism and thiol switches at cysteine residues in ALAD remain to be investigated.

In continuation of our investigations into the redox control in TBS, the aim of this study was to explore *Arabidopsis* ALAD for its thiol-based redox control. To address this question, we investigated the potential redox switches of ALAD in vivo in *Arabidopsis* wild-type as well as TRX f1 and NTRC-deficient mutant seedlings and in vitro by characterization of recombinant ALAD. Additionally, we performed in planta and in vitro protein–protein interaction studies with ALAD.

2. Materials and Methods

2.1. Plant Growth and Mutant Lines

Arabidopsis thaliana wild type (Col-0), *ntrc* (SALK_012208), *trx f1* (SALK_128365) and *ntrc/trx f1* were grown under short-day standard conditions (10 h/14 h light/dark) at 120 µmol photons m^{-2} s^{-1},

22 °C and 70% relative humidity. The *trx f1* [26], the *ntrc* [7] and the *ntrc/trx f1* double mutant [27] were kindly provided by Peter Geigenberger.

2.2. Cloning, Expression and Purification of Recombinant ALAD and TRX f1

A full-length cDNA encoding ALAD1 (*HEMB1*; AT1G69740) was cloned into Novagen pET28a ((Merck Millipore, Burlington, MA, USA) without the predicted sequence for the transit peptide (ChloroP [28]). For expression, the recombinant vectors were transformed into *E. coli* Novagen Rosetta™(DE3) (Merck Millipore) strains. The expression of ALAD was induced by addition of 1 mM isopropyl β-D-1-thiogalactopyranoside (IPTG), the proteins were expressed at 37 °C under continuous shaking for 3 h. TRX f1 expression was performed after induction with 0.2 mM IPTG for 3 h at 37 °C under continuous shaking. The N-terminal 6 x His-tagged fusion proteins were purified by means of nickel-nitrilotriacetic acid (Ni-NTA) agarose beads (Thermo Fisher Scientific, Waltham, MA, USA)) and the protein extracts were concentrated using Amicon® Ultra-4 Centrifugal Filter Units (Merck-Millipore, Burlington, MA, USA).

2.3. Protein Extraction

Leaf tissue was homogenized in liquid nitrogen and resuspended in protein extraction buffer (56 mM Na_2CO_3, 2% (*w/v*) sodium dodecyl sulfate (SDS), 12% (*w/v*) sucrose, 2 mM ethylenediaminetetraacetic acid (EDTA), pH = 8.0) and heated for 20 min at 70 °C. After 10 min centrifugation at room temperature, the protein concentration of the supernatant was determined by Pierce™ Bicinchoninic Acid (BCA) Protein Assay Kit (Thermo Fisher Scientific, Waltham, MA, USA). After addition of 100 mM DTT the samples were briefly boiled before loading. Twenty μg protein of each extract were separated by 12% SDS polyacrylamide gels and subsequently blotted on nitrocellulose membranes. The membranes were probed with specific antibodies according to [29]. The anti-ALAD antibody was generated in rabbits using purified, recombinant His-tagged ALAD. The serum was diluted 1:2500 in TBS buffer containing 1% milk powder (*w/v*). As second antibody, a horseradish peroxidase (HRP)-conjugated anti-rabbit antibody was subjected to the protein-containing membranes (Agrisera, 1:10,000 dilution). The Clarity Western ECL™ Blotting Substrate (Bio-Rad, Hercules, CA, USA) was used for immune detection. The quantification of immune-blots based on two biological replicates was performed using the image analysis software GelAnalyzer 2010a (Istvan Lazar, www.gelanalyzer.com).

2.4. Gel-Shift Assays

To visualize the redox-dependent oligomerization and redox state of monomeric ALAD, 0.5 μM of the purified protein was preincubated in phosphate-buffered saline (PBS) (150 mM NaCl, 20 mM Na_2HPO_4, pH = 7.4, untreated = UT), oxidized in PBS containing H_2O_2 (1 mM) or reduced with dithiothreitol (DTT) (1–100 mM). The samples were subsequently separated in 10% SDS polyacrylamide gels under non-reducing conditions and blotted on nitrocellulose membranes. The membranes were probed with a 6 × His-Tag specific antibody conjugated to HRP (Sigma-Aldrich, St. Louis, MO, USA) and probed as previously described. The labeling of free thiols with methoxypolyethylene glycol maleimide-5000 (mPEG-MAL, Sigma-Aldrich (St. Louis, MO, USA)) was performed under denaturing conditions with minor adjustments according to [30]. Aliquots of recombinant ALAD were prepared and either preincubated with PBS containing H_2O_2 (1 mM), DTT (1–100 mM) or normal PBS (UT). After trichloroacetic acid (TCA) precipitation, the protein was alkylated with 10 mM mPEG-MAL. Then, the samples were resuspended in 1 × Laemmli buffer, separated on 8% SDS-polyacrylamide gel electrophoresis (PAGE), blotted and analyzed as previously described. To label the oxidized and buried thiols, free thiols were first blocked after the preincubation using 100 mM N-ethylmaleimide (NEM, Sigma-Aldrich). Following TCA precipitation, the proteins were reduced using 100 mM DTT. At this step, all reversible cysteine modifications were reduced and the proteins were subsequently precipitated with TCA. Then the mPEG-MAL labeling was performed as described above.

2.5. ALAD Activity Assay with Recombinant Protein and Plant Extracts

The ALAD enzymatic assay was performed with total leaf extracts of the soluble fraction and recombinant proteins. For recombinant ALAD, 475 µL assay buffer (50 mM K_2HPO_4, 2 mM $MgCl_2$, pH = 8.0) including 0.5 µg recombinant ALAD was preincubated with DTT (0.1–10 mM), $CuCl_2$ (5–20 µM) and TRX f1 (0.5 µM) for 10 min at 37 °C. The reaction was started by the addition of 25 µL of 100 mM ALA and stopped after 10 min at 37 °C and 600 rpm by the addition of 1 volume 10% ice-cold TCA including 10 mM $HgCl_2$. After centrifugation, 1 volume of Ehrlich's reagent was added and porphobilinogen was quantified at λ 555 nm photometrically [31]. For the activity measurement of plant ALAD, leaf material was ground in liquid nitrogen and the powder was resuspended in extraction buffer (25 mM Tris-HCl, pH = 8.2). After centrifugation, the supernatant was collected in a fresh tube and used for the ALAD assay. The total protein amount of the extract was quantified using the Pierce™ BCA Protein Assay Kit (Thermo Fisher Scientific). The reaction was performed by adding 1 volume of 2 × reaction buffer (25 mM Tris-HCl pH 8.2, 10 mM ALA, 12 mM $MgCl_2$, and the addition of 2 mM DTT for reducing conditions) to the extract. Samples were incubated for 90 min at 37 °C and constant shaking (600 rpm). The reaction was stopped with 1 volume 10% ice-cold TCA, 10 mM $HgCl_2$, and porphobilinogen was quantified as described.

2.6. Bimolecular Fluorescence Complementation Assay

The full-length coding sequences of *ALAD*, *TRX f1*, *TRX m4* and *NTRC* were cloned into the GATEWAY vectors pVYCE and pVYNE (Invitrogen, Carlsbad, CA, USA) and transformed into *Agrobacterium tumefaciens* (GV2260). The fusion proteins include the C- or N-terminal part of yellow fluorescence protein (YFP), respectively. After leaf infiltration in *Nicotiana benthamiana*, the fusion proteins were expressed for 48–72 h in darkness. The YFP fluorescence was detected by a LSM 800 confocal microscope (Zeiss; λex 514 nm, λem (YFP) 530–555 nm, λem (Chl) 600–700 nm).

2.7. Pull-Down Experiments

Purified 6 × His-tagged TRX f2(C112S), TRX m4(C119S) and NTRC(C457S) proteins were used as bait for the pull-down assay. One hundred µg of the purified protein was incubated with Ni-NTA agarose (Thermo Fisher Scientific) in PBS buffer for 1 h at 4 °C. Subsequently, chloroplast extracts were solubilized with 1% (*w/v*) dodecyl maltoside (DM) and incubated (100 µg chlorophyll amount of chloroplast extract) with the recombinant proteins associated with the Ni-NTA agarose for 1.5 h at 4 °C under gentle rotation. Empty Ni-NTA agarose was used as a negative control. After washing the Ni-NTA agarose five times with PBS buffer, the proteins bound to the beads were eluted with PBS buffer containing 250 mM imidazole, separated by 12% SDS-PAGE gel and analyzed by immunoblotting with specific antibodies.

3. Results

3.1. Structural Analysis and Protein Sequence Alignment Reveal Four Highly Conserved Cysteine Residues in Arabidopsis ALAD

Two X-ray structures of Mg^{2+} dependent ALAD have been published, revealing a high structural homology of *Pseudomonas aeruginosa* and *Chlorobium vibrioforme* ALAD with 1.67 Å and 2.6 Å resolution, respectively [32–35]. *Arabidopsis* ALAD1 (encoded by the *HEMB1* gene) consists of 430 amino acid residues (aa). The 3D-structural analysis and the protein sequence alignment of ALAD1 homologs in the selected plants/photosynthetic organisms (Figure 1) revealed the following peptide domains: a predicted 52-aa N-terminal chloroplast transit peptide, the N-terminus for multimerization and a $(\alpha\beta)8$-barrel domain including the active site between D220 and Y416 (prediction by homology, Basic Local Alignment Search Tool (BLAST), National Center for Biotechnology Information (NCBI)). Two conserved lysine residues (K298 and K351) were in close proximity in the tertiary structure (structure prediction with Phyre2 [36]) and were responsible for the formation of the catalytically essential Schiff base intermediates with one of the two substrate molecules of ALA. These two binding sites were termed A- and P-side and were decisive for the destination of the ALA molecule in the asymmetric PBG as acetate or propionate half [19].

The mature *Arabidopsis* ALAD1 had in total six cysteine residues. Four cysteine residues were highly conserved in higher plants (C152, C251, C404 and C426). C251 and C404 were localized in the active site (Figure 2). C251 is conserved even in green algae, like *Chlamydomonas reinhardtii*. Interestingly, a second *Arabidopsis* ALAD isoform (ALAD2, encoded by *HEMB2*) showed a cysteine-arginine substitution at the homologous positions of C152 and C251. Notably, C404 was highly conserved in all isoforms in the higher plants, *Chlamydomonas* and *Chlorobium*, indicating its potential importance for catalytic or regulatory function.

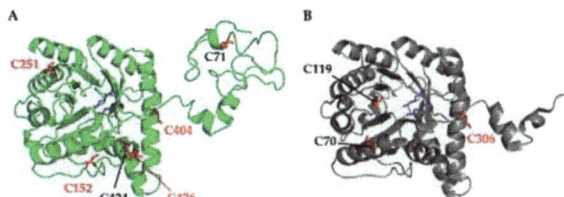

Figure 1. Comparative analysis of the X-ray 3D structure of *Chlorobium vibrioforme* 5-aminolevulinic acid dehydratase (ALAD) and the modelled ALAD from *Arabidopsis thaliana*. (**A**) The structure of mature *Arabidopis* ALAD was visualized with PyMOL (Schrödinger) after structure prediction with Phyre2. The modeled structure shows high structural homology with the two X-ray structures of the Mg^{2+} dependent ALADs from *Pseudomonas aeruginosa* [32,33] and *Chlorobium vibrioforme* [34,35]. (**B**) The *Chlorobium vibrioforme* ALAD structure based on RCSB protein data bank (PDB) entry 1W1Z [34]. Cysteines are highlighted by red sticks, conserved cysteines in higher plants are highlighted in red lettering. The two Schiff base lysine residues (for *A. thaliana* K298 and K351) are highlighted in the two structures by blue sticks.

3.2. Posttranslational Stability of ALAD in TRX and NTRC-Deficient Arabidopsis Seedlings

Leaf material from three-week-old seedlings of *trx f1*, *ntrc* and *ntrc/trx f1* and wild-type control grown under standard conditions was harvested (Figure 3A, showing four-week-old seedlings). The leaves of the three mutants had 15%, 51%, and 68% less chlorophyll, respectively, compared to control leaves. The reduced pigment content was consistent with previous reports [10,27]. The protein content of several TBS enzymes, which were previously proposed to be redox-controlled (see Figure 1 in [1]), was analyzed in the three mutant and wild-type lines. Compared to the wild type, the accumulation of some of these TBS enzymes was compromised in the single reductant-deficient mutants and more severely decreased in the double mutant (Figure 3B). The lower contents of GluTR, CHLM and POR confirmed previous reports showing the redox-dependent stability of the enzymes in TBS [10,14]. Furthermore, we also observed that the amounts of GSAAT and ALAD were decreased in the three mutants compared to control.

Comparative quantification of the contents of the immuno-analysed proteins in the *ntrc/trx f1* double mutant relative to the wild type yielded a 26% decreased amount of GluTR, 65% of CHLM, 19% of GSAAT and 91% of POR. The ALAD content was decreased in *ntrc*, *trx f1* and *ntrc/trx f1* by 33%, 25% and 36%, respectively, compared to the wild type. On the other hand, the levels of protopyrphyrinogen oxidase 1 (PPOX1), chlorophyll synthase (CHLG) and the Mg chelatase subunit CHLI were not negatively affected in all three mutants. We point to the rather diverse decreased contents of the different TBS enzymes under the inadequate reducing power. This requires further studies to explore the variation of protein stability in response to oxidizing conditions. It will be important to clarify whether TRX-mediated redox control of TBS enzymes modulate more protein stability than enzyme activity.

Figure 2. Alignment of the ALAD of different plant species (*Arabidopsis thaliana* ALAD encoded by *HEMB1* and *HEMB2*, *Arabidopsis lyrata*, *Hordeum vulgare*, *Nicotiana tabacum*, *Glycine soja*, *Oryza sativa*, *Cucumis sativus*, *Zea mays*), *Chlamydomonas reinhardtii* and *Chlorobium vibriforme*. The six cysteines of the mature *A. thaliana* protein are highlighted with arrows (red arrows indicate the four conserved cysteines in higher plants). The different domains are annotated by structure prediction (ChloroP, Basic Local Alignment Search Tool (BLAST)). M1-R52: chloroplast transit peptide; A53-G95: N-terminal arm; T96-R430: ($\alpha\beta$)8-barrel domain; D220-Y416 active site; * Schiff base lysine residues (K298, K351); • allosteric magnesium binding site (R107, D272, E336, D340); × aspartate-rich active site metal binding site (D224, D232, D269).

As a result, the immune blots indicate that the lower reducing capacity of the mutants visibly affected the stability of several TBS enzymes. In particular, it is emphasized that the abundance of the first enzymes of the pathways GluTR, GSAAT and ALAD was perturbed in the analyzed mutants, indicating a need for the redox-dependent control of enzyme accumulation at the level of ALA synthesis and the subsequent conversion of ALA into the monopyrrole structure.

To explore the consequence of NTRC and TRX f1 deficiency, leaf extracts of the three mutant and wild-type (Col-0) seedlings were assayed for ALAD activity (Figure 3C). The control extracts contained the highest activity in non-treated and DTT-supplemented extracts. The wild-type ALAD activity was hardly enhanced by the DTT (by 5%). The ALAD activity of the *ntrc*, *trx f1* and *ntrc/trx f1* was decreased by 23% (without DTT, *w/o*) and 14% (with DTT), 31% (*w/o*) and 18% (DTT) as well as 34% (*w/o*) and 24% (DTT), respectively, compared to the wild-type. It is proposed that the decreased enzyme activity in the three mutants was mainly due to the lower ALAD content. The ALAD activity of *ntrc*, *trx f1* and *ntrc/trx f1* could be increased by DTT by 17%, 24% and 21%, respectively, in comparison to the activity of the mutants without additional DTT. Wild-type ALAD activity was scarcely enhanced by DTT, suggesting that ALAD in wild-type seedlings was entirely in the reduced form. In contrast, the weak DTT-driven elevation of ALAD activity in the mutants could be explained by the partially-oxidized state of the ALAD.

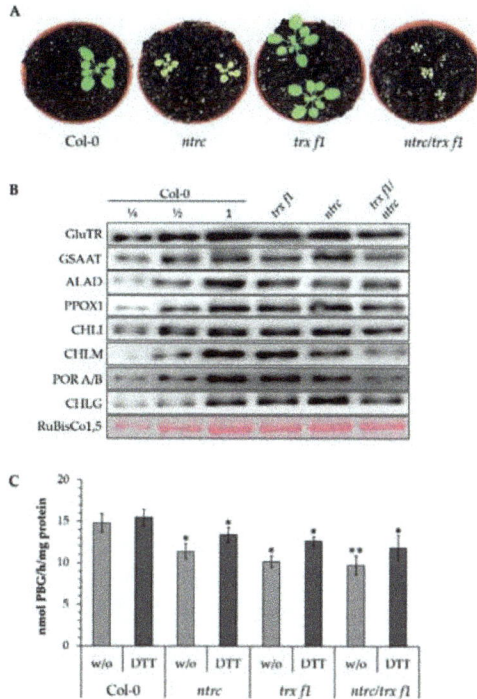

Figure 3. Redox control of tetrapyrrole synthesis (TBS) enzymes in wild type and mutants. (**A**) Four-week-old seedlings of wild type, *ntrc*, *trf f1* and *ntrc/trx f1* mutants. The plants grew under short day conditions (10 h/14 h light/dark) at 120 µmol photons m^{-2} s^{-1} at room temperature. (**B**) Contents of several TBS enzymes in wild-type (Col-0), *trx f1*, *ntrc* and *ntrc/trx f1* plants growing under short day conditions for 21 days. GluTR: glutamyl-tRNA reductase, GSAAT: glutamate-1-semialdehyde aminotransferase, ALAD: 5-aminolevulinic acid dehydratase, PPOX1: protoporphyrinogen IX oxidase, CHLI: subunit of the Mg chelatase, CHLM: Mg protoporphyrin methyltransferase, POR: protochlorophyllide oxidoreductase, CHLG: chlorophyll synthase, RuBisCo1,5: ribulose-bisphosphate carboxylase large subunit as loading control. (**C**) Redox-dependent ALAD activity. ALAD activity of the soluble protein fraction was measured from leaf extracts of four-week-old Col-0, *ntrc*, *trx f1* and *ntrc/trx f1* plants. The assay was performed with or without (w/o) 1 mM DTT. The data correspond to three biological replicates. Statistical significance of the ALAD activity of the mutants compared to Col-0 plants (with or without DTT) is indicated by asterisks (*, $p \leq 0.05$ and **, $p \leq 0.01$, Student's *t* test).

3.3. ALAD Interacts with TRX and NTRC

The mutual protein–protein interaction is the precondition of the TRX-dependent redox regulation of ALAD. To demonstrate the physical interaction of ALAD with TRXs, two different approaches were performed. First, transiently-expressed candidate proteins fused with either the C- or the N-terminal of half of the yellow-fluorescence protein (YFP) in *Nicotiana benthamiana*, resulting in the visible YFP signal observed by confocal laser scanning fluorescence microscopy (Figure 3A). The images of the bimolecular complementation (BiFC) assay revealed the interactions of ALAD with the TRXs f1 and m4, as well as NTRC (Figure 4A).

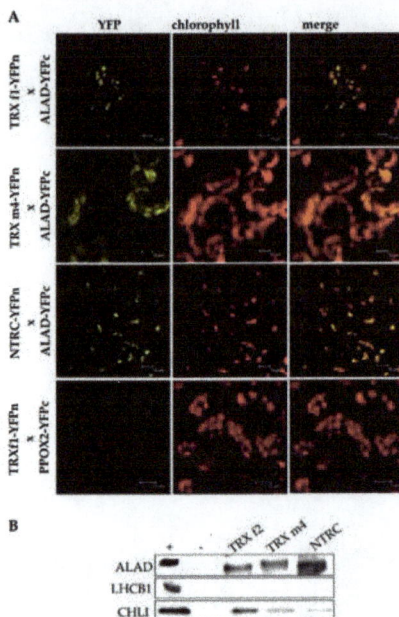

Figure 4. The physical interaction between ALAD and NTRC, TRX f1 and TRX m4. The interaction between ALAD and different thioredoxin (TRX) variants was demonstrated by bimolecular complementation assay (BiFC) (**A**) or a pulldown assay (**B**). (**A**) *Nicotiana benthamiana* leaves were infiltrated with *Agrobacterium tumefaciens* strains. After two days of transient expression of the transgenic gene constructs, images were taken by laser scanning fluorescence microscopy. Left row: YFP; middle row: autofluorescence of the chlorophyll; right: merge. PPOX2 and TRX f2 expression did not lead to mutual protein–protein interaction and was used as a negative control for the BiFC experiment. (**B**) The recombinant, His-tagged bait proteins TRX f2, TRX m4, NTRC were incubated with chloroplast extract. Proteins in the eluate were detected by immune analysis using antibodies against ALAD, LHCB1 and CHLI. The chloroplast extract served as a positive control (+). For the negative control, the pulldown assay was performed with Ni-NTA agarose, lacking the binding of a bait protein (−).

Second, the physical interaction of these proteins was confirmed by an in vitro TRX affinity chromatography approach. Total *Arabidopsis* chloroplast extracts were incubated with recombinant His-tagged TRX f2, TRX m4 and NTRC. After rigorous washing, TRX bound proteins were eluted and detected by immune analysis (Figure 4B). We found that ALAD was reduced by all three reductants. As a positive control, CHLI was reduced by TRX f2, as shown in a previous study [11]. As a negative control, LIGHT-HARVESTING CHLOROPHYLL BINDING PROTEIN 1 OF PHOTOSYSTEM II (LHCB1) could not be detected in the eluate.

3.4. Redox-Dependent Structural Modifications of Recombinant ALAD

Plastid-localized *Arabidopsis* ALAD contains six cysteine residues. Maleimide labeling of recombinant ALAD points to accessible cysteine residues. The mobility shift of the methoxypolyethylene glycol maleimide 5000 (mPEG-MAL)-conjugated proteins constituted around more than 5 kDa per labeled cysteine in a SDS-polyacrylamide gel electrophoresis (SDS-PAGE). Two parallel approaches were applied to test the potential number of reduced cysteines of ALAD under oxidizing and untreated conditions.

When oxidized, the recombinant ALAD was labeled with mPEG-MAL, and a ladder of four additional protein bands appeared in comparison with the entirely reduced maleimide-conjugated ALAD variant (Figure 5A). In addition, a weak ALAD band was detectable with the electrophoretic mobility of ALAD, without additional mPEG-MAL labeling. This result indicated that the cysteine residues of ALAD were accessible to mPEG-MAL to different extents, while at least four cysteine residues formed an intramolecular disulfide bond. It is worth mentioning that the mobility of the PEG-labeled ALAD variants did not allow an estimation of the real protein size. Unlike each amino acid residue, mPEG-MAL did not bind to SDS. This was previously reported for a comparable observation for acyl-coenzyme A:cholesterolacyltransferase 1 upon mPEG-MAL treatment [37].

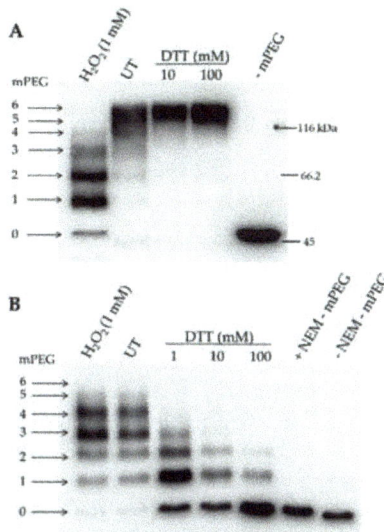

Figure 5. Labeling of the recombinant His-tagged ALAD containing six cysteines by using methoxypolyethylene glycol maleimide (mPEG-MAL)-5000. (**A**) mPEG-MAL labeling after pretreatment of ALAD in oxidizing (CuCl$_2$/hydrogen peroxide) or reducing conditions (DTT). The accessibility of the reduced cysteines in mature ALAD depended on this pretreatment. Each binding of a mPEG-MAL-5000 molecule to a cysteine induced a characteristic mobility shift in an 8% SDS-PAGE. The arrows numbered 0–6 indicate the unlabeled (0) and labeled ALAD (1–6). (**B**) Labeling of the oxidized and buried cysteines. After the pretreatment with oxidizing and reducing agents, all exposed cysteines were blocked irreversibly with N-ethylmaleimide (NEM). Subsequently, all samples were reduced with DTT (100 mM) and subsequently labeled with mPEG-MAL-5000. UT: untreated.

A similar observation was made regarding the susceptibility of cysteine residues in ALAD to mPEG-MAL when the free cysteine residues of the oxidized and reduced ALAD were initially blocked with N-ethylmaleimide (NEM) prior to the successive reduction of the protein and labeling with mPEG-MAL. Initially, two cysteine residues were oxidized and did not react with NEM, while the subsequent reduction of ALAD made up to four cysteines accessible to mPEG-MAL (Figure 5B). Consequentially, with the increasing amount of supplemented DTT, the enzyme became more accessible to NEM prior to mPEG-MAL-treatment. Taken together, these data imply that the four cysteine residues in ALAD were sensitive to the changes in redox potential and that ALAD has a possibility to form two disulfide bounds in fully oxidized conditions.

Consistent with the indication of the modified ALAD redox status with an increasing amount of DTT, as visualized after mPEG labeling, it was shown in a non-reducing polyacrylamide gel that monomeric ALAD migrated more slowly than the oxidized form under reducing conditions (Figure 6A). At least two additional bands can be differentiated apart from the reduced ALAD form (red, ox1 and ox2). It is assumed that one (ox1) or two (ox2) intramolecular disulfide bonds could lead to a compact ALAD structure compared to the reduced state of the protein, resulting in the enhanced mobility of the oxidized forms in the polyacrylamide gel (Figure 6A,C). Additionally, under oxidizing conditions the formation of dimeric and oligomeric structures was observed.

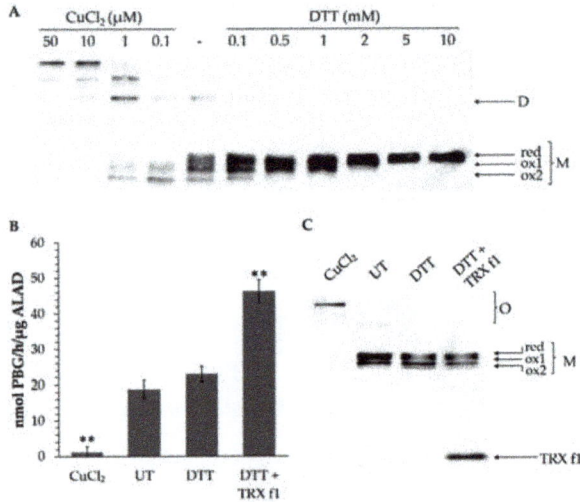

Figure 6. Redox-dependent structure and activity of recombinant ALAD. (**A**) Formation of mono-, di- and oligomeric ALAD and the redox-state of the monomer under oxidized, untreated and reduced conditions. The samples were separated by a non-reducing SDS-10% polyacrylamide gel. The black arrows indicate the different redox-states of an ALAD monomer (red, ox1, ox2) due to changes in protein mobility after the formation of internal disulfide bonds (M = monomer, D = dimer, O = oligomers). (**B**) ALAD activity assay after the preincubation of recombinant ALAD (50 nM) either with CuCl$_2$ (5 μM), DTT (0.1 mM), DTT (0.1 mM) + 0.5 μM recombinant TRX f1 or untreated (UT) conditions. The activity is presented as the nmol porphobilinogen (PBG) formed in one hour per μg ALAD. Statistical significance compared with the activity of the untreated ALAD (UT) is indicated by asterisks (**, $p \leq 0.01$, Student's t test). (**C**) ALAD preincubated together with CuCl$_2$, 0.1 mM DTT, DTT (0.1 mM) + recombinant TRX f1 or without any additives to the buffer (UT) was separated by a non-reducing SDS-12% polyacrylamide gel.

We can summarize that the redox status of ALAD differed under oxidized, reduced and untreated conditions. The ALAD activity was determined under these three conditions and upon the additional supply of TRX (Figure 6B). It seems that the enzyme activity increased with the reduced state of ALAD. The untreated recombinant ALAD was not entirely reduced and additional reducing power enhanced the enzyme activity. These results were in line with the observed redox state of recombinant ALAD in Figure 6A. It was important to note that the ALAD activity was able to be increased when TRX was supplied to the enzyme assay, confirming that ALAD is usually reduced in planta by TRX, while no additional reduction of ALAD was observed under this condition (Figure 6C).

4. Discussion

4.1. Redox-Dependent Modification of Alad Stability and Activity in Arabidopsis

The ALAD-catalyzed formation of the monopyrrole PBG directed into the porphyrin-synthesizing pathway of TBS. Porphyrins and subsequently Mg porphyrins are light-absorbing intermediates of the pathway. Tight control of the metabolic pathway prevents the accumulation of the free porphyrins and consequently the photosensitization of foliar cells. Thiol-based redox control is one of the mechanisms to rapidly adjust the enzyme activity in TBS.

In planta ALAD activity was lower in the *trx f1*, *ntrc* and *trx f1/ntrc* knock-out mutants in comparison to wild-type plants. This could be explained by a decreased accumulation of ALAD in the *trx f1* and *ntrc* mutants (Figure 3B). The ALAD activity of the mutant extracts was stimulated upon DTT supply to 17% (*ntrc*), 24% (*trx f1*) and 21% (*ntrc/trx f1*), respectively, compared to a minor increased of activity (5%) for the wild-type extract. It is proposed that the ALAD pool of the wild-type plant seems to be mainly reduced under illumination, whereas the ALAD pool of the *ntrc*- and *trx f1*-deficient mutants is partially oxidized as result of the absent reductants. Thus it is concluded that TRX and NTRC are important for ALAD stability and enzymatic activity in planta. It is likely that the deficiency of TRX and NTRC is at least partially compensated by other reductants, so that only a low (non-detectable) amount of ALAD was found to be oxidized. The entire loss of reducing power to adjust the redox state of metabolic enzymes in chloroplasts was also not observed in previous reports [14,27].

4.2. Redox-Dependent Modification of Recombinant ALAD

The accessible thiol groups of cysteine residues were detected by the conjugation of mPEG-MAL to ALAD (Figure 5). It seems that not all cysteine residues were accessible upon oxidation and one or two cysteines were constantly accessible to mPEG-MAL and, therefore, not involved in in vitro disulfide bonding. After the reduction of ALAD and blocking of free cysteine residues, most of the protein was not accessible to mPEG-MAL labeling. When oxidized ALAD was blocked by NEM, 3–4 mPEG-MAL molecules were bound to ALAD. We did not observe ALAD conjugated with 5 or 6 mPEG-MAL molecules, indicating again that two out of six cysteines were not redox sensitive (Figure 5). Future studies should elucidate which cysteine residues contribute to the redox-dependent thiol switch.

The transfer of the redox status of purified ALAD in oxidized and reduced conditions was displayed in non-reducing gels (Figure 6). It is suggested that at least four of the six cysteine residues contributed to the structural modifications leading to a mobility shift in the non-reducing gel. Figure 6 also indicated that ALAD di- and oligomers were dismantled to monomers under reducing conditions. Moreover, the activity of the reduced and oxidized forms of ALAD differed, and TRX f1 further stimulated ALAD activity (Figure 6).

However, it remains speculative whether the in planta ALAD requires the monomeric structure for its maximum enzyme activity. Structural analysis revealed the stable oligomeric structures for pea ALAD [23]. A Mg^{2+}-dependent shift of a hexameric to an more active octameric ALAD was reported. This regulatory mechanism for light-dependent chlorophyll synthesis might be sensitive when a light-stimulated uptake of Mg^{2+} into chloroplasts is taken into account.

Based on our results, future experiments are promising to identify the redox-dependent cysteine residues of ALAD and to explore the physiological consequences of impaired thiol-based redox control on the stability and activity of wild-type and each of the six cysteine-substitution mutant ALAD variants, respectively, expressed in the *A. thaliana* hemb1 mutant background.

5. Conclusions

In conclusion, the thiol-disulfide cycling of recombinant ALAD modifies the enzyme activity. Reductants such as TRX f1 stimulate the ALAD activity. Due to the lack of reducing power also affecting the stability of ALAD in planta, it will be challenging in the future to assign the cysteine residues for the regulatory adjustment of these redox switches, resulting in modified ALAD stability and enzymatic activity.

Author Contributions: The conception was based on B.G.'s grant applications; D.W. and S.K. performed the investigations and validated the results; B.G. and D.W. wrote the manuscript; B.G. and P.W. supervised the project; and P.W., D.W. and B.G. corrected and revised the manuscript.

Funding: This work was supported by the Deutsche Forschungsgemeinschaft grant (DFG 936 17-1 in the SPP 1710 "Thiol Switches") to Bernhard Grimm.

Conflicts of Interest: The authors decre no conflict of interest.

Abbreviations

Aa	amino acid residues
ALA	5-aminolevulinic acid
ALAD	5-aminolevulinic acid dehydratase
BiFC	bimolecular fluorescence complementation
BLAST	Basic Local Alignment Search Tool
CHLG	chlorophyll synthase
CHLI	subunit of the Mg chelatase
CHLM	Mg protoporphyrin methyltransferase
DTT	dithiothreitol
GluTR	glutamyl-tRNA reductase
GSAAT	glutamate-1-semialdehyde aminotransferase
mPEG-MAL	methoxypolyethylene glycol maleimide 5000
NEM	N-ethylmaleimide
NTRC	NADPH-dependent thioredoxin reductase C
PBG	porphobilinogen
POR	protochlorophyllide oxidoreductase
PPOX1/2	protoporphyrinogen IX oxidase
ROS	reactive oxygen species
RuBisCo1,5	ribulose-bisphosphate carboxylase large subunit
TBS	tetrapyrrole biosynthesis
TRX	thioredoxin
YFP	yellow-fluorescence protein

References

1. Richter, A.S.; Grimm, B. Thiol-based redox control of enzymes involved in the tetrapyrrole biosynthesis pathway in plants. *Front. Plant Sci.* **2013**, *4*, 371. [CrossRef] [PubMed]
2. Stenbaek, A.; Jensen, P.E. Redox regulation of chlorophyll biosynthesis. *Phytochemistry* **2010**, *71*, 853–859. [CrossRef] [PubMed]
3. Busch, A.W.; Montgomery, B.L. Interdependence of tetrapyrrole metabolism, the generation of oxidative stress and the mitigative oxidative stress response. *Redox Biol.* **2015**, *4*, 260–271. [CrossRef] [PubMed]
4. Buchanan, B.B. The Path to Thioredoxin and Redox Regulation Beyond Chloroplasts. *Plant Cell Physiol.* **2017**, *58*, 1826–1832. [CrossRef] [PubMed]
5. Michelet, L.; Zaffagnini, M.; Morisse, S.; Sparla, F.; Perez-Perez, M.E.; Francia, F.; Danon, A.; Marchand, C.H.; Fermani, S.; Trost, P.; et al. Redox regulation of the Calvin-Benson cycle: Something old, something new. *Front. Plant Sci.* **2013**, *4*, 470. [CrossRef] [PubMed]
6. Serrato, A.J.; Fernandez-Trijueque, J.; Barajas-Lopez, J.D.; Chueca, A.; Sahrawy, M. Plastid thioredoxins: A "one-for-all" redox-signaling system in plants. *Front. Plant Sci.* **2013**, *4*, 463. [CrossRef] [PubMed]
7. Serrato, A.J.; Perez-Ruiz, J.M.; Spinola, M.C.; Cejudo, F.J. A novel NADPH thioredoxin reductase, localized in the chloroplast, which deficiency causes hypersensitivity to abiotic stress in *Arabidopsis thaliana*. *J. Biol. Chem.* **2004**, *279*, 43821–43827. [CrossRef] [PubMed]
8. Perez-Ruiz, J.M.; Naranjo, B.; Ojeda, V.; Guinea, M.; Cejudo, F.J. NTRC-dependent redox balance of 2-Cys peroxiredoxins is needed for optimal function of the photosynthetic apparatus. *Proc. Natl. Acad. Sci. USA* **2017**, *114*, 12069–12074. [CrossRef] [PubMed]
9. Spinola, M.C.; Perez-Ruiz, J.M.; Pulido, P.; Kirchsteiger, K.; Guinea, M.; Gonzalez, M.; Cejudo, F.J. NTRC new ways of using NADPH in the chloroplast. *Physiol. Plant.* **2008**, *133*, 516–524. [CrossRef] [PubMed]

10. Richter, A.S.; Peter, E.; Rothbart, M.; Schlicke, H.; Toivola, J.; Rintamaki, E.; Grimm, B. Posttranslational influence of NADPH-dependent thioredoxin reductase C on enzymes in tetrapyrrole synthesis. *Plant. Physiol.* **2013**, *162*, 63–73. [CrossRef] [PubMed]

11. Luo, T.; Fan, T.; Liu, Y.; Rothbart, M.; Yu, J.; Zhou, S.; Grimm, B.; Luo, M. Thioredoxin redox regulates ATPase activity of magnesium chelatase CHLI subunit and modulates redox-mediated signaling in tetrapyrrole biosynthesis and homeostasis of reactive oxygen species in pea plants. *Plant. Physiol.* **2012**, *159*, 118–130. [CrossRef] [PubMed]

12. Richter, A.S.; Perez-Ruiz, J.M.; Cejudo, F.J.; Grimm, B. Redox-control of chlorophyll biosynthesis mainly depends on thioredoxins. *FEBS Lett.* **2018**, *592*, 3111–3115. [CrossRef] [PubMed]

13. Richter, A.S.; Wang, P.; Grimm, B. Arabidopsis Mg-Protoporphyrin IX Methyltransferase Activity and Redox Regulation Depend on Conserved Cysteines. *Plant Cell Physiol.* **2016**, *57*, 519–527. [CrossRef] [PubMed]

14. Da, Q.; Wang, P.; Wang, M.; Sun, T.; Jin, H.; Liu, B.; Wang, J.; Grimm, B.; Wang, H.B. Thioredoxin and NADPH-Dependent Thioredoxin Reductase C Regulation of Tetrapyrrole Biosynthesis. *Plant Physiol.* **2017**, *175*, 652–666. [CrossRef] [PubMed]

15. Jensen, P.E.; Reid, J.D.; Hunter, C.N. Modification of cysteine residues in the ChlI and ChlH subunits of magnesium chelatase results in enzyme inactivation. *Biochem. J.* **2000**, *352 Pt 2*, 435–441. [CrossRef]

16. Stenbaek, A.; Hansson, A.; Wulff, R.P.; Hansson, M.; Dietz, K.J.; Jensen, P.E. NADPH-dependent thioredoxin reductase and 2-Cys peroxiredoxins are needed for the protection of Mg-protoporphyrin monomethyl ester cyclase. *FEBS Lett.* **2008**, *582*, 2773–2778. [CrossRef] [PubMed]

17. Balmer, Y.; Koller, A.; del Val, G.; Manieri, W.; Schurmann, P.; Buchanan, B.B. Proteomics gives insight into the regulatory function of chloroplast thioredoxins. *Proc. Natl. Acad. Sci. USA* **2003**, *100*, 370–375. [CrossRef] [PubMed]

18. Lindahl, M.; Florencio, F.J. Thioredoxin-linked processes in cyanobacteria are as numerous as in chloroplasts, but targets are different. *Proc. Natl. Acad. Sci. USA* **2003**, *100*, 16107–16112. [CrossRef] [PubMed]

19. Spencer, P.; Jordan, P.M. Characterization of the two 5-aminolaevulinic acid binding sites, the A- and P-sites, of 5-aminolaevulinic acid dehydratase from Escherichia coli. *Biochem. J.* **1995**, *305 Pt 1*, 151–158. [CrossRef]

20. Tang, W.; Wang, W.; Chen, D.; Ji, Q.; Jing, Y.; Wang, H.; Lin, R. Transposase-derived proteins FHY3/FAR1 interact with PHYTOCHROME-INTERACTING FACTOR1 to regulate chlorophyll biosynthesis by modulating HEMB1 during deetiolation in Arabidopsis. *Plant Cell* **2012**, *24*, 1984–2000. [CrossRef] [PubMed]

21. Boese, Q.F.; Spano, A.J.; Li, J.M.; Timko, M.P. Aminolevulinic acid dehydratase in pea (*Pisum sativum* L.). Identification of an unusual metal-binding domain in the plant enzyme. *J. Biol. Chem.* **1991**, *266*, 17060–17066. [PubMed]

22. Kervinen, J.; Dunbrack, R.L., Jr.; Litwin, S.; Martins, J.; Scarrow, R.C.; Volin, M.; Yeung, A.T.; Yoon, E.; Jaffe, E.K. Porphobilinogen synthase from pea: Expression from an artificial gene, kinetic characterization, and novel implications for subunit interactions. *Biochemistry* **2000**, *39*, 9018–9029. [CrossRef] [PubMed]

23. Kokona, B.; Rigotti, D.J.; Wasson, A.S.; Lawrence, S.H.; Jaffe, E.K.; Fairman, R. Probing the oligomeric assemblies of pea porphobilinogen synthase by analytical ultracentrifugation. *Biochemistry* **2008**, *47*, 10649–10656. [CrossRef] [PubMed]

24. Jaffe, E.K. The Remarkable Character of Porphobilinogen Synthase. *ACC Chem. Res.* **2016**, *49*, 2509–2517. [CrossRef] [PubMed]

25. Balange, A.P.; Lambert, C. In vitro activation of δ-aminolevulinate dehydratase from far-red irradiated radish (*Raphanus sativus* L.) seedlings by thioredoxin f. *Plant Sci. Lett.* **1983**, *32*, 253–259. [CrossRef]

26. Thormahlen, I.; Ruber, J.; von Roepenack-Lahaye, E.; Ehrlich, S.M.; Massot, V.; Hummer, C.; Tezycka, J.; Issakidis-Bourguet, E.; Geigenberger, P. Inactivation of thioredoxin f1 leads to decreased light activation of ADP-glucose pyrophosphorylase and altered diurnal starch turnover in leaves of Arabidopsis plants. *Plant Cell Environ.* **2013**, *36*, 16–29. [CrossRef] [PubMed]

27. Thormahlen, I.; Meitzel, T.; Groysman, J.; Ochsner, A.B.; von Roepenack-Lahaye, E.; Naranjo, B.; Cejudo, F.J.; Geigenberger, P. Thioredoxin f1 and NADPH-Dependent Thioredoxin Reductase C Have Overlapping Functions in Regulating Photosynthetic Metabolism and Plant Growth in Response to Varying Light Conditions. *Plant Physiol.* **2015**, *169*, 1766–1786. [CrossRef] [PubMed]

28. Emanuelsson, O.; Nielsen, H.; von Heijne, G. ChloroP, a neural network-based method for predicting chloroplast transit peptides and their cleavage sites. *Protein Sci.* **1999**, *8*, 978–984. [CrossRef] [PubMed]

29. Sambrook, J.; Russell, D.W. *Molecular Cloning: A Laboratory Manual*; Cold Spring Harbor Laboratory Press: New York, NY, USA, 2001.

30. Muthuramalingam, M.; Dietz, K.J.; Stroher, E. Thiol-disulfide redox proteomics in plant research. *Methods Mol. Biol.* **2010**, *639*, 219–238. [CrossRef] [PubMed]

31. Mauzerall, D.; Granick, S. The occurrence and determination of delta-amino-levulinic acid and porphobilinogen in urine. *J. Biol. Chem.* **1956**, *219*, 435–446. [PubMed]

32. Frankenberg, N.; Erskine, P.T.; Cooper, J.B.; Shoolingin-Jordan, P.M.; Jahn, D.; Heinz, D.W. High resolution crystal structure of a Mg2+-dependent porphobilinogen synthase. *J. Mol. Biol.* **1999**, *289*, 591–602. [CrossRef] [PubMed]

33. Frere, F.; Schubert, W.D.; Stauffer, F.; Frankenberg, N.; Neier, R.; Jahn, D.; Heinz, D.W. Structure of porphobilinogen synthase from Pseudomonas aeruginosa in complex with 5-fluorolevulinic acid suggests a double Schiff base mechanism. *J. Mol. Biol.* **2002**, *320*, 237–247. [CrossRef]

34. Coates, L.; Beaven, G.; Erskine, P.T.; Beale, S.I.; Avissar, Y.J.; Gill, R.; Mohammed, F.; Wood, S.P.; Shoolingin-Jordan, P.; Cooper, J.B. The X-ray structure of the plant like 5-aminolaevulinic acid dehydratase from Chlorobium vibrioforme complexed with the inhibitor laevulinic acid at 2.6 A resolution. *J. Mol. Biol.* **2004**, *342*, 563–570. [CrossRef] [PubMed]

35. Coates, L.; Beaven, G.; Erskine, P.T.; Beale, S.I.; Wood, S.P.; Shoolingin-Jordan, P.M.; Cooper, J.B. Structure of Chlorobium vibrioforme 5-aminolaevulinic acid dehydratase complexed with a diacid inhibitor. *Acta Crystallogr. D Biol. Crystallogr.* **2005**, *61*, 1594–1598. [CrossRef] [PubMed]

36. Kelley, L.A.; Mezulis, S.; Yates, C.M.; Wass, M.N.; Sternberg, M.J. The Phyre2 web portal for protein modeling, prediction and analysis. *Nat. Protoc.* **2015**, *10*, 845–858. [CrossRef] [PubMed]

37. Guo, Z.Y.; Chang, C.C.; Lu, X.; Chen, J.; Li, B.L.; Chang, T.Y. The disulfide linkage and the free sulfhydryl accessibility of acyl-coenzyme A:cholesterol acyltransferase 1 as studied by using mPEG5000-maleimide. *Biochemistry* **2005**, *44*, 6537–6546. [CrossRef] [PubMed]

![antioxidants logo] *antioxidants*

MDPI

Article

New Insights into the Potential of Endogenous Redox Systems in Wheat Bread Dough

Nicolas Navrot [1,2,*], Rikke Buhl Holstborg [3], Per Hägglund [1,4], Inge Lise Povlsen [3,5] and Birte Svensson [1]

[1]   Enzyme and Protein Chemistry (EPC), Department of Systems Biology, Technical University of Denmark, Building 224, DK-2800 Kgs Lyngby, Denmark; pmh@sund.ku.dk (P.H.); bis@bio.dtu.dk (B.S.)
[2]   Institute of Plant Molecular Biology, CNRS UPR 2357, University of Strasbourg, 12 rue du Général Zimmer, 67084 Strasbourg CEDEX, France
[3]   DuPont Nutrition Biosciences ApS, Edwin Rahrs Vej 38, DK-8220 Brabrand, Denmark; Rikke.Buhl.Holstborg@dupont.com (R.B.H.); inge.lise.povlsen@arlafoods.com (I.L.P.)
[4]   Department of Biomedical Sciences, University of Copenhagen, Blegdamsvej 3, DK-2200 København N, Denmark
[5]   Arla Food Ingredients, Sønderhøj 12, DK-8260 Viby J, Denmark
*   Correspondence: nicolas.navrot@ibmp-cnrs.unistra.fr

Received: 2 October 2018; Accepted: 7 December 2018; Published: 12 December 2018

Abstract: Various redox compounds are known to influence the structure of the gluten network in bread dough, and hence its strength. The cereal thioredoxin system (NTS), composed of nicotinamide adenine dinucleotide phosphate (NADPH)-dependent thioredoxin reductase (NTR) and thioredoxin (Trx), is a major reducing enzymatic system that is involved in seed formation and germination. NTS is a particularly interesting tool for food processing due to its heat stability and its broad range of protein substrates. We show here that barley NTS is capable of remodeling the gluten network and weakening bread dough. Furthermore, functional wheat Trx that is present in the dough can be recruited by the addition of recombinant barley NTR, resulting in dough weakening. These results confirm the potential of NTS, especially NTR, as a useful tool in baking for weakening strong doughs, or in flat product baking.

Keywords: wheat; thioredoxin; thioredoxin reductase; baking; redox; dough rheology

---

1. Introduction

Bread is a basic element of the daily diet all over the world, and standard production procedures are important for achieving reproducible quality in various production sites. As the raw material shows natural differences in many qualitative aspects, such as protein content, it has been crucial to develop tools to manipulate the bread properties to compensate for natural variations between wheat cultivars, or to adapt flour for specific applications. As an example, the variation in high molecular weight (HMW) glutenin contents in different flours influences the elasticity of the dough [1]. An accurate assessment of the variations found in flours from different wheat cultivars is thus essential in order to understand and predict dough properties [2]. Sulfur-containing molecules such as glutathione and cysteine are known to influence the redox state of the gluten network in the dough, and hence its strength [3]. In this context, small redox active molecules such as ascorbate, glutathione, or cysteine, and enzymes such as oxidases have been investigated in the past. Berland and Launey [4] showed that ascorbate and glutathione (GSH) strengthen and weaken bread dough, respectively. GSH is thought to suppress the disulfide formation of gluten proteins by binding to thiol groups of cysteine residues, mostly in glutenins [5]. Ascorbate acts indirectly by causing GSH depletion through the activity of the endogenous dehydroascorbate reductase enzyme (DHAR, also referred to as glutathione dehydrogenase), which uses GSH to reduce ascorbate that is oxidized by H_2O_2 and other reactive oxygen species [6] (Figure 1A). In contrast, the addition of GSH in gluten-free systems, e.g., rice batters, yielded an increased bread volume, due to the reduction of disulfides in the protein network [7]. The addition of cysteine reduces dough strength and augments its fluidity, and cysteine is considered by the FDA to be a GRAS ingredient of up to concentrations of 90 ppm in flour, although its chemical or animal origin is a slight drawback to its use [8]. Plant thiol-based reducing and oxidizing enzymes are thus to be considered when looking for indicators or tools for dough improvement, especially when looking for natural, non-animal origin additives [9].

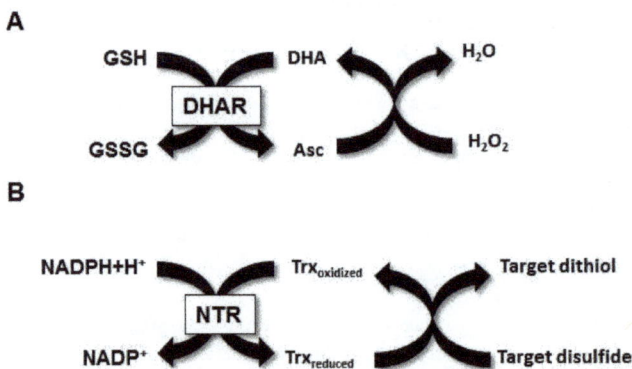

A

GSH   DHA H_2O

DHAR

GSSG   Asc H_2O_2

B

$NADPH+H^+$ $Trx_{oxidized}$ Target dithiol

NTR

$NADP^+$ $Trx_{reduced}$ Target disulfide

Figure 1. Ascorbate- and thioredoxin-dependent reducing pathways. (**A**) Dehydroascorbate reductase (DHAR) regenerates reduced ascorbate by oxidizing reduced glutathione (GSH) to disulfide-linked oxidized glutathione (GSSG). Ascorbate (Asc) can then be oxidized back to dehydroascorbate (DHA) upon reaction with a molecule H_2O_2 or other oxidants. (**B**) Reduced Trx regenerates an oxidized target disulfide bond to a dithiol. Trx itself is then reduced to its active form by a nicotinamide adenine dinucleotide phosphate (NADPH)-dependent thioredoxin reductase.

Nicotinamide adenine dinucleotide phosphate (NADPH)-dependent thioredoxin reductase (NTR) and thioredoxin (Trx) constitute a versatile multi-enzymatic NTS that has been extensively studied in plants. NTS functions by transferring reducing power from NADPH to various target proteins via a Trx-mediated disulfide bridge reduction [10] (Figure 1B). Plant genomes typically contain two NTR-encoding genes and more (up to 20) Trx isoforms, which in all cell types are involved in the regulation of enzymes from the normal metabolism, as well as stress-related responses [11,12]. Proteomics studies identified numerous such potential targets, of which several are only observed at the protein level [13]. Several Trx targets are known in cereal seeds, e.g., barley α-amylase/subtilisin inhibitor (BASI), and proteomics methods have been developed to expand knowledge of these systems [14]. Trx received interest in the field of food chemistry due to its broad range of protein substrates, and to its robustness. NTS, in particular, was investigated with regard to its ability to reduce the allergenicity of bread and other food products [15,16]. Indeed, redox biochemistry in food is far from being well understood, and most likely, these multi-enzyme systems behave differently according to the process studied.

Plant Trx are highly expressed in seeds, and these very stable proteins likely remain functional in raw dough. In contrast, upstream reducing enzymes, e.g., the flavoprotein NTR, are rather unstable enzymes, and are likely to be lost during industrial flour processing. We therefore hypothesize that endogenous Trx that is present in wheat dough could be used as a natural reducing agent to modify dough properties, if supplemented with reducing equivalents provided by NADPH and barley NTR. We examined these hypotheses here in an experimental dough- and bread-making system.

2. Materials and Methods

Raw material. Wheat flour (Reform flour, batch No. 1009160013, protein content 10.9%) and the yeast strain (*Saccharomyces cerevisiae*) were provided by Danisco.

Chemicals. NADPH and Ellman's reagent were purchased from Sigma-Aldrich (St. Louis, MO, USA). Rabbit antibodies raised against wheat Trx h1 were a kind gift from Bob B. Buchanan (UC Berkeley, CA, USA).

Enzymes. Recombinant barley NTR (HvNTR2) and Trx (HvTrxh1) were produced and purified as previously described [17,18]. The specific activity and detailed kinetic properties of these proteins have been described previously [17–19], and it has been demonstrated that HvNTR2 and the homologue HvNTR1 show similar catalytic efficiencies with the two Trx h isozymes, HvTrxh1 and HvTrxh2.

Dough preparation. Flour (50 g for rheological experiment, 300 g for baking experiments) was mixed with water to 55% water absorption using method AACC 54-21 and 2% (bakers' percentage) NaCl on flour weigh. NADPH and either or both NTR and Trx were then added, and the dough was mixed for 6 min. Dough strength was monitored until a stable value of ca. 400 Brabender units (BU) was reached, as measured by farinograph. Yeast (6% Malterserkors yeast on flour) and glucose (2% on flour) were included in the 300 g samples used in the baked products experiments. NADPH, NTR, Trx, and 10 mM Na-phosphate buffer (pH 6) were kept separately on ice, and mixed just before their addition to the flour in the mixing chamber, immediately after water addition.

Kiffer rig measurements. After mixing, the dough was shaped into a bun, of which samples were cut out for Kiffer rig measurements. Eight dough strips were shaped in a specifically designed mold, and left to rest at 25 °C for 30 min in a tightly closed plastic bag (unless stated otherwise). Eight strips per sample were then assayed by measuring Kieffer uniaxial extension using a Stable Micro Systems model TA-XT2i texture analyzer. The resistance of the dough strip (in g) and the maximum elongation (length in mm, just before the strip broke) were recorded using the Texture Expert Exceed version 2.64 software.

Baked products. Minibreads and buns from yeast-containing samples were prepared by mixing 15 and 50 g of initial flour, respectively, to 450 BU. After fitting into molds, minibreads and buns were proofed at 34 °C, 85% relative humidity (RH) for 55 min, to allow for dough development, then minibread dough stability was tested by sliding half of them down a custom-made slide ("shock treatment"), and finally the products were baked at 220 °C for 8 min (MIWE aeromat CS oven). For enzyme testing, to the flour mix (300 g) was added either buffer (10 mM sodium phosphate pH 6) or a mix of 16 mg NADPH and 2 mg NTR in buffer. The weights and volumes of the minibreads and buns were measured after cooling down. The section and bottom surfaces of four minibreads and buns for each treatment were measured from scanned sections, using ImageJ software.

Dough protein extraction and thiol titration. Dough samples were frozen in liquid N_2 after mixing and development. Samples were freeze-dried overnight and ground to powder using a mortar and pestle. Sixty milligrams of powder was used for protein extraction (0.5 mL of extraction buffer = 30 mM Tris-HCl pH 8, 200 mM NaCl, 10% glycerol, 0.2% Tween 20) was added. Samples were incubated for 2 hr at 4 °C with shaking, centrifuged at 14,000 rpm for 20 min, and supernatants were collected (soluble protein fraction). Twenty milligrams of protein were loaded and run on NuPAGE Novex 12–15% Bis-Tris minigels (Invitrogen, Carlsbad, CA, USA), and stained with Coomassie blue.

Volumes of 20 and 200 μL of extract were used to measure free thiols with Ellman's reagent. Briefly, the sample and 5 μL 20 mM DTNB in ethanol were mixed in 10 mM Tris pH 8 buffer (total volume 500 μL). After 30 min in the dark, A_{412} was recorded, and the thiol concentration was calculated from $\varepsilon_M = 13,600$ $M^{-1}cm^{-1}$ for free TNB anion, and the results were normalized by setting a value of 100 for the control sample.

Western blot. A total of 30 mg soluble protein were loaded and run on NuPAGE Novex 12–15% Bis-Tris minigels (Invitrogen), and blotted onto Hybond-ECL membranes (GE Healthcare, Chicago, IL, USA). The quality of the protein transfer was checked by Ponceau staining, and membranes were afterwards incubated overnight in blocking solution consisting of 2% bovine serum albumine (BSA)

Antioxidants **2018**, *7*, 190

in 30 mM Tris pH 8, 200 mM NaCl, 0.05% Tween 20 buffer (TBST). Membranes were incubated with 1/2500 diluted anti-wheat Trx rabbit primary antibody in blocking solution for 2 hr, washed three times for 10 min in TBST, and incubated for 1 hr in 1/2500 diluted secondary goat anti-rabbit horseradish peroxidase-coupled antibody (Dako Danmark A/S, Glostrup, Denmark). After three washes with TBST, membranes were put in plastic film, and an ECL Plus kit (GE Healthcare) and a UVP-Bioimaging system (AH Diagnostics, Helsinki, Finland) were used for detection.

3. Results and Discussion

3.1. Effect of NTS on Dough Rheological Properties

As a starting point, a concentration of thiols was chosen based on the 90 ppm cysteine FDA regulation, taking into account that our enzymatic system would probably be less active in the non-optimal dough environment (pH, water content). An amount equivalent to 300 ppm thiol was added in the form of 120 mg Trx, and 13 mg NTR with 140 mg NADPH, for a 50 g flour assay. The amounts of each component were then progressively decreased and mixed in various combinations to assess the effects of the different compounds on the dough system. NADPH and NTR were observed to exert strong effects on the rheological properties of the dough (Figure 2). Control experiments with only buffer or NTR in the absence of NADPH gave reproducible values of ca. 60 g and 40 mm for dough resistance and maximal elongation, respectively (samples 1 and 2). The first assay using amounts corresponding to 300 ppm free thiol (sample 10: 25 mg NADPH plus 2.75 mg NTR) resulted in very soft and sticky dough with barely measurable properties (only two samples in this experiment gave non-aberrant values). Reproducible and measurable values were obtained for the other combinatorial assays, with variable amounts of NADPH and NTR, and as little as 1.25 mg NADPH and 5.5 mg NTR for 50 g flour significantly reduced the resistance and augmented the extensibility of the dough. This effect appeared to be limited by NADPH as well as NTR, as a stronger effect was observed by increasing the amount of either of these two. Additional softening was found when including the remaining NTS component, barley thioredoxin h1 (HvTrxh1), which is known to be regenerated by NTR.

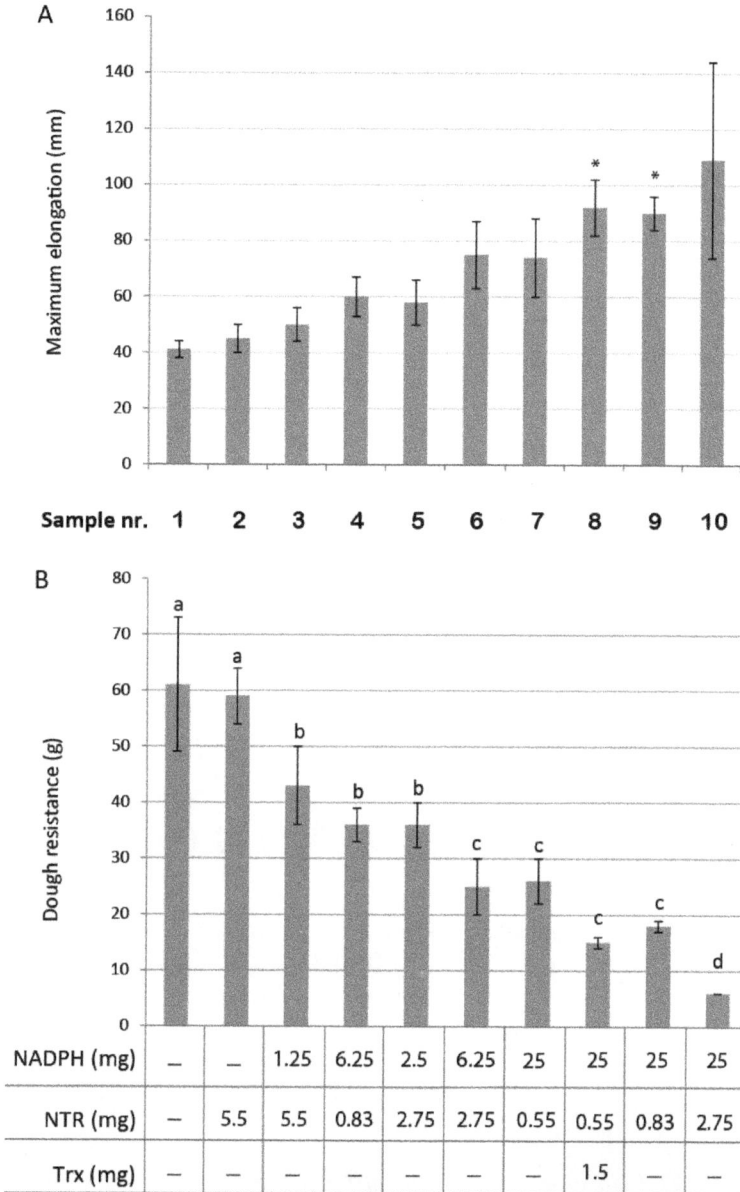

Figure 2. Kieffer rig rheological measurements of dough upon treatment with various amounts of nicotinamide adenine dinucleotide phosphate (NADPH), NADPH-dependent thioredoxin reductase (NTR), thioredoxin (Trx). (**A**) Maximal elongation obtained in mm. Asterisks indicate values that are significantly different from sample 1 (control) (Tukey test, $n = 6$, $p < 0.05$). Sample 10 was excluded from statistical analysis due to high variance on the only two measured values. (**B**) Dough resistance in g. Values are the average of 6–8 measurements ± standard deviation, except sample 10, for which $n = 2$. Different letters indicate significant differences (Tukey test, $n = 6$, $p \leq 0.05$).

3.2. Mini Breads and Buns

Further tests were conducted using yeast-containing dough. The influence of NTS was first assessed on minibreads, i.e., 15 g fresh dough baked in tin molds, yielding 11.6 g $\pm$ 0.2 g of baked product. Addition of NTR and NADPH to the dough clearly increased the minibread volume, due to increased height (Figure 3A). This indicates that our exogenous NTS was active and effective during dough resting, in addition to yeast metabolism. The increase in volume was explained by the relaxation of the gluten network, allowing for further expansion of CO_2 bubbles. This effect was suppressed when a light shock treatment was applied, resulting in dramatic collapse of the dough, due to the weakened gluten network. In this case, the measured height dropped below those seen for the control dough. When no enzyme was added to the dough, the shock treatment had no significant effect on minibread volume (Figure 3A,D).

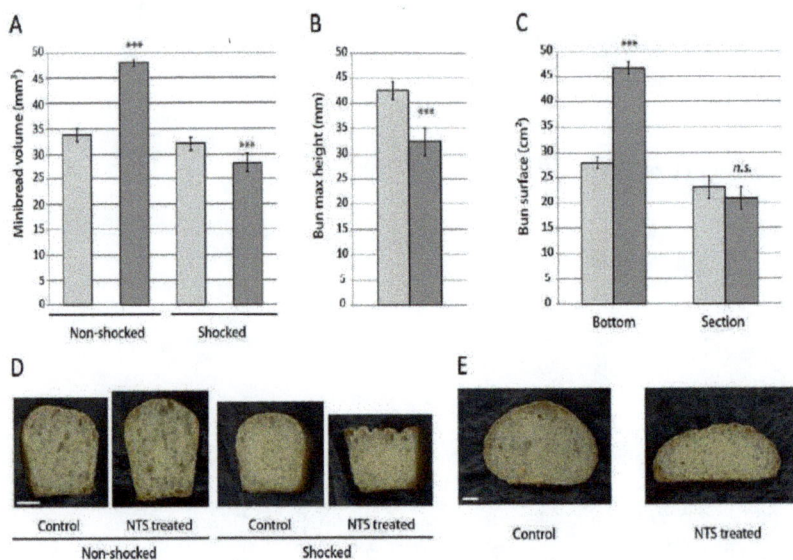

Figure 3. Shape parameters of buns and minibreads treated with NADPH and NTR. The average values for four buns or five minibreads $\pm$ standard error are shown. Control: light grey. NTS-treated samples: dark grey. (**A**) Bottom and cross-section surfaces of buns (cm^2). (**B**) Height of buns (mm). (**C**) Minibread volume (mm^3). (**D**) Cross-section photography of representative buns treated with phosphate buffer only (control) or 16 mg NADPH and 2 mg NTR in buffer (treated). (**E**) Cross-section photography of representative minibreads treated with phosphate buffer only (control) or 16 mg NADPH and 2 mg NTR in buffer (treated), with or without shock treatment. Asterisks indicate significant difference to control (Student *t*-test, $p \leq 0.01$). Scale bar = 10 mm.

In case of free-standing rolls (50 g fresh weight, 39 g $\pm$ 1.1 g baked product) we only analyzed the effect of enzyme addition. This addition caused an increase in the bottom surface of the rolls, while the maximum height decreased (Figure 3B,C). This confirmed our rheological tests on dough, i.e., the dough is softened and more extensible upon the addition of NTS. The loss in height was somehow compensated by an increased bottom surface, resulting in a change in the shape of the roll, which was flatter upon enzyme treatment, and a slight loss of volume (Figure 3E).

3.3. Protein Extraction and Dough Thiol Content

Dough samples were snap-frozen in liquid nitrogen, and ground with a mortar and pestle. After the addition of phosphate buffer, the soluble fraction was analyzed by sodium dodecyl sulfate–polyacrylamide gel electrophoresis (SDS-PAGE) and Western blot, and free thiols were also quantified. Quantification showed a higher content of free thiols that were extractible from the dough when it was treated with high amounts of NTS (Figure 4A). Unfortunately, no statistical significant difference could be calculated, due to the small number ($n = 2$) of measurements done. In yeast-containing dough, the free thiol content was higher than in samples without yeast, and NTS seemed to induce a slight decrease in thiol content (Figure 4B). Western blots using anti-wheat Trxh1 detected recombinant barley Trxh1 in those dough samples to which it was added (Figure 4C). This cross-species reactivity was previously observed [17,20–22]. A strong background signal was observed, and it prevented the detection of endogenous wheat Trx that was present in the flour.

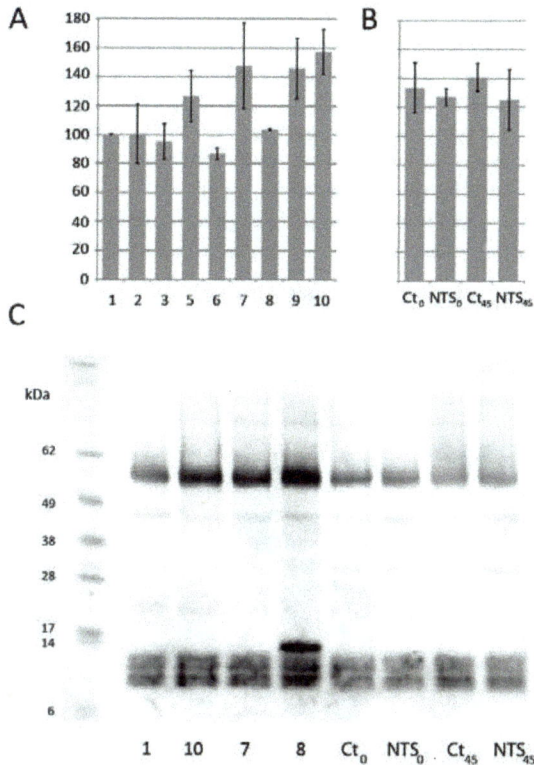

Figure 4. (**A**) Dough soluble thiol content measured as by Ellmann reagent in samples 1 to 10, corresponding to the conditions detailed in Figure 2. (**B**) Yeast-containing dough soluble thiol content measured as by Ellmann reagent in dough used for baking tests. Time point 0 corresponds to dough sampled just after mixing, and $t = 45$ min to dough just before baking, after development. Control sample value was set to 100%. (**C**) Western blot profile of dough extract with anti-wheat Trxh1 polyclonal antibodies. A marker in the first lane (size in kDa) was added for reference from the picture of the Ponceau-stained membrane before immunodetection.

4. Conclusions

We showed in this work that NTS is a powerful tool for the remodeling of the gluten network and the weakening of the dough, likely through functional wheat Trx being recruited in the dough upon addition of recombinant NTR. The addition of NTS induced dough weakening, as expected from reducing agents. It was possible to extract protein from dough and detect recombinant Trx spiked in the dough, together with a background signal from other proteins including most probably wheat Trx. Soluble dough extracts were also used to estimate the contents of soluble free thiols in the dough, which likely increased when NTS was added. Furthermore, this softening both of dough with and without yeast showed that added NTR and Trx activity were not lost due to elevated yeast metabolism in the developing dough. It is suggested that the optimization of the systems can involve increasing amounts of NTR added to the mix. The synergistic effect of Trx will allow for the reduction of the levels of NADPH needed for efficacy, and lead to a cost-effective procedure to soften strong flour systems using endogenous cereal NTR and Trx, where dough weakening is required, e.g., in flat product baking or laminated dough. Relaxation of the gluten network by enzymatic reduction is also increasing the volume of developed dough, although the shock sensibility observed here is a severe issue for any application in this direction. Using the NTS system, the costly addition of NADPH could be circumvented via the development of yeast strains containing high endogenous NTS system activity and/or to strains designed for specific applications, such as pizza or laminated dough making. The development of yeast GSH biosynthetic pathway mutants has already shown increases in their contents of GSH and derivatives [23], opening tracks for the development of industrial strains and the reduction of the use of chemicals such as cysteine. On the contrary, yeast strains with low NTS activity could help bread-makers to increase dough strength and bread volume. The shelf-life and alimentary properties of "NTS boosted" bread could also reveal further interesting aspects of the Trx system in food, since Trx is not only a regulator of the global redox state of the cell, but also regulates many enzymes that are involved in seed metabolism, such as starch hydrolase limit dextrinase [24].

Author Contributions: N.N., P.H., I.L.P., B.S. designed the experiments. N.N., R.B.H. performed the experiments. N.N. designed the manuscript draft. N.N., B.S., P.H., I.L.P. edited and reviewed the manuscript.

Funding: This work was supported by the Danish Council for Technology and Production Sciences (FTP, Grant 274-08-0413) and the Carlsberg Foundation.

Acknowledgments: Aida Curovic is acknowledged for technical help with the recombinant expression and purification of the barley NTR and Trx enzymes.

Conflicts of Interest: The authors declare no conflicts of interest.

Abbreviations

Trx	thioredoxin
NTR	NADPH-dependent thioredoxin reductase
NADPH	Nicotinamide adenine dinucleotide phosphate
GSH	glutathione
NTS	NADPH-dependent NTR and Thioredoxin system

References

1. Shewry, P.R.; Halford, N.G.; Belton, P.S.; Tatham, A.S. The structure and properties of gluten: An elastic protein from wheat grain. *Philos. Trans. R. Soc. Lond. B Biol. Sci.* **2002**, *357*, 133–142. [CrossRef] [PubMed]
2. Aamodt, A.; Magnus, E.M.; Faergestad, E.M. Effect of flour quality, ascorbic acid, and DATEM on dough rheological parameters and hearth loaves characteristics. *J. Food Sci.* **2003**, *68*, 2201–2210. [CrossRef]
3. Reinbold, J.; Rychlik, M.; Asam, S.; Wieser, H.; Koehler, P. Concentrations of total glutathione and cysteine in wheat flour as affected by sulfur deficiency and correlation to quality parameters. *J. Agric. Food Chem.* **2008**, *56*, 6844–6850. [CrossRef] [PubMed]
4. Berland, S.; Launay, B. Rheological properties of wheat flour doughs in steady and dynamic shear: Effect of water content and some additives. *Cereal Chem.* **1995**, *72*, 48–52.
5. Grosch, W.; Wieser, H. Redox reactions in wheat dough as affected by ascorbic acid. *J. Cereal Sci.* **1999**, *29*, 1–16. [CrossRef]
6. Dong, W.; Hoseney, R.C. Effects of certain breadmaking oxidants and reducing agents on dough rheological properties. *Cereal Chem.* **1995**, *72*, 58–64.
7. Yano, H. Improvements in the Bread-Making Quality of Gluten-Free Rice Batter by Glutathione. *J. Agric. Food Chem.* **2010**, *58*, 7949–7954. [CrossRef]

8. L-Cysteine. Code of Federal Regulations, 21CFR184.1271. Available online: https://www.ecfr.gov/cgi-bin/text-idx?SID=dc75a9bfd74137ed4b5dafeedd8db200&mc=true&node=se21.3.184_11271&rgn=div8 (accessed on 1 October 2018).
9. Every, D.; Simmons, L.D.; Ross, M.P. Distribution of redox enzymes in millstreams and relationships to chemical and baking properties of flour. *Cereal Chem.* **2006**, *83*, 62–68. [CrossRef]
10. Gütle, D.D.; Roret, T.; Hecker, A.; Reski, R.; Jacquot, J.P. Dithiol disulphide exchange in redox regulation of chloroplast enzymes in response to evolutionary and structural constraints. *Plant Sci.* **2017**, *255*, 1–11. [CrossRef]
11. Gelhaye, E.; Rouhier, N.; Navrot, N.; Jacquot, J.P. The plant thioredoxin system. *Cell Mol. Life Sci.* **2005**, *62*, 24–35. [CrossRef]
12. Geigenberger, P.; Thormählen, I.; Daloso, D.M.; Fernie, A.R. The Unprecedented Versatility of the Plant Thioredoxin System. *Trends Plant Sci.* **2017**, *22*, 249–262. [CrossRef]
13. Montrichard, F.; Alkhalfioui, F.; Yano, H.; Vensel, W.H.; Hurkman, W.J.; Buchanan, B.B. Thioredoxin targets in plants: The first 30 years. *J. Proteom.* **2009**, *72*, 452–474. [CrossRef]
14. Hägglund, P.; Bunkenborg, J.; Maeda, K.; Svensson, B. Identification of thioredoxin targets using a quantitative proteomics approach based on isotope-coded affinity tags—The ICAT switch. *J. Proteom. Res.* **2008**, *7*, 5270–5276. [CrossRef]
15. Del Val, G.; Yee, B.C.; Lozano, R.M.; Buchanan, B.B. Thioredoxin treatment increases digestibility and lowers allergenicity of milk. *J. Allergy Clin. Immunol.* **1999**, *103*, 690–697. [CrossRef]
16. Li, Y.C.; Ren, J.P.; Cho, M.J.; Zhou, S.M.; Kim, Y.B.; Guo, H.X.; Wong, J.H.; Niu, H.B.; Kim, H.K.; Morigasaki, S.; et al. The level of expression of thioredoxin is linked to fundamental properties and applications of wheat seeds. *Mol. Plant* **2009**, *2*, 430–441. [CrossRef]
17. Maeda, K.; Hägglund, P.; Finnie, C.; Svensson, B.; Henriksen, A. Structural basis for target protein recognition by the protein disulfide reductase thioredoxin. *Structure* **2006**, *14*, 1701–1710. [CrossRef] [PubMed]
18. Kirkensgaard, K.; Hägglund, P.; Finnie, C.; Svensson, B.; Henriksen, A. Crystal structure of Hordeum vulgare NADPH-dependent thioredoxin reductase 2. Unwinding the reaction mechanism. *Acta Crystallogr. D* **2009**, *65*, 932–941. [CrossRef] [PubMed]
19. Shahpiri, A.; Svensson, B.; Finnie, C. The NADPH-dependent thioredoxin reductase/thioredoxin system in germinating barley seeds: Gene expression, protein profiles, and interactions between isoforms of thioredoxin h and thioredoxin reductase. *Plant Physiol.* **2008**, *146*, 789–799. [CrossRef]
20. Cecere, F.; Iuliano, A.; Albano, F.; Zappelli, C.; Castellano, I.; Grimaldi, P.; Masullo, M.; De Vendittis, E.; Ruocco, M.R. Diclofenac-induced apoptosis in the neuroblastoma cell line SH-SY5Y: Possible involvement of the mitochondrial superoxide dismutase. *J. Biomed. Biotechnol.* **2010**, *2010*, 801726–801737. [CrossRef] [PubMed]
21. Andoh, T.; Chock, P.B.; Chiueh, C.C. The roles of thioredoxin in protection against oxidative stress-induced apoptosis in SH-SY5Y cells. *J. Biol. Chem.* **2002**, *277*, 9655–9660. [CrossRef] [PubMed]
22. Das, K.C.; Lewis-Molock, Y.; White, C.W. Elevation of manganese superoxide dismutase gene expression by thioredoxin. *Am. J. Respir. Cell Mol. Biol.* **1997**, *17*, 713–726. [CrossRef] [PubMed]
23. Orumets, K.; Kevvai, K.; Nisamedtinov, I.; Tamm, T.; Paalme, T. YAP1 over-expression in *Saccharomyces cerevisiae* enhances glutathione accumulation at its biosynthesis and substrate availability levels. *Biotechnol. J.* **2012**, *7*, 566–568. [CrossRef] [PubMed]
24. Jensen, J.M.; Hägglund, P.; Christensen, H.E.M.; Svensson, B. Inactivation of barley limit dextrinase inhibitor by thioredoxin-catalysed disulfide reduction. *FEBS Lett.* **2012**, *586*, 2479–2482. [CrossRef] [PubMed]

MDPI

St. Alban-Anlage 66

4052 Basel

Switzerland

Tel. +41 61 683 77 34

Fax +41 61 302 89 18

www.mdpi.com

Antioxidants Editorial Office

E-mail: antioxidants@mdpi.com

www.mdpi.com/journal/antioxidants

www.ingramcontent.com/pod-product-compliance
Lightning Source LLC
Chambersburg PA
CBHW051721210326
41597CB00032B/5562